MW00717216

TRANSFORMER DESIGN PRINCIPLES

TRANSFORMER DESIGN PRINCIPLES

With Applications to Core-Form Power Transformers

Robert M. Del Vecchio
Bertrand Poulin
Pierre T. Feghali
Dilipkumar M. Shah
and
Rajendra Ahuja

Published in 2002 by
Taylor & Francis
29 West 35th Street
New York, NY 10001

Printed in the United States of America on acid-free paper.

1 2 3 4 5 6 7 8 9 10

British Library Cataloguing in Publication Data

Transformer design principles : with application to
core-form power transformers
1.Electric transformers - Design
I.Del Vecchio, Robert M.
621.3'14

ISBN 90-5699-703-3

CONTENTS

PREFACE

Many of the standard texts on power transformers are now over ten years old and some much older. Much has changed in transformer design since these books were written. Newer and better materials are now available for core and winding construction. Powerful computers now make it possible to produce more detailed models of the electrical, mechanical and thermal behavior of transformers than previously possible. Although many of these modern approaches to design and construction are found scattered in the literature, there is a need to have this information available in a single source as a reference for the designer or power engineer and as a starting point for the student or novice.

It is hoped that the present work can serve both purposes. As a text for beginners, we emphasize the physical basis of transformer operation. We also discuss the physical effects which result from various fault conditions and their implications for design. Physical principles and mathematical techniques are presented in a reasonably self-contained manner, although references are provided to additional material. For the specialist such as a power or transformer design engineer, detailed models are presented which focus on various aspects of a transformer under normal or abnormal conditions. Cost minimization techniques, which form the starting point for most designs, are also presented.

Although this book primarily deals with power transformers, many of the physical principles discussed or mathematical modeling techniques presented apply equally well to other types of transformers. The presentation is kept as general as possible so that designers or users of other transformer types will have little difficulty applying many of the results to their own designs. The emphasis on fundamentals should make this process easier and should also foster the development of new and more powerful design tools in the future.

The International System of Units (SI) is used throughout the text. However, an occasional figure, graph, or table may show quantities in the British system of units. Sometimes a quantity is given in British units in parentheses after its metric value.

References are referred to generally by the first three letters of the first author's name followed by the last two digits of the publication date, e.g. [Abc98]. In cases where this format cannot be followed, an appropriate substitute is made. They are listed alphabetically at the end of the book.

We wish to thank Harral Robin for guidance throughout the course of this work. We would also like to acknowledge many helpful suggestions from power industry representatives and consultants over the years.

1. INTRODUCTION TO TRANSFORMERS

Summary Beginning with the principle of induction discovered by Faraday, the transformer slowly evolved to fill a need in electrical power systems. The development of 3 phase a.c. power has led to a great variety of transformer types. We discuss some of these types and their use in power systems. We also discuss and contrast some of the main construction methods. The principle components of a transformer are highlighted with special emphasis on core-form power transformers. Some of the basic considerations which determine the design of these components are presented. A look at some newer technologies is given which could impact the future development of transformers.

1.1 HISTORICAL BACKGROUND

Transformers are electrical devices which change or transform voltage levels between two circuits. In the process, current values are also transformed. However, the power transferred between the circuits is unchanged, except for a typically small loss which occurs in the process. This transfer only occurs when alternating current (a.c.) or transient electrical conditions are present. Transformer operation is based on the principle of induction discovered by Faraday in 1831. He found that when a changing magnetic flux links a circuit, a voltage or electromotive force (emf) is induced in the circuit. The induced voltage is proportional to the number of turns linked by the changing flux. Thus when two circuits are linked by a common flux and there are different linked turns in the two circuits, there will be different voltages induced. This situation is shown in Fig. 1.1 where an iron core is shown carrying the common flux. The induced voltages V_1 and V_2 will differ since the linked turns N_1 and N_2 differ.

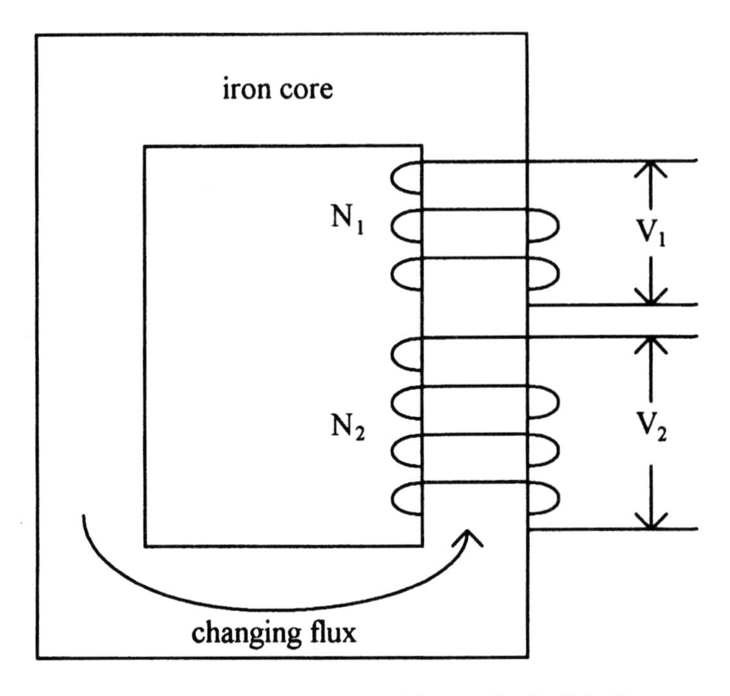

Figure 1.1 Transformer principle illustrated for two circuits linked by a common changing flux

Devices based on Faraday's discovery, such as inductors, were little more than laboratory curiosities until the advent of a.c. electrical systems for power distribution which began towards the end of the nineteenth century. Actually the development of a.c. power systems and transformers occurred almost simultaneously since they are closely linked. The invention of the first practical transformer is attributed to the Hungarian engineers Karoly Zipernowsky, Otto Blathy, and Miksa Deri in 1885 [Jes97]. They worked for the Hungarian Ganz factory. Their device had a closed toroidal core made of iron wire. The primary voltage was a few kilo volts and the secondary about 100 volts. It was first used to supply electric lighting.

Modern transformers differ considerably from these early models but the operating principle is still the same. In addition to transformers used in power systems which range in size from small units which are attached to the tops of telephone poles to units as large as a small house and weighing hundreds of tons, there are a myriad of transformers used in the electronics industry. These latter range in size from units weighing a few pounds and used to convert electrical outlet voltage to lower values required by transistorized circuitry to micro-transformers which are deposited directly onto silicon substrates via lithographic techniques.

Needless to say, we will not be covering all these transformer types here in any detail, but will instead focus on the larger power transformers. Nevertheless, many of the issues and principles discussed are applicable to all transformers.

1.2 USES IN POWER SYSTEMS

The transfer of electrical power over long distances becomes more efficient as the voltage level rises. This can be seen by considering a simplified example. Suppose we wish to transfer power P over a long distance. In terms of the voltage V and line current I, this power can be expressed as

$$P = VI \tag{1.1}$$

where rms values are assumed and the voltage and current are assumed to be in phase. For a line of length L and cross-sectional area A, its resistance is given by

$$R = \rho \frac{L}{A} \tag{1.2}$$

where ρ is the electrical resistivity of the line conductor. The electrical losses are therefore

$$\text{Loss} = I^2 R = \frac{\rho L I^2}{A} \tag{1.3}$$

and the voltage drop is

$$\text{Voltage drop} = IR = \frac{\rho L I}{A} \tag{1.4}$$

Substituting for I from (1.1), we can rewrite the loss and voltage drop as

$$\text{Loss} = \frac{\rho L P^2}{A V^2} \quad , \quad \text{Voltage drop} = \frac{\rho L P}{A V} \tag{1.5}$$

Since P, L, ρ are assumed given, the loss and voltage drop can be made as small as desired by increasing the voltage V. However, there are limits to increasing the voltage, such as the availability of adequate and safe insulation structures and the increase of corona losses.

We also notice from (1.5) that increasing the cross-section area of the line conductor A can lower the loss and voltage drop. However as A increases, the weight of the line conductor and therefore its cost also increase so that a compromise must be reached between the cost of losses and acceptable voltage drop and the conductor material costs.

In practice, long distance power transmission is accomplished with voltages in the range of 100 - 500 kV and more recently with voltages as high as 765 kV. These high voltages are, however, incompatible with safe usage in households or factories. Thus the need for transformers is apparent to convert these to lower levels at the receiving end. In addition, generators are, for practical reasons such as cost and efficiency, designed to produce electrical power at voltage levels of ~ 10 to 40 kV. Thus there is also a need for transformers at the sending end of the line to boost the generator voltage up to the required transmission levels. Fig. 1.2 shows a simplified version of a power system with actual voltages indicated. GSU stands for generator step-up transformer.

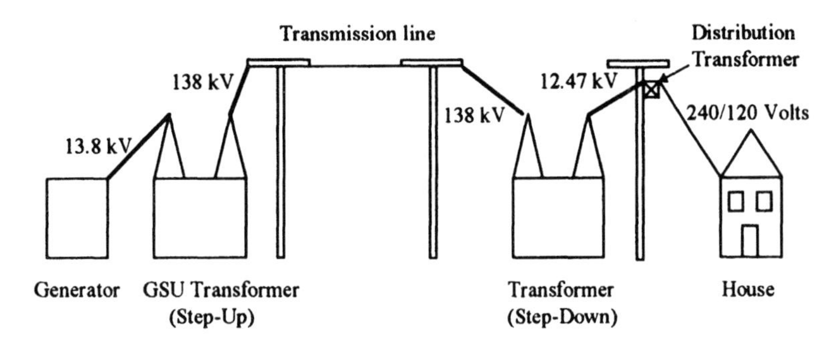

Figure 1.2 Schematic drawing of a power system

In modern power systems, there is usually more than one voltage step-down from transmission to final distribution, each step down requiring a transformer. Fig. 1.3 shows a transformer situated in a switch yard. The transformer takes input power from a high voltage line and converts it to lower voltage power for local use. The secondary power could be further stepped down in voltage before reaching the final consumer. This transformer could supply power to a large number of these smaller step down transformers. A transformer of the size shown could support a large factory or a small town.

Figure 1.3 Transformer located in a switching station, surrounded by auxiliary equipment

There is often a need to make fine voltage adjustments to compensate for voltage drops in the lines and other equipment. These voltage drops depend on the load current so they vary throughout the day. This is accomplished by equipping transformers with tap changers. These are devices which add or subtract turns from a winding, thus altering its voltage. This process can occur under load conditions or with the power disconnected from the transformer. The corresponding devices are called respectively load or no-load tap
changers. Load tap changers are typically sophisticated mechanical devices which can be remotely controlled. The tap changes can be made to occur automatically when the voltage levels drop below or rise above certain predetermined values. Maintaining nominal or expected voltage levels is highly desirable since much electrical equipment is designed to operate efficiently and sometimes only within a certain voltage range. This is particularly true for solid state equipment. No-load tap changing is usually performed manually. This type of tap changing can be useful if long term drifts are occurring in the voltage level. Thus it is done infrequently. Fig. 1.4 shows three load tap changers and their

connections to three windings of a power transformer. The same transformer can be equipped with both types of tap changers.

Figure 1.4 Three load tap changers attached to three windings of a power transformer. These tap changers were made by the Maschinenfabrik Reinhausen Co., Germany.

Most power systems today are three phase systems, i.e. they produce sinusoidal voltages and currents in three separate circuits which are displaced in time relative to each other by 1/3 of a cycle or 120 electrical degrees as shown in Fig. 1.5. Note that, at any instant of time, the 3 voltages sum to zero. Such a system made possible the use of generators and motors without commutators which were cheaper and safer to operate. Thus transformers were required which transformed all 3 phase voltages. This could be accomplished by using 3 separate transformers, one for each phase, or more commonly by combining all 3 phases within a single unit, permitting some economies particularly in the core structure. A sketch of such a unit is shown in Fig. 1.6. Note that the three fluxes produced by the different phases are, like the voltages and currents, displaced in time by 1/3 of a cycle relative to each other. This means that, when they overlap in the top or bottom yokes of the core, they cancel each other out. Thus the yoke steel does not have to be designed to carry more flux than is produced by a single phase.

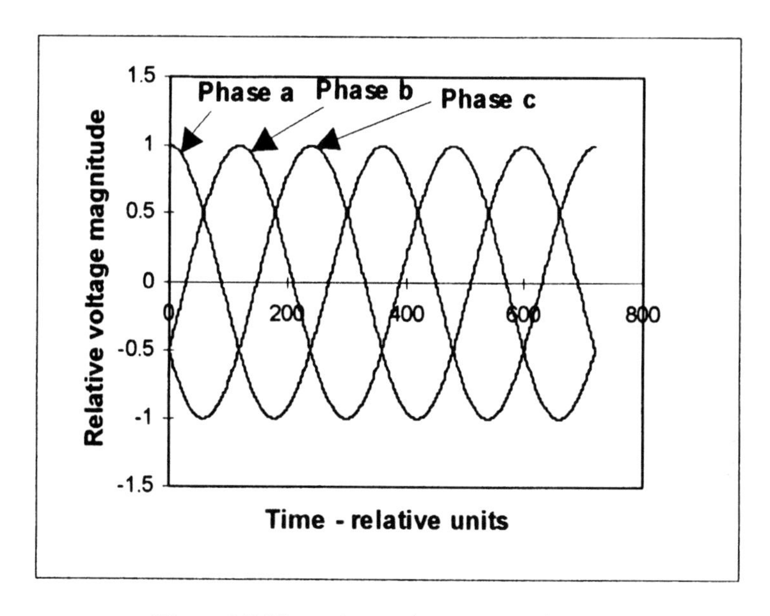

Figure 1.5 Three phase voltages versus time.

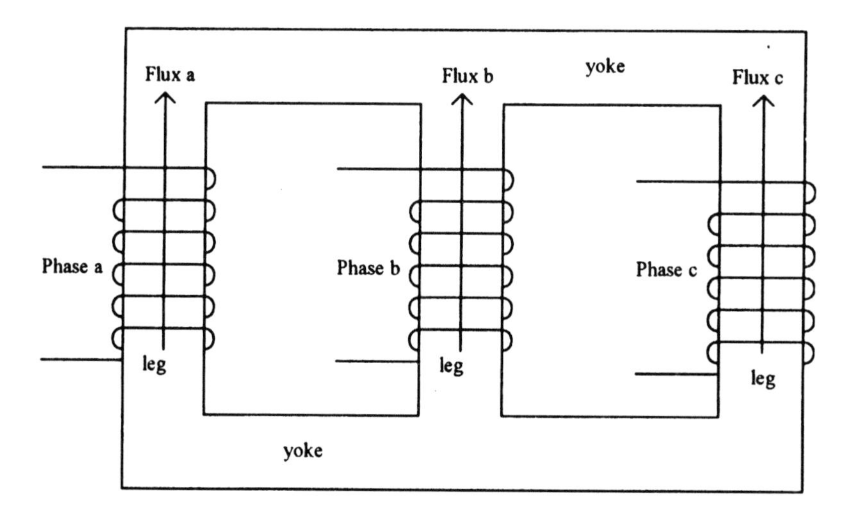

Figure 1.6 Three phase transformer utilizing a 3 phase core

At some stages in the power distribution system, it is desirable to furnish single phase power. For example, this is the common form of household power. To accomplish this, only one of the output circuits of a

3 phase unit is used to feed power to a household or group of households. The other circuits feed similar groups of households. Because of the large numbers of households involved, on average each phase will be equally loaded.

Because modern power systems are interconnected so that power can be shared between systems, sometimes voltages do not match at interconnection points. Although tap changing transformers can adjust the voltage magnitudes, they do not alter the phase angle. A phase angle mismatch can be corrected with a phase shifting transformer. This inserts an adjustable phase shift between the input and output voltages and currents. Large power phase shifters generally require two 3 phase cores housed in separate tanks. A fixed phase shift, usually of 30°, can be introduced by suitably interconnecting the phases of standard 3 phase transformers, but this is not adjustable.

Transformers are fairly passive devices containing very few moving parts. These include the tap changers and cooling fans which are needed on most units. Sometimes pumps are used on oil filled transformers to improve cooling. Because of their passive nature, transformers are expected to last a long time with very little maintenance. Transformer lifetimes of 25 - 50 years are common. Often units will be replaced before their useful life is up because of improvements in losses, efficiency, and other aspects over the years. Naturally a certain amount of routine maintenance is required. In oil filled transformers, the oil quality must be checked periodically and filtered or replaced if necessary. Good oil quality insures sufficient dielectric strength to protect against electrical breakdown. Key transformer parameters such oil and winding temperatures, voltages, currents, and oil quality as reflected in gas evolution are monitored continuously in many power systems. These parameters can then be used to trigger logic devices to take corrective action should they fall outside of acceptable operating limits. This strategy can help prolong the useful operating life of a transformer. Fig. 1.7 shows the end of a transformer tank where a control cabinet is located which houses the monitoring circuitry. Also shown projecting from the sides are radiator banks equipped with fans. This transformer is fully assembled and is being moved to the testing location in the plant.

Figure 1.7 End view of a transformer tank showing the control cabinet which houses the electronics.

1.3 CORE-FORM AND SHELL-FORM TRANSFORMERS

Although transformers are primarily classified according to their function in a power system, they also have subsidiary classifications according to how they are constructed. As an example of the former type of classification, we have generator step-up transformers which are connected directly to the generator and raise the voltage up to the line transmission level or distribution transformers which are the final step in a power system, transferring single phase power directly to the household or customer. As an example of the latter type of classification, perhaps the most important is the distinction between core-form and shell-form transformers.

The basic difference between a core-form and shell-form transformer is illustrated in Fig. 1.8. In a core-form design, the coils are wrapped or stacked around the core. This lends itself to cylindrical shaped coils. Generally high voltage and low voltage coils are wound concentrically, with the low voltage coil inside the high voltage one. In the 2 coil split shown in Fig. 1.8a, each coil group would consist of both high and low voltage windings. This insures better magnetic coupling between the coils. In the shell form design, the core is wrapped or stacked around the coils. This lends itself to flat oval shaped coils called pancake coils, with the high and low voltage windings stacked on top of each other, generally in more than one layer each in an alternating fashion.

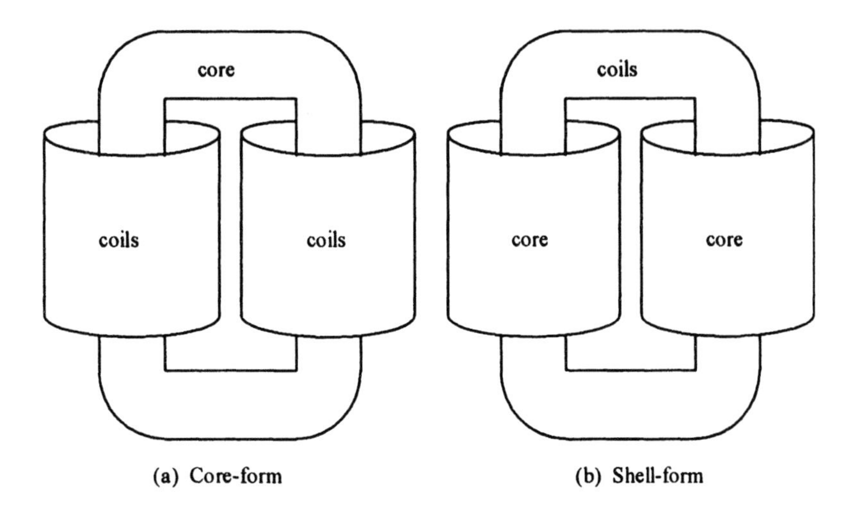

Figure 1.8 Single phase core-form and shell-form transformers contrasted

Each of these types of construction has its advantages and disadvantages. Perhaps the ultimate determination between the two comes down to a question of cost. In distribution transformers, the shell form design is very popular because the core can be economically wrapped around the coils. For moderate to large power transformers, the core-form design is more common, possibly because the short circuit forces can be better managed with cylindrically shaped windings.

1.4 STACKED AND WOUND CORE CONSTRUCTION

In both core-form and shell-form types of construction, the core is made of thin layers or laminations of electrical steel, especially developed for its good magnetic properties. The magnetic properties are best along the rolling direction so this is the direction the flux should naturally want to take in a good core design. The laminations can be wrapped around the coils or stacked. Wrapped or wound cores have few, if any, joints so they carry flux nearly uninterrupted by gaps. Stacked cores have gaps at the corners where the core steel changes direction. This results in poorer magnetic characteristics than for wound cores. In larger power transformers, stacked cores are much more common while in small distribution transformers, wound cores predominate. The laminations for both types of cores are coated with an insulating coating to prevent large eddy current paths from developing which would lead to high losses.

In one type of wound core construction, the core is wound into a continuous "coil". The core is then cut so that it can be inserted around the coils. The cut laminations are then shifted relative to each other and reassembled to form a staggered stepped type of joint. This type of joint allows the flux to make a smoother transition over the cut region than would be possible with a butt type of joint where the laminations are not staggered. Very often, in addition to cutting, the core is reshaped into a rectangular shape to provide a tighter fit around the coils. Because the reshaping and cutting operations introduce stress into the steel which is generally bad for the magnetic properties, these cores need to be re-annealed before use to help restore these properties. A wound core without a joint would need to be wound around the coils or the coils would need to be wound around the core. Techniques for doing this are available but somewhat costly.

In stacked cores for core-form transformers, the coils are circular cylinders which surround the core. Therefore the preferred cross-section of the core is circular since this will maximize the flux carrying area. In practice, the core is stacked in steps which approximates a circular cross-section as shown in Fig. 1.9. Note that the laminations are coming out of the paper and carry flux in this direction which is the sheet rolling direction. The space between the core and innermost coil is needed to provide insulation clearance for the voltage difference between the winding and the core which is at ground potential. It is also used for structural elements.

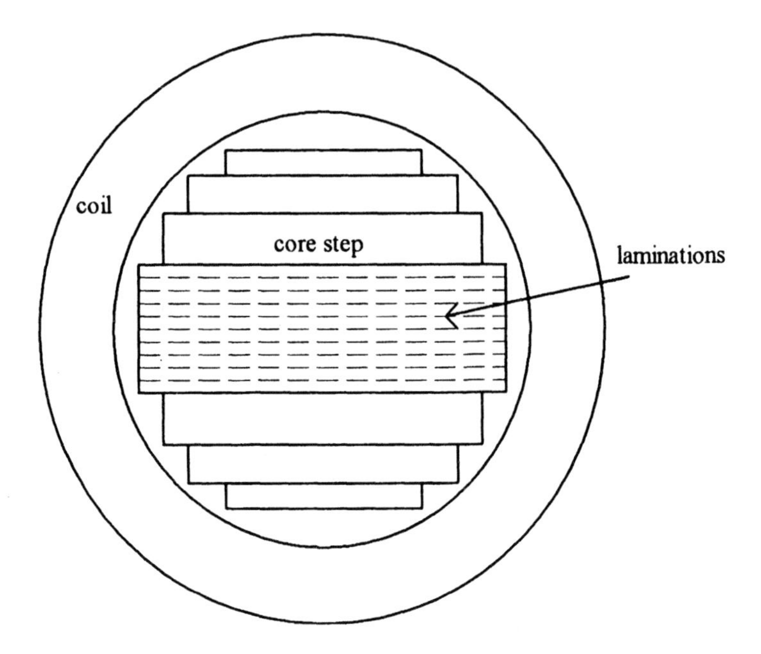

Figure 1.9 Stepped core used in core-form transformers to approximate a circular cross-section

For a given number of steps, one can maximize the core area to obtain an optimal stacking pattern. Fig. 1.10 shows the geometric parameters which can be used in such an optimization, namely the x and y coordinates of the stack corners which touch the circle of radius R. Only 1/4 of the geometry is modeled due to symmetry considerations.

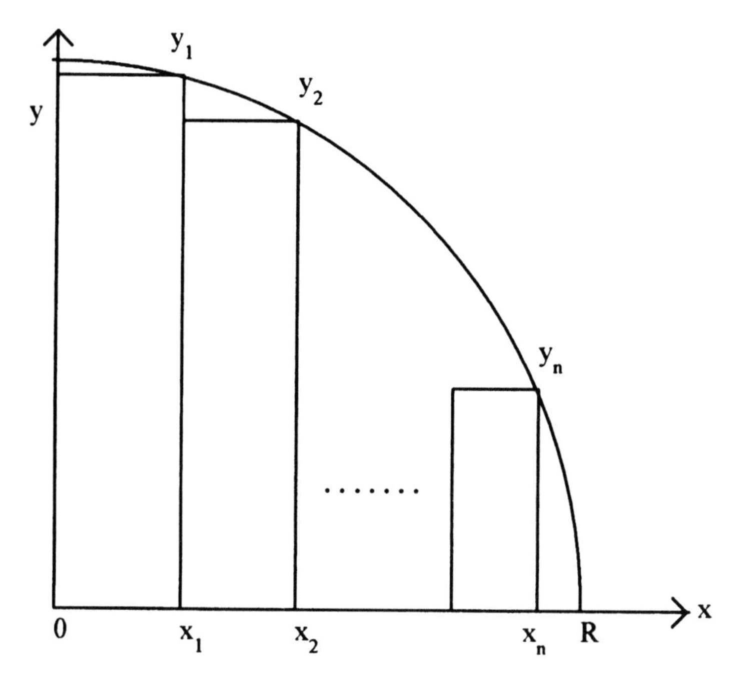

Figure 1.10 Geometric parameters for finding the optimum step pattern

The corner coordinates must satisfy

$$x_i^2 + y_i^2 = R^2 \tag{1.6}$$

For a core with n steps, where n refers to the number of stacks in half the core cross-section, the core area, A_n, is given by

$$A_n = 4\sum_{i=1}^{n}(x_i - x_{i-1})y_i = 4\sum_{i=1}^{n}(x_i - x_{i-1})\sqrt{R^2 - x_i^2} \tag{1.7}$$

where $x_0 = 0$. Thus the independent variables are the x_i since the y_i can be determined from them using (1.6) To maximize A_n, we need to solve the n equations

$$\frac{\partial A_n}{\partial x_i} = 0 \tag{1.8}$$

We can show that

$$\frac{\partial^2 A_n}{\partial x_i^2} < 0 \tag{1.9}$$

so that the solution to (1.8) does represent a maximum. Inserting (1.7) into (1.8), we get after some algebraic manipulation,

$$\left(R^2 - x_i^2\right)\left[x_{i+1}^2 - x_i(3x_i - 2x_{i-1})\right] + x_i^2(x_i - x_{i-1})^2 = 0\,, i = 1,\ldots,n \tag{1.10}$$

In the first and last equations (i = 1 and i = n), we need to use $x_o = 0$ and $x_{n+1} = R$.

Since (1.10) represents a set of non-linear equations, an approximate solution scheme such as a Newton-Raphson iteration can be used to solve them. Note that these equations can be normalized by dividing by R^4 so that the normalized solution coordinates x_i/R are independent of R. Table 1.1 gives the normalized solution for various numbers of steps.

Table 1.1 Normalized x coordinates which maximize the core area for a given number of steps.

# Steps n	Fract of circle occupied $A_n/\pi R^2$	Normalized x coordinates, x_i/R
1	.6366	.7071
2	.7869	.5257, .8506
3	.8510	.4240, .7070, .9056
4	.8860	.3591, .6064, .7951, .9332
5	.9079	.3138, .5336, .7071, .8457, .9494
6	.9228	.2802, .4785, .6379, .7700, .8780, .9599
7	.9337	.2543, .4353, .5826, .7071, .8127, .9002, .9671
8	.9419	.2335, .4005, .5375, .6546, .7560, .8432, .9163, .9723
9	.9483	.2164, .3718, .4998, .6103, .7071, .7921, .8661, .9283, .9763
10	.9534	.2021, .3476, .4680, .5724, .6648, .7469, .8199, .8836, .9376, .9793

In practice, because only a limited number of standard sheet widths are kept in inventory and because stack heights are also descretized, at least by the thickness of an individual sheet, it is not possible to achieve the ideal coverage given in the table. Fig. 1.11 shows a 3 phase stepped core for a core-form transformer without the top yoke. This is added after the coils are inserted over the legs. The bands around the legs are made of a high strength non-conducting material. They help hold the laminations together and prevent them from vibrating in service. Such vibrations are a source of noise.

Figure 1.11 Three phase stepped core for a core-form transformer without the top yoke.

1.5 TRANSFORMER COOLING

Because power transformers are greater than 99% efficient, the input and output power are nearly the same. However because of the small inefficiency, there are losses inside the transformer. The sources of these losses are I^2R losses in the conductors, losses in the electrical steel due to the changing flux which it carries, and losses in metallic tank walls and other metallic structures caused by the stray time varying flux. These losses lead to temperature rises which must be controlled by cooling. The primary cooling media for transformers are oil and air. In oil cooled transformers, the coils and core are immersed in an oil filled tank. The oil is then circulated through radiators or other types of heat exchanger so that the ultimate cooling medium is the surrounding air or possibly water for some types of heat exchangers. In small distribution transformers, the tank surface in contact with the air provides enough cooling surface so that radiators are not needed. Sometimes in these units the tank surface area is augmented by means of fins or corrugations.

The cooling medium in contact with the coils and core must provide adequate dielectric strength to prevent electrical breakdown or discharge between components at different voltage levels. For this reason, oil immersion is common in higher voltage transformers since oil has a higher breakdown strength than air. Often one can rely on the natural convection of oil through the windings, driven by buoyancy effects, to provide adequate cooling so that pumping isn't necessary. Air is a more efficient cooling medium when it is blown by means of fans through the windings for air cooled units.

In some applications, the choice of oil or air is dictated by safety considerations such as the possibility of fires. For units inside buildings, air cooling is common because of the reduced fire hazard. While transformer oil is combustible, there is usually little danger of fire since the transformer tank is often sealed from the outside air or the oil surface is blanketed with an inert gas such as nitrogen. Although the flash point of oil is quite high, if excessive heating or sparking occurs inside an oil filled tank, combustible gasses could be released.

Another consideration in the choice of cooling is the weight of the transformer. For mobile transformers such as those used on planes or trains or units designed to be transportable for emergency use, air cooling might be preferred since oil adds considerably to the overall weight. For units not so restricted, oil is the preferred cooling medium so that one finds oil cooled transformers in general use from large generator or substation units to distribution units on telephone poles.

There are other cooling media which find limited use in certain applications. Among these is sulfur hexaflouride gas, usually pressurized. This is a relatively inert gas which has a higher breakdown strength than air and finds use in high voltage units where oil is ruled out for reasons such as those mentioned above and where air doesn't provide enough dielectric strength. Usually when referring to oil cooled transformers, one means that the oil is standard transformer oil. However there are other types of oil which find specialized usage. One of these is silicone oil. This can be used at a higher temperature than standard transformer oil and at a reduced fire hazard.

1.6 WINDING TYPES

For core-form power transformers, there are two main methods of winding the coils. These are sketched in Fig. 1.12. Both types are cylindrical coils, having an overall rectangular cross-section. In a disk coil, the turns are arranged in horizontal layers called disks which are wound alternately out-in, in-out, etc. The winding is usually continuous so that the last inner or outer turn gradually transitions between the adjacent layers. When the disks have only one turn, the winding is called a helical winding. The total number of turns will usually dictate whether the winding will be a disk or helical winding. The turns within a disk are usually touching so that a double layer of insulation separates the metallic conductors. The space between the disks is left open, except for structural separators called key spacers. This allows room for cooling fluid to flow between the disks, in addition to providing clearance for withstanding the voltage difference between them.

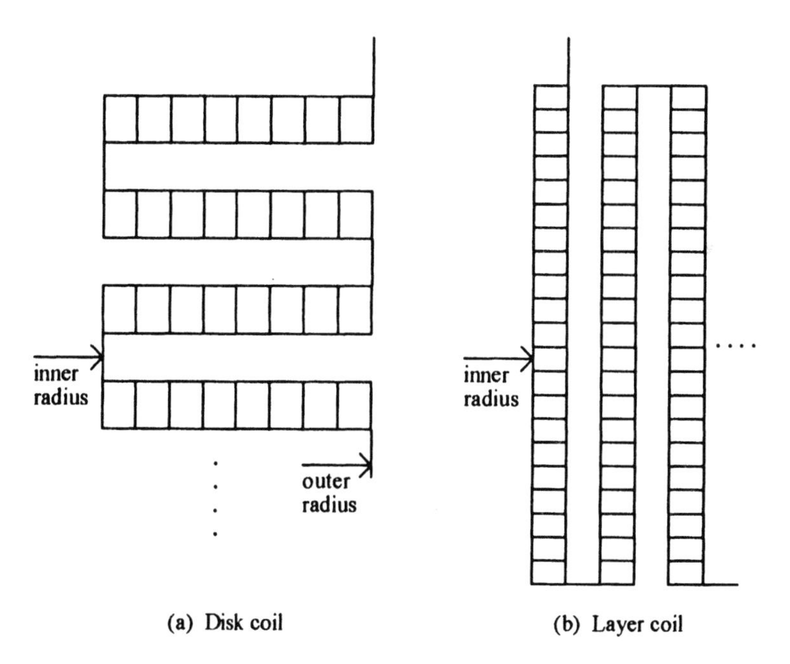

Figure 1.12 Two major types of coil construction for core-form power transformers

In a layer coil, the coils are wound in vertical layers, top-bottom, bottom-top, etc. The turns are typically wound in contact with each other in the layers but the layers are separated by means of spacers so that cooling fluid can flow between them. These coils are also usually continuous with the last bottom or top turn transitioning between the layers.

Both types of winding are used in practice. Each type has its proponents. In certain applications, one or the other type may be more efficient. However, in general they can both be designed to function well in terms of ease of cooling, ability to withstand high voltage surges, and mechanical strength under short circuit conditions.

If these coils are wound with more than one wire or cable in parallel, then it is necessary to insert cross-overs or transpositions which interchange the positions of the parallel cables at various points along the winding. This is done to cancel loop voltages induced by the stray flux. Otherwise such voltages would drive currents around the loops formed when the parallel turns are joined at either end of the winding, creating extra losses.

The stray flux also causes localized eddy currents in the conducting wire whose magnitude depends on the wire cross-sectional dimensions. These eddy currents and their associated losses can be reduced by subdividing the wire into strands of smaller cross-sectional dimensions. However these strands are then in parallel and must therefore be transposed to reduce the loop voltages and currents. This can be done during the winding process when the parallel strands are wound individually. Wire of this type, consisting of individual strands covered with an insulating paper wrap, is called magnet wire. The transpositions can also be built into the cable. This is called continuously transposed cable and generally consists of a bundle of 5 - 83 strands, each covered with a thin enamel coating. One strand at a time is transposed along the cable every 12 to 16 times its width so that all the strands are eventually transposed approximately every 25 - 50 cm (10 - 20 in) along the length of the cable. The overall bundle is then sheathed in a paper wrap.

Fig. 1.13 shows a disk winding situated over inner windings and core and clamped at either end via the insulating blocks and steel structure shown. Leads emerging from the top and bottom of one of the inner windings are also visible on the right. The staggered short horizontal gaps shown are transition points between disks. Vertical columns of key spacer projections are also barely visible. This outer high voltage winding is center fed so that the top and bottom halves are connected in parallel. The leads feeding this winding are on the left.

Figure 1.13 Disk winding shown in position over inner windings and core. Clamping structures and leads are also shown.

1.7 INSULATION STRUCTURES

Transformer windings and leads must operate at high voltages relative to the core, tank, and structural elements. In addition, different windings and even parts of the same winding operate at different voltages. This requires that some form of insulation between these various parts be provided to prevent voltage breakdown or corona discharges. The surrounding oil or air which provides cooling has some insulating value. The oil is of a special composition and must be purified to remove small particles and moisture. The type of oil most commonly used, as mentioned previously, is called transformer oil. Further insulation is provided by paper covering over the wire or cables. When saturated with oil, this paper has a high insulation value. Other types of wire covering besides paper are sometimes used, mainly for specialty applications. Other insulating structures which are generally present in sheet form, often wrapped into a cylindrical shape, are made of pressboard. This is a material made of cellulose fibers which are compacted together into a fairly dense and rigid matrix. Key spacers, blocking material, and lead support structures are also commonly made of pressboard.

Although normal operating voltages are quite high, 10 - 500 kV, the transformer must be designed to withstand even higher voltages which can occur when lightning strikes the electrical system or when power is suddenly switched on or off in some part of the system. However infrequently these occur, they could permanently damage the insulation, disabling the unit, unless the insulation is designed to withstand them. Usually such events are of short duration. There is a time dependence to how the insulation breaks down. A combination of oil and pressboard barriers can withstand higher voltages for shorter periods of time. In other words, a short duration high voltage pulse is no more likely to cause breakdown than a long duration low voltage pulse. This means that the same insulation that can withstand normal operating voltages which are continuously present can also withstand the high voltages arising from lightning strikes or switching operations which are present only briefly. In order to insure that the abnormal voltages do not exceed the breakdown limits determined by their expected durations, lightning or surge arrestors are used to limit them. These arrestors thus guarantee that the voltages will not rise above a certain value so that breakdown will not occur, assuming their durations remain within the expected range.

Because of the different dielectric constants of oil or air and paper, the electric stresses are unequally divided between them. Since the oil dielectric constant is about half that of paper and air is even a smaller fraction of papers', the electric stresses are generally higher in oil or air than in the paper insulation. Unfortunately, oil or air has a lower breakdown stress than paper. In the case of oil, it has been found that subdividing the oil gaps by means of thin insulating barriers, usually made of pressboard, can raise the breakdown stress in the oil. Thus large oil gaps between the windings are usually subdivided by multiple pressboard barriers as shown schematically in Fig. 1.14. This is referred to as the major insulation structure. The oil gap thicknesses are maintained by means of long vertical narrow sticks glued around the circumference of the cylindrical pressboard barriers. Often the barriers are extended by means of end collars which curve around the ends of the windings to provide subdivided oil gaps at either end of the windings to strengthen these end oil gaps against voltage breakdown.

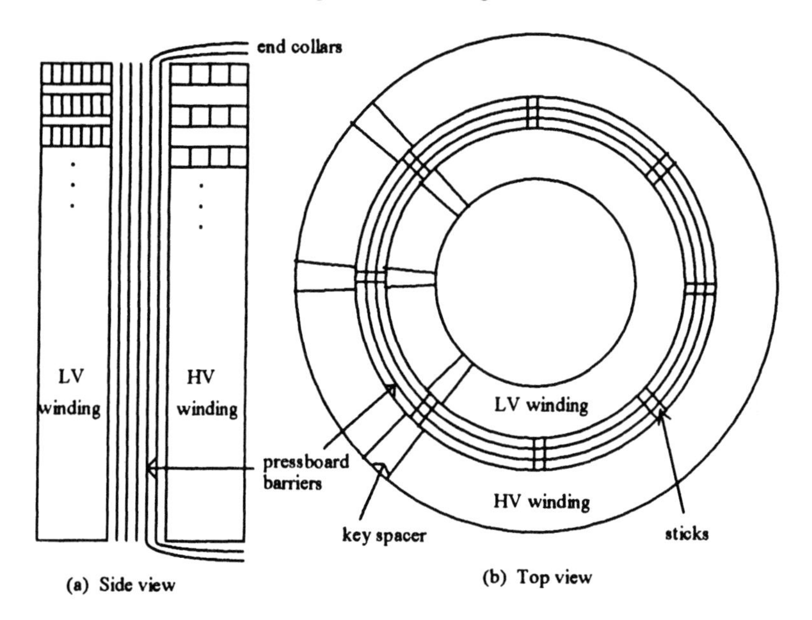

Figure 1.14 Major insulation structure consisting of multiple barriers between windings. Not all the keyspacers or sticks are shown.

The minor insulation structure consists of the smaller oil gaps separating the disks and maintained by the key spacers which are narrow insulators, usually made of pressboard, spaced radially around the disk's circumference as shown in Fig. 1.14b. Generally these oil gaps are small enough that subdivision is not required. In addition the turn to turn insulation, usually made of paper, can be considered as part of the minor insulation structure. Fig. 1.15 shows a pair of windings as seen from the top. The finger is pointing to the major insulation structure between the windings. Key spacers and vertical sticks are also visible.

Figure 1.15 Top view of two windings showing the major insulation structure, key spacers, and sticks

The leads which connect the windings to the bushings or tap changers or to other windings must also be properly insulated since they are at high voltage and pass close to tank walls or structural supports which are grounded. They also can pass close to other leads at different voltages. High stresses can be developed at bends in the leads, particularly if they are sharp, so that additional insulation may be required in these areas. Fig. 1.16 shows a rather extensive set of leads along with structural supports made of pressboard. The leads pass close to the metallic clamps at the top and bottom and will also be near the tank wall when the core and coil assembly is inserted into the tank.

Figure 1.16 Leads and their supporting structure emerging from the coils on one side of a 3 phase transformer

Although voltage breakdown levels in oil can be increased by means of barrier subdivisions, there is another breakdown process which must be guarded against. This is breakdown due to creep. It occurs along the surfaces of the insulation. It requires sufficiently high electric stresses directed along the surface as well as sufficiently long uninterrupted paths over which the high stresses are present. Thus the barriers themselves, sticks, key spacers, and lead supports can be a source of breakdown due to creep. Ideally one should position these insulation structures so that their surfaces conform to voltage equipotential surfaces to which the electric field is perpendicular. Thus there would be no electric fields directed along the surface. In practice, this is not always possible so that a compromise must be reached.

The major and minor insulation designs, including overall winding to winding separation and number of barriers as well as disk to disk separation and paper covering thickness, are often determined by design rules based on extensive experience. However, in cases of newer or unusual designs, it is often desirable to do a field calculation using a finite-element program or other numerical procedure. This can be especially helpful when the potential for creep breakdown exists. Although these methods can provide accurate calculations of electric stresses, the breakdown process is not as well understood so that there is usually some judgment involved in deciding what level of electrical stress is acceptable.

1.8 STRUCTURAL ELEMENTS

Under normal operating conditions, the electromagnetic forces acting on the transformer windings are quite modest. However, if a short circuit fault occurs, the winding currents can increase 10 - 30 fold, resulting in forces of 100 - 900 times normal since the forces increase as the square of the current. The windings and supporting structure must be designed to withstand these fault current forces without permanent distortion of the windings or supports. Because current protection devices are usually installed, the fault currents are interrupted after a few cycles.

Faults can be caused by falling trees which hit power lines, providing a direct current path to ground or by animals or birds bridging across two lines belonging to different phases, causing a line to line short. These should be rare occurrences but over the 20 - 50 year lifetime of a transformer, their probability increases so that sufficient mechanical strength to withstand these is required.

The coils are generally supported at the ends with pressure rings. These are thick rings of pressboard or other material which cover the winding ends. The center opening allows the core to pass through. The rings are in the range of 3 - 10 cm (1 -4 in) for large power transformers. Some blocking made of pressboard or wood is required between the tops of the windings and the rings since all of the windings are not of the same height. Additional blocking is usually placed between the ring and the top yoke and clamping structure to provide some clearance between the high winding voltages and the grounded core and clamp. These structures can be seen in Fig. 1.13. The metallic clamping structure can also be seen.

The top and bottom clamps are joined by vertical tieplates which pass along the sides of the core. The tieplates have threaded ends so they pull the top and bottom clamps together by means of tightening bolts, compressing the windings. These compressive forces are transmitted along the windings via the key spacers which must be strong enough in compression to accommodate these forces. The clamps and tieplates are made of steel. Axial forces which tend to elongate the windings when a fault occurs will put the tieplates in tension. The tieplates must also be strong enough to carry the gravitational load when the core and coils are lifted as a unit since the lifting hooks are attached to the clamps. The tieplates are typically about 1 cm (3/8 in) thick and of varying width depending on the expected short circuit forces and transformer weight. The width is often subdivided to reduce eddy current losses. Fig. 1.17 shows a top view of the clamping structure. The unit shown is being lifted by means of the lifting hooks.

The radial fault forces are countered inwardly by means of the sticks which separate the oil barriers and by means of additional support next to the core. The windings themselves, particularly the innermost one, are often made of hardened copper or bonded cable to provide additional resistance to the inward radial forces. The outermost winding is usually subjected to an outer radial force which puts the wires or cables in tension. The material itself must be strong enough to resist these tensile forces since there is no supporting structure on the outside to counter these forces. A measure of the material's strength is its proof stress. This is the stress required to produce a permanent elongation of 0.2% (sometimes 0.1% is used). Copper of specified proof stress can be ordered from the wire or cable supplier.

The leads are also acted on by extra forces during a fault. These are produced by the stray flux from the coils or from nearby leads interacting with the lead's current. The leads are therefore braced by means of wooden or pressboard supports which extend from the clamps. This lead support structure can be quite complicated, especially if there are many leads and interconnections. It is usually custom made for each unit. Fig. 1.16 is an example of such a structure.

Figure 1.17 Top view of clamping structure for a 3 phase transformer.

The assembled coil, core, clamps, and lead structure is placed in a transformer tank. The tank serves many functions, one of which is to contain the oil for an oil filled unit. It also provides protection not only for the coils and other transformer structures but for personnel from the high voltages present. If made of soft (magnetic) steel, it keeps stray flux from getting outside the tank. The tank is usually made airtight so that air doesn't enter and oxidize the oil.

Aside from being a containment vessel, the tank also has numerous attachments such as bushings for getting the electrical power into and out of the unit, an electronic control and monitoring cabinet for recording and transferring sensor information to remote processors and receiving control signals, and radiators with or without fans to provide cooling. On certain units, there is a separate tank compartment for tap changing equipment. Also some units have a conservator attached to the tank cover or to the top of the radiators. This is a large, usually cylindrical, structure which contains oil in communication with the main tank oil. It also has an air space which is separated from the oil by a sealed diaphragm. Thus, as the tank oil expands and contracts due to temperature changes, the flexible diaphragm accommodates these volume changes while maintaining a sealed oil environment. Fig 18 shows a large power transformer installed in a switchyard. The cylindrical conservator is visible on top of the radiator bank. The high and low voltage bushings which are mounted on the tank cover are visible. Also shown are the surge arrestors which in this case are mounted on top of the conservator.

Figure 1.18 Large power transformer showing tank and attachments.

1.9 THREE PHASE CONNECTIONS

There are two basic types of 3 phase connections in common use, the Y (Wye) and Δ (Delta) connections as illustrated schematically in Fig. 1.19. In the Y connection, all 3 phases are connected to a common point which may or may not be grounded. In the Δ connection, the phases are connected end to end with each other. In the Y connection, the line current flows directly into the winding where it is called the winding or phase current. Note that in a balanced 3 phase system, the currents sum to zero at the common node in Fig. 1.19a. Therefore, under balanced conditions, even if this point were grounded, no current would flow to ground. In the Δ connection, the line and phase currents are different. On the other hand, the line to line voltages in the Y connection differ from the voltages across the windings or phase voltages whereas they are the same in the Δ connection. Note that the coils are shown at angles to each other in the figure to emphasize the type of interconnection whereas in practice they are side by side and vertically oriented as shown for example in Fig. 1.16. The leads are snaked about to handle the interconnections.

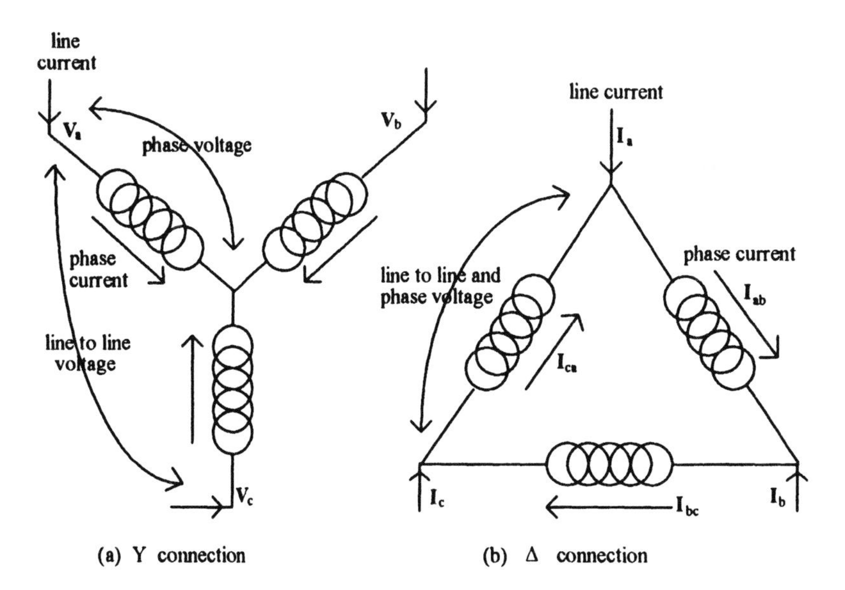

Figure 1.19 Basic 3 phase connections

To be more quantitative about the relationship between line and phase quantities, we must resort to phasor notation. The 3 phase voltages as shown in Fig. 1.5 can be written

$$V_a = V_o \cos(\omega t),\ V_b = V_o \cos(\omega t + 240°),\ V_c = V_o \cos(\omega t + 120°) \quad (1.11)$$

where V_o is the peak voltage, t is the time, and ω the angular frequency ($\omega = 2\pi f$, f the frequency in Hz). Actually the 240° and 120° above should be expressed in radians to be consistent with the expression for ω. When using degrees, $\omega = 360°f$ should be understood. Using the identity

$$e^{j\theta} = \cos\theta + j\sin\theta \quad (1.12)$$

where j is the imaginary unit $(j^2 = -1)$, (1.11) can be written

$$V_a = V_o \operatorname{Re}\{e^{j\omega t}\},\ V_b = V_o \operatorname{Re}\{e^{j(\omega t + 240°)}\},\ V_c = V_o \operatorname{Re}\{e^{j(\omega t + 120°)}\} \quad (1.13)$$

where Re denotes taking the real part of.

Complex quantities are more easily visualized as vectors in the complex plane. Thus (1.12) can be described as a vector of unit magnitude in the complex plane with real component $\cos\theta$ and imaginary component $\sin\theta$. This can be visualized as a unit vector starting at the origin and making an angle θ with the real axis. Any complex number can be described as such a vector but having, in general, a magnitude different from unity. Fig. 1.20 shows this pictorial description. As θ increases, the vector rotates in a counter clockwise fashion about the origin. If we let $\theta = \omega t$, then as time increases, the vector rotates with a uniform angular velocity ω about the origin. When dealing with complex numbers having the same time dependence, $e^{j\omega t}$, as in (1.13), it is customary to drop this term or to simply set $t = 0$. Since these vectors all rotate with the same angular velocity, their relative positions with respect to each other in the complex plane remain unchanged, i.e. if the angle between two such vectors is 120° at $t = 0$, it will remain 120° for all subsequent times. The resulting vectors, with common time dependence removed, are called phasors.

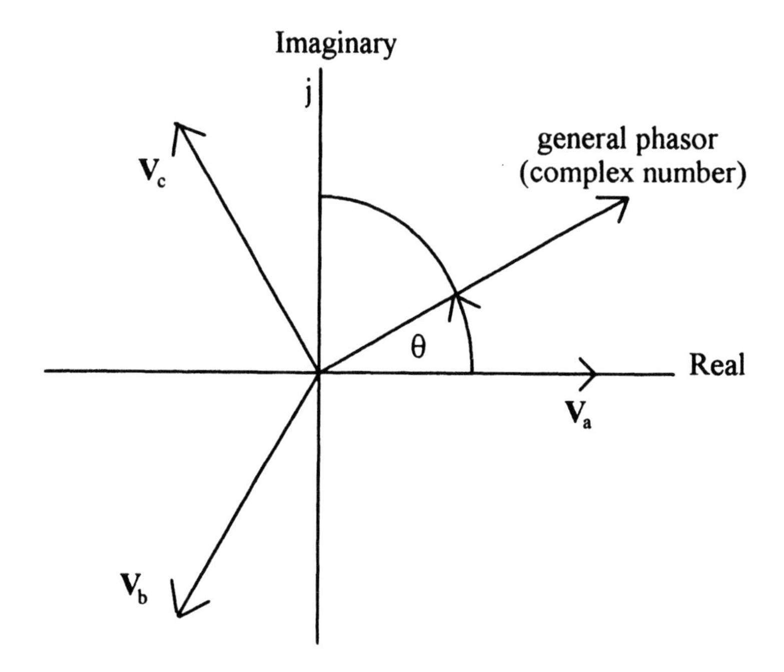

Figure 1.20 Phasors as vectors in the complex plane

Using bold faced type to denote phasors, from (1.13) define

$$\mathbf{V}_a = V_o \quad , \quad \mathbf{V}_b = V_o e^{j240^\circ} \quad , \quad \mathbf{V}_c = V_o e^{j120^\circ} \qquad (1.14)$$

Then to recover (1.13), multiply by the complex time dependence and take the real part. These phasors are also shown in Fig. 1.20. Note that the process of recovering (1.13), which is equivalent to (1.11), can be visualized as multiplying by the complex time dependence and taking their projections on the real axis as the phasors rotate counterclockwise while maintaining their relative orientations. Thus $\mathbf{V}_a$ peaks (has maximum positive value) at $t = 0$, $\mathbf{V}_b$ achieves its positive peak value next at $\omega t = 120°$ followed by $\mathbf{V}_c$ at $\omega t = 240°$. This ordering is called positive sequence ordering and $\mathbf{V}_a$, $\mathbf{V}_b$, $\mathbf{V}_c$ are referred to as a positive sequence set of voltages. A similar set of phasors can also be used to describe the currents. Note that, using vector addition, this set of phasors adds to zero as is evident in Fig. 1.20.

If $\mathbf{V}_a$, $\mathbf{V}_b$, $\mathbf{V}_c$ are the phase voltages in a Y connected set of transformer windings, let $\mathbf{V}_{ab}$ denote the line to line voltage between phases a and b , etc. for the other line to line voltages. Then these voltages are given by,

$$V_{ab} = V_a - V_b \quad , \quad V_{bc} = V_b - V_c \quad , \quad V_{ca} = V_c - V_a \tag{1.15}$$

Using a phasor description, these can be readily calculated. The line to line phasors are shown graphically in Fig. 1.21. Fig. 1.21a shows the vector subtraction process explicitly and Fig. 1.21b shows the set of 3 line to line voltage phasors. Note that these form a positive sequence set that is rotated 30° relative to the phase voltages. The line to line voltage magnitude can be found geometrically as the diagonal of the parallelogram formed by equal length sides making an angle of 120° with each other. Thus we have

$$\begin{aligned} |\mathbf{V}_{ab}| = |\mathbf{V}_a - \mathbf{V}_b| &= \sqrt{|\mathbf{V}_a|^2 + |\mathbf{V}_b|^2 - 2|\mathbf{V}_a||\mathbf{V}_b|\cos 120^\circ} \\ &= \sqrt{|\mathbf{V}_a|^2 + |\mathbf{V}_b|^2 + |\mathbf{V}_a||\mathbf{V}_b|} = \sqrt{3}|\mathbf{V}_a| \end{aligned} \tag{1.16}$$

where $|\ |$ denotes taking the magnitude. We have used the fact that the different phase voltages have the same magnitude.

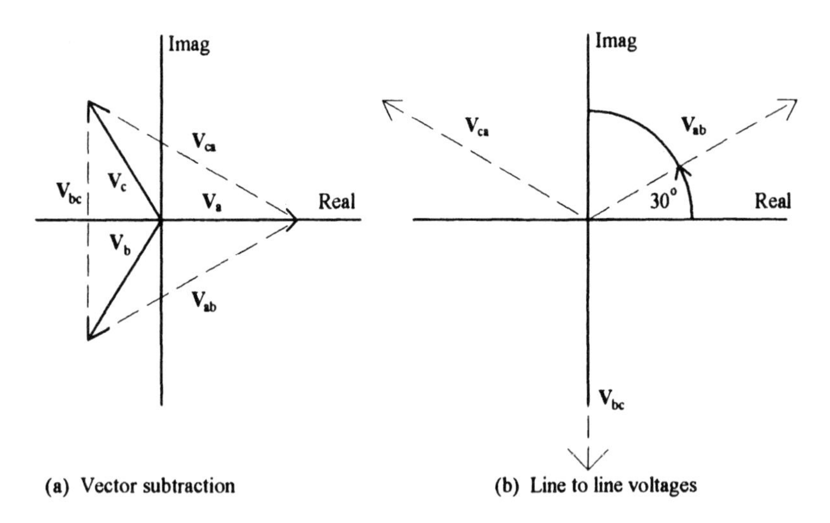

Figure 1.21 Phasor representation of line to line voltages and their relation to the phase voltages in a Y connected set of 3 phase windings

Thus the magnitude of the line to line voltage is √3 times the phase voltage magnitude in Y connected transformer coils. Since the phase voltages are internal to the transformer and therefore impact the winding insulation structure, a more economical design is possible if these can be lowered. Hence, a Y connection is often used for the high voltage coils of a 3 phase transformer.

The relationship between the phase and line currents of a delta connected set of windings can be found similarly. From Fig. 1.19, we see that the line currents are given in terms of the phase currents by

$$\mathbf{I}_a = \mathbf{I}_{ab} - \mathbf{I}_{ca} \quad , \quad \mathbf{I}_b = \mathbf{I}_{bc} - \mathbf{I}_{ab} \quad , \quad \mathbf{I}_c = \mathbf{I}_{ca} - \mathbf{I}_{bc} \tag{1.17}$$

These are illustrated graphically in Fig. 1.22. Note that, as shown in Fig. 1.22b, the line currents form a positive sequence set rotated -30° relative to the phase currents. One could also say that the phase currents are rotated +30° relative to the line currents. The magnitude relationship between the phase and line currents follows similarly as for the voltages in a Y connection. However, let us use phasor subtraction directly in (1.17) to show this. We have

$$\begin{aligned}\mathbf{I}_a &= |\mathbf{I}_{ab}|e^{j0^\circ} - |\mathbf{I}_{ca}|e^{j120^\circ} = |\mathbf{I}_{ab}| - |\mathbf{I}_{ca}|(\cos 120^\circ + j\sin 120^\circ)\\ &= |\mathbf{I}_{ab}| - |\mathbf{I}_{ca}|\left(-\frac{1}{2} + j\frac{\sqrt{3}}{2}\right) = |\mathbf{I}_{ab}|\left[\left(1+\frac{1}{2}\right) - j\frac{\sqrt{3}}{2}\right]\\ &= |\mathbf{I}_{ab}|\sqrt{\frac{9}{4}+\frac{3}{4}}e^{-j\tan^{-1}(1/\sqrt{3})} = \sqrt{3}|\mathbf{I}_{ab}|e^{-j30^\circ}\end{aligned} \tag{1.18}$$

where we have used $|\mathbf{I}_{ab}| = |\mathbf{I}_{ca}|$.

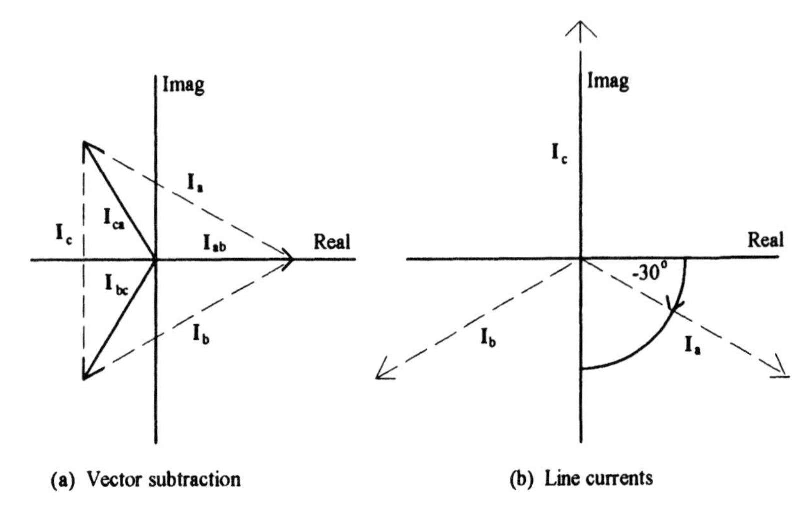

Figure 1.22 Phasor representation of line currents and their relation to the phase currents in a delta connected set of 3 phase windings

Thus the magnitude of the line current into or out of a delta connected set of windings are √3 times the winding or phase current magnitude. Since low voltage windings have higher phase currents than high voltage windings, it is common to connect these in delta because lowering the phase currents can produce a more economical design.

The rated power into a phase of a 3 phase transformer is the terminal voltage to ground times the line current. We can deal with magnitudes only here since rated power is at unity power factor. Thus for a Y connected set of windings, the terminal voltage to ground equals the voltage across the winding and the line current equals the phase current. Hence the total rated power into all three phases is 3 × the phase voltage × the phase current which is the total winding power. In terms of the line to line voltages and line current, we have, using (1.16), that the total rated power is √3 × the line to line voltage × the line current.

For a delta connected set of windings, there is considered to be a virtual ground at the center of the delta. From the geometrical relationships we have developed above, the line voltage to ground is therefore the line to line voltage ÷ √3. Thus the total rated power is 3 × the line voltage to ground × the line current = √3 × line to line voltage × the line current. Since the phase voltage equals the line to line voltage and the phase current = 1/√3 × the line current according to (1.18), the total rated power can also be expressed as 3 × the phase voltage × the phase current or the total winding power. Thus the total rated power, whether expressed in terms of line quantities or phase quantities, is the same for Y and delta connected windings. We also note that the rated input power is the same as the power flowing through the windings. This may be obvious here, but there are some connections where this is not the case, in particular autotransformer connections.

An interesting 3 phase connection is the open delta connection. In this connection, one of the windings of the delta is missing although the terminal connections remain the same. This is illustrated in Fig. 1.23. This could be used especially if the 3 phases consist of separate units and if one of the phases is missing either because it is intended for future expansion or it has been disabled for some reason. Thus instead of (1.17) for the relationship between line and phase currents, we have

$$I_a = I_{ab} - I_{ca} \quad , \quad I_b = -I_{ab} \quad , \quad I_c = I_{ca} \tag{1.19}$$

But this implies

$$I_a + I_b + I_c = 0 \tag{1.20}$$

so that the terminal currents form a balanced 3 phase system. Likewise the terminal voltages form a balanced 3 phase system. However only 2 phases are present in the windings. As far as the external electrical system is concerned, the 3 phases are balanced. The total rated input or output power is, as before, 3 × line voltage to ground × line current = √3 × line to line voltage × line current. But as (1.19) shows, the line current has the same magnitude as the phase or winding current. Also, as before, the line to line voltage equals the voltage across the winding or phase voltage. Hence the total rated power is √3 × phase voltage × phase current. Previously for a full delta connection, we found that the total rated power was 3 × phase voltage × phase current. Thus for the open delta, the rated power is only 1/√3 = 0.577 times that of a full delta connection. As far as winding utilization goes, in the full delta connection 2 windings carry 2/3 × 3 × phase voltage × phase current = 2 × phase voltage × phase current. Thus the winding power utilization in the open delta connection is √3/2 = 0.866 times that of a full delta connection. Thus this connection is not as efficient as a full delta connection.

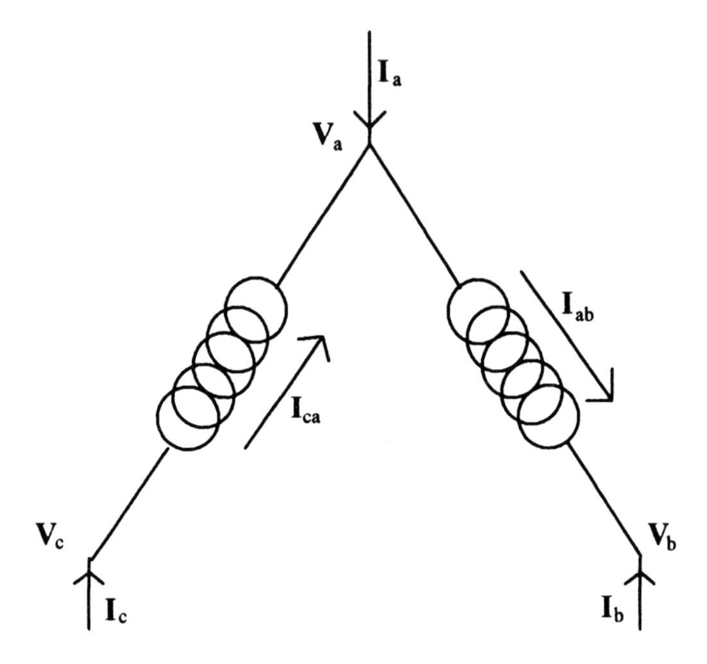

Figure 1.23 Open delta connection

1.10 MODERN TRENDS

Changes in power transformers tend to occur very slowly. Issues of reliability over long periods of time and compatibility with existing systems must be addressed by any new technology. A major change which has been ongoing since the earliest transformers is the improvement in core steel. The magnetic properties, including losses, have improved dramatically over the years. Better stacking methods, such as stepped lapped construction, have resulted in lower losses at the joints. The use of laser or mechanical scribing has also helped lower the losses in these steels. Further incremental improvements in all of these areas can be expected.

The development of amorphous metals as a core material is relatively new. Although these materials have very low losses, lower than the best rolled electrical steels, they also have a rather low saturation induction (~1.5 Tesla versus 2.1 Tesla for rolled steels). They are also rather brittle and difficult to stack. This material has tended to be more expensive than rolled electrical steel and, since expense is always an issue, has limited their use. However, this could change with the cost of no-load losses to the utilities. Amorphous metals have found use as wound cores in distribution transformers. However, their use as stacked cores in large power transformers is problematic.

The development of improved wire types, such as transposed cable with epoxy bonding, is an ongoing process. Newer types of wire insulation covering such as Nomex are being developed. Nomex is a synthetic material which can be used at higher temperatures than paper. It also has a lower dielectric constant than paper so it produces a more favorable stress level in the adjacent oil than paper. Although it is presently a more expensive material than paper, it has found a niche in air cooled transformers or in the rewinding of older transformers. Its thermal characteristics would probably be underutilized in transformer oil filled transformers because of the limitations on the oil temperatures.

Pressboard insulation has undergone improvements over time such as precompressing to produce higher density material which results in greater dimensional stability in transformer applications. This is especially helpful in the case of key spacers which bear the compressional forces acting on the winding. Also pre-formed parts made of pressboard, such as collars at the winding ends and high voltage lead insulation assemblies, are becoming more common and are facilitating the development of higher voltage transformers.

Perhaps the biggest scientific breakthrough which could revolutionize future transformers is the discovery of high temperature superconductors. These materials are still in the early stage of development. They could operate at liquid nitrogen temperatures which is a big improvement over the older superconductors which operate at liquid helium temperatures. It has been exceedingly difficult to make these new superconductors into wires of the lengths required in transformers. Nevertheless, prototype units are being built and technological improvements can be expected [Meh98].

A big change which is occurring in newer transformers is the increasing use of on-line monitoring devices. Fiber optic temperature sensors are being inserted directly into the windings to monitor the hottest winding temperature. This can be used to keep the transformer's loading or overloading within appropriate bounds so that acceptable insulation and adjacent oil temperatures are not exceeded and the thermal life is not too negatively impacted. Gas analysis devices are being developed to continuously record the amounts and composition of gasses in the cover gas or dissolved in the oil. This can provide an early indication of overheating or of arcing so that corrective action can be taken before the situation deteriorates too far. Newer fiber optic current sensors based on the Faraday effect are being developed. These weigh considerably less than present current sensors and are much less bulky. Newer miniaturized voltage sensors are also being developed. Sensor data in digitized form can be sent from the transformer to a remote computer for further processing. Newer software analysis tools should help to more accurately analyze fault conditions or operational irregularities.

Although tap changers are mechanical marvels which operate very reliably over hundreds of thousands of tap changing operations, as with any mechanical device, they are subject to wear and must be replaced or refurbished from time to time. Electronic tap changers, using solid state components, have been developed. Aside from essentially eliminating the wear problem, they also have a much faster response time than mechanical tap changers which could be very useful in some applications. Their expense relative to mechanical tap changers has been one factor limiting their use. Further developments perhaps resulting in lower cost can be expected in this area.

As mentioned previously, there are incentives to transmit power at higher voltages. Some of the newer high voltage transmission lines operate in a d.c. mode. In this case, the conversion equipment at the ends of the line which change a.c. to d.c. and vice versa requires a transformer. However this transformer does not need to operate at the line voltage. For high voltage a.c. lines, however, the transformer must operate at these higher voltages. At present, transformers which operate in the range of 750 - 800 kV have been built. Even higher voltage units have been developed, but this technology is still somewhat experimental. A better understanding of high voltage breakdown mechanisms, especially in oil, is needed to spur growth in this area.

2. TRANSFORMER CIRCUIT MODELS, INCLUDING MAGNETIC CORE CHARACTERISTICS, AND APPLICATIONS

Summary The characteristics of transformer cores are discussed in terms of their basic magnetic properties and how these influence transformer design. Special emphasis is placed on silicon steel cores since these are primarily used in power transformers. However, the magnetic concepts discussed are applicable to all types of cores. The magnetic circuit approximation is introduced and its use in obtaining the properties of cores with joints or gaps is discussed. Basic features of the magnetization process are used to explain inrush current and to calculate its magnitude. The inclusion of the transformer core in electrical circuit models is discussed. Although non-linearities in the magnetic characteristics can be included in these models, for inductions well below saturation a linear approximation is adequate. For many purposes, the circuit models can be further approximated by eliminating the core. As an application, an approximate circuit model is used to calculate the voltage regulation of a two winding transformer.

2.1 INTRODUCTION

Transformer cores are constructed predominantly of ferromagnetic material. The most common material used is iron, with the addition of small amounts of silicon and other elements which help improve the magnetic properties and/or lower losses. Other materials which find use in electronic transformers are the nickel-iron alloys (permalloys) and the iron-oxides (ferrites). The amorphous metals, generally consisting of iron, boron, and other additions, are also finding use as cores for distribution transformers. These materials are all broadly classified as ferromagnetic and, as such, share many properties in common. Among these are saturation magnetization or induction, hysteresis, and a Curie temperature above which they cease to be ferromagnetic.

Cores made of silicon steel (~ 3% Si) are constructed of multiple layers of the material in sheet form. The material is fabricated in rolling mills from hot slabs or ingots. Through a complex process of multiple rolling, annealing, and coating stages, it is formed into thin sheets of from 0.18 - 0.3 mm (7 - 11 mil) thickness and up to a meter (39 in) wide. The material has its best magnetic properties along the rolling direction and a well constructed core will take advantage of this. The good rolling direction magnetic properties are due to the underlying crystalline orientation which is called a Goss or cube-on-edge texture as shown in Fig. 2.1. The cubic crystals have the highest permeability along the cube edges. The visible edges pointing along the rolling direction are highlighted in the figure. Modern practice can achieve crystal alignments of > 95%. The permeability is much lower along the cube diagonals or cube face diagonals. The latter are pointing in the sheet width direction.

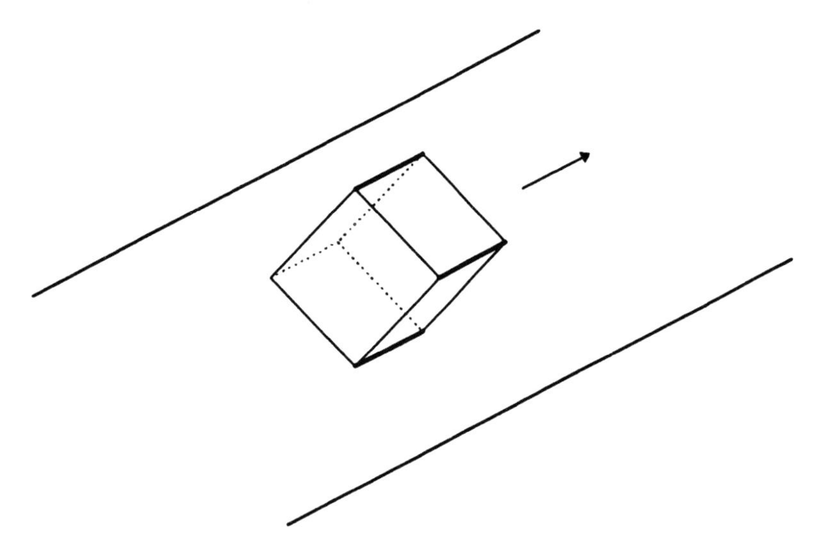

Figure 2.1 Goss or cube-on-edge crystalline texture for silicon steel.

In addition to its role in aiding crystal alignment, the silicon helps increase the resistivity of the steel from about 25 μΩ-cm for low carbon magnetic steel to about 50 μΩ-cm for 3% Si-Fe. This higher resistivity leads to lower eddy current losses. Silicon also lowers the saturation induction from about 2.1 T for low carbon steel to about 2.0 T for 3% Si-Fe. Silicon confers some brittleness on the material, which is an obstacle to rolling to even thinner sheet thicknesses. At higher silicon levels, the brittleness increases to the point where it becomes difficult to roll. This is unfortunate because at 6% silicon content, the magnetostriction of the steel disappears. Magnetostriction is a length change or strain which is produced by the induction in the material. This contributes to the noise level in a transformer.

The nickel-iron alloys or permalloys are also produced in sheet form. Because of their malleability, they can be rolled extremely thin. The sheet thinness results in very low eddy current losses so that these materials find use in high frequency applications. Their saturation induction is lower than that for silicon steel.

Ferrite cores are made of sintered power. They generally have isotropic magnetic properties. They can be cast directly into the desired shape or machined after casting. They have extremely high resistivities which permits their use in high frequency applications. However, they have rather low saturation inductions.

Amorphous metals are produced by directly casting the liquid melt onto a rotating, internally cooled, drum. The liquid solidifies extremely rapidly, resulting in the amorphous (non-crystalline) texture of the final product. The material comes off the drum in the form of a thin ribbon with controlled widths which can be as high as ~ 25 cm (10 in). The material has a magnetic anisotropy determined by the casting direction and subsequent magnetic anneals so that the best magnetic properties are along the casting direction. Their saturation induction is about 1.5 T. Because of their thinness and composition, they have extremely low losses. These materials are very brittle which has limited their use to wound cores. Their low losses make them attractive for use in distribution transformers, especially when no-load loss evaluations are high.

Ideally a transformer core would carry the flux along a direction of highest permeability and in a closed path. Path interruptions caused by joints, which are occupied by low permeability air or oil, lead to poorer overall magnetic properties. In addition, the cutting or slitting operations can introduce localized stresses which degrade the magnetic properties. In stacked cores, the joints are often formed by overlapping the laminations in steps to facilitate flux transfer across the joint. Nevertheless, the corners result in regions of higher loss. This can be accounted for in design by multiplying the ideal magnetic circuit losses, usually provided by the manufacturer on a per unit weight basis, by a building factor > 1. Another, possibly better, way to account for the extra loss is to apply a loss multiplying factor to the steel occupying the corner or joint region only. More fundamental methods to account for these extra losses have been proposed but these tend to be too elaborate for routine use. Joints also give rise to higher exciting current, i.e. the current in the coils necessary to drive the required flux around the core.

2.2 BASIC MAGNETISM

The discovery by Oersted that currents give rise to magnetic fields led Ampere to propose that material magnetism results from localized currents. He proposed that large numbers of small current loops, appropriately oriented, could create the magnetic fields associated with magnetic materials and permanent magnets. At the time, the atomic nature of matter was not understood. With the Bohr model of the atom, where electrons are in orbit around a small massive nucleus, the localized currents could be associated with the moving electron. This gives rise to an orbital magnetic moment which persists even though a quantum description has replaced the Bohr model. In addition to the orbital magnetism, the electron itself was found to possess a magnetic moment which cannot be understood simply from the circulating current point of view. Atomic magnetism results from a combination of both orbital and electron moments.

In some materials, the atomic magnetic moments either cancel or are very small so that little material magnetism results. These are known as paramagnetic or diamagnetic materials, depending on whether an applied field increases or decreases the magnetization. Their permeabilities relative to vacuum are nearly equal to 1. In other materials, the atomic moments are large and there is an innate tendency for them to align due to quantum mechanical forces. These are the ferromagnetic materials. The alignment forces are very short range, operating only over atomic distances. Nevertheless, they create regions of aligned magnetic moments, called domains, within a magnetic material. Although each domain has a common orientation, this orientation differs from domain to domain. The narrow separations between domains are regions where the magnetic moments are transitioning from one orientation to another. These transition zones are referred to as domain walls.

In non-oriented magnetic materials, the domains are typically very small and randomly oriented. With the application of a magnetic field, the domain orientation tends to align with the field direction. In addition, favorably orientated domains tend to grow at the expense of unfavorably oriented ones. As the magnetic field increases, the domains eventually all point in the direction of the magnetic field, resulting in a state of magnetic saturation. Further increases in the field cannot orient more domains so the magnetization does not increase but is said to saturate. From this point on, further increases in induction are due to increases in the field only.

The relation between induction, **B**, magnetization, **M**, and field, **H**, in SI units, is

$$\mathbf{B} = \mu_o(\mathbf{H} + \mathbf{M}) \tag{2.1}$$

For many materials, M is proportional to H,

$$\mathbf{M} = \chi \mathbf{H} \tag{2.2}$$

where χ is the susceptibility which need not be a constant. Substituting into (2.1)

$$\mathbf{B} = \mu_o(1+\chi)\mathbf{H} = \mu_o \mu_r \mathbf{H} \tag{2.3}$$

where $\mu_o = 1 + \chi$ is the relative permeability. We see directly in (2.1) that, as **M** saturates because all the domains are similarly oriented, **B** can only increase due to increases in **H**. This occurs at fairly high **H** or exciting current values, since **H** is proportional to the exciting current. At saturation, since all the domains have the same orientation, there are no domain walls. Since **H** is generally small compared to **M** for high permeability ferromagnetic materials up to saturation, the saturation magnetization and saturation induction are nearly the same and will be used interchangeably.

As the temperature increases, the thermal energy begins to compete with the alignment energy and the saturation magnetization begins to fall until the Curie point is reached where ferromagnetism completely disappears. For 3 % Si-Fe, the saturation magnetization or induction at 20 °C is 2.0 T and the Curie temperature is 746 °C. This should be compared with pure iron where the saturation induction at 20 °C is 2.1 T and the Curie temperature is 770 °C. The fall off with temperature follows fairly closely a theoretical relationship between ratios of saturation induction at absolute temperature T to saturation induction at T = 0 °K to the ratio of absolute temperature T to the Curie temperature expressed in °K. For pure iron, this relationship is shown graphed in Fig. 2.2 [Ame57]. This same graph also applies rather closely to other iron containing magnetic materials such as Si-Fe as well as to nickel and cobalt based magnetic materials.

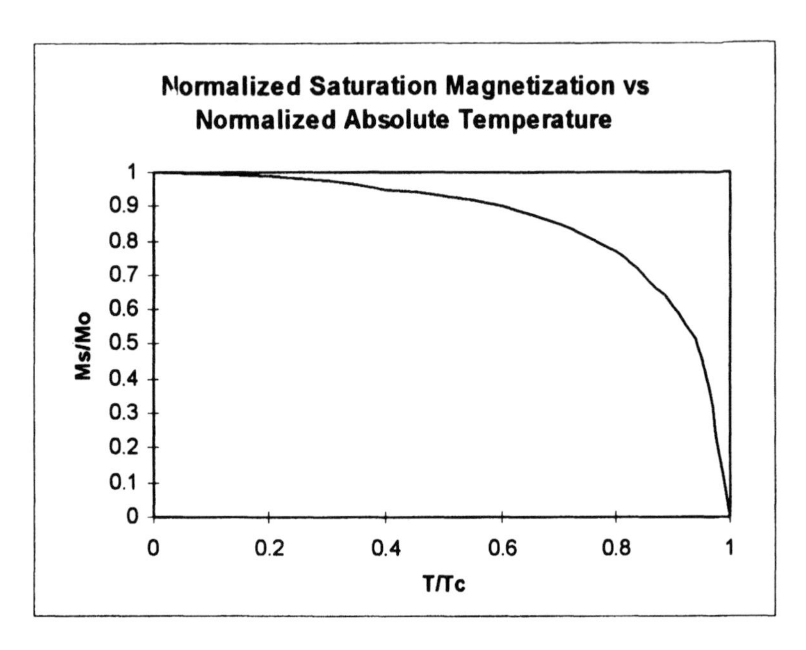

Figure 2.2 Relationship between saturation magnetization and absolute temperature, expressed in relative terms, for pure iron. This also applies reasonably well to other ferromagnetic materials containing predominately iron, nickel, or cobalt.

Thus to find the saturation magnetization of 3 % Si-Fe at a temperature of 200 °C = 473 °K, take the ratio $T/T_C = 0.464$. From the graph, this corresponds to $M_S/M_0 = 0.94$. On the other hand, we know that at 20 °C where $T/T_C = 0.287$ that $M_S/M_0 = 0.98$. Thus $M_0 = 2.0/0.98 = 2.04$. Thus $M_S(T{=}200\ °C) = 0.94(2.04) = 1.92$ T. This is only a 4 % drop in saturation magnetization. Considering that core temperatures are unlikely to reach 200 °C, temperature effects on magnetization should not be a problem in transformers under normal operating conditions.

Ferromagnetic materials typically exhibit the phenomenon of magnetostriction, i.e. a length change or strain resulting from the induction or flux density which they carry. Since this length change is independent of the sign of the induction, for an a.c. induction at frequency f, the length oscillations occur at frequency 2f. These length vibrations contribute to the noise level in transformers. Magnetostriction is actually a fairly complex phenomenon and can exhibit hysteresis as well as anisotropy.

Another source of noise in transformers, often overlooked, results from the transverse vibrations of the laminations at unsupported free ends. This can occur at the outer surfaces of the core and shunts if these are not constrained. This can be shown qualitatively by considering the situation shown in Fig. 2.3. In Fig. 2.3a, we show a leakage flux density vector $\mathbf{B}_1$ impinging on a packet of tank shunt laminations which are flat against the tank wall and rigidly constrained. After striking the lamination packet, the flux is diverted into the packet and transported upward since we are looking at the bottom end. In Fig. 2.3b, the outer lamination is constrained only up to a certain distance from the end, d, beyond which it is free to move. We show its loose end at a distance x from the rest of the packet. Part of the flux density B_1 is diverted along this outer packet and a reduced flux density B_2 impinges on the remaining packets. We can assume the magnetic shunts are linear since the flux density they carry is usually far below saturation. The magnetic energy for linear materials is given by

$$E_m = \frac{1}{2}\int \mathbf{H}\cdot\mathbf{B}\,dV \tag{2.4}$$

Using $\mathbf{B} = \mu_o\mu_r\,\mathbf{H}$, this becomes

$$E_m = \frac{1}{2\mu_o\mu_r}\int B^2 dV \tag{2.5}$$

In highly oriented Si-Fe, μ_r is generally quite large ($\mu_r > 5000$) so that most of the magnetic energy resides in the oil or air where $\mu_r = 1$. Thus we can ignore the energy stored in the laminations. For the situation shown in Fig. 2.3, let ℓ be a distance along which B_1 is parallel and reasonably constant and let B_1 also be constant through a cross-sectional area A determined by d and a unit distance into the paper. Then the magnetic energy associated with Fig. 2.3b is

$$E_m = \frac{A}{2\mu_o}\left[\left(\ell - \frac{x}{2}\right)B_1^2 + \frac{x}{2}B_2^2\right] = \frac{A}{2\mu_o}\left[\ell B_1^2 - \frac{x}{2}\left(B_1^2 - B_2^2\right)\right] \tag{2.6}$$

The magnetic force component in the x direction is

$$F_{m,x} = -\frac{\partial E_m}{\partial x} = \frac{A}{4\mu_o}\left(B_1^2 - B_2^2\right) \tag{2.7}$$

Since $B_2 < B_1$, this force acts to pull the outer lamination outward in the direction of x. It is independent of the sign of B so that if B is sinusoidal of frequency f, $F_{m,x}$ will have a frequency of 2f. Thus it contributes to the transformer noise at the same frequency as magnetostriction.

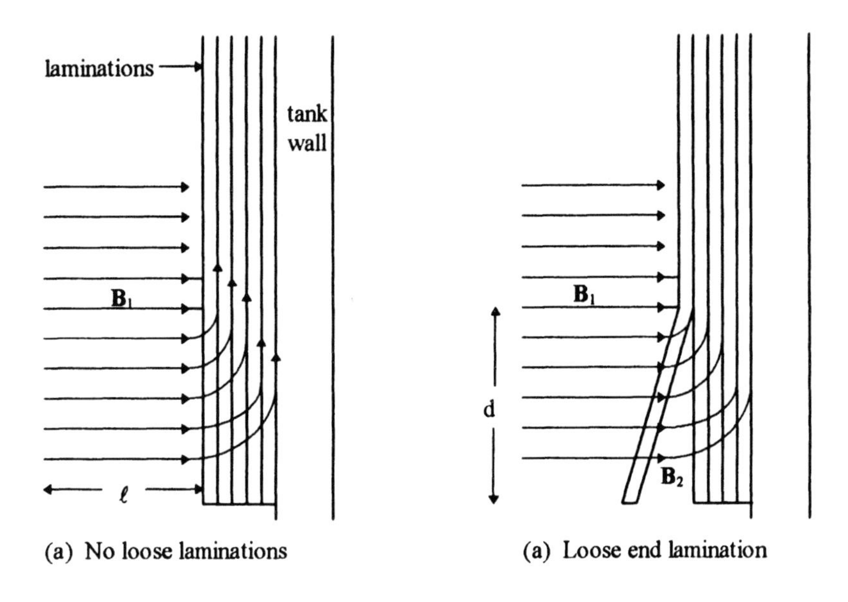

Figure 2.3 Geometry for simplified force calculation for a loose end lamination.

2.3 HYSTERESIS

Hysteresis, as the name implies, means that the present state of a ferromagnetic material depends on its past magnetic history. This is usually illustrated by means of a B-H diagram. Magnetic field changes are assumed to occur slowly enough that eddy current effects can be ignored. We assume the B and H fields are collinear although, in general, they need not be. Thus we can drop the vector notation. If we start out with a completely demagnetized specimen (this state requires careful preparation) and increase the magnetic field from 0, the material will follow the initial curve as shown in Fig. 2.4a. This curve can be continued to saturation, B_s. If at some point along the initial curve, the field is reversed and decreased, the material will follow a normal hysteresis loop as shown in Fig. 2.4b. If the magnetic field is cycled repeatedly between $\pm H_{max}$, the material will stay on a normal hysteresis loop determined by H_{max}, their being a whole family of such loops as shown in Fig. 2.4c. The largest loop occurs when B_{max} reaches B_s. This is called the major loop. These loops are symmetrical about the origin. If at some point along a normal loop, other than the extreme points, the field is reversed and cycled through a smaller cycle back to its original value before the reversal occurred, a minor or incremental loop is traced as shown in Fig. 2.4d. Considering the many other possibilities for field reversals, the resulting hysteresis paths can become quite complicated.

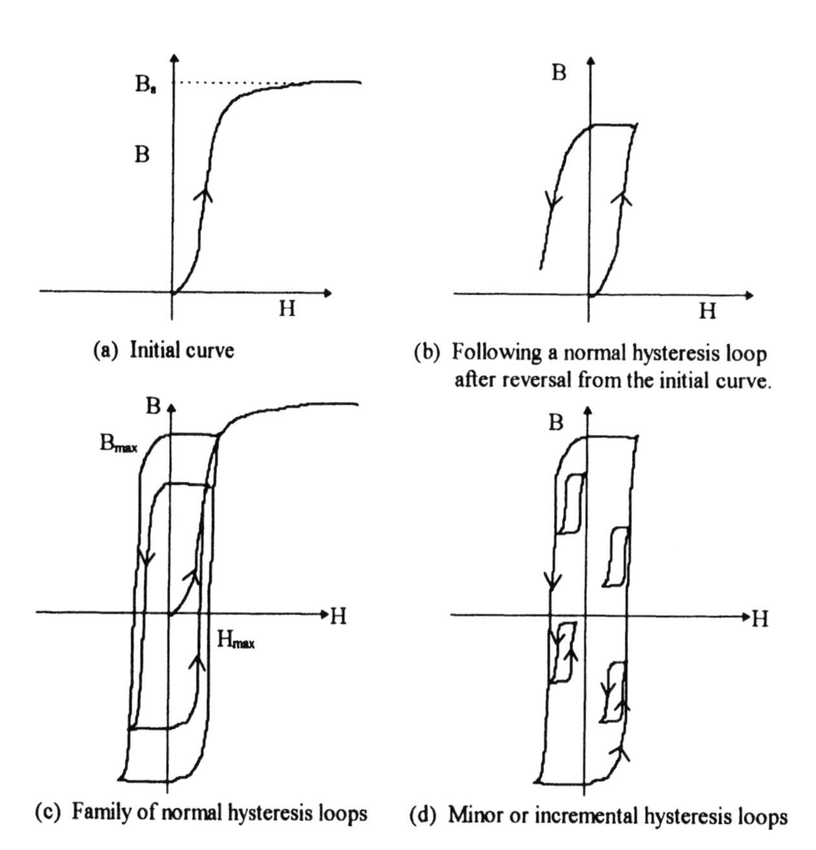

Figure 2.4 Hysteresis processes

Perhaps the most important magnetic path is a normal hysteresis loop since this is traced in a sinusoidal cyclic magnetization process. Several key points along such a path are shown in Fig. 2.5. As the field is lowered from H_{max} to zero, the induction remaining is called the remanence, B_r. As the field is further lowered into negative territory, the absolute value of the field at which the induction drops to zero is called the coercivity, H_c. . Because the loop is symmetrical about the origin, there are corresponding points on the negative branches. At any point on a magnetization path, the ratio of B to H is called the permeability while the slope of the B-H curve at that point is called the differential permeability. Other types of permeability can be defined. The area of the hysteresis loop is the magnetic energy per unit volume and per cycle dissipated in hysteresis processes.

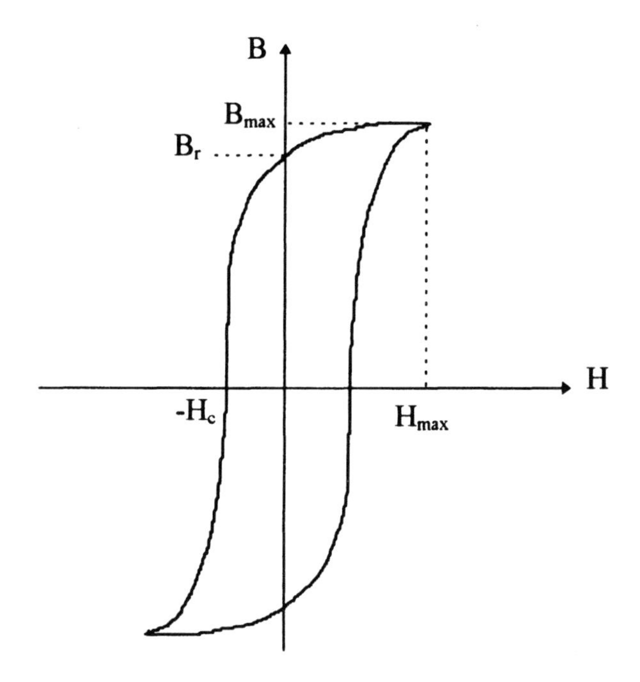

Figure 2.5 Key points along a normal hystersis loop

In oriented Si-Fe, the relative permeability for inductions reasonably below saturation is so high that the initial curve is close to the B axis if B and H are measured in the same units as they are in the Gaussian system or if B vs $\mu_o H$ is plotted in the SI system. In addition, the hysteresis loops are very narrow in these systems of units and closely hug the initial curve. Thus for all practical purposes, we can assume a single valued B-H characteristic for these high permeability materials coinciding with the initial curve. Since inductions in transformer cores are kept well below saturation in normal operation to avoid high exciting currents, the effects of saturation are hardly noticeable and the core, for many purposes, can be assumed to have a constant permeability.

2.4 MAGNETIC CIRCUITS

In stacked cores and especially in cores containing butt joints as well as in gapped reactor cores, the magnetic path for the flux is not through a homogeneous magnetic material. Rather there are gaps occupied by air or oil or other non-magnetic materials of relative permeability equal to 1. In such cases, it is possible to derive effective permeabilities by using a magnetic circuit approximation. This approximation derives from the mathematical similarity of magnetic and electrical laws. In resistive electrical circuits, the conductivities of the wires and circuit elements are usually so much higher than the surrounding medium (usually air) that little current leaks away from the circuit. However, in magnetic circuits where the flux corresponds to the electric current, the circuit permeability, which is the analog of conductivity, is not orders of magnitude higher than that of the surrounding medium so that flux leakage does occur. Thus, whereas the circuit approach is nearly exact for electric circuits, it is only approximate for magnetic circuits [Del94].

Corresponding to Kirchoff's current law at a node

$$\sum_i I_i = 0 \tag{2.8}$$

where I_i is the current into a node along a circuit branch i (positive if entering the node, negative if leaving), we have the approximate magnetic counterpart

$$\sum_i \Phi_i = 0 \tag{2.9}$$

where Φ_i is the flux into a node. Whereas (2.8) is based on conservation of current and ultimately charge, (2.9) is based on conservation of flux. Kirchoff's voltage law can be expressed

$$\sum_i V_i = \oint \mathbf{E} \cdot d\boldsymbol{\ell} = \text{emf} \tag{2.10}$$

where the voltage drops V_i are taken around the loop in which an emf is induced. The electric field path integral in (2.10) is an alternate way of expressing the voltage drops. With $\mathbf{E} \leftrightarrow \mathbf{H}$, the magnetic analogy is

$$\oint \mathbf{H} \cdot d\boldsymbol{\ell} = \text{mmf} = \text{NI} \tag{2.11}$$

In this case, (2.11) is exact.

In terms of the current density $\mathbf{J}$ and flux density $\mathbf{B}$, we have

$$I = \int \mathbf{J} \cdot d\mathbf{A} \quad , \quad \Phi = \int \mathbf{B} \cdot d\mathbf{A} \tag{2.12}$$

so that $\mathbf{J}$ and $\mathbf{B}$ are corresponding quantities in the two systems. Ohm's law in its basic form can be written

$$\mathbf{J} = \sigma \mathbf{E} \tag{2.13}$$

where σ is the conductivity. This corresponds to

$$\mathbf{B} = \mu \mathbf{H} \tag{2.14}$$

in the magnetic system, where μ is the permeability.

To obtain resistance and corresponding reluctance expressions for simple geometries, consider a resistive or reluctive element of length L and uniform cross-sectional area A as shown in Fig. 2.6. Let a uniform current density J flow in the resistive element and a uniform flux density B flow through the magnetic element. Then

$$I = JA \quad , \quad \Phi = BA \tag{2.15}$$

Let a uniform electric field E drive the current and a uniform magnetic field H drive the flux. Then

$$V = EL \quad , \quad NI = HL \tag{2.16}$$

Using (2.13), we have

$$\frac{I}{A} = \sigma \frac{V}{L} \quad \Rightarrow \quad V = \left(\frac{L}{\sigma A}\right) I \tag{2.17}$$

where the quantity in parenthesis is recognized as the resistance. Similarly from (2.14),

$$\frac{\Phi}{A} = \mu \frac{NI}{L} \quad \Rightarrow \quad NI = \left(\frac{L}{\mu A}\right) \Phi \tag{2.18}$$

where the quantity in parenthesis is called the reluctance. A similar analysis could be carried out for other geometries using the basic field correspondences.

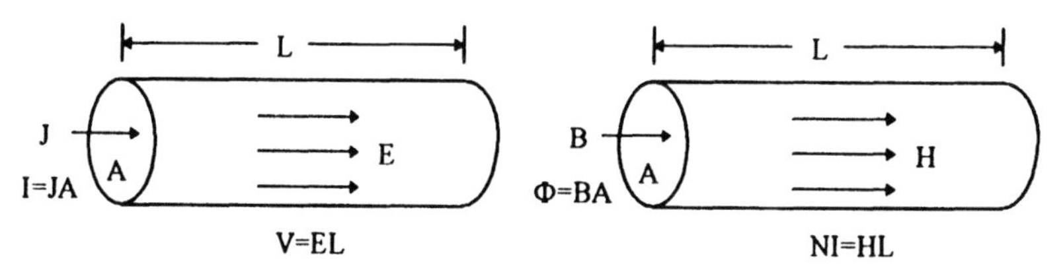

(a) Simple resistive electrical element (a) Simple reluctive magnetic element

Figure 2.6 Geometries for simple resistance and reluctance calculation

As an application of the circuit approach, consider a simple magnetic core with an air gap as shown in Fig. 2.7. A flux Φ is driven around the circuit by a coil generating an mmf of NI. The path through the magnetic core of permeability $\mu = \mu_o \mu_r$ has mean length L and the path in the air gap of permeability μ_o has length L_o. The reluctances in the magnetic material and air gap are

$$R_{mag} = \frac{L}{\mu_o \mu_r A} \quad , \quad R_{air} = \frac{L_o}{\mu_o A} \tag{2.19}$$

Since these reluctances are in series, we have

$$NI = \Phi\left(\frac{L}{\mu_o \mu_r A} + \frac{L_o}{\mu_o A}\right) = \Phi \frac{L}{\mu_o \mu_r A}\left(1 + \mu_r \frac{L_o}{L}\right) \tag{2.20}$$

Thus since μ_r can be quite large, the reluctance with air gap can be much larger than that without so that more mmf is required to drive a given flux. Note that we ignored fringing in the air gap. This could be approximately accounted for by letting the area A be larger in the air gap than in the core material.

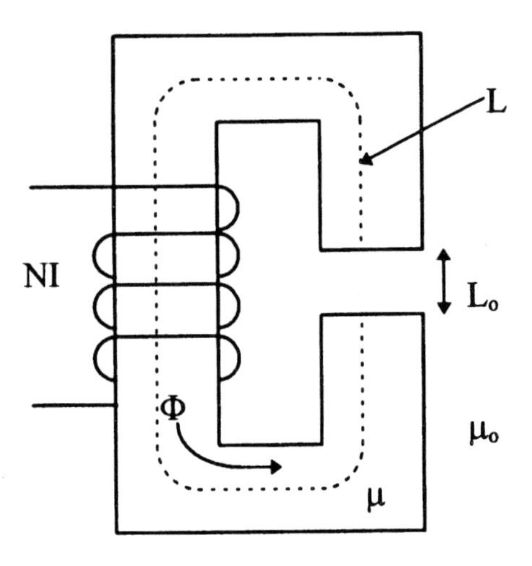

Figure 2.7 Air gap magnet driven by mmf = NI

In general, for two reluctances in series having the same cross-sectional area A but lengths L_1, L_2 and permeabilities μ_1, μ_2, the total reluctance is

$$R_{tot} = \frac{L_1}{\mu_1 A} + \frac{L_2}{\mu_2 A} = \frac{L}{A}\left(\frac{f_1}{\mu_1} + \frac{f_2}{\mu_2}\right) \tag{2.21}$$

where $L = L_1 + L_2$, $f_1 = L_1/L$, $f_2 = L_2/L$. Thus the effective permeability of the combination is given by

$$\frac{1}{\mu_{eff}} = \frac{f_1}{\mu_1} + \frac{f_2}{\mu_2} \tag{2.22}$$

This could be extended to more elements in series. An identical relationship holds for the effective conductivity of a series of equal cross-sectional area resistive elements.

Thus in a transformer core with joints, the effective permeability is reduced relative to that of an ideal core without joints. This requires a higher exciting current to drive a given flux through the core. The core losses will also increase mainly due to flux distortion near the joint region. For linear materials, the slope of the B-H curve will decrease for a jointed core relative to an unjointed core. For materials which follow a hysteresis loop, the joints will have the effect of skewing or tilting the effective hysteresis loop of the jointed core away from the vertical compared with the unjointed ideal core. This can be seen by writing (2.11) as

$$H_1\ell_1 + H_2\ell_2 = NI = H_{eff}\ell \tag{2.23}$$

where H_1 is the field in the core steel and H_2 the field in the gap. ℓ_1 and ℓ_2 are core and gap lengths, ℓ the total length, and H_{eff} an effective applied field. Assuming the gap is linear with permeability μ_o (relative permeability 1), we have

$$B_2 = \mu_o H_2 \tag{2.24}$$

Also assume that there is no leakage, that the core and gap have the same cross-sectional area, and the flux is uniformly distributed across it. Then $B_1 = B_2 = B$ from flux continuity. Thus we get for the H field in the core material

$$H_1 = H_{eff}\frac{\ell}{\ell_1} - \frac{B}{\mu_o}\frac{\ell_2}{\ell_1} \tag{2.25}$$

Since $\ell \approx \ell_1$, H_1 is more negative than H_{eff} when B is positive. To see the effect this has on the hysteresis loop of the jointed core, refer to Fig. 2.8. On the ascending part of the loop, we show the H_{eff} value which applies to the jointed core. Since $B > 0$, $H_1 < H_{eff}$ by (2.25) so the B value associated with H_{eff} corresponds to a B value lower down on the loop for the intrinsic core material. The same situation applies on the descending part of the loop. The points on the $B = 0$ axis have $H_1 \approx H_{eff}$ so the loop is fixed at these points. The net effect is that the loop is skewed towards the right on the top (and towards the left on the bottom). Notice that the remanence falls for the jointed loop because according to (2.25), when $H_{eff} = 0$, $H_1 < 0$ and the associated B value is lower down on the loop.

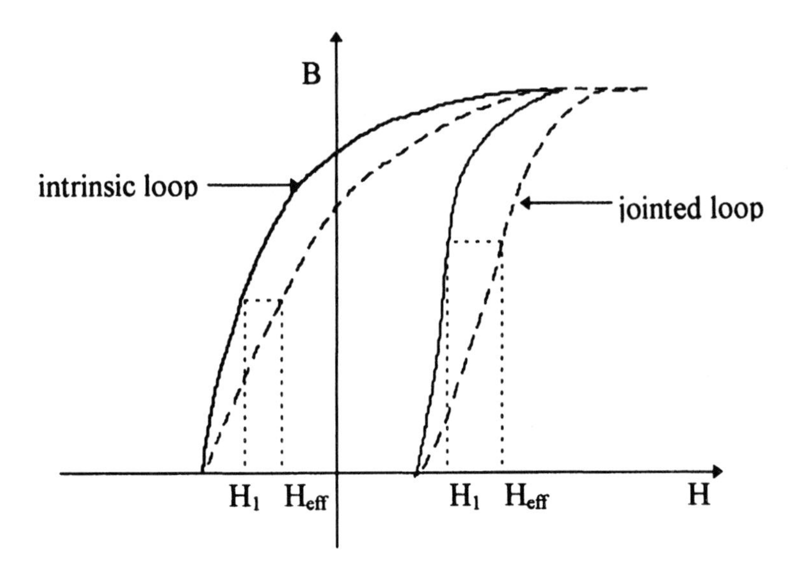

Figure 2.8 Effect of air or oil gaps on the hysteresis loop

2.5 INRUSH CURRENT

When a transformer is disconnected from a power source, the current is interrupted and the magnetic field or mmf driving flux through the core is reduced to zero. As we have seen in the preceding section, the core retains a residual induction which is called the remanence when the hysteresis path is on the positive descending (negative ascending) branch of a normal loop. In other cases, there could also be a residual induction but it would not have as high a magnitude. In order to drive the core to the zero magnetization state, it would be necessary to gradually lower the peak induction while cycling the field. Since the intrinsic normal hysteresis loops for oriented Si-Fe have fairly flat tops (or bottoms), the remanence is close to the peak induction. However, the presence of gaps in the core reduces this somewhat. When the unit is re-energized by a voltage source, the flux change must match the voltage change according to Faraday's law,

$$V = -N\frac{d\Phi}{dt} \tag{2.26}$$

For a sinusoidal voltage source, the flux is also sinusoidal,

$$\Phi = \Phi_p \sin(\omega t + \varphi) \tag{2.27}$$

where Φ_p is the peak flux. Assuming uniform flux density, we have $\Phi_p = B_p A_c$, where A_c is the core cross-sectional area. Substituting into (2.26), we get

$$V = -N\omega B_p A_c \cos(\omega t + \varphi) \tag{2.28}$$

Hence the peak voltage, using $\omega = 2\pi f$, where f is the frequency in Hz, is

$$V_p = 2\pi f N B_p A_c \tag{2.29}$$

Thus to follow the voltage change, the induction must change by $\pm B_p$ over a cycle. If, in a worst case scenario, the voltage source is turned on when the voltage is at a value which requires a $-B_p$ value and the remanent induction has a positive value of nearly B_p, then the induction will triple to nearly $3\ B_p$ when the voltage reaches a value corresponding to $+B_p$. Since B_p is usually ~ 10 to 20 % below saturation in typical power transformers, this means that the core will be driven strongly into saturation, which requires a very high exciting current. This exciting current is called the inrush current and can be many times the normal load current in a transformer.

Actually as saturation is approached, the flux will no longer remain confined to the core but will spill into the air or oil space inside the coil which supplies the exciting current. Thus, beyond saturation, the entire area inside the exciting coil, including the core, must be considered the flux carrying area and the incremental relative permeability is 1. Let B_r be the residual induction in the core which, without loss of generality, we can take to be positive. In the following, we will assume that this is the remanence. Thus the residual flux is $\Phi_r = B_r A_c$. Let $\Delta\Phi$ be the flux change required to bring the voltage from its turn-on point up to its maximum value in the same sense as the residual flux. $\Delta\Phi$ could be positive, negative or zero, depending on the turn-on point. We assume it is positive here.

Part of the increase in $\Delta\Phi$ will simply bring the induction up to the saturation level, entailing the expenditure of little exciting power or current. This part is given approximately by $(B_s - B_r)A_c$, where B_s is the saturation induction. Beyond this point further increases in $\Delta\Phi$ occur with the expenditure of high exciting current since the incremental relative permeability is 1. Since beyond saturation, the core and air or oil have the same permeability, the incremental flux density will be the same throughout the interior of the coil, ignoring end effects. Letting the interior coil area up the mean radius R_m be A ($A = \pi R_m$), the incremental flux density is

$$B_{inc} = \frac{\Delta\Phi - (B_s - B_r)}{A} \tag{2.30}$$

The incremental magnetic field inside the coil, ignoring end effects, is $H_{inc} = NI/h$, where NI are the exciting amp-turns and h the coil height. If we ignore the exciting amp-turns required to reach saturation since these are comparatively small, then $H_{inc} = H$, where H is the total field. Since the permeability for this flux is μ_o, we have $B_{inc} = \mu_o H_{inc} = \mu_o H$ which implies

$$NI = \frac{h}{\mu_o A}\left[\Delta\Phi - (B_s - B_r)A_c\right] \tag{2.31}$$

For transformers with stacked cores and step-lapped joints, $B_r \approx 0.9\ B_p$. Assuming, in the worst case, that the voltage is turned on at a correspond flux density that is at the most negative point in the cycle, we have, using (2.29)

$$\Delta\Phi_{max} = \frac{2V_p}{2\pi fN} = 2B_p A_c \tag{2.32}$$

In this expression, it is the flux change that matters since this is directly related to the voltage change as indicated in the first equality. In the second equality, we are measuring this flux change mathematically as if it all occurred in the core even though we know this is not true physically. Thus (2.31) becomes

$$NI = \frac{hA_c}{\mu_o A}(2.9B_p - B_s) \tag{2.33}$$

For $h = 2$ m, $A_c/A = 0.5$, $B_p = 1.7$ T, $B_s = 2$ T, we obtain $NI = 2.33 \times 10^6$ A-t. For $N = 500$, $I = 4660$ Amps. This is quite high for an exciting current.

While the voltage is constrained to be sinusoidal, the exciting current will be distorted due mainly to saturation effects. Even below saturation there is some distortion due to non-linearities in the B-H curve. Fig. 2.9a illustrates the situation on inrush. The sinusoidal voltage is proportional to the incremental induction which is shown displaced by the remanent induction. It requires high peak H values near its peak and comparatively small to zero H values near its trough. This is reflected in the exciting current which is proportional to H. This current appears as a series of positive pulses separated by broad regions of near zero value as shown in Fig. 2.9b. The high exciting inrush current will damp out with time as suggested in Fig. 2.9b. This is due to resistive effects.

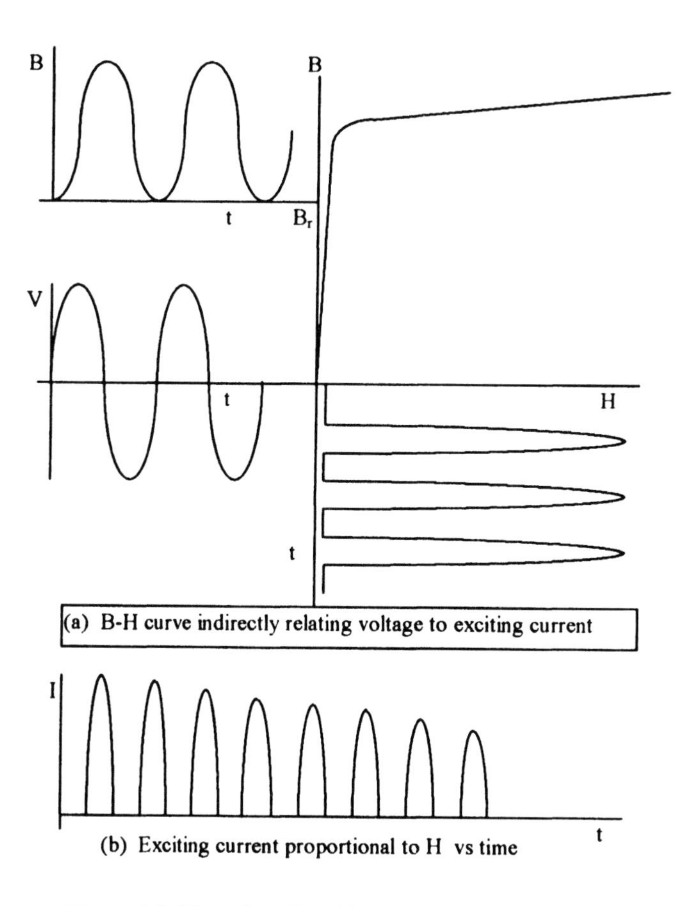

Figure 2.9 Distortion of exciting current due to saturation

Because the inrush current can be as large as a fault current, it is important to find some way of discriminating them so that false alarms are not set off when the unit is turned on. To this end we examine the time dependence of the two types of current. Using (2.27), we can write

$$\Delta\Phi = \Phi_p \sin(\omega t + \varphi) - \Phi_p \sin\varphi = B_p A_c\left[\sin(\omega t + \varphi) - \sin\varphi\right] \quad (2.34)$$

Thus, using (2.31) and ignoring the resistive damping since we are only interested in examining the first few cycles,

$$I(t) = \frac{hA_c}{\mu_o NA}\left[B_p \sin(\omega t + \varphi) - \left(B_s + B_p \sin\varphi - B_r\right)\right] \quad (2.35)$$

This last equation holds as long as I(t) is positive. It is nearly zero for negative values of the expression on the right hand side since then the core is not saturated. Thus (2.35) will remain positive over a cycle for values of ωt between

$$\omega t_1 = \sin^{-1} X - \varphi \quad , \quad \omega t_2 = \pi - \sin^{-1} X - \varphi$$

$$\text{where} \quad X = \frac{B_s + B_p \sin\varphi - B_r}{B_p} \quad (2.36)$$

Thus the interval per cycle over which (2.35) is positive is

$$\omega t_2 - \omega t_1 = \pi - 2\sin^{-1} X \quad (2.37)$$

Shifting the time origin to t_1 , (2.35) becomes

$$I(t) = \begin{cases} I_p\left[\sin\left(\omega t + \sin^{-1} X\right) - X\right] & , \quad 0 \le \omega t \le \pi - 2\sin^{-1} X \\ 0 & , \quad \pi - 2\sin^{-1} X \le \omega t \le 2\pi \end{cases} \quad (2.38)$$

$$\text{where} \quad I_p = \frac{hA_c B_p}{\mu_o NA}$$

Extreme values of X are determined by $\sin\varphi = \pm 1$,

$$X_{min} = -1 + \frac{(B_s - B_r)}{B_p} \quad , \quad X_{max} = 1 + \frac{(B_s - B_r)}{B_p} \quad (2.39)$$

Thus X cannot be < -1. If $X > 1$, this simply means there is no noticeable inrush current. (2.38) is graphed in Fig. 2.10 for $X = -0.5$.

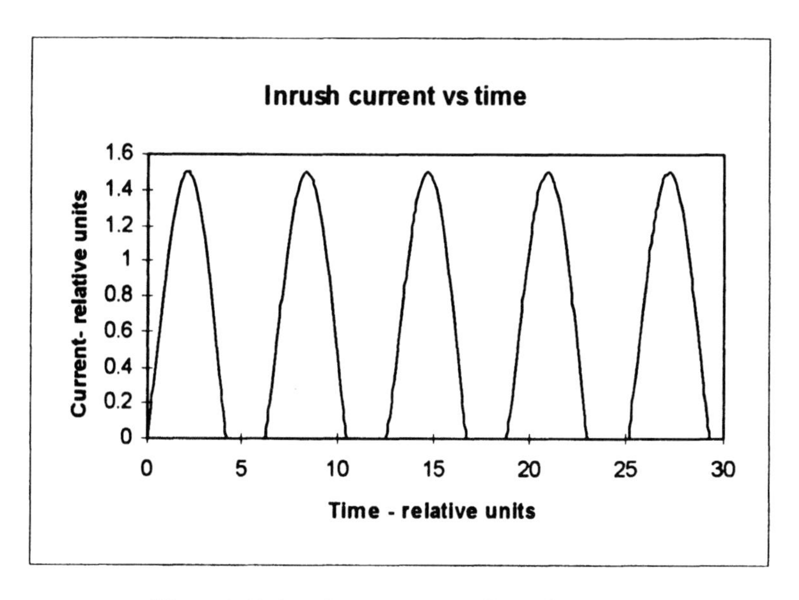

Figure 2.10 Inrush current (normalized) for $X = -0.5$

Let us find the harmonic content of (2.38) by performing a Fourier analysis. To facilitate this, rewrite (2.38) as

$$I(t) = \begin{cases} I_{max}\left[\sqrt{1-X^2}\sin\omega t + X\cos\omega t - X\right], & 0 \le \omega t \le \pi - 2\sin^{-1} X \\ 0 & , \pi - 2\sin^{-1} X \le \omega t \le 2\pi \end{cases} \tag{2.40}$$

We use the expansion,

$$I(t) = a_0 + a_1 \cos\omega t + a_2 \sin\omega t + \cdots + b_1 \sin\omega t + b_2 \sin\omega t + \cdots$$

where

$$a_0 = \frac{1}{2\pi}\int_0^{2\pi} I(\omega t)d(\omega t) \tag{2.41}$$

$$a_n = \frac{1}{\pi}\int_0^{2\pi} I(\omega t)\cos(n\omega t)d(\omega t)$$

$$b_n = \frac{1}{\pi}\int_0^{2\pi} I(\omega t)\sin(n\omega t)d(\omega t)$$

Evaluating the first few harmonics and letting $\alpha = \pi - 2\sin^{-1}X$, we get

$$\frac{a_0}{I_p} = \frac{1}{2\pi}\left\{\sqrt{1-X^2}(1-\cos\alpha) + X\sin\alpha - X\alpha\right\}$$

$$\frac{b_1}{I_p} = \frac{1}{\pi}\left\{\sqrt{1-X^2}\,\frac{\sin^2\alpha}{2} + X\left(\frac{\alpha}{2} + \frac{\sin 2\alpha}{4}\right) - X\sin\alpha\right\}$$

$$\frac{a_2}{I_p} = \frac{1}{\pi}\left\{\sqrt{1-X^2}\left[\frac{\cos\alpha - 1}{2} + \frac{1-\cos 3\alpha}{6}\right] + X\left[\frac{\sin\alpha}{2} + \frac{\sin 3\alpha}{6}\right] - X\frac{\sin 2\alpha}{2}\right\}$$

$$\frac{b_2}{I_p} = \frac{1}{\pi}\left\{\sqrt{1-X^2}\left[\frac{\sin\alpha}{2} - \frac{\sin 3\alpha}{6}\right] + X\left[\frac{(1-\cos\alpha)}{2} + \frac{(1-\cos 3\alpha)}{6}\right] - \frac{X}{2}(1-\cos 2\alpha)\right\} \tag{2.42}$$

The ratio of second harmonic amplitude to fundamental amplitude is given by

$$\frac{2^{nd}}{1^{st}} = \frac{\sqrt{a_2^2 + b_2^2}}{\sqrt{a_1^2 + b_1^2}} \tag{2.43}$$

This is tabulated in Table 2.1 for a range of X values.

Table 2.1 Ratio of second to first harmonic amplitudes for inrush current as a function of the parameter X

X	2nd / 1st	X	2nd / 1st
-0.9	0.018	0.1	0.479
-0.8	0.048	0.2	0.534
-0.7	0.085	0.3	0.591
-0.6	0.127	0.4	0.647
-0.5	0.171	0.5	0.705
-0.4	0.219	0.6	0.763
-0.3	0.268	0.7	0.822
-0.2	0.318	0.8	0.881
-0.1	0.371	0.9	0.940
0	0.424		

Thus we see that the second harmonic is a significant fraction of the fundamental for most X values. In fact, the lowest practical X value can be determined from (2.39) by setting $B_s = 2\ T$, $B_p \approx 0.85\ B_s = 1.7\ T$, $B_r = 0.9\ B_p = 1.53\ T$. We obtain $X = -0.72$. At this value, the ratio of second to first harmonic is > 8 % which can be considered a lower limit.

For comparison purposes, we now examine the time dependence of the fault current. We will ignore the load current at the time of the fault and assume the transformer is suddenly grounded at $t = 0$. The equivalent circuit is shown in Fig. 2.11 which will be derived later. Here R and L are the resistance and leakage reactance of the transformer, including any contributions from the system. The voltage is given by

$$V = V_p \sin(\omega t + \varphi) \tag{2.44}$$

where φ is the phase angle which can have any value, in general, since the fault can occur at any time during the voltage cycle. The circuit equation is

$$V_p \sin(\omega t + \varphi) = \begin{cases} RI + L\dfrac{dI}{dt} & , \quad t \geq 0 \\ 0 & , \quad t < 0 \end{cases} \tag{2.45}$$

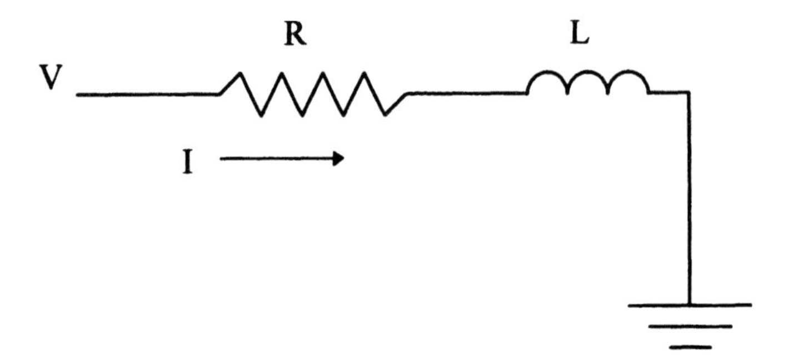

Figure 2.11 Circuit for fault current analysis

Using Laplace transforms, the current transform is given by [Hue72]

$$I(s) = \frac{V_p(s\sin\varphi + \omega\cos\varphi)}{(R + Ls)(s^2 + \omega^2)} \tag{2.46}$$

Taking the inverse transform, we obtain

$$I(t) = \frac{V_p}{R\sqrt{1+\left(\frac{\omega L}{R}\right)^2}}\left[\sin(\beta - \varphi)e^{-\frac{R}{L}t} + \sin(\omega t + \varphi - \beta)\right] \tag{2.47}$$

where $\beta = \tan^{-1}\left(\frac{\omega L}{R}\right)$

The steady state peak current amplitude is given by

$$I_{p,ss} = \frac{V_p}{R\sqrt{1+\left(\frac{\omega L}{R}\right)^2}} \tag{2.48}$$

Using this and letting $\tau = \omega t$, $\nu = \omega L/R$, we can rewrite (2.47)

$$I(\tau) = I_{p,ss}\left[\sin(\beta - \varphi)e^{-\frac{\tau}{\nu}} + \sin(\tau + \varphi - \beta)\right]$$

$$= I_{p,ss}\left[\sin(\beta - \varphi)\left(e^{-\frac{\tau}{\nu}} - \cos\tau\right) + \cos(\beta - \varphi)\sin\tau\right] \qquad (2.49)$$

where $\beta = \tan^{-1}\nu$

To find the maximum amplitude for a given φ, we need to solve

$$\frac{\partial I}{\partial \tau} = \sin(\beta - \varphi)\left(\frac{e^{-\frac{\tau}{\nu}}}{\nu} - \sin\tau\right) - \cos(\beta - \varphi)\cos\tau = 0 \qquad (2.50)$$

In addition, if we wish to determine the value of φ which produces the largest fault current, we need to solve

$$\frac{\partial I}{\partial \varphi} = \cos(\beta - \varphi)\left(e^{-\frac{\tau}{\nu}} - \cos\tau\right) - \sin(\beta - \varphi)\sin\tau = 0 \qquad (2.51)$$

Solving (2.50) and (2.51) simultaneously, we find

$$\tan(\beta - \varphi) = \nu = \tan\beta \qquad (2.52)$$

$$e^{-\frac{\tau}{\nu}} - \nu\sin\tau - \cos\tau = 0 \qquad (2.53)$$

Equation (2.52) shows that $\varphi = 0$ produces the maximum amplitude and the time at which this maximum occurs is given by the solution of (2.53) for $\tau > 0$. Substituting into (2.49), we obtain

$$\frac{I_{max}}{I_{p,ss}} = \sqrt{1 + \nu^2}\,\sin\tau \qquad (2.54)$$

This is the asymmetry factor over the steady state peak amplitude with τ obtained by solving (2.53).

The asymmetry factor is generally considered to be with respect to the steady state rms current value. This new ratio is called K in the literature and is thus given by substituting $I_{p,ss} = \sqrt{2}I_{rms,ss}$ in (2.54),

$$K = \frac{I_{max}}{I_{rms,ss}} = \sqrt{2\left(1+v^2\right)}\sin\tau \quad , \quad v = \frac{\omega L}{R} = \frac{x}{r} \tag{2.55}$$

with τ obtained by solving (2.53). We have expressed the ratio of leakage impedance to resistance as x/r. x and r are normalized quantities, i.e. the leakage reactance and resistance divided by a base impedance value which cancels out in the ratio. (2.55) has been parametrized as [IEE57]

$$K = \sqrt{2}\left\{1 + e^{-\frac{r}{x}\left(\phi+\frac{\pi}{2}\right)}\sin\phi\right\} \quad , \quad \phi = \tan^{-1}\left(\frac{x}{r}\right) = \beta \text{ and } \frac{x}{r} = v \tag{2.56}$$

This parametrization agrees with (2.55) to within 0.7 %. Table 2.2 shows some of the K values obtained by the two methods.

Table 2.2 Comparison of exact with parametrized K values

$v = x / r$	K (exact)	K (parametrized)
1	1.512	1.509
2	1.756	1.746
5	2.192	2.184
10	2.456	2.452
50	2.743	2.743
1000	2.828	2.824

At $\varphi = 0$ where the asymmetry is greatest, (2.49) becomes

$$I(\tau) = \frac{\sqrt{2}I_{rms,ss}}{\sqrt{1+v^2}}\left[v\left(e^{-\frac{\tau}{v}} - \cos\tau\right) + \sin\tau\right] \tag{2.57}$$

This is graphed in Fig. 2.12 for $v = 10$.

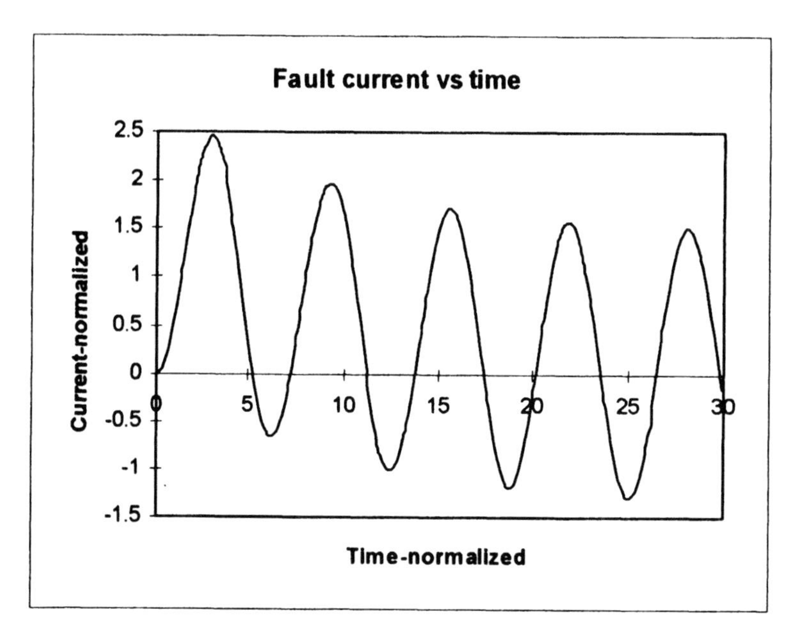

Figure 2.12 Fault current versus time for the case of maximum offset and $\nu = x/r = 10$.

The lowest Fourier coefficients of (2.49) are

$$
\begin{aligned}
\frac{a_0}{I_{p,ss}} &= \sin(\beta-\varphi)\frac{\nu}{2\pi}\left(1-e^{-\frac{2\pi}{\nu}}\right)\\
\frac{a_1}{I_{p,ss}} &= \sin(\beta-\varphi)\left[\frac{\nu}{\pi\left(1+\nu^2\right)}\left(1-e^{-\frac{2\pi}{\nu}}\right)-1\right]\\
\frac{b_1}{I_{p,ss}} &= \sin(\beta-\varphi)\frac{\nu^2}{\pi\left(1+\nu^2\right)}\left(1-e^{-\frac{2\pi}{\nu}}\right)+\cos(\beta-\varphi)\\
\frac{a_2}{I_{p,ss}} &= \sin(\beta-\varphi)\frac{\nu}{\pi\left(1+4\nu^2\right)}\left(1-e^{-\frac{2\pi}{\nu}}\right)\\
\frac{b_2}{I_{p,ss}} &= \sin(\beta-\varphi)\frac{2\nu^2}{\pi\left(1+4\nu^2\right)}\left(1-e^{-\frac{2\pi}{\nu}}\right)
\end{aligned}
\tag{2.58}
$$

The ratio of second harmonic amplitude to fundamental, (2.43), is tabulated in Table 2.3 for a range of φ and ν values.

Table 2.3 Ratio of second harmonic to fundamental amplitude of fault current for various values of the reactance to resistance ratio and voltage phase angle

ν=x/r	$\varphi = 0°$	$\varphi = 30°$	$\varphi = 45°$	$\varphi = 90°$	φ=120°	φ=135°
1	0.0992	0.0357	0	0.119	0.168	0.166
2	0.128	0.0752	0.0440	0.0752	0.152	0.167
3	0.127	0.0831	0.0566	0.0475	0.123	0.145
4	0.118	0.0819	0.0589	0.0323	0.101	0.124
5	0.109	0.0778	0.0576	0.0232	0.0543	0.107
6	0.0996	0.0731	0.0550	0.0175	0.0735	0.0941
7	0.0916	0.0684	0.0521	0.0137	0.0644	0.0835
8	0.0845	0.0641	0.0493	0.0110	0.0572	0.0750
9	0.0784	0.0601	0.0465	0.0090	0.0515	0.0680
10	0.0730	0.0565	0.0440	0.0075	0.0467	0.0622
15	0.0540	0.0432	0.0341	0.0036	0.0320	0.0434
20	0.0427	0.0347	0.0276	0.0022	0.0242	0.0333
25	0.0352	0.0290	0.0232	0.0014	0.0195	0.0270
50	0.0188	0.0158	0.0128	0.0004	0.0099	0.0138
100	0.0097	0.0083	0.0067	0.0001	0.0050	0.0070
500	0.0020	0.0017	0.0014	0	0.0010	0.0014
1000	0.0010	0.0008	0.0007	0	0.0005	0.0007

For power transformers x/r is usually > 20. We see from Table 2.3 that the second harmonic content relative to the fundamental is < 4.3 % for x/r > 20. On the other hand, we found in Table 2.1 that the second to first harmonic ratio is > 8 % under virtually all conditions for the inrush current. Thus a determination of this ratio can distinguish inrush from fault currents. In fact x/r would need to fall below 10 before this method breaks down.

2.6 TRANSFORMER CIRCUIT MODEL WITH CORE

To simplify matters, we will consider a single phase, two winding transformer. For a balanced 3 phase system, the phases can be analyzed separately for most purposes. We wish to develop a circuit model of such a transformer under normal a.c. conditions which includes the effects of the core. Capacitive effects have been traditionally ignored except at higher frequencies. However, under no-load conditions, these are not necessarily negligible in modern transformers made with high permeability steel cores. We will indicate several ways by which capacitive effects can be included.

If the secondary winding were open circuited, then the transformer would behave like an inductor with a high permeability closed iron core. It would therefore have a high inductance so that little exciting current would be required to generate the voltage or back emf. Some I^2R loss will be generated by the exciting current, however this will be small compared with the load current losses. There will, however, be losses in the core due to the changing flux. These losses are to a good approximation proportional to the square of the induction, B^2. Hence they are also proportional to the square of the voltage across the core. Thus these losses can be accounted for by putting an equivalent resistor across the transformer voltage and ground, where the resistor has the value

$$R_c = \frac{V_{rms}^2}{W_c} \tag{2.59}$$

where V_{rms} is the rms phase voltage and W_c is the core loss. The open circuited inductance can be obtained from

$$V = L_c \frac{dI_{ex}}{dt} \quad \Rightarrow \quad L_c = \frac{V_{rms}}{\omega I_{ex,rms}} \tag{2.60}$$

where I_{ex} is the inductive component of the exciting current, which we assume to be sinusoidal with angular frequency ω. Thus the circuit so far will look like Fig. 2.13. The resistance R_p is the resistance of the primary (or excited) winding. Note that as saturation is approached, the inductance L_c as well as the resistance R_c will become non-linear. If necessary, capacitive effects can be included by putting an equivalent capacitance in parallel with the core inductance and resistance. Since the core losses are supplied by the input power source, there is a component of the total excitation current, $I_{ex,\,tot}$, which generates the core loss. We labeled it I_c in the figure. It will be in quadrature with I_{ex}.

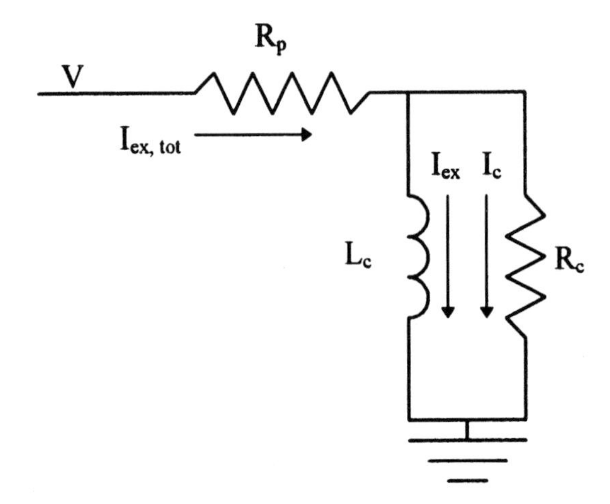

Figure 2.13 Transformer circuit model with secondary open circuited

When the secondary circuit is connected to a load, the emf generated in the secondary winding by the changing core flux will drive a current through the secondary circuit. This additional current (amp-turns) would alter the core flux unless equal and opposite amp-turns flow in the primary winding. Since the core flux is determined by the impressed primary voltage, the net amp-turns must equal the small exciting amp-turns. Hence the primary and secondary amp-turns due to load current must cancel out.

Fig. 2.14 shows a schematic of the flux pattern in a two winding transformer under load. The currents are taken as positive when they flow into a winding and the dots on the terminals indicate that the winding sense is such that the induced voltage is positive at that terminal relative to the terminal at the other end of the winding when positive exciting current flows into the transformer. Notice that the bulk of the flux flows through the core and links both windings. However some of the flux links only one winding. When referring to flux linkages, we assume partial linkages are included. Some of these can be seen in the figure.

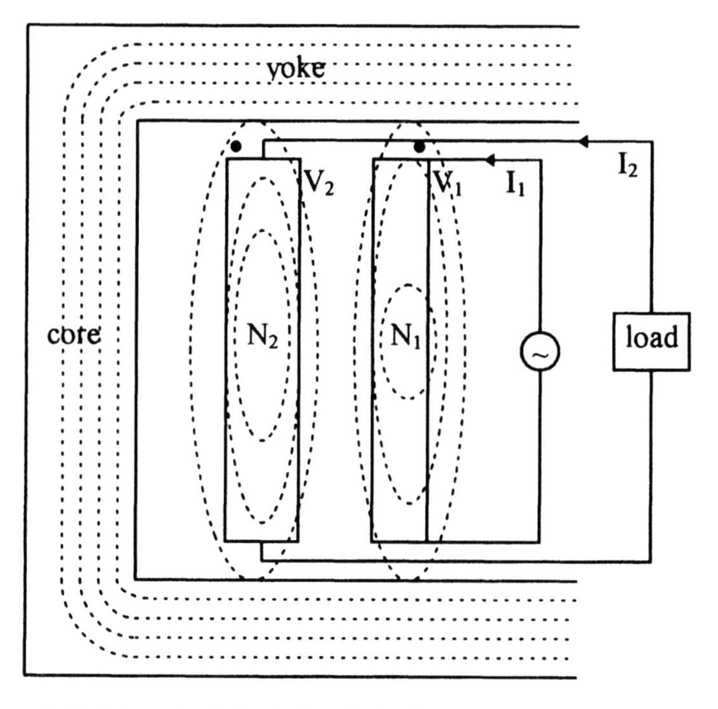

Figure 2.14 Schematic of a 2 winding single plase transformer with leakage flux

The voltage equations for the two windings are

$$V_1 = R_1 I_1 + \frac{d\lambda_1}{dt} \quad , \quad V_2 = R_2 I_2 + \frac{d\lambda_2}{dt} \tag{2.61}$$

where λ_1 and λ_2 are the total flux linkages for windings 1 and 2 and R_1 , R_2 their resistances. We are ignoring core loss here as that would complicate matters unnecessarily.

The traditional way to further develop (2.61) is to subdivide the total flux linkages into contributions from the two windings. Thus

$$\lambda_1 = \lambda_{11} + \lambda_{12} \quad , \quad \lambda_2 = \lambda_{22} + \lambda_{21} \tag{2.62}$$

where λ_{11} are the flux linkages contributed by winding 1 to itself and λ_{12} are the flux linkages contributed by winding 2 to winding 1 and similarly for λ_{22} and λ_{21}. Further, for linear materials, we can write

$$\lambda_{11} = L_1 I_1 \quad , \quad \lambda_{12} = M_{12} I_2$$
$$\lambda_{22} = L_2 I_2 \quad , \quad \lambda_{21} = M_{21} I_1 \tag{2.63}$$

where L and M are self and mutual inductances and $M_{12} = M_{21} = M$. Thus (2.61) can be written

$$V_1 = R_1 I_1 + L_1 \frac{dI_1}{dt} + M \frac{dI_2}{dt} \quad , \quad V_2 = R_2 I_2 + L_2 \frac{dI_2}{dt} + M \frac{dI_1}{dt} \tag{2.64}$$

Assuming sinusoidal voltages and currents and reverting to phasor notation (bold faced type), (2.64) becomes

$$\mathbf{V}_1 = R_1 \mathbf{I}_1 + j\omega L_1 \mathbf{I}_1 + j\omega M \mathbf{I}_2 \quad , \quad \mathbf{V}_2 = R_2 \mathbf{I}_2 + j\omega L_2 \mathbf{I}_2 + j\omega M \mathbf{I}_1 \tag{2.65}$$

Separating the exciting current component out of $\mathbf{I}_1$ so that $\mathbf{I}_1 = \mathbf{I}_{ex} + \mathbf{I}_1'$ and using amp-turn balance,

$$N_1 \mathbf{I}_1' = -N_2 \mathbf{I}_2 \tag{2.66}$$

we obtain by substituting into (2.65),

$$\mathbf{V}_1 = R_1 \mathbf{I}_1 + j\omega\left(L_1 - \frac{N_1}{N_2} M\right)\mathbf{I}_1 + j\omega \frac{N_1}{N_2} M \mathbf{I}_{ex}$$
$$\mathbf{V}_2 = R_2 \mathbf{I}_2 + j\omega\left(L_2 - \frac{N_2}{N_1} M\right)\mathbf{I}_2 + j\omega M \mathbf{I}_{ex} \tag{2.67}$$

Defining single winding leakage inductances as

$$L_{\ell 1} = L_1 - \frac{N_1}{N_2} M \quad , \quad L_{\ell 2} = L_2 - \frac{N_2}{N_1} M \tag{2.68}$$

the circuit model corresponding to (2.67) is given in Fig. 2.15. These single winding leakage inductance are for one winding with respect to the other and thus depend on both windings.

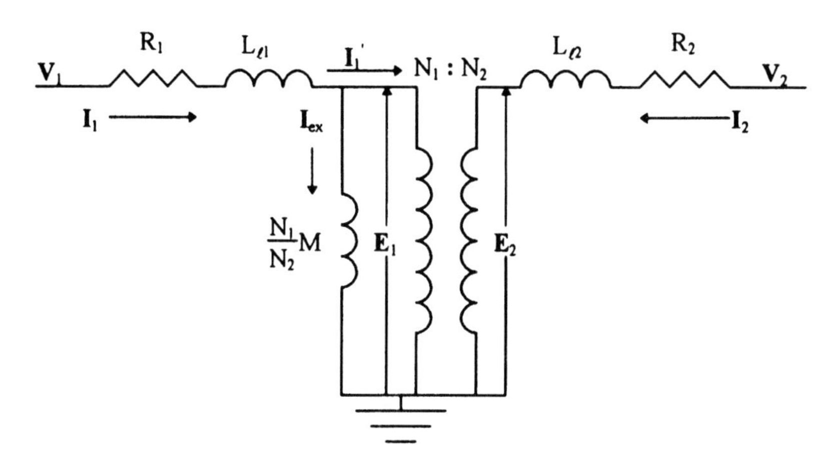

Figure 2.15 Circuit model of a 2 winding transformer under load

We have introduced an ideal transformer in Fig. 2.15. In such a device the currents into it are related by (2.66) and the voltages by

$$\frac{\mathbf{E}_1}{\mathbf{E}_2} = \frac{N_1}{N_2} \tag{2.69}$$

Using (2.66) - (2.69), we can write

$$\mathbf{V}_2 = \mathbf{E}_2 + (R_2 + j\omega L_{\ell 2})\mathbf{I}_2 = \left(\frac{N_2}{N_1}\right)\mathbf{E}_1 - \left(\frac{N_1}{N_2}\right)(R_2 + j\omega L_{\ell 2})\mathbf{I}_1'$$

$$= \left(\frac{N_2}{N_1}\right)\left[\mathbf{E}_1 - \left(\frac{N_1}{N_2}\right)^2 (R_2 + j\omega L_{\ell 2})\mathbf{I}_1'\right] \tag{2.70}$$

or

$$\left(\frac{N_1}{N_2}\right)\mathbf{V}_2 = \mathbf{E}_1 - \left(\frac{N_1}{N_2}\right)^2 (R_2 + j\omega L_{\ell 2})\mathbf{I}_1'$$

Thus the circuit diagram in Fig. 2.15 can be redrawn as shown in Fig. 2.16. The missing element in Fig. 2.15 or 2.16 is the core loss and possibly capacitive effects. This could be added by putting a resistance and capacitor in parallel with the voltage $\mathbf{E}_1$.

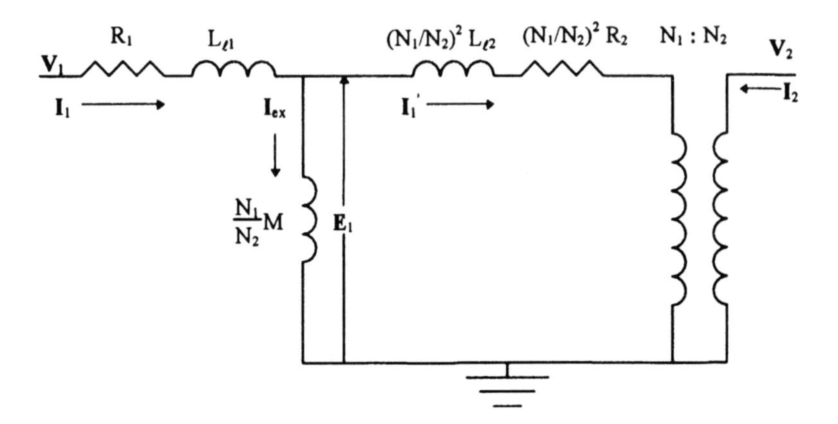

Figure 2.16 Circuit model of a 2 winding transformer under load referred to the primary side

It is instructive to develop a circuit model using a more fundamental approach which does not neglect the core non-linearities [MIT43]. We will utilize the concept of flux linkages per turn so that if λ is a flux linkage, N the number of turns linked, then $\Phi = \lambda/N$ is the flux linkage/turn. If the flux linked every turn of a circuit 100 %, then Φ would be the total flux passing through the circuit.

Assume only coil 1 has current flowing through it. Let Φ_{11} be the flux linkage/turn in coil 1 produced by this current. Let Φ_{21} be the flux linkage/turn linking coil 2 produced by the current in coil 1. Define the leakage flux of coil 1 with respect to coil 2 as

$$\Phi_{\ell 1} = \Phi_{11} - \Phi_{21} = \frac{\lambda_{11}}{N_1} - \frac{\lambda_{21}}{N_2} \tag{2.71}$$

Similarly, if only coil 2 is carrying current, we define the leakage flux of coil 2 with respect to coil 1 as

$$\Phi_{\ell 2} = \Phi_{22} - \Phi_{12} = \frac{\lambda_{22}}{N_2} - \frac{\lambda_{12}}{N_1} \tag{2.72}$$

where Φ_{22} is the flux linkage/turn linking coil 2 and Φ_{12} is the flux linkage/turn linking coil 1 produced by current in coil 2. Referring to Fig. 2.14, we note that much of the flux produced by coil 1 or 2 flows in the core and thus links both coils. Most of the remaining flux will have paths in air or only partially in the core. Thus the leakage flux associated with either coil consists of this remaining flux which has a large part of its path in air or oil. Therefore, the mmf required to produce it will be almost entirely in oil or air which means that it, as well as the current, will be linearly related to the leakage flux. In other words, non-linearities in the core will have little impact on the leakage flux. Thus the single winding leakage inductances defined as

$$L_{\ell 1} = \frac{N_1 \Phi_{\ell 1}}{I_1} \quad , \quad L_{\ell 2} = \frac{N_2 \Phi_{\ell 2}}{I_2} \tag{2.73}$$

will be essentially constant.

From (2.62) we see that the total flux linking winding 1 and 2 can be expressed as

$$\lambda_1 = N_1(\Phi_{11} + \Phi_{12}) \quad , \quad \lambda_2 = N_2(\Phi_{22} + \Phi_{21}) \tag{2.74}$$

Substituting from (2.71) and (2.72),

$$\lambda_1 = N_1(\Phi_{\ell 1} + \Phi_{21} + \Phi_{12}) \quad , \quad \lambda_2 = N_2(\Phi_{\ell 2} + \Phi_{12} + \Phi_{21}) \tag{2.75}$$

Letting $\Phi = \Phi_{12} + \Phi_{21}$ and using (2.73), (2.75) becomes

$$\lambda_1 = L_{\ell 1} I_1 + N_1 \Phi \quad , \quad \lambda_2 = L_{\ell 2} I_2 + N_2 \Phi \tag{2.76}$$

so that the voltage equations (2.61) become

$$V_1 = R_1 I_1 + L_{\ell 1} \frac{dI_1}{dt} + N_1 \frac{d\Phi}{dt} \quad , \quad V_2 = R_2 I_2 + L_{\ell 2} \frac{dI_2}{dt} + N_2 \frac{d\Phi}{dt} \tag{2.77}$$

Φ is the sum of the fluxes linking one coil and produced by the other. In order to add these legitimately, we need to assume that the material characteristics are linear. However, we can, at this stage, assume that these mutual fluxes are produced by the combined action of the currents acting simultaneously. This will allow non-linear effects to be included and Φ can therefore no longer be regarded as the sum of separate fluxes.

As before, we expect the core excitation current, I_{ex} , to be supplied by the primary coil so that

$$I_1 = I_{ex} + I_1' \tag{2.78}$$

where I_1 satisfies the amp-turn balance condition (2.66) when load current flows in coil 2. Thus, from (2.77) and (2.78) we obtain the equivalent circuit shown in Fig. 2.17. This model is identical with that derived earlier and shown in Fig. 2.15 where linearity was assumed except that now it is clear that the core characteristics may be non-linear. Also the constancy of the leakage inductances as circuit elements is apparent. As before, we can reposition the ideal transformer and derive a circuit model equivalent to that shown in Fig. 2.16 except that the linear inductive element $(N_1/N_2)M$ is replaced by the possibly nonlinear $N_1 d\Phi/dt$. This is shown in Fig. 2.18.

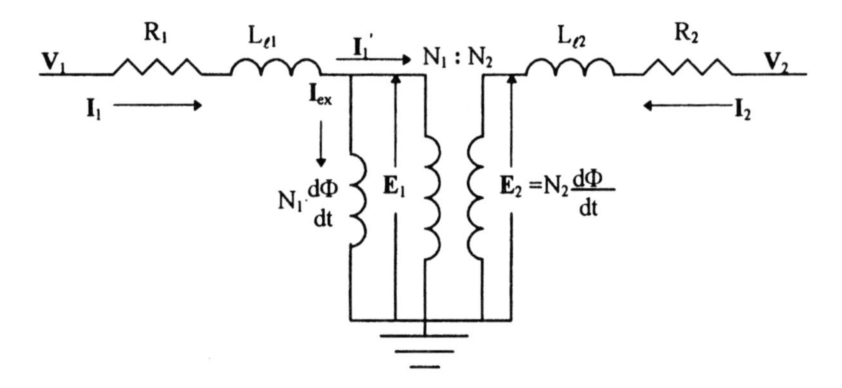

Figure 2.17 Circuit model of a 2 winding transformer under load, including a possibly non-linear core

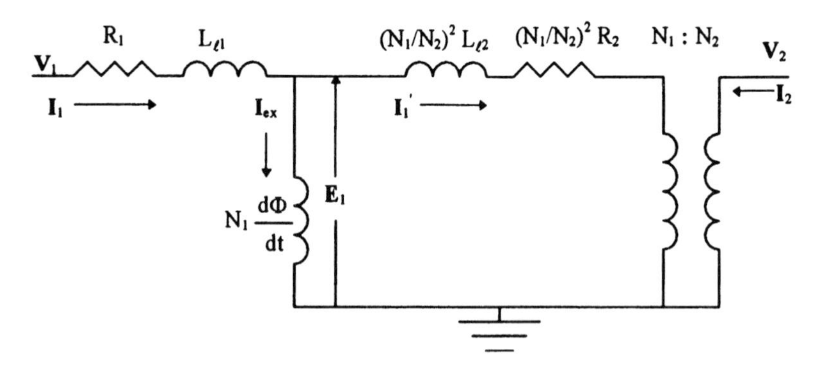

Figure 2.18 Circuit model of a 2 winding transformer under load referred to the primary side and including a possibly non-linear core

With the addition of core loss and possibly capacitive effects which can be accounted for by placing a resistance and capacitance in parallel with $\mathbf{E}_1$, Figs. 17 or 18 can be regarded as virtually exact circuit models of a 2 winding transformer phase under load. In practice, because the exciting current is small, it is usually permissible to transfer the shunt branch to either side of the impedances as shown in Fig. 2.19, where the resistances and reactances have been combined to give equivalent quantities

$$R_{1,eq} = R_1 + \left(\frac{N_1}{N_2}\right)^2 R_2 \quad , \quad L_{1,eq} = L_{\ell 1} + \left(\frac{N_1}{N_2}\right)^2 L_{\ell 2} \qquad (2.79)$$

These are the 2 winding resistance and leakage inductance referred to the primary side. Although the equivalent resistance includes the I^2R losses in the windings, it can also include losses caused by the stray flux since these are proportional to the stray induction squared to a good approximation which, in turn, is proportional to the square of the current. For sinusoidal currents, we can define an equivalent impedance by

$$Z_{1,eq} = R_{1,eq} + j\omega L_{1,eq} \qquad (2.80)$$

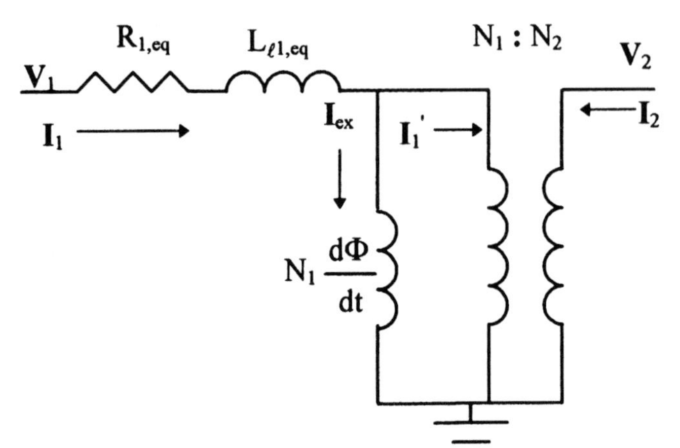

(a) Shunt branch transferred to the right

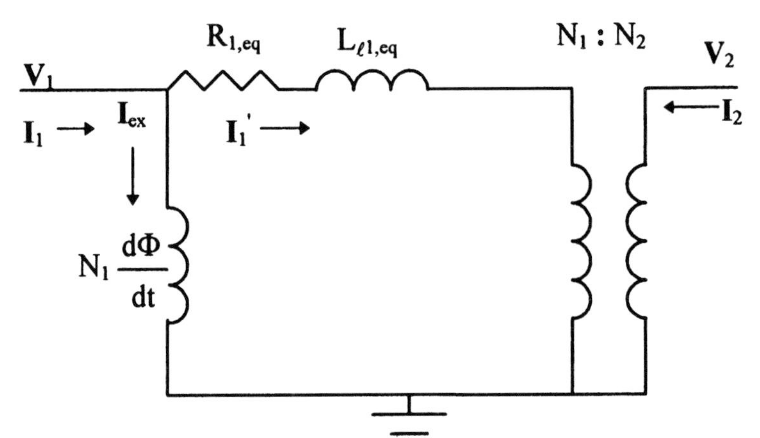

(b) Shunt branch transferred to the left

Figure 2.19 Approximate circuit models of a 2 winding transformer under load referred to the primary side and including a possibly non-linear core

For many purposes, it is even possible to dispense with the exciting current branch so that the approximate equivalent circuit reduces to Fig. 2.20, with the simplified circuit equations

$$\mathbf{V}_1 = \mathbf{I}_1 Z_{1,eq} + \mathbf{E}_1 \quad , \quad \mathbf{V}_2 = \mathbf{E}_2 \tag{2.81}$$

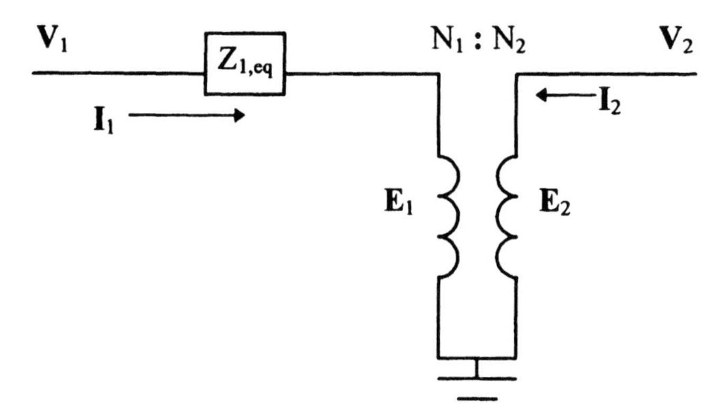

Figure 2.20 Approximate circuit model of a 2 winding transformer under load referred to the primary side and ignoring core excitation

Using (2.66), with $I_1' = I_1$ since we are ignoring exciting current, and (2.69), we can write for (2.81)

$$\mathbf{V}_1 = -\left(\frac{N_2}{N_1}\right)\mathbf{I}_2 Z_{1,eq} + \left(\frac{N_1}{N_2}\right)\mathbf{E}_2 = \left(\frac{N_1}{N_2}\right)\left[\mathbf{E}_2 - \left(\frac{N_2}{N_1}\right)^2 Z_{1,eq}\mathbf{I}_2\right]$$

or (2.82)

$$\left(\frac{N_2}{N_1}\right)\mathbf{V}_1 = \mathbf{E}_2 - \left(\frac{N_2}{N_1}\right)^2 Z_{1,eq}\mathbf{I}_2 = \mathbf{V}_2 - \left(\frac{N_2}{N_1}\right)^2 Z_{1,eq}\mathbf{I}_2$$

Letting $E_1' = V_1$ and $E_2' = (N_2/N_1)V_1$, (2.82) becomes

$$\mathbf{V}_2 = \left(\frac{N_2}{N_1}\right)^2 Z_{1,eq}\mathbf{I}_2 + \mathbf{E}_2' \quad , \quad \mathbf{V}_1 = \mathbf{E}_1' \tag{2.83}$$

where $\mathbf{E}_1'/\mathbf{E}_2' = N_1/N_2$. The circuit model for (2.83) is shown in Fig. 2.21 where

$$Z_{2,eq} = \left(\frac{N_2}{N_1}\right)^2 Z_{1,eq} = \left[\left(\frac{N_2}{N_1}\right)^2 R_1 + R_2\right] + j\omega\left[\left(\frac{N_2}{N_1}\right)^2 L_{\ell 1} + L_{\ell 2}\right] \tag{2.84}$$

$Z_{2,eq}$ is the equivalent impedance referred to the secondary side. It should noted that the equivalent circuits with the core excitation branch included could be transformed to ones having the impedances on the secondary side by similar methods.

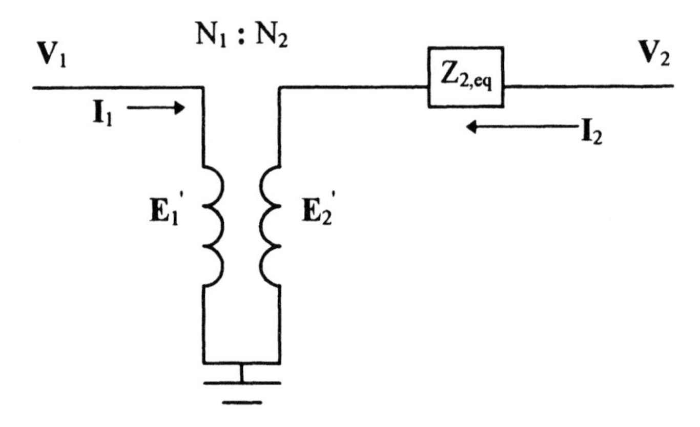

Figure 2.21 Approximate circuit model of a 2 winding transformer under load referred to the secondary side and ignoring core excitation

We should note that the exciting current may not be restricted to the primary winding when load current flows but may be shared by both windings. This can lead to legitimate circuit models with shunt branches on either side. These are all mathematically equivalent since impedances can be transferred across the ideal transformer present in the circuit. We demonstrated this above for a series impedance. We show this now for a shunt impedance by means of Fig. 2.22.

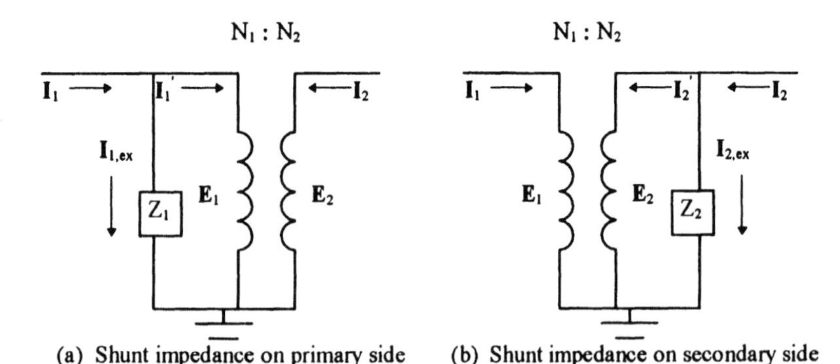

(a) Shunt impedance on primary side (b) Shunt impedance on secondary side

Figure 2.22 Transfer of shunt impedances across the ideal transformer

In Fig 22a, we have the relationships

$$\mathbf{E}_1 = \mathbf{I}_{1,ex} Z_1 \quad , \quad \frac{\mathbf{E}_1}{\mathbf{E}_2} = \frac{N_1}{N_2} \quad , \quad \frac{\mathbf{I}_1^{'}}{\mathbf{I}_2} = -\frac{N_2}{N_1} \quad , \quad \mathbf{I}_1 = \mathbf{I}_1^{'} + \mathbf{I}_{1,ex} \tag{2.85}$$

Manipulating these equations, we obtain

$$\mathbf{E}_2 = \left(\frac{N_2}{N_1}\right)\left[\mathbf{I}_1 + \left(\frac{N_2}{N_1}\right)\mathbf{I}_2\right] Z_1 \tag{2.86}$$

From Fig. 2.22b, we have $\mathbf{I}_2 = \mathbf{I}_2^{'} + \mathbf{I}_{2,ex}$ and $\mathbf{I}_1/\mathbf{I}_2^{'} = -N_2/N_1$. Using these, (2.86) becomes

$$\mathbf{E}_2 = \left(\frac{N_2}{N_1}\right)^2 Z_1 \mathbf{I}_{2,ex} = Z_2 \mathbf{I}_{2,ex} \tag{2.87}$$

Thus the equivalent shunt impedance on the secondary side is obtained in the same manner as for series impedances as is seen by comparing (2.84). The voltage and currents transform in the usual way across the ideal transformer in both circuits of Fig. 2.22.

We see from Fig. 2.20 and 2.21 that if the shunt branch is ignored so that the exciting current is zero and equal and opposite amp-turns flow in the 2 coils, then the inductance of this configuration is the 2 winding leakage inductance. This can then be obtained from the energy in the magnetic field by methods to be described later.

We mentioned earlier that in modern power transformers, which typically have very low core exciting current, capacitive effects can be important in situations where the core characteristics play a role. We indicated that one way of including these effects was by placing a shunt capacitance across the core. However, for greater accuracy, a better method is to use the circuit model shown in Fig. 2.23 or one of its equivalents. This model also includes the core losses by means of a core resistance. There are shunt capacitances across the input and output so that even if one of the coils is open circuited, the shunt capacitance would act as a load so that current would flow in both windings. This model has been found to accurately reproduce test results over a frequency range near the power frequency.

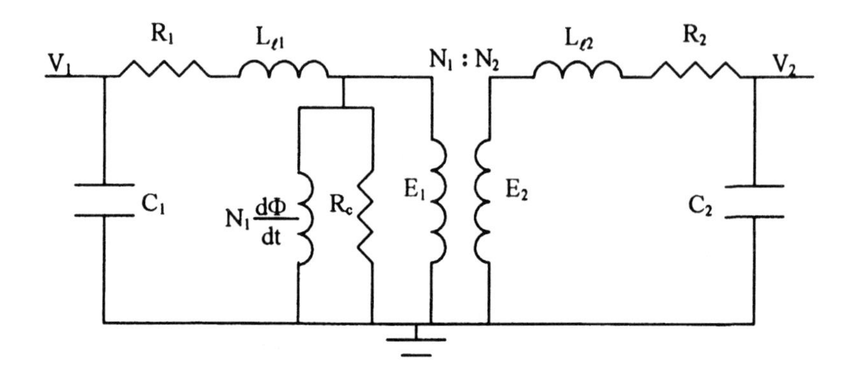

Figure 2.23 Circuit model of a 2 winding transformer under load, including a possibly non-inear core, core losses, and capacitative effects

When the core has non-linear Φ vs I_{ex} characteristics, since $E_1 = N_1\, d\Phi/dt$, a sinusoidal E_1 will result in a non-sinusoidal I_{ex}. For Φ reasonably below saturation, the non-linearities are mild enough that I_{ex} can be considered sinusoidal for most purposes when the primary circuit is driven by a sinusoidal voltage. In this case $N_1\, d\Phi/dt$ can be replaced by a magnetizing inductance or reactance. As the transformer is driven into saturation, the exciting current becomes nonsinusoidal and also constitutes a greater fraction of the load current. In this case, it begins to matter which coil carries it. As was seen in the extreme case of inrush current, the exciting current depended on the mean area of the coil through which it flowed. In this extreme, the circuit modals we have developed here are no longer applicable. One would need to work directly with flux linkages. One could define non-linear inductances and mutual inductances as in (2.63) but it would no longer be true that $M_{12} = M_{21}$. Also the L's and M's would depend on possibly both I_1 and I_2.

2.7 PER UNIT SYSTEM

Transformer impedances, along with other quantities such as voltages and currents are often expressed in the per unit (p.u.) system. In this system, these quantities are expressed as a ratio with respect to the transformer's nominal or rated phase quantities. Thus, if the rated or base phase voltages are V_{b1} , V_{b2} and the base currents are I_{b1} , I_{b2} where 1 and 2 refer to the primary and secondary sides, then the base impedances are

$$Z_{b1} = \frac{V_{b1}}{I_{b1}} \quad , \quad Z_{b2} = \frac{V_{b2}}{I_{b2}} \tag{2.88}$$

The rated or base voltages, currents, and impedances are assumed to transfer from one side to the other by means of the ideal transformer relationships among voltages, currents, and impedances. These base quantities are all taken to be positive. Thus the minus sign is neglected in the base current transfer across sides. Hence, it can be shown that the base power, P_b , is the same on both sides of a transformer.

$$P_b = V_{b1} I_{b1} = V_{b2} I_{b2} \tag{2.89}$$

In an actual transformer, the real power into a transformer is nearly the same as that leaving it on the secondary side. This is because the transformer losses are a small fraction of the power transferred.

The primary side voltage, V_1 , current, I_1 , and equivalent impedance, $Z_{1,eq}$, are expressed in the per unit system by

$$V_{1,pu} = v_1 = \frac{V_1}{V_{b1}} \quad , \quad I_{1,pu} = i_1 = \frac{I_1}{I_{b1}} \quad , \quad Z_{1,eq,pu} = z_{1,eq} = \frac{Z_{1,eq}}{Z_{b1}} \tag{2.90}$$

and similarly for the secondary quantities. Often the p.u. values are multiplied by 100 and expressed as a percentage. However, it is best not to use the percentage values in calculations since this can lead to errors. As indicated in (2.90), we use lower case letters to denote p.u. quantities.

Since the base quantities transfer across the ideal transformer in the same manner as their corresponding circuit quantities, the p.u. values of the circuit quantities are the same on both the primary and secondary sides. Thus the 1 or 2 subscripts can be dropped when referring to p.u. quantities. Although voltages, currents, and impedances for transformers of greatly different power ratings can differ considerably, their per unit values tend to be very similar. This can facilitate calculations since one has a pretty good idea of the magnitudes of the quantities being calculated. Thus, the 2 winding leakage impedances when expressed in the p.u system are generally in the range of 5 - 15 % for all power transformers. The exciting currents of modern power transformers are typically $\sim 0.1\%$ in the p.u. system. This is also their percentage of the rated load current since this has the value of 1 or 100 % in the p.u. system. The 2 winding resistances which account for the transformers losses can be obtained in the p.u. system by noting that modern power transformers are typically > 99.5 % efficient. This means that < 0.5 % of the rated input power goes into losses. Thus we have

$$\frac{\text{Losses}}{\text{Rated Power}} = \frac{R_{1,eq} I_{b1}^2}{V_{b1} I_{b1}} = \frac{R_{1,eq}}{V_{b1}/I_{b1}} = \frac{R_{1,eq}}{Z_{b1}} = R_{1,eq,pu} = r < 0.5\% \qquad (2.91)$$

We omitted a 1 subscript on the per unit equivalent resistance, r , since it is the same on both sides of the transformer. Using this and the above estimate of the leakage impedance in the p.u. system, x , for power transformers, we can estimate the x/r ratio,

$$\frac{x}{r} = \frac{0.05 - 0.15}{< 0.005} > 10 - 30 \qquad (2.92)$$

Thus 10 is probably a lower limit and, as previously shown, is high enough to allow discriminating inrush from fault current on the basis of second harmonic analysis.

2.8 VOLTAGE REGULATION

At this point, it is useful to discuss the topic of voltage regulation as an application of the transformer circuit model just developed. In this context, the core characteristics do not play a significant role so we will use the simplified circuit model of Fig. 2.21. Voltage regulation is defined as the change in the magnitude of the secondary voltage between its open circuited value and its value when loaded divided by the value when loaded with the primary voltage held constant. We can represent the load by an equivalent impedance, Z_L , and the relevant circuit is shown in Fig. 2.24. In the figure, we have shown a load current, I_L , where $I_L = -I_2$.

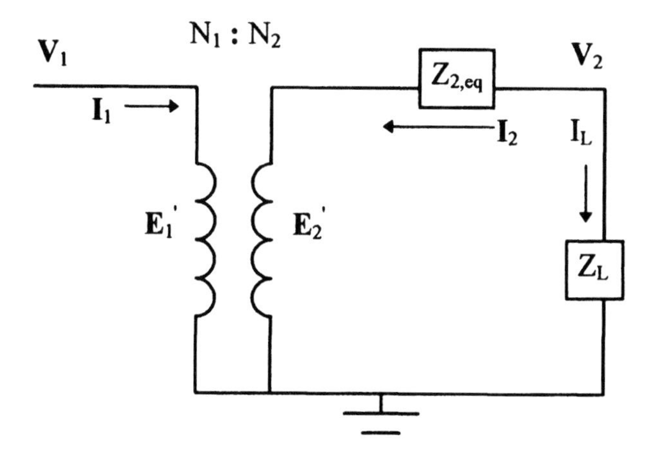

Figure 2.24 Approximate circuit models of a 2 winding transformer with load included and referred to the secondary side

It is convenient to perform this calculation in the per unit system. We will use lower case letters to represent per unit quantities. Thus since base quantities transfer across the ideal transformer like their corresponding physical quantities, we have

$$e_1' = \frac{E_1'}{V_{b1}} = \frac{E_2'}{V_{b2}} = e_2' = e' \quad , \quad i_1 = \frac{I_1}{I_{b1}} = -\frac{I_2}{I_{b2}} = -i_2 = i_L \tag{2.93}$$

Therefore the ideal transformer can be eliminated from the circuit in the per unit system. Also, since $z_{2,eq} = z_{1,eq}$, we can drop the numerical subscript and denote the equivalent transformer p.u. impedance by

$$z = r + jx \tag{2.94}$$

where r is the p.u. equivalent resistance and x the p.u. 2 winding leakage reactance. Similarly we write

$$z_L = \frac{Z_L}{Z_{b2}} = r_L + jx_L \tag{2.95}$$

The p.u. circuit is shown in Fig. 2.25.

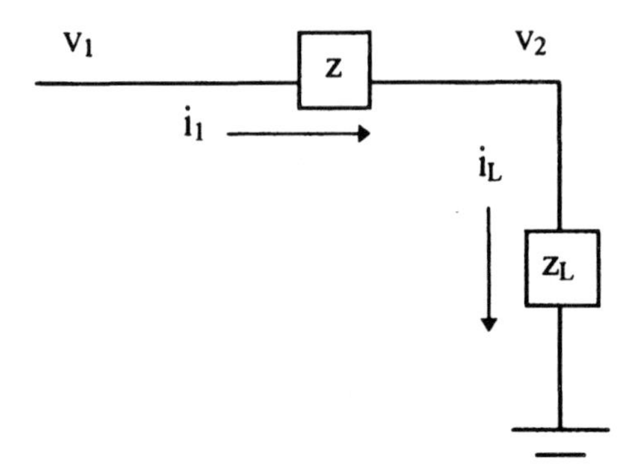

Figure 2.25 The circuit of Fig. 2.24 shown in the p.u. system

Thus when the secondary terminal is open circuited so z_L and its ground connection are missing in Fig. 2.25, we obtain for the open circuited value of v_2,

$$v_{2,oc} = v_1 \tag{2.96}$$

We assume that $\mathbf{v}_1$ is a reference phasor (zero phase angle) and v_1 is therefore its magnitude. When the load is present, we have

$$\mathbf{v}_2 = \mathbf{i}_L \mathbf{z}_L \quad , \quad \mathbf{v}_1 = \mathbf{i}_L (\mathbf{z} + \mathbf{z}_L) \tag{2.97}$$

Thus, from (2.97),

$$\mathbf{v}_2 = \frac{\mathbf{z}_L \mathbf{v}_1}{(\mathbf{z} + \mathbf{z}_L)} = \frac{v_1}{(1 + \mathbf{z}/\mathbf{z}_L)} \quad \Rightarrow \quad |\mathbf{v}_2| = \frac{v_1}{|1 + \mathbf{z}/\mathbf{z}_L|} \tag{2.98}$$

Hence the voltage regulation is given by

$$\text{Regulation} = \frac{v_{2,oc} - |\mathbf{v}_2|}{|\mathbf{v}_2|} = |1 + \mathbf{z}/\mathbf{z}_L| - 1 \tag{2.99}$$

It is customary to express $\mathbf{z}_L$ in terms of the load voltage and current via (2.97),

$$z_L = \frac{\mathbf{v}_2}{\mathbf{i}_L} = \frac{|\mathbf{v}_2|e^{j\theta_v}}{|\mathbf{i}_L|e^{j\theta_i}} = \frac{|\mathbf{v}_2|e^{j(\theta_v-\theta_i)}}{|\mathbf{i}_L|} = \frac{|\mathbf{v}_2|e^{j\theta}}{|\mathbf{i}_L|} \tag{2.100}$$

where $\theta = \theta_v - \theta_i$ is the angle by which the voltage leads the current. For an inductive load, θ is positive. Letting v_2 and i_L represent the magnitudes of the load voltage and current, i.e. dropping the magnitude signs, the impedance ratio in (2.99) can be written

$$\begin{aligned}\frac{z}{z_L} &= (r+jx)\frac{i_L}{v_2}e^{-j\theta} = \frac{i_L}{v_2}(r+jx)(\cos\theta - j\sin\theta) \\ &= \frac{i_L}{v_2}\left[(r\cos\theta + x\sin\theta) + j(x\cos\theta - r\sin\theta)\right]\end{aligned} \tag{2.101}$$

Substituting (2.101) into (2.99), we get

$$\begin{aligned}\text{Regulation} &= \sqrt{\left[1+\frac{i_L}{v_2}(r\cos\theta + x\sin\theta)\right]^2 + \left[\frac{i_L}{v_2}(x\cos\theta - r\sin\theta)\right]^2} - 1 \\ &= \sqrt{1+2\frac{i_L}{v_2}(r\cos\theta + x\sin\theta) + \left(\frac{i_L}{v_2}\right)^2\left(r^2+x^2\right)} - 1\end{aligned} \tag{2.102}$$

Since z is generally small compared with z_L, the terms other than unity in (2.102) are small compared to unity. Using the approximation for small ε,

$$\sqrt{1+\varepsilon} \approx 1 + \frac{1}{2}\varepsilon - \frac{1}{8}\varepsilon^2 \tag{2.103}$$

the regulation is given to second order in z/z_L by

$$\text{Regulation} = \frac{i_L}{v_2}(r\cos\theta + x\sin\theta) + \frac{1}{2}\left(\frac{i_L}{v_2}\right)^2(x\cos\theta - r\sin\theta)^2 \tag{2.104}$$

3. REACTANCE CALCULATIONS

Summary Leakage reactances are calculated for transformers having up to six separate windings per leg but with 2 or 3 terminals per phase. These calculations may be applied to single phase units or to one phase of a 3 phase unit. Windings may be connected in series or auto-connected. In addition, tap windings or series taps within a winding may be identified. Terminal-terminal and positive/negative sequence leakage reactances are calculated, as well as the T-equivalent circuit model leakage reactances for 3 terminal transformers. Expressions for per-unit quantities are also given. When taps are present within a winding or as separate windings, calculations are performed for the all in, all out, and center or neutral position and for all combinations if more than one type of tap is present. The calculations are based on a 2 winding reactance formula which assumes that the windings are uniform along their length, with a correction for end fringing flux. This has been found to be very accurate for most purposes but, if greater accuracy is desired, a better two winding reactance calculation could be substituted without changing most of the formulas presented here.

3.1 INTRODUCTION

The reactance calculations performed here are on a per phase basis so they would apply to a single phase unit or to one phase of a three phase transformer. The phase can have up to six windings, interconnected in such a way that only 2 or 3 terminals (external or buried) result. Thus auto-transformers, with or without tertiary, are included as well as transformers with tap windings. it should be noted that these reactances are positive or negative sequence reactances. Zero sequence reactances are somewhat sensitive to the three phase connection and to whether the transformer is core-form or shell-form, but this does not appear to be true for the positive/negative sequence reactances. Thus, in the following discussion, we will be dealing with a single phase system which may or may not be part of a 3 phase system.

It is usually desirable to design some reactance into a transformer in order to limit any fault current. In addition, these reactances determine the voltage regulation of the unit. Hence it is desirable at the design stage to be able to calculate these reactances based on the geometry of the coils and core and the nature of the (single phase) winding interconnections.

We begin by discussing ideal transformers, i.e. having no reactance or resistance, since real transformers are usually described by adding lumped resistance and reactance circuit elements to a model of an ideal transformer. The following references have been used in this chapter: [MIT43], [Lyo37], [Blu51], [Wes64].

3.2 IDEAL TRANSFORMERS

In an ideal 2 winding transformer as depicted in Fig. 3.1, the entire flux ϕ links both windings so that, by Faraday's law, the induced emf's are given by

$$E_1 = -N_1 \frac{d\phi}{dt} \quad , \quad E_2 = -N_2 \frac{d\phi}{dt} \tag{3.1}$$

Hence

$$\frac{E_1}{E_2} = \frac{N_1}{N_2} \tag{3.2}$$

We use the convention that the current is positive when entering the positive terminal and that the fluxes generated by these positive currents add.

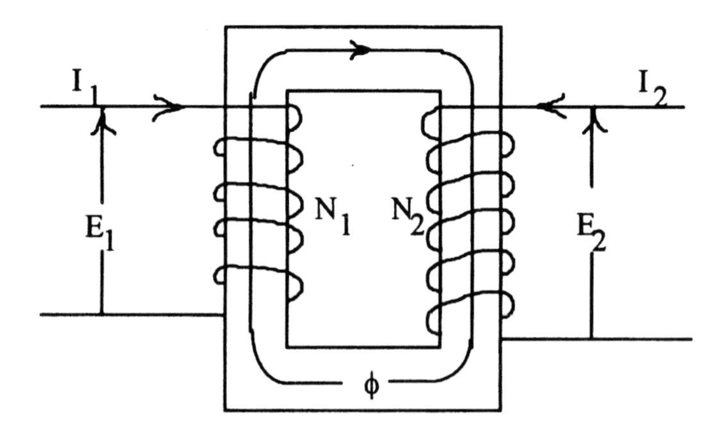

Figure 3.1 Ideal 2 winding transformer

In an ideal transformer, zero mmf is required to generate the flux so that

$$N_1I_1 + N_2I_2 = 0 \tag{3.3}$$

which implies

$$\frac{I_1}{I_2} = -\frac{N_2}{N_1} \tag{3.4}$$

i.e., the secondary current I_2 is leaving terminal 2 (negative) when the primary current I_1 enters terminal 1. Also, from the above equations

$$\frac{E_1I_1}{E_2I_2} = -1 \tag{3.5}$$

i.e. the instantaneous power entering terminal 1 equals the power leaving terminal 2 so that the unit is lossless as expected.

For a 3 winding unit, the above formulas become

$$\frac{E_1}{N_1} = \frac{E_2}{N_2} = \frac{E_3}{N_3} \tag{3.6}$$

and

$$N_1I_1 + N_2I_2 + N_3I_3 = 0 \tag{3.7}$$

which imply

$$E_1I_1 + E_2I_2 + E_3I_3 = 0 \tag{3.8}$$

i.e. the net power into the unit is 0, or the power entering equals the power leaving. The generalization to more than 3 windings is straightforward.

If the ideal transformer has a load of impedance Z_2 connected to its secondary terminals as shown in Fig. 3.2, we have the additional equation

$$E_2 = -I_2Z_2 \tag{3.9}$$

where $-I_2$ is the load current, usually denoted I_L.

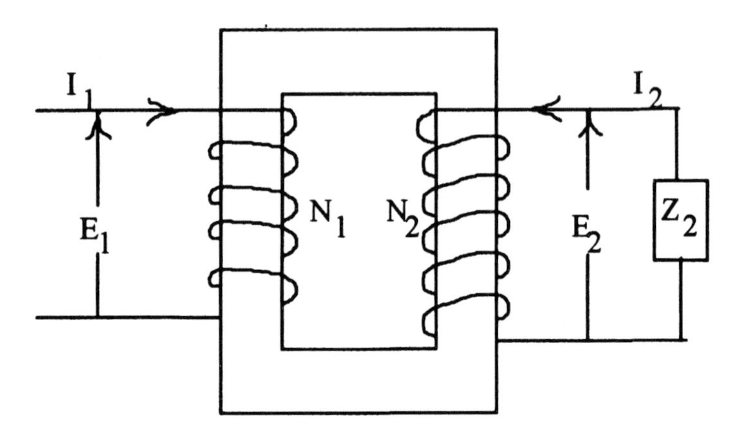

Figure 3.2 Ideal 2 winding transformer with a load on its secondary terminals

But from the above equations,

$$\frac{E_1}{I_1} = \left(\frac{N_1}{N_2}\right)E_2 \Big/ \left(-\frac{N_2}{N_1}\right)I_2 = -\left(\frac{N_1}{N_2}\right)^2 \frac{E_2}{I_2} = \left(\frac{N_1}{N_2}\right)^2 Z_2 \tag{3.10}$$

Thus, as seen from the primary terminals, the load appears as an impedance Z_1,

$$E_1 = I_1 Z_1 \tag{3.11}$$

where

$$Z_1 = \left(\frac{N_1}{N_2}\right)^2 Z_2 \tag{3.12}$$

At this point we introduce per-unit values. That is, we choose base values for voltages, currents, etc. and express the actual voltages, currents, etc. as ratios with respect to these base values. Only two independent base values need to be chosen and these are usually taken as power or VI and voltage V. The base power is normally taken as the VI rating of the unit per phase and the voltage taken as the open circuit rated phase voltage of each terminal. Thus we have a $(VI)_b$ and V_{1b}, V_{2b}, ... where b denotes base value. From these we derive base currents of

$$I_{1b} = \frac{(VI)_b}{V_{1b}} \quad , \quad \text{etc.} \tag{3.13}$$

and base impedances

$$Z_{1b} = \frac{(V_{1b})^2}{(VI)_b} \quad , \quad \text{etc.} \tag{3.14}$$

Letting small letters denote per-unit quantities, we have

$$v_1 = \frac{V_1}{V_{1b}} \ ,\text{etc.} \qquad i_1 = \frac{I_1}{I_{1b}} \ ,\text{etc.} \qquad z_1 = \frac{Z_1}{Z_{1b}} \ ,\text{etc.} \tag{3.15}$$

Ignoring core excitation, at no load we have $V_1 = E_1$, etc. so that from (3.6), (3.13), (3.14),

$$\frac{V_{1b}}{V_{2b}} = \frac{N_1}{N_2}\ ,\text{etc.} \qquad \frac{I_{1b}}{I_{2b}} = \frac{N_2}{N_1}\ ,\text{etc.} \qquad \frac{Z_{1b}}{Z_{2b}} = \left(\frac{V_{1b}}{V_{2b}}\right)^2 = \left(\frac{N_1}{N_2}\right)^2 \ ,\text{etc.} \tag{3.16}$$

since everything is on a common VI base.

In terms of per-unit quantities, we have for an ideal transformer

$$e_1 = \frac{E_1}{V_{1b}} = 1 \quad , \quad \text{etc.} \tag{3.17}$$

and, from (3.7) and (3.16)

$$N_1 \frac{I_1}{I_{1b}} + N_2 \frac{I_2}{I_{1b}} + N_3 \frac{I_3}{I_{1b}}$$

$$= N_1\left(\frac{I_1}{I_{1b}}\right) + N_2\left(\frac{I_2}{I_{2b}}\right)\left(\frac{I_{2b}}{I_{1b}}\right) + N_3\left(\frac{I_3}{I_{3b}}\right)\left(\frac{I_{3b}}{I_{1b}}\right)$$

$$= N_1(i_1 + i_2 + i_3) = 0$$

Therefore

$$i_1 + i_2 + i_3 = 0 \tag{3.18}$$

Thus an ideal 3-circuit transformer can be represented by a one circuit description as shown in Fig. 3.3, if per-unit values are used. This also holds if more than 3 circuits are present.

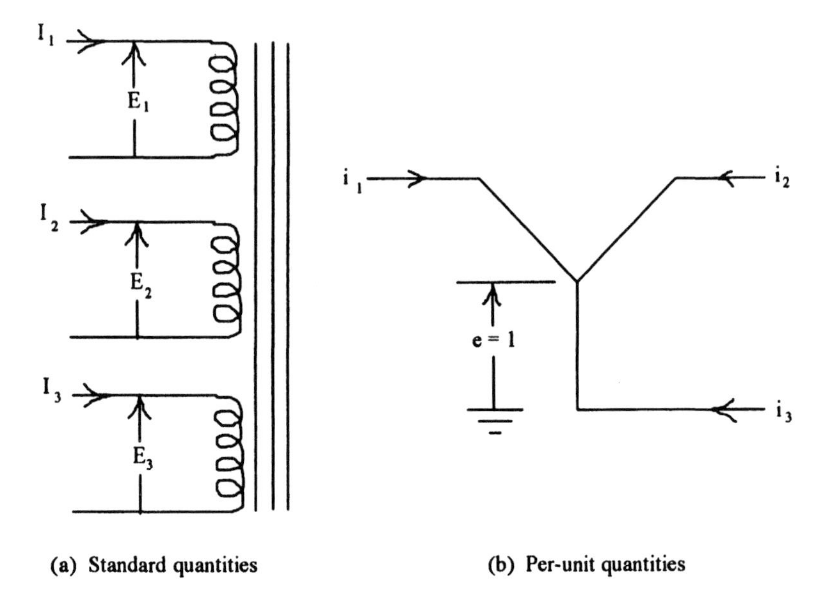

Figure 3.3 Ideal 3 circuit transformer schematic

3.2.1 Ideal Autotransformer

The ideal 2-terminal autotransformer is shown in Fig. 3.4. The two coils, labeled s for series and c for common, are connected together so that their voltages add to produce the high voltage terminal voltage E_1. Thus $E_1 = E_s + E_c$. The secondary or low voltage terminal voltage is $E_2 = E_c$. Similarly, from the figure, $I_1 = I_s$ and $I_2 = I_c - I_s$. Using the expressions for a two winding unit, which are true regardless of the interconnections involved,

$$\frac{E_s}{N_s} = \frac{E_c}{N_c}$$

$$I_s N_s + I_c N_c = 0$$

we find

$$E_1 = E_c\left(1+\frac{N_s}{N_c}\right) = E_2\left(\frac{N_c+N_s}{N_c}\right)$$

$$I_1N_s+(I_2+I_1)N_c = I_1(N_c+N_s)+I_2N_c = 0$$

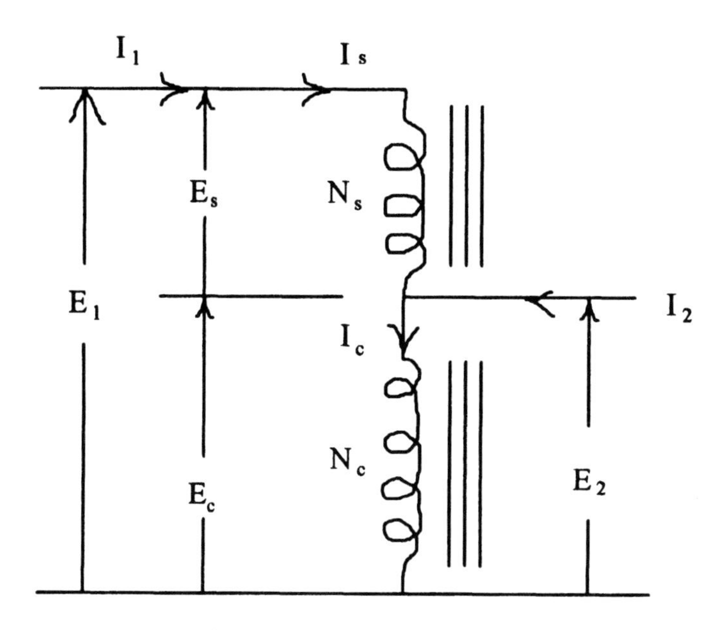

Figure 3.4 Ideal autotransformer

Thus

$$\frac{E_1}{E_2} = \left(\frac{N_c+N_s}{N_c}\right) \quad \text{and} \quad \frac{I_1}{I_2} = -\left(\frac{N_c}{N_c+N_s}\right) \tag{3.19}$$

so that the effective turns ratio as seen by the terminals is $n = (N_C + N_S) / N_C$. Any impedances on the low voltage terminal could be transferred to the high voltage circuit by the square of this turns ratio.

The co-ratio is defined as

$$r = \frac{E_1-E_2}{E_1} = 1-\frac{E_2}{E_1} = 1-\frac{1}{n} = \frac{n-1}{n} \tag{3.20}$$

where r < 1. The terminal power rating of an autotransformer is E_1I_1 (= $-E_2I_2$) but the power rating of each coil is E_sI_s (= $-E_cI_c$). The ratio of these power ratings is

$$\frac{E_1I_1}{E_sI_s} = \frac{E_1}{(E_1 - E_2)} = \frac{1}{1 - 1/n} = \frac{1}{r} \tag{3.21}$$

Thus, since r < 1, the terminal rating is always greater that the single coil rating. Since the single coil rating is the same as the terminal rating of a conventional 2-winding transformer, this shows an important advantage of using the auto connection.

3.3 LEAKAGE IMPEDANCE FOR 2-WINDING TRANSFORMERS

In real transformers not all the flux links the windings. In addition, there are resistive losses in the windings as well as core losses. The basic equations for the terminal voltages are now given by

$$V_1 = I_1R_1 - \frac{d\lambda_1}{dt} \quad \text{and} \quad V_2 = I_2R_2 - \frac{d\lambda_2}{dt} \tag{3.22}$$

where λ_1 is the flux linkages of coil 1 and R_1 its resistance and similarly for coil 2. If we let
ϕ_c = the common flux linking all turns of both coils, which is mainly core flux, we can write

$$\lambda_1 = (\lambda_1 - N_1\phi_c) + N_1\phi_c \quad \text{and} \quad \lambda_2 = (\lambda_2 - N_2\phi_c) + N_2\phi_c \tag{3.23}$$

The quantity $(\lambda_i - N_i\phi_c)$ is the leakage flux of coil i. It exists mainly in the oil or air and conductor material but not in the core to any great extent. Thus it exists in non-magnetic (or linear) materials and therefore should depend linearly on the currents. Thus we can write very generally

$$-\frac{d(\lambda_1 - N_1\phi_c)}{dt} = L_1\frac{dI_1}{dt} + M_{12}\frac{dI_2}{dt}$$

or, assuming sinusoidal quantities,

$$-\frac{d(\lambda_1 - N_1\phi_c)}{dt} = jI_1X_{11} + jI_2X_{12} \tag{3.24}$$

where I_1 and I_2 are phasors, $X_{11} = 2\pi f L_1$ and $X_{12} = 2\pi f M_{12}$, where f is the frequency, and j is the imaginary unit. Using similar expressions for λ_2 , (3.22) becomes

$$\begin{aligned} V_1 &= I_1R_1 + jI_1X_{11} + jI_2X_{12} + E_1 \\ V_2 &= I_2R_2 + jI_2X_{22} + jI_1X_{12} + E_2 \end{aligned} \tag{3.25}$$

where $E_1 = -N_1\, d\phi_c/dt$ is the no-load terminal voltage of terminal 1, etc. for E_2. We have $E_1/E_2 = N_1/N_2$. We have also used the fact that $X_{12} = X_{21}$ for linear systems.

We are going to ignore the exciting current of the core since this is normally much smaller than the load currents. Thus (3.3) applies so we can rewrite (3.25)

$$\begin{aligned} V_1 &= I_1\left[R_1 + j\left(X_{11} - \frac{N_1}{N_2}X_{12}\right)\right] + E_1 \\ V_2 &= I_2\left[R_2 + j\left(X_{22} - \frac{N_2}{N_1}X_{12}\right)\right] + E_2 \end{aligned} \tag{3.26}$$

or, more succinctly,

$$\begin{aligned} V_1 &= I_1Z_1 + E_1 \\ V_2 &= I_2Z_2 + E_2 \end{aligned} \tag{3.27}$$

where Z_1 and Z_2 are single winding leakage impedances. This can be visualized by means of Fig. 3.5a.

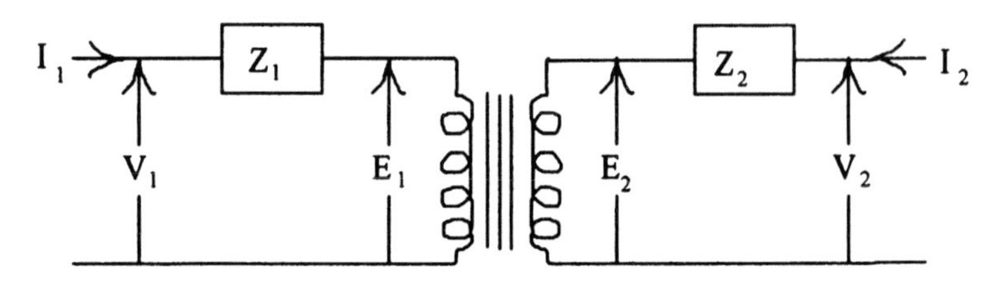

(a) Separate circuits and leakage

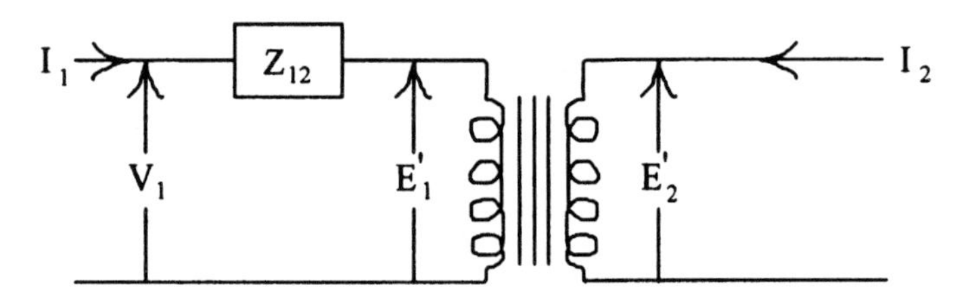

(b) Separate circuits with a single effective 2-winding leakage

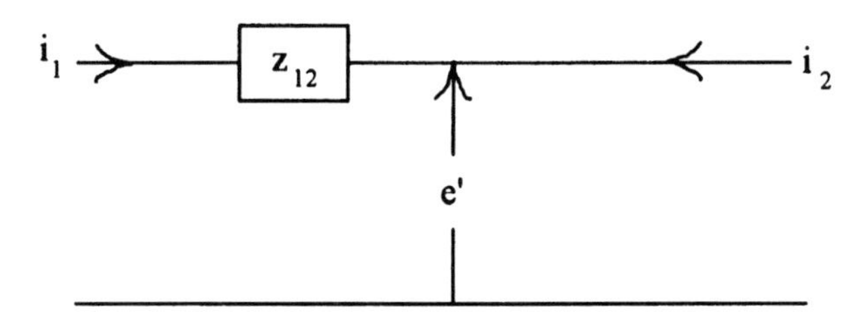

(c) Single circuit using per-unit

Figure 3.5 Circuit models of a 2 winding transformer with leakage impedance

Let $V_2 = E'_2 = I_2 Z_2 + E_2$. Then, since $E_1/E_2 = N_1/N_2$ and $I_1/I_2 = -N_2/N_1$, $E'_2 = -(N_1/N_2)I_1 Z_2 + (N_2/N_1)E_1$ or rewriting

$$E_1 = \frac{N_1}{N_2} E'_2 + \left(\frac{N_1}{N_2} \right)^2 I_1 Z_2 \tag{3.28}$$

Substituting into (3.27), we get

$$V_1 = I_1\left[Z_1 + \left(\frac{N_1}{N_2}\right)^2 Z_2\right] + \frac{N_1}{N_2}E_2'$$

$$V_2 = E_2' \tag{3.29}$$

We see that (3.29) can be expressed in terms of a single effective impedance $Z_{12} = Z_1 + (N_1/N_2)^2 Z_2$,

$$V_1 = I_1 Z_{12} + E_1'$$

$$V_2 = E_2' \tag{3.30}$$

where $E'_1/E'_2 = N_1/N_2$. The circuit model for this is shown in Fig. 3.5b. Using per-unit values, the picture in Fig. 3.5c applies. Note that $E_1/E_2 = E'_1/E'_2 = N_1/N_2$ but that $E_1 \neq E'_1$, $E_2 \neq E'_2$ except at no load.

Thus a two winding transformer is characterized by a single value of leakage impedance Z_{12} which has both resistive and reactive components,

$$Z_{12} = R_{12} + jX_{12}$$

where

$$R_{12} = R_1 + \left(\frac{N_1}{N_2}\right)^2 R_2 \tag{3.31}$$

and

$$X_{12} = X_1 + \left(\frac{N_1}{N_2}\right)^2 X_2$$

referred to the primary winding. We can obtain expressions for referring quantities to the secondary winding by interchanging 1 and 2 in the above formulas. In large power transformers $X_{12} >> R_{12}$ so we are normally concerned with obtaining leakage reactances.

In terms of previously defined quantities,

$$X_1 = X_{11} - \frac{N_1}{N_2} X_{12}$$
$$X_2 = X_{22} - \frac{N_2}{N_1} X_{12} \qquad (3.32)$$

(Note that X_{12} in (3.32) is not the same as that in (3.31). We will usually use the symbol Z_{12} when referring to leakage impedances and ignore the resistive component so no confusion should arise.) Z_{12} will be calculated by more direct methods later so (3.32) is rarely used.

We should note that another method of obtaining the effective 2-winding leakage impedance, which corresponds with how it is measured, is to short circuit terminal 2 and perform an impedance measurement using terminal 1. Thus

$$Z_{12} = \left.\frac{V_1}{I_1}\right|_{V_2=0} \qquad (3.33)$$

Using (3.27), this implies that $E_2 = -I_2 Z_2$. But, using (3.2) and (3.3), we get $E_1 = I_1(N_1/N_2)^2 Z_2$. Substituting into the V_1 equation of (3.27), we find

$$V_1 = I_1\left[Z_1 + \left(\frac{N_1}{N_2}\right)^2 Z_2\right]$$

so that, from (3.33), we get $Z_{12} = Z_1 + (N_1/N_2)^2 Z_2$ as before.

3.3.1 Leakage Impedance for a 2-Winding Autotransformer

The circuit model for a 2-winging autotransformer can be constructed from separate windings as shown in Fig. 3.6. From the definition of leakage impedance (3.33), we measure the impedance at the H terminal with the X terminal shorted. But this will yield the same leakage impedance as that of an ordinary 2-winding transformer so that $Z_{HX} = Z_{12}$, i.e. the terminal leakage impedance of a two winding autotransformer is the same as that of a transformer with the same windings but not auto-connected.

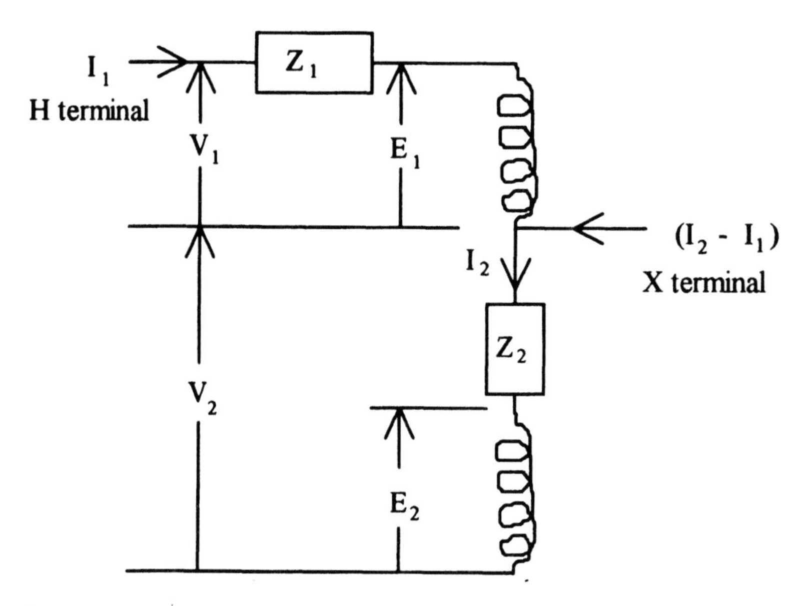

Figure 3.6 Circuit model of a 2 winding autotransformer with leakage impedance, based on a seperate windings circuit model

3.4 LEAKAGE IMPEDANCES FOR 3-WINDING TRANSFORMERS

We can go through the same arguments for a 3-winding transformer, isolating the flux common to all coils, ϕ_C , and expressing the leakage flux for each coil, which exists in non-magnetic materials, in terms of self and mutual inductances which are constants. We obtain

$$V_1 = I_1R_1 + jI_1X_{11} + jI_2X_{12} + jI_3X_{13} + E_1$$

$$V_2 = I_2R_2 + jI_2X_{22} + jI_3X_{23} + jI_1X_{12} + E_2 \qquad (3.34)$$

$$V_3 = I_3R_3 + jI_3X_{33} + jI_1X_{13} + jI_2X_{23} + E_3$$

The order of the suffixes is irrelevant, since for linear materials (constant permeability) $X_{ij} = X_{ji}$.

Using the same assumptions as before concerning the neglect of core excitation, equation (3.7), we substitute $I_3 = -(N_1/N_3)\, I_1 - (N_2/N_3)\, I_2$

in the first 2 equations and $I_2 = -(N_1/N_2)\, I_1 - (N_3/N_2)\, I_3$ in the third equation of (3.34) to obtain

$$V_1 = I_1\left(R_1 + jX_{11} - j\frac{N_1}{N_3}X_{13}\right) + jI_2\left(X_{12} - \frac{N_2}{N_3}X_{13}\right) + E_1$$

$$V_2 = I_2\left(R_2 + jX_{22} - j\frac{N_2}{N_3}X_{23}\right) + jI_1\left(X_{12} - \frac{N_1}{N_3}X_{23}\right) + E_2 \qquad (3.35)$$

$$V_3 = I_3\left(R_3 + jX_{33} - j\frac{N_3}{N_2}X_{23}\right) + jI_1\left(X_{13} - \frac{N_1}{N_2}X_{23}\right) + E_3$$

Now add and subtract $jI_1 \frac{N_1}{N_2}\left(X_{12} - \frac{N_1}{N_3}X_{23}\right)$ from the first equation, add and subtract $jI_2 \frac{N_2}{N_1}\left(X_{12} - \frac{N_2}{N_3}X_{13}\right)$ from the second, and add and subtract $jI_3 \frac{N_3}{N_1}\left(X_{13} - \frac{N_3}{N_2}X_{12}\right)$ from the third equation of (3.35) to obtain

$$V_1 = I_1\left(R_1 + jX_{11} - j\frac{N_1}{N_3}X_{13} - j\frac{N_1}{N_2}X_{12} + j\frac{N_1^2}{N_2N_3}X_{23}\right)$$
$$+\left[jI_1\frac{N_1}{N_2}\left(X_{12} - \frac{N_1}{N_3}X_{23}\right) + jI_2\left(X_{12} - \frac{N_2}{N_3}X_{13}\right) + E_1\right]$$

$$V_2 = I_2\left(R_2 + jX_{22} - j\frac{N_2}{N_3}X_{23} - j\frac{N_2}{N_1}X_{12} + j\frac{N_2^2}{N_1N_3}X_{13}\right) \qquad (3.36)$$
$$+\left[jI_1\left(X_{12} - \frac{N_1}{N_3}X_{23}\right) + jI_2\frac{N_2}{N_1}\left(X_{12} - \frac{N_2}{N_3}X_{13}\right) + E_2\right]$$

$$V_3 = I_3\left(R_3 + jX_{33} - j\frac{N_3}{N_2}X_{23} - j\frac{N_3}{N_1}X_{13} + j\frac{N_3^2}{N_1N_2}X_{12}\right)$$
$$+\left[jI_1\left(X_{13} - \frac{N_1}{N_2}X_{23}\right) + jI_3\frac{N_3}{N_1}\left(X_{13} - \frac{N_3}{N_2}X_{12}\right) + E_3\right]$$

Substituting for $I_3 = -(N_1/N_3)\, I_1 - (N_2/N_3)\, I_2$ in the term in brackets in the last equation, we obtain

$$V_3 = I_3\left(R_3 + jX_{33} - j\frac{N_3}{N_2}X_{23} - j\frac{N_3}{N_1}X_{13} + j\frac{N_3^2}{N_1N_2}X_{12}\right)$$
$$+\left[jI_1\left(\frac{N_3}{N_2}X_{12} - \frac{N_1}{N_2}X_{23}\right) + jI_2\left(\frac{N_3}{N_1}X_{12} - \frac{N_2}{N_1}X_{13}\right) + E_3\right] \quad (3.37)$$

Comparing the terms in brackets of the resulting V_1, V_2, V_3 equations, we find, using (3.6),

$$\frac{[\ \ ]_1}{N_1} = \frac{[\ \ ]_2}{N_2} = \frac{[\ \ ]_3}{N_3} \quad (3.38)$$

Therefore, labeling the terms in brackets E'_1, E'_2, E'_3, we obtain

$$V_1 = I_1Z_1 + E_1' \quad , \quad V_2 = I_2Z_2 + E_2' \quad , \quad V_3 = I_3Z_3 + E_3' \quad (3.39)$$

where

$$Z_1 = R_1 + jX_{11} - j\left(\frac{N_1}{N_3}X_{13} + \frac{N_1}{N_2}X_{12} - \frac{N_1^2}{N_2N_3}X_{23}\right)$$

$$Z_2 = R_2 + jX_{22} - j\left(\frac{N_2}{N_3}X_{23} + \frac{N_2}{N_1}X_{12} - \frac{N_2^2}{N_1N_3}X_{13}\right) \quad (3.40)$$

$$Z_3 = R_3 + jX_{33} - j\left(\frac{N_3}{N_2}X_{23} + \frac{N_3}{N_1}X_{13} - \frac{N_3^2}{N_1N_2}X_{12}\right)$$

and

$$\frac{E_1'}{N_1} = \frac{E_2'}{N_2} = \frac{E_3'}{N_3} \quad (3.41)$$

Here Z_1, Z_2, Z_3 are the single winding leakage impedances and the applicable multi-circuit model is shown in Fig. 3.7a. If we express quantities in terms of per unit values, the single circuit description of Fig. 3.7b applies. This is possible because (3.41) implies $e'_1 = e'_2 = e'_3 = e'$ and by equation (3.18). This figure should be compared with Fig. 3.3.

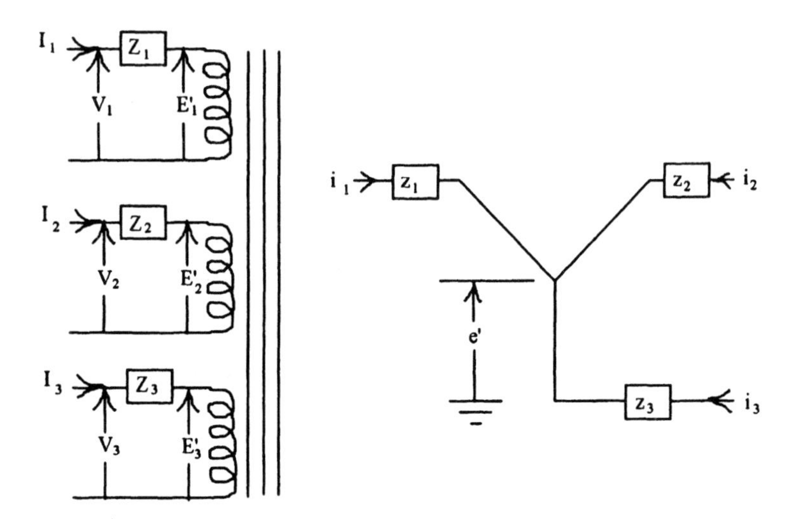

(a) Separate circuit description (b) Single circuit description in terms of per-unit values

Figure 3.7 Circuit models of a 3 winding transformer with leakage impedance

Since the single winding leakage impedances are not directly measured or easily calculated, it is desirable to express these in terms of 2-winding leakage impedances. We will refer to these 2-winding leakage impedances as Z_{12}, Z_{13}, Z_{23} which correspond to the notation Z_{12} used previously for the two winding case. Thus to measure the 2-winding leakage impedance between winding 1 and 2, we short circuit 2, open circuit 3, and measure the impedance at terminal 1,

$$Z_{12} = \left.\frac{V_1}{I_1}\right|_{\substack{V_2=0 \\ I_3=0}} \tag{3.42}$$

From (3.39) we see that this implies $I_2 Z_2 + E'_2 = 0$. Using (3.6) and (3.7), we find for Z_{12} and similarly for Z_{13} and Z_{23}

$$Z_{12} = Z_1 + \left(\frac{N_1}{N_2}\right)^2 Z_2 \quad , \quad Z_{13} = Z_1 + \left(\frac{N_1}{N_3}\right)^2 Z_3 \quad , \quad Z_{23} = Z_2 + \left(\frac{N_2}{N_3}\right)^2 Z_3 \tag{3.43}$$

The subscript ordering is chosen so that the second subscript refers to the shorted winding. The expression changes if we reverse subscripts, according to

$$Z_{12} = \left(\frac{N_1}{N_2}\right)^2 Z_{21} \quad , \text{ etc.} \tag{3.44}$$

Solving (3.43) for the Z_i's, we get

$$\begin{aligned}
Z_1 &= \frac{1}{2}\left[Z_{12} + Z_{13} - \left(\frac{N_1}{N_2}\right)^2 Z_{23}\right] \\
Z_2 &= \frac{1}{2}\left(\frac{N_2}{N_1}\right)^2\left[Z_{12} + \left(\frac{N_1}{N_2}\right)^2 Z_{23} - Z_{13}\right] \\
Z_3 &= \frac{1}{2}\left(\frac{N_3}{N_1}\right)^2\left[Z_{13} + \left(\frac{N_1}{N_2}\right)^2 Z_{23} - Z_{12}\right]
\end{aligned} \tag{3.45}$$

Using per-unit values, where Z_{1b} is the base impedance of circuit 1 so that $z_{12} = Z_{12}/Z_{1b}$, etc. and $z_1 = Z_1/Z_{1b}$, etc. we find, using (3.16), that (3.45) can be expressed as

$$\begin{aligned}
z_1 &= (z_{12} + z_{13} - z_{23})/2 \\
z_2 &= (z_{12} + z_{23} - z_{13})/2 \\
z_3 &= (z_{13} + z_{23} - z_{12})/2
\end{aligned} \tag{3.46}$$

Similarly (3.43) becomes, in per-unit terms,

$$z_{12} = z_1 + z_2 \quad , \quad z_{13} = z_1 + z_3 \quad , \quad z_{23} = z_2 + z_3 \tag{3.47}$$

3.4.1 Leakage Impedances for an Autotransformer with Tertiary

The autotransformer with tertiary circuit model can be obtained by interconnecting elements of the 3-winding transformer circuit model as shown in Fig. 3.8. Here the notation corresponds to that of Fig. 3.7a. The problem is to re-express this in terms of terminal quantities. Thus the appropriate terminal 1 voltage is $V_1 + V_2$ and the appropriate terminal 2 current is $I_2 - I_1$. Using (3.39),

$$V_1 + V_2 = I_1 Z_1 + I_2 Z_2 + E_1' + E_2' \tag{3.48}$$

Substituting from (3.7) into this equation and the V_2 and V_3 equations of (3.39), we obtain after some algebraic manipulations,

$$\begin{aligned}
V_1 + V_2 &= I_1\left(Z_1 - \frac{N_1}{N_2} Z_2\right) + \left[-\frac{N_3}{N_2} I_3 Z_2 + E_1' + E_2'\right] \\
V_2 &= (I_2 - I_1)\frac{Z_2}{(1 + N_2/N_1)} + \left[-\frac{N_3}{N_1 + N_2} I_3 Z_2 + E_2'\right] \\
V_3 &= I_3\left(Z_3 + \frac{N_3^2 Z_2}{N_2(N_1 + N_2)}\right) + \left[-\frac{N_3^2}{N_2(N_1 + N_2)} I_3 Z_2 + E_3'\right]
\end{aligned} \tag{3.49}$$

The term $\dfrac{N_3^2 I_3 Z_2}{N_2(N_1 + N_2)}$ was added and subtracted from the V_3 equation above.

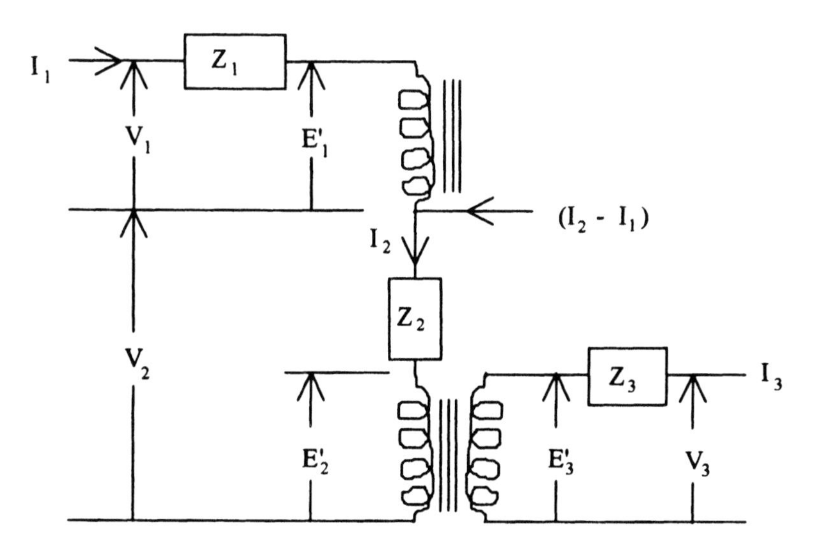

Figure 3.8 3 winding autotransformer circuit model, derived from the 3 separate winding circuit model

We see that the terms in brackets satisfy

$$\frac{[\ ]_1}{N_1 + N_2} = \frac{[\ ]_2}{N_2} = \frac{[\ ]_3}{N_3} \tag{3.50}$$

Labeling these terms in brackets E_H, E_X, E_Y , we can rewrite (3.49)

$$V_H = Z_H I_H + E_H \quad , \quad V_X = Z_X I_X + E_X \quad , \quad V_Y = Z_Y I_Y + E_Y \tag{3.51}$$

where $V_H = V_1 + V_2$, $V_X = V_2$, $V_Y = V_3$, $I_H = I_1$, $I_X = I_2 - I_1$, $I_Y = I_3$, and

$$Z_H = Z_1 - \frac{N_1}{N_2} Z_2$$

$$Z_X = \frac{Z_2}{(1 + N_2/N_1)} \tag{3.52}$$

$$Z_Y = Z_3 + \frac{N_3^2}{N_2(N_1 + N_2)} Z_2$$

Thus the circuit model shown in Fig. 3.9a looks like that of Fig. 3.7a with H, X, Y substituted for 1, 2, 3.

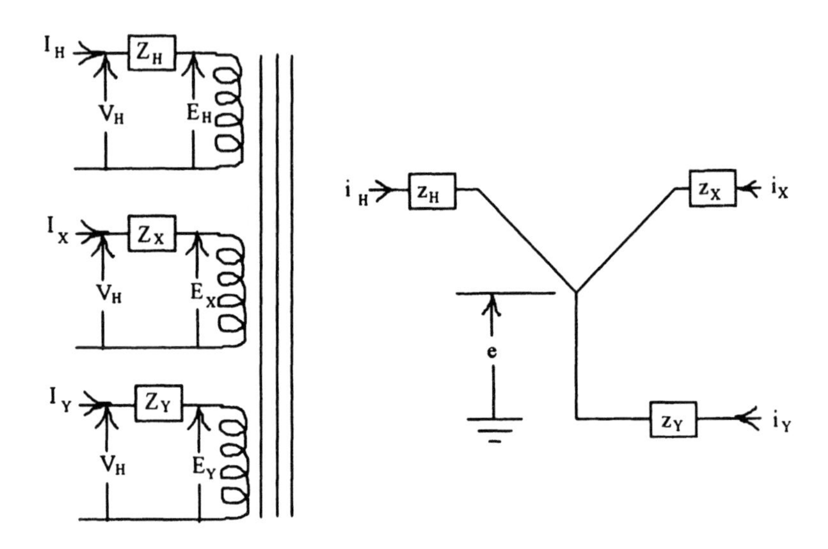

(a) Separate circuit description (b) Single circuit description in terms of per-unit

Figure 3.9 Circuit models of a 3 winding autotransformer based on terminal parameters

In terms of the measured two terminal impedances, we have as before

$$Z_{HX} = \left.\frac{V_H}{I_H}\right|_{\substack{V_X=0\\ I_Y=0}} \tag{3.53}$$

Rewriting (3.7), we find

$$\begin{aligned} &N_1I_1 + N_2(I_2 - I_1) + N_2I_1 + N_3I_3 \\ &\quad = (N_1 + N_2)I_H + N_2I_X + N_3I_Y = 0 \end{aligned} \tag{3.54}$$

From (3.53) with (3.50), (3.51) and (3.54), we obtain

$$Z_{HX} = Z_H + \left(\frac{N_1 + N_2}{N_2}\right)^2 Z_X$$

$$Z_{HY} = Z_H + \left(\frac{N_1 + N_2}{N_3}\right)^2 Z_Y \qquad (3.55)$$

$$Z_{XY} = Z_X + \left(\frac{N_2}{N_3}\right)^2 Z_Y$$

At this point, it is worthwhile to revert to per-unit quantities. Because of the auto connection, we have, again choosing the VI per phase rating of the unit and the rated (no-load) terminal voltages per phase, $(VI)_b$, V_{Hb}, V_{Xb}, V_{Yb},

$$\frac{V_{Hb}}{V_{Xb}} = \frac{N_1 + N_2}{N_2} \quad , \quad \frac{V_{Hb}}{V_{Yb}} = \frac{N_1 + N_2}{N_3} \quad , \quad \frac{V_{Xb}}{V_{Yb}} = \frac{N_2}{N_3}$$

$$\frac{I_{Hb}}{I_{Xb}} = \frac{N_2}{N_1 + N_2} \quad , \quad \frac{I_{Hb}}{I_{Yb}} = \frac{N_3}{N_1 + N_2} \quad , \quad \frac{I_{Xb}}{I_{Yb}} = \frac{N_3}{N_2} \qquad (3.56)$$

$$\frac{Z_{Hb}}{Z_{Xb}} = \left(\frac{N_1 + N_2}{N_2}\right)^2 \quad , \quad \frac{Z_{Hb}}{Z_{Yb}} = \left(\frac{N_1 + N_2}{N_3}\right)^2 \quad , \quad \frac{Z_{Xb}}{Z_{Yb}} = \left(\frac{N_2}{N_3}\right)^2$$

since $(VI)_b$ is the same for each terminal. From (3.54), we have on a per unit basis,

$$i_H + i_X + i_Y = 0 \qquad (3.57)$$

Similarly from (3.50), on a per-unit basis

$$e_H = e_X = e_Y = e \qquad (3.58)$$

On a per-unit basis, (3.55) becomes, using (3.56)

$$z_{HX} = z_H + z_X \quad , \quad z_{HY} = z_H + z_Y \quad , \quad z_{XY} = z_X + z_Y \qquad (3.59)$$

where the same VI base is used for all the terminals. Solving these for z_H, z_X, z_Y, we obtain a set of equations similar to (3.46)

$$z_H = (z_{HX} + z_{HY} - z_{XY})/2$$
$$z_X = (z_{HX} + z_{XY} - z_{HY})/2 \tag{3.60}$$
$$z_Y = (z_{HY} + z_{XY} - z_{HX})/2$$

Equation (3.52) contains terminal impedances and single coil impedances and the bases are different for these. Keeping $(VI)_b$ the same for both, we have

$$\frac{Z_{1b}}{Z_{Hb}} = \left(\frac{N_1}{N_1 + N_2}\right)^2, \frac{Z_{2b}}{Z_{Hb}} = \left(\frac{N_2}{N_1 + N_2}\right)^2, \frac{Z_{3b}}{Z_{Hb}} = \left(\frac{N_3}{N_1 + N_2}\right)^2$$

$$\frac{Z_{1b}}{Z_{Xb}} = \left(\frac{N_1}{N_2}\right)^2 \quad , \quad \frac{Z_{2b}}{Z_{Xb}} = 1 \quad , \quad \frac{Z_{3b}}{Z_{Xb}} = \left(\frac{N_3}{N_2}\right)^2 \tag{3.61}$$

$$\frac{Z_{1b}}{Z_{Yb}} = \left(\frac{N_1}{N_3}\right)^2 \quad , \quad \frac{Z_{2b}}{Z_{Yb}} = \left(\frac{N_2}{N_3}\right)^2 \quad , \quad \frac{Z_{3b}}{Z_{Yb}} = 1$$

Thus, on a per-unit basis, (3.52) becomes

$$z_H = \left(\frac{N_1}{N_1 + N_2}\right)^2 z_1 - \frac{N_1}{N_2}\left(\frac{N_2}{N_1 + N_2}\right)^2 z_2$$

$$z_X = \frac{z_2}{(1 + N_2/N_1)} \tag{3.62}$$

$$z_Y = z_3 + \left(\frac{N_2}{N_1 + N_2}\right) z_2$$

In terms of the terminal turns ratio $n = (N_1 + N_2)/N_2$, (3.62) can be written

$$z_H = \left(\frac{n-1}{n}\right)^2 z_1 - \frac{(n-1)}{n^2} z_2$$

$$z_X = \left(\frac{n-1}{n}\right) z_2 \tag{3.63}$$

$$z_Y = z_3 + \frac{z_2}{n}$$

The per-unit circuit model is depicted in Fig. 3.9b. Note that from (3.45) and (3.46), the autotransformer circuit parameters can be derived from 2-winding leakage impedance values which, we will see, can be obtained with reasonable accuracy, using an analytic formula.

3.4.2 Leakage Impedance between 2 Windings Connected in Series and a Third Winding

It is useful to calculate the impedance between a pair of windings connected in series and a third winding in terms of 2-winding leakage impedances, as shown in Fig. 3.10. This can be regarded as a special case of an autotransformer with the X-terminal open. But, by definition, this leakage impedance is just Z_{HY}. From (3.55) and (3.52), we obtain

$$Z_{HY} = Z_1 + Z_2 + \left(\frac{N_1 + N_2}{N_3}\right)^2 Z_3 \tag{3.64}$$

or, in per-unit terms, using (3.61)

$$z_{HY} = \left(\frac{N_1}{N_1 + N_2}\right)^2 z_1 + \left(\frac{N_2}{N_1 + N_2}\right)^2 z_2 + z_3 \tag{3.65}$$

These can be expressed in terms of 2 winding leakage reactances by means of (3.45) and (3.46)

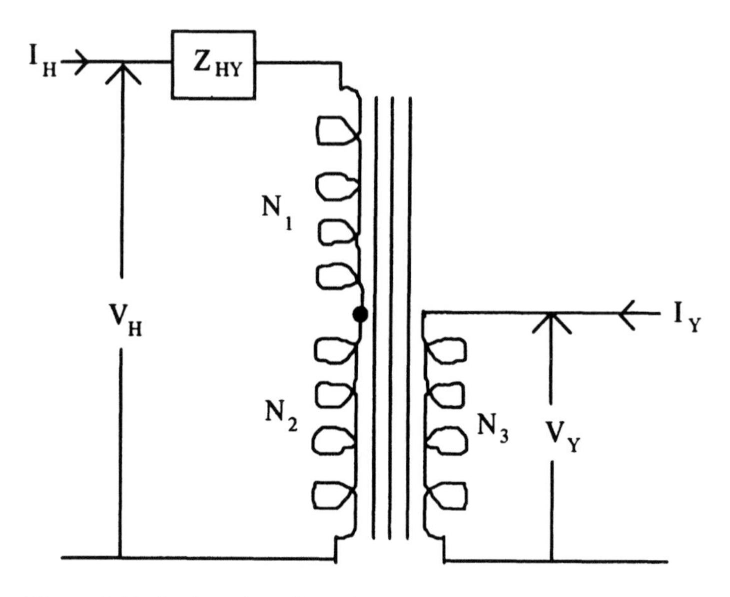

Figure 3.10 Leakage impedance between two series connected windings and a third

3.4.3 Leakage Impedance of a 2-Winding Autotransformer with X-Line Taps

A circuit model of a 2-winding autotransformer with X-line taps constructed from the 3 separate winding circuit model is shown in Fig. 3.11. The derivation of its 2-terminal leakage impedance uses (3.39) together with (3.6) and (3.7). In terms of terminal parameters, we have $V_{HT} = V_1 + V_2$, $V_{XT} = V_2 + V_3$, $I_{HT} = I_1$, $I_{XT} = I_3$, where T is appended to indicate the presence of a tap winding. We also require that $I_2 = I_1 + I_3$.

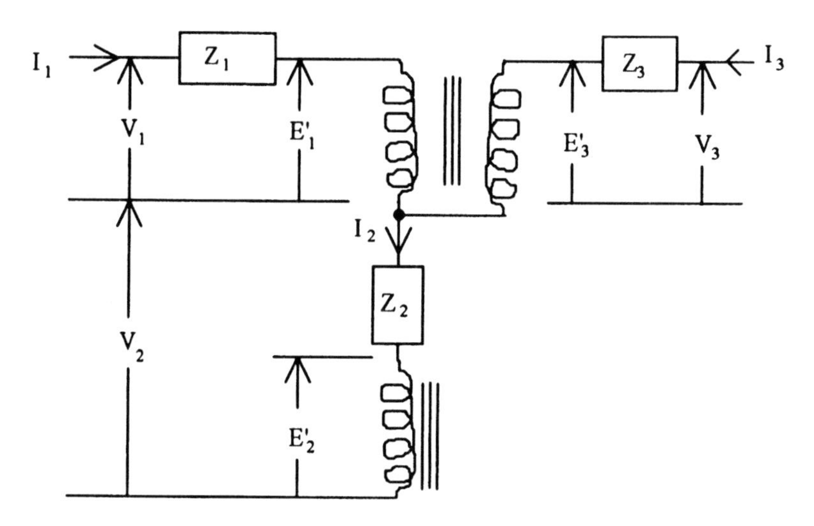

Figure 3.11 Circuit model of a 2 winding autotransformer with X-line taps, derived from the 3 separate winding circuit

Thus

$$
\begin{aligned}
V_{HT} = V_1 + V_2 &= I_1 Z_1 + (I_1 + I_3) Z_2 + E_1' + E_2' \\
&= I_1 (Z_1 + Z_2) + I_3 Z_2 + E_1' + E_2' \\
V_{XT} = V_2 + V_3 &= I_3 Z_3 + (I_1 + I_3) Z_2 + E_2' + E_3' \\
&= I_3 (Z_2 + Z_3) + I_1 Z_2 + E_2' + E_3'
\end{aligned}
\tag{3.66}
$$

Substitute $I_3 = -(N_1/N_3)I_1 - (N_2/N_3)I_2$ into the V_{HT} equation and $I_1 = -(N_2/N_1)I_2 - (N_3/N_1)I_3$ into the V_{XT} equation to obtain

$$
\begin{aligned}
V_{HT} &= I_1\left[Z_1 + \left(1 - \frac{N_1}{N_3}\right)Z_2\right] - \frac{N_2}{N_3} Z_2 I_2 + E_1' + E_2' \\
V_{XT} &= I_3\left[Z_3 + \left(1 - \frac{N_3}{N_1}\right)Z_2\right] - \frac{N_2}{N_1} Z_2 I_2 + E_2' + E_3'
\end{aligned}
\tag{3.67}
$$

Add and subtract $\left(\frac{N_2}{N_2+N_3}\right)Z_2I_1$ from the V_{HT} equation and add and subtract $\left(\frac{N_2}{N_1+N_2}\right)Z_2I_3$ from the V_{XT} equation to obtain, after some algebraic manipulations,

$$V_{HT} = I_1\left[Z_1 + \left(\frac{N_3-N_1}{N_2+N_3}\right)Z_2\right] + E_1' + E_2'$$

$$V_{XT} = I_3\left[Z_3 + \left(\frac{N_1-N_3}{N_1+N_2}\right)Z_2\right] + E_2' + E_3' \tag{3.68}$$

which can be rewritten

$$V_{HT} = I_{HT}Z_{HT} + E_{HT} \quad , \quad V_{XT} = I_{XT}Z_{XT} + E_{XT} \tag{3.69}$$

where

$$Z_{HT} = Z_1 + \left(\frac{N_3-N_1}{N_2+N_3}\right)Z_2 \quad , \quad Z_{XT} = Z_3 + \left(\frac{N_1-N_3}{N_1+N_2}\right)Z_2 \tag{3.70}$$

To obtain the 2 terminal leakage impedance, we use the definition

$$Z_{HT,XT} = \left.\frac{V_{HT}}{I_{HT}}\right|_{V_{XT}=0} \tag{3.71}$$

We have

$$\frac{E_{HT}}{N_1+N_2} = \frac{E_{XT}}{N_2+N_3} \tag{3.72}$$

and from (3.7) we obtain

$$\begin{aligned} N_1I_1 + N_2(I_1+I_3) + N_3I_3 &= (N_1+N_2)I_1 + (N_2+N_3)I_3 \\ &= (N_1+N_2)I_{HT} + (N_2+N_3)I_{XT} = 0 \end{aligned} \tag{3.73}$$

Using (3.69) together with (3.72) and (3.73), we obtain from (3.71)

$$Z_{HT,XT} = Z_{HT} + \left(\frac{N_1 + N_2}{N_2 + N_3}\right)^2 Z_{XT} \tag{3.74}$$

Using (3.70), this becomes

$$Z_{HT,XT} = Z_1 + \left(\frac{N_1 - N_3}{N_2 + N_3}\right)^2 Z_2 + \left(\frac{N_1 + N_2}{N_2 + N_3}\right)^2 Z_3 \tag{3.75}$$

On a per-unit basis, using a fixed $(VI)_b$ and, noting that

$$\frac{Z_{1b}}{Z_{HTb}} = \left(\frac{N_1}{N_1 + N_2}\right)^2 \quad , \quad \frac{Z_{2b}}{Z_{HTb}} = \left(\frac{N_2}{N_1 + N_2}\right)^2 \quad , \quad \frac{Z_{3b}}{Z_{HTb}} = \left(\frac{N_3}{N_1 + N_2}\right)^2$$

we find

$$\begin{aligned} z_{HT,XT} = &\left(\frac{N_1}{N_1 + N_2}\right)^2 z_1 + \left(\frac{N_2}{N_1 + N_2}\right)^2 \left(\frac{N_1 - N_3}{N_2 + N_3}\right)^2 z_2 \\ &+ \left(\frac{N_3}{N_2 + N_3}\right)^2 z_3 \end{aligned} \tag{3.76}$$

This is the two terminal leakage impedance of a 2 winding autotransformer with X-line taps. It can be expressed in terms of the 2 winding leakage impedances via equations (3.45) and (3.46).

3.4.4 More General Leakage Impedance Calculations

The cases considered so far cover most of the configurations encountered in practice. However, other situations can be covered by using these results as building blocks. For example, to obtain the leakage impedance between three windings connected in series and a fourth winding, use formula (3.64) or (3.65) to obtain the leakage impedance between the fourth winding and two of the other windings connected in series. Then add the third winding. Considering the two series winding as a single winding, calculate the single winding impedances for this new "three winding" system and reapply (3.64) or (3.65).

Another example would be an autotransformer with tertiary and X-line taps. The circuit model for this can be derived from the 2-winding autotransformer with X-line taps by adding a tertiary winding. Then the 2-terminal leakage reactance, equation (3.75) or (3.76), between the series + common and X-line tap + common, together with the 2-terminal leakage reactances between the tertiary and the series + common and between the tertiary and the X-line tap + common can be used to construct a 3-terminal impedance model similar to Fig. 3.9.

3.5 TWO WINDING LEAKAGE REACTANCE FORMULA

All of the reactance circuit parameters obtained here for 2 or 3 terminal transformers can be expressed in terms of 2-winding leakage reactances. These can be calculated by advanced analytical techniques or finite element methods which solve Maxwell's equations directly. These methods are especially useful if the distribution of amp-turns along the winding is non-uniform, due, for example, to tapped out sections or thinning in sections of windings adjacent to taps in neighboring windings. However, simpler idealized calculations have proven adequate in practice, particularly at the early design stage. These simpler calculations will be discussed here but, however these 2-winding reactances are obtained, they can be used directly in the formulas derived previously.

The simple reactance calculation assumes that the amp-turns are uniformly distributed along the windings. It also treats the windings as if they were infinitely long insofar as the magnetic field is concerned, although a correction for fringing at the ends is included in the final formula. The parameters of interest are shown in Fig. 3.12. If the windings were infinitely long and the amp-turns per unit length were equal for the two windings then the magnetic field as a function of radius would be proportional to the amp-turn distribution shown in Fig. 3.12b, i.e. in the SI system,

$$H(r) = NI(r)/h \tag{3.77}$$

where NI(r) is the function of r shown in the figure, linearly increasing from 0 through winding 1, remaining constant in the gap, and decreasing to 0 through winding 2. H is independent of the z-coordinate in this model and points vectorially in the z-direction. The flux density is therefore

$$B(r) = \mu_o H(r) = \mu_o NI(r)/h \tag{3.78}$$

since the permeabilities of the materials in or between the winding are essentially that of vacuum, $\mu_0 = 4\pi\times10^{-7}$ in the MKS system. B also points in the z-direction. We can take $h = (h_1 + h_2)/2$ as an approximation.

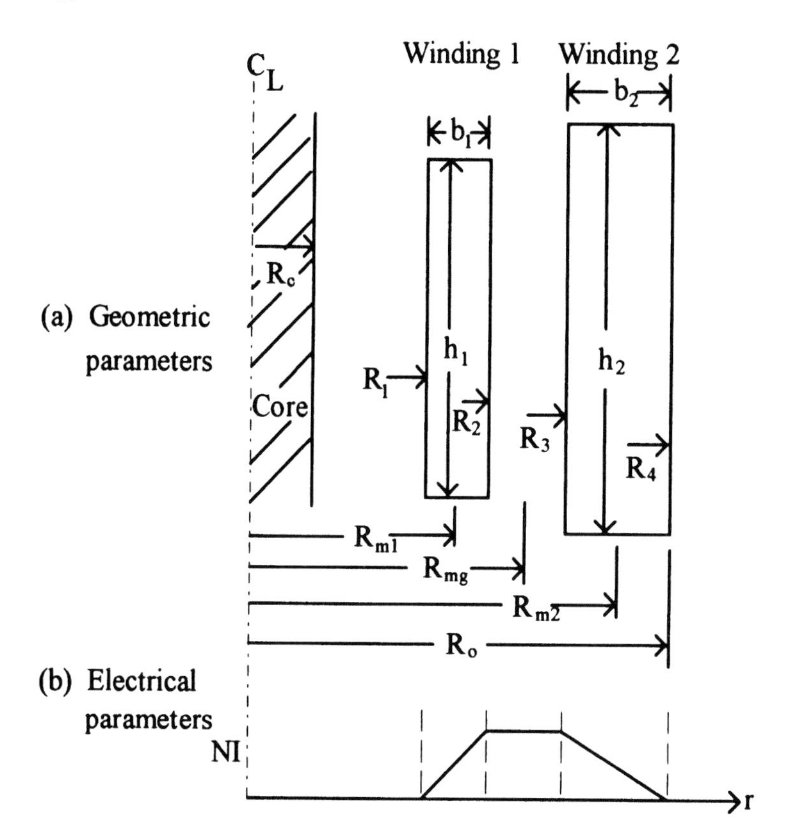

Figure 3.12 Parameters used in 2 winding leakage reactance calculation

For calculation purposes, we need to express B as a function of r analytically

$$B(r) = \mu_o \frac{NI}{h} \begin{cases} \dfrac{(r - R_1)}{(R_2 - R_1)} & , \quad R_1 < r < R_2 \\ 1 & , \quad R_2 < r < R_3 \\ \dfrac{(R_4 - r)}{(R_4 - R_3)} & , \quad R_3 < r < R_4 \end{cases} \tag{3.79}$$

where the R_i are indicated in Fig. 3.12. The leakage inductance, L, can be obtained from the magnetic energy in the leakage field by means of the expression

$$\frac{1}{2}LI^2 = \frac{1}{2\mu_o}\int_{\text{space}} B^2\,dV \tag{3.80}$$

Substituting (3.79) into the integral, we get

$$\frac{1}{2\mu_o}\int B^2\,dV = \frac{1}{2\mu_o}\left(\frac{\mu_o NI}{h}\right)^2 2\pi h\left\{\int_{R_1}^{R_2}\left(\frac{r-R_1}{R_2-R_1}\right)^2 r\,dr + \int_{R_2}^{R_3} 1\cdot r\,dr + \int_{R_3}^{R_4}\left(\frac{R_4-r}{R_4-R_3}\right)^2 r\,dr\right\}$$

$$= \frac{\pi\mu_o(NI)^2}{h}\left\{\frac{(R_2^2-R_1^2)}{6}+\frac{(R_2-R_1)^2}{12}+\frac{(R_3^2-R_2^2)}{2}+\frac{(R_4^2-R_3^2)}{6}-\frac{(R_4-R_3)^2}{12}\right\} \tag{3.81}$$

In terms of mean radii R_m, thicknesses b, and gap g shown in Fig. 3.12, (3.81) can be written

$$\frac{1}{2\mu_o}\int B^2\,dV = \frac{\pi\mu_o(NI)^2}{h}\left\{\frac{R_{m1}b_1}{3}+\frac{b_1^2}{12}+R_{mg}g+\frac{R_{m2}b_2}{3}-\frac{b_2^2}{12}\right\} \tag{3.82}$$

Because the terms in b_1^2 and b_2^2 are so much smaller that the others, we drop them and find, using (3.80)

$$L = \frac{2\pi\mu_o N^2}{h}\left\{\frac{R_{m1}b_1}{3}+\frac{R_{m2}b_2}{3}+R_{mg}g\right\} \tag{3.83}$$

The leakage reactance is $X_L = 2\pi fL$ so we get

$$X_L = \frac{(2\pi)^2\mu_o f N^2}{h}\left\{\frac{R_{m1}b_1}{3}+\frac{R_{m2}b_2}{3}+R_{mg}g\right\} \tag{3.84}$$

On a per-unit basis,

$$X_b = \frac{V_b^2}{(VI)_b} = \frac{N^2(V_b/N)^2}{(VI)_b}$$

where $(VI)_b$ is the base volt-amps per phase and V_b/N is the base Volts/turn. Letting x denote the per-unit reactance, where $x = X_L/X_b$, we get

$$x = \frac{(2\pi)^2 \mu_o f(VI)_b}{(V_b/N)^2 h}\left\{\frac{R_{m1}b_1}{3} + \frac{R_{m2}b_2}{3} + R_{mg}g\right\}$$
$$= 2.977 \times 10^{-3} \frac{(VI)_b}{(V_b/N)^2 h}\left\{\frac{R_{m1}b_1}{3} + \frac{R_{m2}b_2}{3} + R_{mg}g\right\} \tag{3.85}$$

in MKS units at 60 Hz or

$$x = 7.5606 \times 10^{-5} \frac{(VI)_b}{(V_b/N)^2 h}\left\{\frac{R_{m1}b_1}{3} + \frac{R_{m2}b_2}{3} + R_{mg}g\right\} \tag{3.86}$$

when lengths are measured in inches. This same result could have been obtained by consideration of flux-linkages but the effort required would have been greater.

In order to correct for fringing, it has been found that a good approximation is to increase h by the amount

$$s = 0.32(R_o - R_c) \tag{3.87}$$

where R_o is the outer radius of the outermost coil and R_c is the core radius. Thus we obtain

$$x = 7.5606 \times 10^{-5} \frac{(VI)_b}{(V_b/N)^2 (h+s)}\left\{\frac{R_{m1}b_1}{3} + \frac{R_{m2}b_2}{3} + R_{mg}g\right\} \tag{3.88}$$

4. FAULT CURRENT CALCULATIONS

Summary Fault currents at the terminals and in the windings are calculated for transformers subjected to the standard faults: 3-phase line to ground, single phase line to ground, line to line, and double line to ground. Since these faults result in unbalanced currents in the 3 phase system, except for 3-phase faults, the method of symmetrical components is introduced and used in the fault analysis. In this method, the unbalanced voltages and currents are replaced by balanced systems of positive, negative, and zero sequence quantities. The circuits associated with each of these sequences can differ. After solving the sequence circuit equations, the unbalanced quantities are obtained by a reverse transformation. A 2-terminal per phase transformer is modeled using a single leakage impedance and a 3-terminal per phase unit is modeled with 3 leakage impedances (T-equivalent circuit). The leakage impedances are the same for the positive and negative sequence circuits but can differ for the zero sequence circuit. The systems attached to the external terminals of the transformer are treated simply as a voltage source in series with an impedance. An asymmetry factor is included to account for an initial transient surge which usually accompanies a fault.

4.1 INTRODUCTION

It is necessary to design transformers to withstand various possible faults, such as a short to ground of one or more phases. The high currents accompanying these faults, approximately 10 to 30 times normal, produce high forces and stresses in the windings and support structure. Also, depending on the fault duration, significant amounts of heat may be generated inside the unit. The design must accommodate the worst case fault which can occur from both the mechanical and thermal standpoints.

The first step in designing to withstand faults is to determine the fault currents in all the windings, which is the subject of this report. Since this is an electrical problem, it requires a circuit model which includes leakage impedances of the transformer and also relevant system impedances. The system is typically represented by a voltage source in series with an impedance, since we are not interested here in detailed fault currents within the system external to the transformer. The transformer circuit model considered here is that of a 2 or 3 terminal per phase unit with all pairs of terminal leakage reactances given either from calculations or measurement (from these the T-equivalent reactances can be obtained). We ignore core excitation since, for modern power transformers, its effects on the fault currents are negligible.

The transformers dealt with here are 3-phase units and the fault types considered are: 3-phase line to ground, single phase line to ground, line to line, and double line to ground. These are the standard fault types and are important because they are most likely to occur on actual systems. The transformer must be designed to withstand the worst of these fault types, or rather each coil must be designed to withstand the worst (highest current) fault it can experience. Note that each fault type refers to a fault on any of the single phase terminals. For example, a 3-phase fault can occur on all the high voltage terminals (H_1, H_2, H_3), all the low voltage terminals (X_1, X_2, X_3), or all the tertiary voltage terminals (Y_1, Y_2, Y_3). Etc. for the other fault types. Note also that faults on a single phase system can be considered as 3-phase faults on a 3-phase system so that these are included automatically in the analysis of faults on 3-phase systems.

Since the fault types considered include faults which produce unbalanced conditions in a 3-phase system, probably the most efficient way of treating them is by the method of symmetrical components. In this method, an unbalanced set of voltages or currents can be represented mathematically by sets of balanced voltages or currents, called sequence voltages or currents. These latter can then be analyzed by means of sequence circuit models. The final results are then obtained by transforming the voltages and currents from the sequence analysis into the voltages and currents of the real system. We will discuss this method in greater detail before proceeding with the specific fault analyses.

It should also be noted that the circuit model calculations to be discussed are for steady-state conditions, whereas actual faults would have a transient phase where the currents can exceed their steady-state values for short periods of time. These enhancement effects are included by means of an asymmetry factor. This factor takes into account the resistance and reactance present at the faulted terminal and is considered to be conservative from a design point of view. The following references have been used: [Ste62a], [Lyo37], [Blu51].

4.2 SYMMETRICAL COMPONENTS

In a balanced 3-phase electrical system, the voltage or current phasors are of equal magnitude and separated by 120° as shown in Fig. 4.1a. They are labeled V_{a1}, V_{b1}, V_{c1} where the order a,b,c corresponds to the order in which the phasors would pass a point, say on the horizontal axis, as they rotate in the counter-clockwise direction (called a positive sequence ordering). (The actual time dependent voltages are found by projecting these rotating vectors onto the horizontal axis, assuming they are rotating with angular velocity $\omega = 2\pi f$.) However, we usually ignore the time dependence and assume the phasors are stationary at some time snapshot. If the tips of the voltage vectors in Fig. 4.1a are represented by complex numbers then, with V_{a1} along the positive real axis, where the subscript 1 refers to positive sequence by convention, we have

$$
\begin{aligned}
V_{a1} &= |V_{a1}|(\cos 0^\circ + j\sin 0^\circ) \\
V_{b1} &= |V_{a1}|(\cos 240^\circ + j\sin 240^\circ) \\
V_{c1} &= |V_{a1}|(\cos 120^\circ + j\sin 120^\circ)
\end{aligned}
\tag{4.1}
$$

or

$$
\begin{aligned}
V_{a1} &= |V_{a1}|(1 + j0) &&= |V_{a1}|\angle 0^\circ \\
V_{b1} &= |V_{a1}|(-0.5 - 0.866j) &&= |V_{a1}|\angle 240^\circ \\
V_{c1} &= |V_{a1}|(-0.5 + 0.866j) &&= |V_{a1}|\angle 120^\circ
\end{aligned}
\tag{4.2}
$$

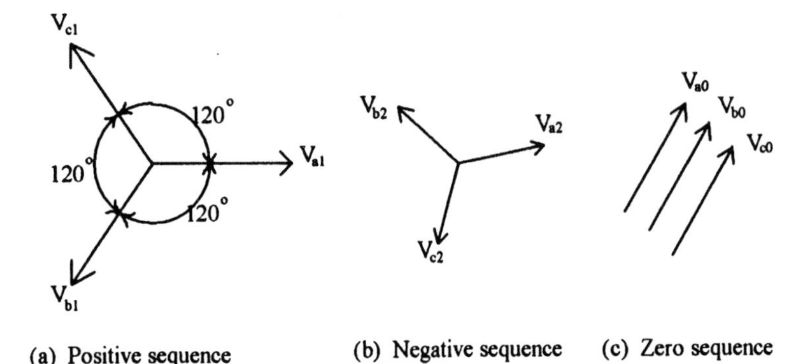

Figure 4.1 Balanced systems of 3 phase phasors

Let $\alpha = \angle 120° = -0.5 + 0.866j$ i.e. a rotation in the positive sense (counterclockwise) by 120°. Using polar notation, $\alpha = e^{j120}$, we see that $\alpha^2 = e^{j240} = \angle 240°$. Thus for a balanced 3-phase positive sequence system

$$V_{a1} = V_{a1} \quad , \quad V_{b1} = \alpha^2 V_{a1} \quad , \quad V_{c1} = \alpha V_{a1} \tag{4.3}$$

Notice that V_{a1} need not be along the positive real axis for (4.3) to hold since α and α^2 are rotation operators which guarantee that V_{b1} and V_{c1} are 240° and 120° from V_{a1} regardless of its position in the complex plane. Also these phasors are of equal magnitude since α and its powers are of unit magnitude.

A negative sequence set of balanced phasors is one with the phase ordering a, c, b as shown in Fig. 4.1b where we see that

$$V_{a2} = V_{a2} \quad , \quad V_{b2} = \alpha V_{a2} \quad , \quad V_{c2} = \alpha^2 V_{a2} \tag{4.4}$$

where 2 refers to negative sequence quantities by convention. These are separated by 120° and have the same magnitude which can differ from the positive sequence magnitude.

A zero sequence set of balanced phasors is shown in Fig. 4.1c. These are all in phase and have equal magnitudes, which can differ from the positive or negative sequence magnitudes. Thus

$$V_{a0} = V_{b0} = V_{c0} \tag{4.5}$$

with 0 used to label zero sequence quantities.

We now show that it is possible to represent any unbalanced set of 3 phasors by means of these balanced sequence sets. Let V_a, V_b, V_c be such an unbalance set as shown in Fig. 4.2. Since the positive, negative, and zero sequence balanced sets are determined once V_{a1}, V_{a2}, and V_{a0} are specified, we need to find only these phase a components of the balanced sets in terms of the original phasors to prove that this representation is possible. Write

$$\begin{aligned} V_a &= V_{a0} + V_{a1} + V_{a2} \\ V_b &= V_{b0} + V_{b1} + V_{b2} \\ V_c &= V_{c0} + V_{c1} + V_{c2} \end{aligned} \tag{4.6}$$

Using (4.3), (4.4), and (4.5), this can be written

$$\begin{aligned} V_a &= V_{a0} + V_{a1} + V_{a2} \\ V_b &= V_{a0} + \alpha^2 V_{a1} + \alpha V_{a2} \\ V_c &= V_{a0} + \alpha V_{a1} + \alpha^2 V_{a2} \end{aligned} \tag{4.7}$$

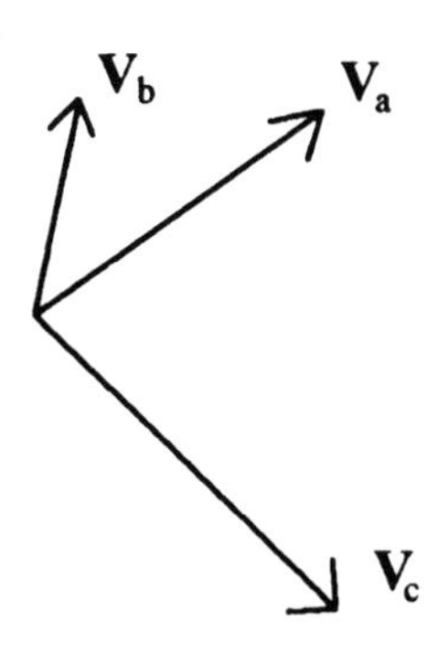

Figure 4.2 Unbalanced set of 3 phasors

In matrix notation, we have

$$\begin{pmatrix} V_a \\ V_b \\ V_c \end{pmatrix} = \begin{pmatrix} 1 & 1 & 1 \\ 1 & \alpha^2 & \alpha \\ 1 & \alpha & \alpha^2 \end{pmatrix} \begin{pmatrix} V_{a0} \\ V_{a1} \\ V_{a2} \end{pmatrix} \tag{4.8}$$

which is often abbreviated

$$\mathbf{V}_{abc} = \mathbf{A}\mathbf{V}_{012} \tag{4.9}$$

where $\mathbf{V}_{abc}$ and $\mathbf{V}_{012}$ are column vectors and **A** is the matrix in (4.8).

V_{a0}, V_{a1}, V_{a2} can be found uniquely if **A** has an inverse. This can be shown to be the case and the result is

$$\begin{pmatrix} V_{a0} \\ V_{a1} \\ V_{a2} \end{pmatrix} = \frac{1}{3} \begin{pmatrix} 1 & 1 & 1 \\ 1 & \alpha & \alpha^2 \\ 1 & \alpha^2 & \alpha \end{pmatrix} \begin{pmatrix} V_a \\ V_b \\ V_c \end{pmatrix} \tag{4.10}$$

as can be verified by direct computation, using the identities

$$\begin{aligned} 1+\alpha+\alpha^2 &= 0 \\ \alpha^3 &= 1 \\ \alpha^4 &= \alpha \end{aligned} \tag{4.11}$$

Equation (4.10) can be abbreviated to

$$\mathbf{V}_{012} = \mathbf{A}^{-1}\mathbf{V}_{abc} \tag{4.12}$$

where $\mathbf{A}^{-1}$ is the matrix in (4.10) including the factor 1/3.

Thus given any unbalanced set of phasors, the balanced positive, negative, and zero sequence sets can be found or conversely, given the balanced sets or just one phasor from each balanced set, chosen customarily to be the a phasor, the unbalanced set can be obtained.

The virtue of this decomposition is that, using symmetrical components, an unbalanced 3 phase system can be analyzed as 3 single phase systems, each applying to one balanced sequence, in the same manner that we need only consider one phase of a conventional balanced 3 phase system. However, the circuit model applying to each sequence may differ from the normal 3 balanced phases circuit model. This latter circuit model applies generally only to the positive sequence. The negative sequence circuit may differ from the positive, particularly if there are generators or motors in the system. Not only can the positive and negative impedances differ but, since generated voltages are usually of positive sequence, voltage sources are absent from the negative sequence circuit. For transformers, the impedances are independent of phase order so that the positive and negative sequence circuit models of transformers are identical.

The zero sequence circuit model can differ considerably from the positive or negative one. For example, in a balanced 3-phase line, the currents add up to zero so there is no need for a return path or ground for the currents to flow. Thus an effective ground point can be assumed. Since zero sequence currents are of equal magnitude and phase, they cannot sum to zero unless they are all zero. Thus for zero sequence currents to flow, a return path or ground is necessary. Thus in the zero sequence circuit model without ground or return path, an infinite impedance must be placed in the circuit.

In order to justify this approach, assume that the phase a, b, and c, circuits are identical as would be typical of a balanced 3-phase system. Let V_{ia}, V_{ib}, V_{ic} be corresponding branch voltages and I_{ia}, I_{ib}, I_{ic} corresponding branch currents for branch i in the phase a, b, and c circuits. Then Kirchoff's voltage and current laws can be expressed in the form

$$\sum_i K_i \mathbf{V}_{iabc} = 0 \quad , \quad \sum_i B_i \mathbf{I}_{iabc} = 0$$

where the K_i and $B_i = \pm 1$ or 0 and the column vector notation is employed so that each of the above vector equations represents 3 scalar equations with the same K_i or B_i coefficients. Then multiplying on the left by $\mathbf{A}^{-1}$, these are transformed into

$$\sum_i K_i \mathbf{V}_{i012} = 0 \quad , \quad \sum_i B_i \mathbf{I}_{i012} = 0$$

Thus we see that each set of sequence voltages or currents satisfies the same Kirchoff equations. This means that the sequence networks behave like ordinary electrical networks.

With regards to the circuit elements, the situation is a bit more complicated. Let the corresponding branch voltages and currents obey an equation of the form, omitting the i subscript for simplicity,

$$\begin{pmatrix} V_a \\ V_b \\ V_c \end{pmatrix} = \begin{pmatrix} Z_{aa} & Z_{ab} & Z_{ac} \\ Z_{ba} & Z_{bb} & Z_{bc} \\ Z_{ca} & Z_{cb} & Z_{cc} \end{pmatrix} \begin{pmatrix} I_a \\ I_b \\ I_c \end{pmatrix}$$

or, more compactly

$$\mathbf{V}_{abc} = \mathbf{Z}\mathbf{I}_{abc}$$

where $\mathbf{Z}$ is an impedance matrix. Then, pre-multiply by $\mathbf{A}^{-1}$ and, substituting $\mathbf{V}_{012} = \mathbf{A}^{-1}\,\mathbf{V}_{abc}$ and $\mathbf{I}_{abc} = \mathbf{A}\mathbf{I}_{012}$, we get

$$\mathbf{V}_{012} = (\mathbf{A}^{-1}\mathbf{Z}\mathbf{A})\mathbf{I}_{012} = \mathbf{Z}_{seq}\mathbf{I}_{012}$$

In general the sequence impedance matrix $\mathbf{Z}_{seq} = \mathbf{A}^{-1}\mathbf{Z}\mathbf{A}$ which relates the sequence voltages and currents is not diagonal even if $\mathbf{Z}$ is. This means that there can be coupling between the different sequence circuits.

If the original circuit branches satisfy current-voltage relationships of the form

$$\begin{pmatrix} V_a \\ V_b \\ V_c \end{pmatrix} = \begin{pmatrix} Z_a & 0 & 0 \\ 0 & Z_a & 0 \\ 0 & 0 & Z_a \end{pmatrix} \begin{pmatrix} I_a \\ I_b \\ I_c \end{pmatrix}$$

i.e. the corresponding branches have the same impedance and are uncoupled, then the sequence equations have the same form. This follows since the above $\mathbf{Z}$ matrix is a multiple of the identity and $\mathbf{Z}_{seq} = Z_a\mathbf{A}^{-1}\mathbf{I}\mathbf{A} = Z_a\mathbf{I}$ reduces to the same multiple of the identity. In this example, there would be no need to use symmetrical components, since the original uncoupled phase circuits could be solved with no more effort.

The method becomes more useful when, as is often the case in practice, each sequence needs to be described by means of its own set of sequence impedances, as for example when there is no return path for zero sequence currents. We assume these sequence impedances are uncoupled and equal for the 3 members of the balanced set. Thus, for the positive sequence circuit we would have the branch current relationship,

$$\begin{pmatrix} V_{a1} \\ V_{b1} \\ V_{c1} \end{pmatrix} = \begin{pmatrix} Z_1 & 0 & 0 \\ 0 & Z_1 & 0 \\ 0 & 0 & Z_1 \end{pmatrix} \begin{pmatrix} I_{a1} \\ I_{b1} \\ I_{c1} \end{pmatrix}$$

and similarly for the negative and zero sequences where Z_2 and Z_0 would be substituted for Z_1 above. In this case, the sequence networks have their own impedances and there is no coupling between sequences. In this case, the matrix **Z** connecting the branch phase quantities is non-diagonal in general. However, under normal operation with balanced positive sequence voltages and currents, it reduces to the positive sequence impedance matrix shown above.

The fault analysis performed here assumes uncoupled sequence circuits. We assume that the sequence impedances are known or can be calculated. For example, Z_1 is taken to be the normal impedance to positive sequence current and for static devices such as transformers, $Z_2 = Z_1$. Z_0 is more difficult to calculate, but can be measured by energizing the 3 terminals of the device with voltages of the same phase.

4.3 FAULT ANALYSIS ON 3-PHASE SYSTEMS

We assume that the system is balanced before the fault occurs, that is, each phase has identical impedances and the currents and voltages are positive sequence sets. Here we consider a general electrical system as shown in Fig. 4.3a. The fault occurs at some location on the system where fault phase currents I_a, I_b, I_c flow. They are shown as leaving the system in the figure. The voltages to ground at the fault point are labeled V_a, V_b, V_c. The system, as viewed from the fault point or terminal, is modeled by means of Thevenin's theorem. First, however, we resolve the voltages and currents into symmetrical components so that we need only analyze one phase of the positive, negative, and zero sequence sets. This is indicated in Fig. 4.3b where the a-phase sequence set has been singled out.

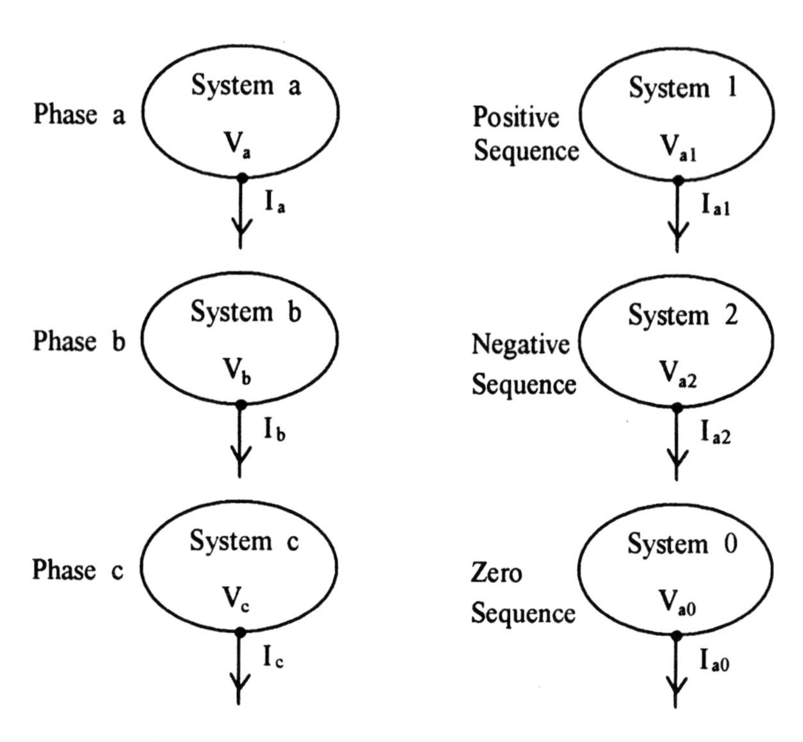

Figure 4.3 Fault at a point on a general electrical system

By Thevenin's theorem, each of the sequence systems can be modeled as a voltage source in series with an impedance, where the voltage source is the open circuit voltage at the fault point and the impedance is found by shorting all voltage sources and measuring or calculating the impedance to ground at the fault terminal. The resulting model is shown in Fig. 4.4. No voltage source is included in the negative and zero sequence circuits, since the standard voltage sources in power systems are positive sequence sources.

The circuit equations for Fig. 4.4 are

$$V_{a1} = E_1 - I_{a1}Z_1 \quad , \quad V_{a2} = -I_{a2}Z_2 \quad , \quad V_{a0} = -I_{a0}Z_0 \qquad (4.13)$$

Since E_1 is the open circuit voltage at the fault terminal, it is the voltage at the fault point before the fault occurs and can be labeled V_{pf} where pf denotes pre-fault. We can omit the label 1 since it is understood to be a positive sequence voltage. Thus

$$V_{a1} = V_{pf} - I_{a1}Z_1 \quad , \quad V_{a2} = -I_{a2}Z_2 \quad , \quad V_{a0} = -I_{a0}Z_0 \qquad (4.14)$$

If there is some resistance in the fault, this could be included in the circuit model. However, because we are interested in the worst case faults (highest fault currents), we assume that the fault resistance is zero.

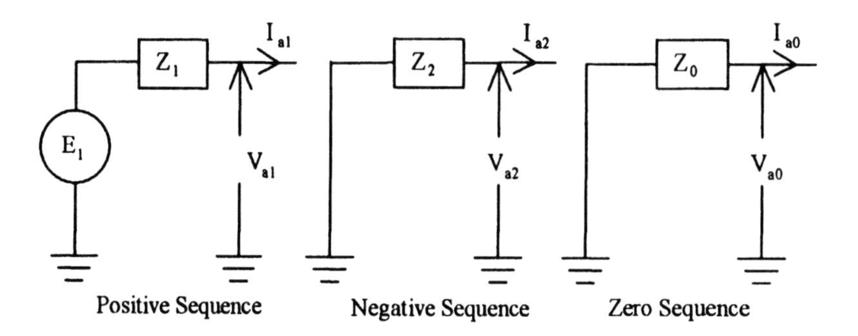

Figure 4.4 Thevenin equivalent sequence circuit models

4.3.1 3-Phase Line to Ground fault

Three phase faults to ground are characterized by

$$V_a = V_b = V_c = 0 \tag{4.15}$$

as shown in Fig. 4.5a. From (4.15), together with (4.10), we find

$$V_{a0} = V_{a1} = V_{a2} = 0 \tag{4.16}$$

Therefore, from (4.14) we get

$$I_{a1} = V_{pf}/Z_1 \quad , \quad I_{a2} = I_{a0} = 0 \tag{4.17}$$

Using (4.17) and (4.8) applied to currents, we find

$$I_a = I_{a1} \quad , \quad I_b = \alpha^2 I_{a1} \quad , \quad I_c = \alpha I_{a1} \tag{4.18}$$

Thus the fault currents, as expected, form a balanced positive sequence set of magnitude V_{pf}/Z_1. This example could have been carried out without the use of symmetrical components since the fault does not unbalance the system.

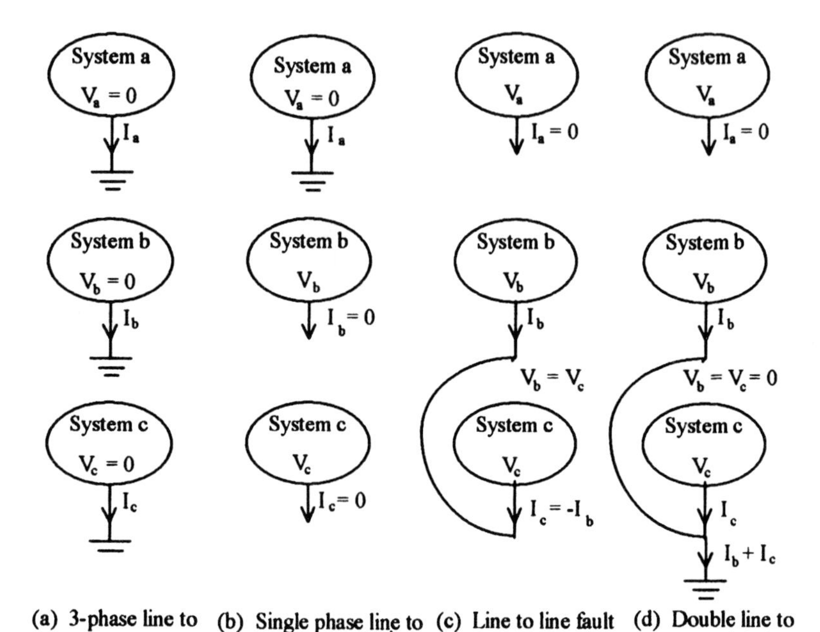

Figure 4.5 Standard fault types on 3 phase systems

4.3.2 Single Phase Line to Ground Fault

For a single phase to ground fault, we assume, without loss of generality, that the a-phase is faulted. Thus we have

$$V_a = 0 \quad , \quad I_b = I_c = 0 \tag{4.19}$$

as indicated in Fig. 4.5b. From (4.10) applied to currents, we get

$$I_{a0} = I_{a1} = I_{a2} = I_a/3 \tag{4.20}$$

From (4.14), (4.19), and (4.20), we find

$$V_a = 0 = V_{a0} + V_{a1} + V_{a2} = V_{pf} - I_{a1}(Z_1 + Z_2 + Z_0)$$

or

$$I_{a1} = \frac{V_{pf}}{(Z_1 + Z_2 + Z_0)} = I_{a2} = I_{a0} \tag{4.21}$$

From (4.19) and (4.20), we get

$$I_a = \frac{3V_{pf}}{(Z_1 + Z_2 + Z_0)} \quad , \quad I_b = I_c = 0 \tag{4.22}$$

4.3.3 Line to Line Fault

A line to line fault can, without loss of generality, be assumed to occur between lines b and c as shown in Fig. 4.5c. The fault equations are

$$V_b = V_c \quad , \quad I_a = 0 \quad , \quad I_c = -I_b \tag{4.23}$$

From (4.10) applied to voltages and currents, we get

$$V_{a1} = V_{a2} \quad , \quad I_{a0} = 0 \quad , \quad I_{a2} = -I_{a1} \tag{4.24}$$

Using (4.14) and (4.24), we find

$$V_{a0} = 0 \quad , \quad V_{a1} - V_{a2} = 0 = V_{pf} - I_{a1}(Z_1 + Z_2)$$

or

$$I_{a1} = \frac{V_{pf}}{(Z_1 + Z_2)} = -I_{a2} \quad , \quad I_{a0} = 0 \tag{4.25}$$

Using (4.8) applied to currents, (4.24), and (4.25), we obtain

$$I_a = 0 \quad , \quad I_c = \frac{j\sqrt{3}V_{pf}}{(Z_1 + Z_2)} = -I_b \tag{4.26}$$

4.3.4 Double Line to Ground Fault

The double line to ground fault, as shown in Fig. 4.5d, can be regarded as involving lines b and c. The fault equations are

$$V_b = V_c = 0 \quad , \quad I_a = 0 \tag{4.27}$$

From (4.27) and (4.10), we find

$$V_{a0} = V_{a1} = V_{a2} = V_a/3 \quad , \quad I_a = I_{a0} + I_{a1} + I_{a2} = 0 \tag{4.28}$$

Using (4.14) and (4.28),

$$I_{a0} + I_{a1} + I_{a2} = 0 = \frac{V_{pf}}{Z_1} - V_{a1}\left(\frac{1}{Z_1} + \frac{1}{Z_2} + \frac{1}{Z_0}\right)$$

or

$$V_{a1} = V_{pf} \Big/ \left(1 + \frac{Z_1(Z_0 + Z_2)}{Z_0 Z_2}\right) \tag{4.29}$$

so that, from (4.14)

$$I_{a1} = \frac{V_{pf}(Z_0 + Z_2)}{Z_0 Z_2 + Z_1(Z_0 + Z_2)}$$

$$I_{a2} = -\frac{V_{pf} Z_0}{Z_0 Z_2 + Z_1(Z_0 + Z_2)} \tag{4.30}$$

$$I_{a0} = -\frac{V_{pf} Z_2}{Z_0 Z_2 + Z_1(Z_0 + Z_2)}$$

Substituting into (4.8) applied to currents, we obtain

$$I_a = 0$$

$$I_b = \frac{V_{pf}\left[-j\sqrt{3}Z_0 - \left(\frac{3}{2} + j\frac{\sqrt{3}}{2}\right)Z_2\right]}{Z_0 Z_2 + Z_1(Z_0 + Z_2)} \tag{4.31}$$

$$I_c = \frac{V_{pf}\left[j\sqrt{3}Z_0 - \left(\frac{3}{2} - j\frac{\sqrt{3}}{2}\right)Z_2\right]}{Z_0 Z_2 + Z_1(Z_0 + Z_2)}$$

4.4 FAULT CURRENTS FOR TRANSFORMERS WITH 2 TERMINALS PER PHASE

A 2 terminal transformer can be modeled by a single leakage reactance which we call z_{HL} where H and L indicate high and low voltage terminals. All electrical quantities from this point on will be taken to mean per-unit quantities and will be written with small letters. This will enable us to describe transformers, using a single circuit.

The high and low voltage systems external to the transformer are described by system impedances z_{SH} , z_{SL} and voltage sources e_{SH} , e_{SL}. The resulting sequence circuit models are shown in Fig. 4.6. A zero subscript is used to label the zero sequence circuit parameters since they can differ considerably from the positive or negative sequence circuit parameters. The positive and negative circuit parameters are equal for transformers and we will assume for the electrical systems also. They bear no distinguishing subscript. We have shown a fault on the H terminal in Fig. 4.6. By interchanging subscripts, the L terminal faults can be obtained.

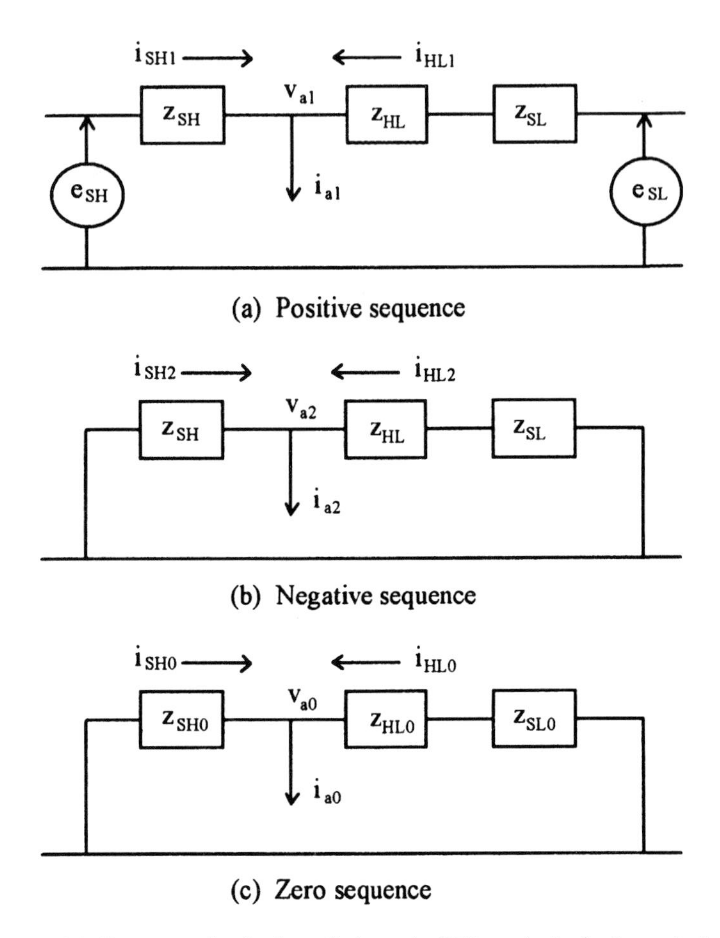

Figure 4.6 Sequence circuits for a fault on the HV terminal of a 2 terminal per phase transformer, using per unit quantities

In order to use the previously developed general results, we need to compute the Thevenin impedances and pre-fault voltage. From Fig. 4.6, we find

$$z_1 = z_2 = \frac{z_{SH}(z_{HL} + z_{SL})}{z_{HL} + z_{SH} + z_{SL}} \quad , \quad z_0 = \frac{z_{SH0}(z_{HL0} + z_{SL0})}{z_{HL0} + z_{SH0} + z_{SL0}} \tag{4.32}$$

and

$$v_{pf} = e_{SH} - i_{SHpf} z_{SH} = e_{SL} - i_{HLpf}(z_{HL} + z_{SL}) \tag{4.33}$$

where the pre-fault (pf) quantities are all positive sequence. Fig. 4.6 and the above formulas assume that both terminals are connected to the HV and LV systems. If either terminal of the transformer is floating, then this is equivalent to setting the system impedance to infinity for that system.

We are interested in obtaining the currents in the transformer during the fault. Thus, according to Fig. 4.6, we need to find i_{HL1}, i_{HL2}, i_{HL0} for the standard faults. Since i_{a1}, i_{a2}, i_{a0} have already been obtained for the standard faults, we must find the transformer currents in terms of these known fault currents. From Fig. 4.6 we see that

$$v_{a1} = e_{SL} - i_{HL1}(z_{HL} + z_{SL}),\ v_{a2} = -i_{HL2}(z_{HL} + z_{SL}),\ v_{a0} = -i_{HL0}(z_{HL0} + z_{SL0})$$

Using (4.33), we can rewrite this as

$$v_{a1} = v_{pf} - i_{HL1}(z_{HL} + z_{SL}) + i_{HLpf}(z_{HL} + z_{SL})$$
$$v_{a2} = -i_{HL2}(z_{HL} + z_{SL}) \quad , \quad v_{a0} = -i_{HL0}(z_{HL0} + z_{SL0}) \tag{4.34}$$

Naturally, if the transformer were not loaded before the fault, we would have $i_{HLpf} = 0$. Substituting the per-unit version of (4.14) into (4.34), we obtain

$$i_{HL1} = i_{a1}\frac{z_1}{(z_{HL} + z_{SL})} + i_{HLpf}$$
$$i_{HL2} = i_{a2}\frac{z_2}{(z_{HL} + z_{SL})} \tag{4.35}$$
$$i_{HL0} = i_{a0}\frac{z_0}{(z_{HL0} + z_{SL0})}$$

4.4.1 3-Phase Line to Ground Fault

For this fault case, we substitute (4.17), expressed in per-unit terms, into (4.35) to obtain

$$i_{HL1} = \frac{v_{pf}}{(z_{HL} + z_{SL})} + i_{HLpf} \quad , \quad i_{HL2} = i_{HL0} = 0 \tag{4.36}$$

Then, using (4.8) applied to currents, we find

$$i_{HLa} = i_{HL1} \quad , \quad i_{HLb} = \alpha^2 i_{HL1} \quad , \quad i_{HLc} = \alpha i_{HL1} \qquad (4.37)$$

i.e. the fault currents in the transformer form a positive sequence set as expected.

4.4.2 Single Phase Line to Ground Fault

For this type of fault, substitute the per-unit versions of (4.21) into (4.35), using $z_1 = z_2$ to obtain

$$i_{HL1} = \frac{v_{pf}}{(z_{HL} + z_{SL})}\left(\frac{z_1}{2z_1 + z_0}\right) + i_{HLpf}$$

$$i_{HL2} = \frac{v_{pf}}{(z_{HL} + z_{SL})}\left(\frac{z_1}{2z_1 + z_0}\right) \qquad (4.38)$$

$$i_{HL0} = \frac{v_{pf}}{(z_{HL0} + z_{SL0})}\left(\frac{z_0}{2z_1 + z_0}\right)$$

Substituting (4.38) into (4.8) applied to currents, we obtain the phase currents,

$$i_{HLa} = v_{pf}\left\{\frac{1}{(z_{HL} + z_{SL})}\left(\frac{2z_1}{2z_1 + z_0}\right) + \frac{1}{(z_{HL0} + z_{SL0})}\left(\frac{z_0}{2z_1 + z_0}\right)\right\} + i_{HLpf}$$

$$i_{HLb} = v_{pf}\left\{\frac{-1}{(z_{HL} + z_{SL})}\left(\frac{z_1}{2z_1 + z_0}\right) + \frac{1}{(z_{HL0} + z_{SL0})}\left(\frac{z_0}{2z_1 + z_0}\right)\right\} + \alpha^2 i_{HLpf}$$

$$i_{HLc} = v_{pf}\left\{\frac{-1}{(z_{HL} + z_{SL})}\left(\frac{z_1}{2z_1 + z_0}\right) + \frac{1}{(z_{HL0} + z_{SL0})}\left(\frac{z_0}{2z_1 + z_0}\right)\right\} + \alpha i_{HLpf}$$

(4.39)

Note that there is fault current in phases b and c inside the transformer even though the fault is on phase a. These b and c fault currents are of lower magnitude than the phase a fault current.

4.4.3 Line to Line Fault

For this type of fault, we substitute the per-unit versions of (4.25) into (4.35) using $z_1 = z_2$, to obtain

$$i_{HL1} = \frac{v_{pf}}{2(z_{HL} + z_{SL})} + i_{HLpf}\ ,\ i_{HL2} = -\frac{v_{pf}}{2(z_{HL} + z_{SL})}\ ,\ i_{HL0} = 0 \qquad (4.40)$$

Using (4.8) applied to currents, we obtain

$$i_{HLa} = i_{HLpf}$$

$$i_{HLb} = -\frac{j\sqrt{3}v_{pf}}{2(z_{HL} + z_{SL})} + \alpha^2 i_{HLpf} \qquad (4.41)$$

$$i_{HLc} = \frac{j\sqrt{3}v_{pf}}{2(z_{HL} + z_{SL})} + \alpha i_{HLpf}$$

In this case, with the fault between phases b and c, phase a is unaffected.

4.4.4 Double Line to Ground Fault

For this fault, we substitute (4.30) expressed in per-unit terms, into (4.35) , using $z_1 = z_2$, to obtain

$$i_{HL1} = \frac{v_{pf}}{(z_{HL} + z_{SL})}\left(\frac{z_1 + z_0}{z_1 + 2z_0}\right) + i_{HLpf}$$

$$i_{HL2} = -\frac{v_{pf}}{(z_{HL} + z_{SL})}\left(\frac{z_0}{z_1 + 2z_0}\right) \qquad (4.42)$$

$$i_{HL0} = -\frac{v_{pf}}{(z_{HL0} + z_{SL0})}\left(\frac{z_0}{z_1 + 2z_0}\right)$$

Using (4.8) applied to currents, we get for the phase currents

$$i_{HLa} = \frac{v_{pf}}{(z_1 + 2z_0)}\left\{\frac{z_1}{(z_{HL} + z_{SL})} - \frac{z_0}{(z_{HL0} + z_{SL0})}\right\} + i_{HLpf}$$

$$i_{HLb} = \frac{v_{pf}}{(z_1 + 2z_0)}\left\{\frac{\alpha^2 z_1 - j\sqrt{3}z_0}{(z_{HL} + z_{SL})} - \frac{z_0}{(z_{HL0} + z_{SL0})}\right\} + \alpha^2 i_{HLpf} \quad (4.43)$$

$$i_{HLc} = \frac{v_{pf}}{(z_1 + 2z_0)}\left\{\frac{\alpha z_1 + j\sqrt{3}z_0}{(z_{HL} + z_{SL})} - \frac{z_0}{(z_{HL0} + z_{SL0})}\right\} + \alpha i_{HLpf}$$

4.4.5 Zero Sequence Impedances

Zero sequence impedances require special consideration since certain transformer 3-phase connections, such as the delta connection, block the flow of zero sequence currents and hence provide an essentially infinite impedance to their passage. This is also true of the ungrounded Y connection. The reason is that the zero sequence currents, being all in phase, require a return path in order to flow. The delta connection provides an internal path for the flow of these currents, circulating around the delta, but blocks their flow through the external lines. These considerations do not apply to positive or negative sequence currents which sum to zero vectorially and so require no return path.

For transformers, since the amp-turns must be balanced for each sequence, in order for zero sequence currents to be present, they must flow in both windings. Thus in a grounded Y-Delta unit, for example, zero sequence currents can flow within the transformer but cannot flow in the external circuit connected to the delta side. Similarly, zero sequence currents cannot flow in either winding if one of them is an ungrounded Y.

Fig. 4.7 shows some examples of zero sequence impedance diagrams for different transformer connections. These should be compared with Fig. 4.6c which applies to a grounded Y/grounded Y connection. Where a break in a line occurs, imagine that an infinite impedance is inserted. Mathematically, one needs to let the impedance approach infinity as a limiting process in the formulas.

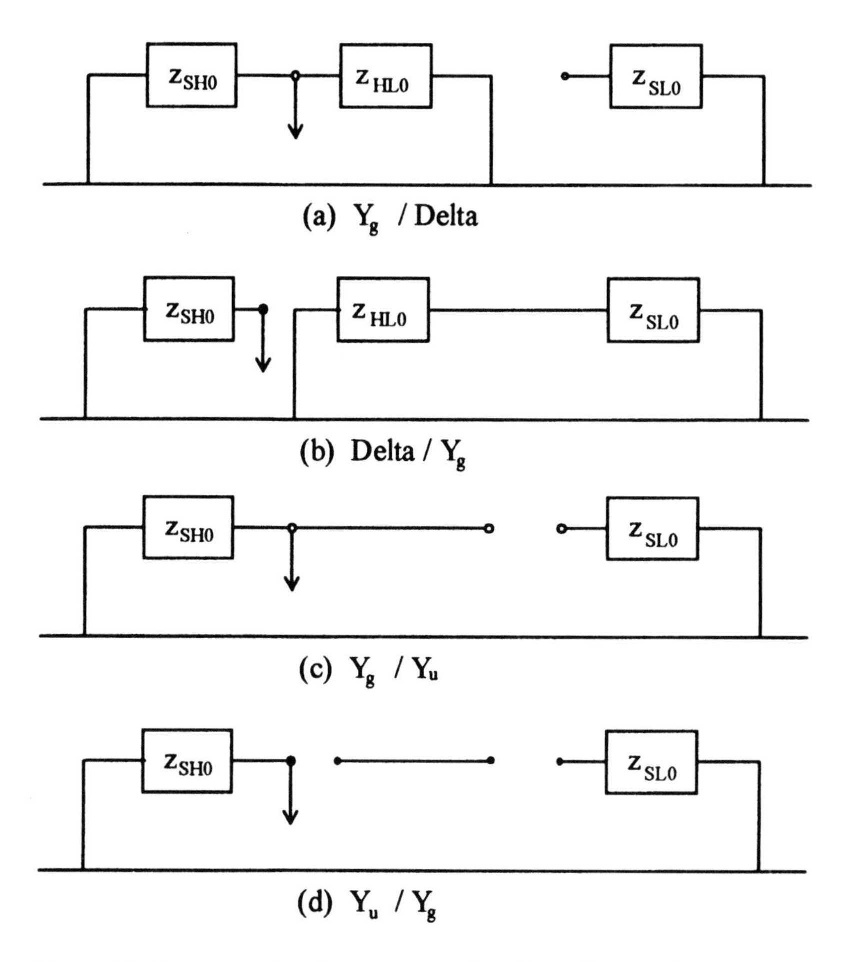

Figure 4.7 Some examples of zero sequence impedance diagrams for 2 terminal transformers. The arrow indicates the fault point. Y_g = grounded Y, Y_u = ungrounded Y

For the Connection in Fig. 4.7a, the Thevenin impedance, looking in from the fault point, is $z_0 = z_{SH0}\, z_{HL0} / (z_{SH0} + z_{HL0})$ i.e. the parallel combination of z_{SH0} and z_{HL0}. In this case, the impedance z_{SL0} is effectively removed from the circuit and replaced by $z_{SL0} = 0$. This substitution should be made in all the formulas.

For the connection in Fig. 4.7b, we find $z_0 = z_{SH0}$. In this case no zero sequence current can flow into the fault from the transformer side of the fault point so effectively $z_{SL0} \rightarrow \infty$. Figs. 7c and 7d are similar. Because of the ungrounded Y connection, no zero sequence current flows in the transformer.

Another issue is the value of the zero sequence impedances themselves when they are fully in the circuit. These values tend to differ from the positive sequence impedances in transformers because the magnetic flux patterns associated with them can be quite different from the positive sequence flux distribution. This difference is taken into account by multiplying factors which multiply the positive sequence impedances to produce the zero sequence values. For 3 phase core form transformers, these multiplying factors tend to be ≈ 0.85, however they can differ for different 3 phase connections and are usually found by experimental measurements.

4.5 FAULT CURRENTS FOR TRANSFORMERS WITH 3 TERMINALS PER PHASE

A 3-terminal transformer can be represented in terms of 3 Y (or T) connected impedances if per-unit quantities are used. Fig. 4.8 shows the sequence circuits for such a transformer where H, X, Y label the transformer impedances and SH, SX, SY label the associated system impedances. The systems are represented by impedances in series with voltage sources. The positive sense of the currents is into the transformer terminals. Although the fault is shown on the H-terminal, by interchanging subscripts, the formulas which follow can apply to faults an any terminal. As before, we have not labeled the positive and negative sequence impedances with subscripts since they are equal for transformers and we assume also for the systems. The zero sequence impedances are distinguished with subscripts because they can differ from their positive sequence counterparts.

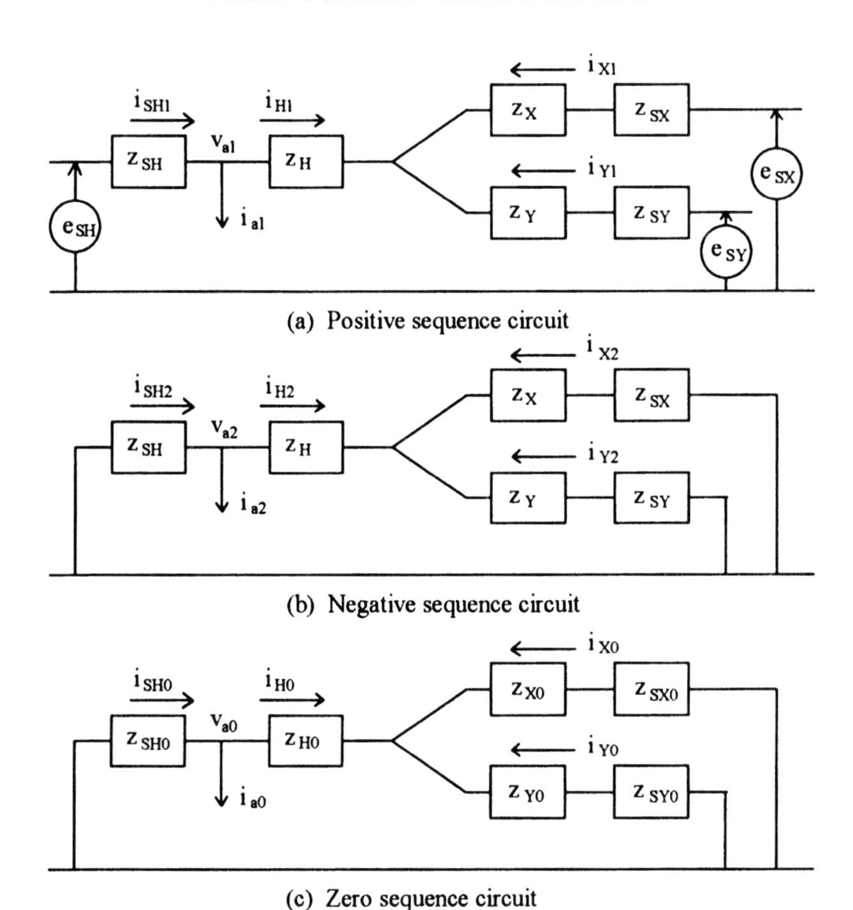

Figure 4.8 Sequence circuits for a fault on the H terminal of a 3 terminal per phase transformer, using per unit quantities

From Fig. 4.8, the Thevenin impedances, looking into the circuits from the fault point, are

$$z_1 = z_2 = \frac{z_{SH}(z_H + w_1)}{z_H + z_{SH} + w_1} \quad \text{where} \quad w_1 = \frac{(z_X + z_{SX})(z_Y + z_{SY})}{z_X + z_Y + z_{SX} + z_{SY}}$$

and (4.44)

$$z_0 = \frac{z_{SH0}(z_{H0} + w_0)}{z_{H0} + z_{SH0} + w_0} \quad \text{where} \quad w_0 = \frac{(z_{X0} + z_{SX0})(z_{Y0} + z_{SY0})}{z_{X0} + z_{Y0} + z_{SX0} + z_{SY0}}$$

The pre-fault voltage is given by

$$\begin{aligned} v_{pf} &= e_{SH} - i_{SHpf} z_{SH} = e_{SX} - i_{Xpf}(z_X + z_{SX}) + i_{Hpf} z_H \\ &= e_{SY} - i_{Ypf}(z_Y + z_{SY}) + i_{Hpf} z_H \end{aligned} \tag{4.45}$$

where pf labels pre-fault quantities which are all positive sequence. We also have

$$i_H + i_X + i_Y = 0 \tag{4.46}$$

which applies to the sequence and pre-fault currents. If a terminal is unloaded, then the corresponding pre-fault current should be set to zero.

From Fig. 4.8,

$$\begin{aligned} v_{a1} &= e_{SX} - i_{X1}(z_X + z_{SX}) + i_{H1} z_H = e_{SY} - i_{Y1}(z_Y + z_{SY}) + i_{H1} z_H \\ v_{a2} &= -i_{X2}(z_X + z_{SX}) + i_{H2} z_H = -i_{Y2}(z_Y + z_{SY}) + i_{H2} z_H \\ v_{a0} &= -i_{X0}(z_{X0} + z_{SX0}) + i_{H0} z_{H0} = -i_{Y0}(z_{Y0} + z_{SY0}) + i_{H0} z_{H0} \end{aligned} \tag{4.47}$$

Solving (4.47), together with (4.44), (4.45), (4.46), and (4.14) expressed in per-unit terms, we obtain

$$i_{H1} = -i_{a1}\frac{z_1}{(z_H + w_1)} + i_{Hpf} \quad , \quad i_{H2} = -i_{a2}\frac{z_2}{(z_H + w_1)} \quad , \quad i_{H0} = -i_{a0}\frac{z_0}{(z_{H0} + w_0)}$$

$$i_{X1} = i_{a1}\frac{z_1}{(z_H + w_1)}\left(\frac{z_Y + z_{SY}}{z_X + z_Y + z_{SX} + z_{SY}}\right) + i_{Xpf}$$

$$i_{X2} = i_{a2}\frac{z_2}{(z_H + w_1)}\left(\frac{z_Y + z_{SY}}{z_X + z_Y + z_{SX} + z_{SY}}\right), \; i_{X0} = i_{a0}\frac{z_0}{(z_{H0} + w_0)}\left(\frac{z_{Y0} + z_{SY0}}{z_{X0} + z_{Y0} + z_{SX0} + z_{SY0}}\right)$$

$$i_{Y1} = i_{a1}\frac{z_1}{(z_H + w_1)}\left(\frac{z_X + z_{SX}}{z_X + z_Y + z_{SX} + z_{SY}}\right) + i_{Ypf}$$

$$i_{Y2} = i_{a2}\frac{z_2}{(z_H + w_1)}\left(\frac{z_X + z_{SX}}{z_X + z_Y + z_{SX} + z_{SY}}\right), \; i_{Y0} = i_{a0}\frac{z_0}{(z_{H0} + w_0)}\left(\frac{z_{X0} + z_{SX0}}{z_{X0} + z_{Y0} + z_{SX0} + z_{SY0}}\right) \tag{4.48}$$

We now use these equations, together with the fault current equations to obtain the currents in the transformer for the various types of fault. We will only list the equations for the sequence currents. Equation (4.8), applied to currents, may be used to obtain the phase currents in terms of the sequence currents.

4.5.1 3-Phase line to ground fault

For this type of fault, we substitute the per-unit version of (4.17) into (4.48) to obtain

$$i_{H1} = -\frac{v_{pf}}{(z_H + w_1)} + i_{Hpf} \quad , \quad i_{H2} = i_{H0} = 0$$

$$i_{X1} = \frac{v_{pf}}{(z_H + w_1)}\left(\frac{z_Y + z_{SY}}{z_X + z_Y + z_{SX} + z_{SY}}\right) + i_{Xpf} \quad , \quad i_{X2} = i_{X0} = 0 \qquad (4.49)$$

$$i_{Y1} = \frac{v_{pf}}{(z_H + w_1)}\left(\frac{z_X + z_{SX}}{z_X + z_Y + z_{SX} + z_{SY}}\right) + i_{Ypf} \quad , \quad i_{Y2} = i_{Y0} = 0$$

4.5.2 Single Phase Line to Ground Fault

For this fault, we substitute the per-unit version of (4.21) into (4.48), using $z_1 = z_2$, to get

$$i_{H1} = -\frac{v_{pf}}{(z_H + w_1)}\left(\frac{z_1}{2z_1 + z_0}\right) + i_{Hpf}$$

$$i_{H2} = -\frac{v_{pf}}{(z_H + w_1)}\left(\frac{z_1}{2z_1 + z_0}\right) \quad , \quad i_{H0} = -\frac{v_{pf}}{(z_{H0} + w_0)}\left(\frac{z_0}{2z_1 + z_0}\right)$$

$$i_{X1} = \frac{v_{pf}}{(z_H + w_1)}\left(\frac{z_1}{2z_1 + z_0}\right)\left(\frac{z_Y + z_{SY}}{z_X + z_Y + z_{SX} + z_{SY}}\right) + i_{Xpf}$$

$$i_{X2} = \frac{v_{pf}}{(z_H + w_1)}\left(\frac{z_1}{2z_1 + z_0}\right)\left(\frac{z_Y + z_{SY}}{z_X + z_Y + z_{SX} + z_{SY}}\right)$$

$$i_{X0} = \frac{v_{pf}}{(z_{H0} + w_0)}\left(\frac{z_0}{2z_1 + z_0}\right)\left(\frac{z_{Y0} + z_{SY0}}{z_{X0} + z_{Y0} + z_{SX0} + z_{SY0}}\right)$$

$$i_{Y1} = \frac{v_{pf}}{(z_H + w_1)}\left(\frac{z_1}{2z_1 + z_0}\right)\left(\frac{z_X + z_{SX}}{z_X + z_Y + z_{SX} + z_{SY}}\right) + i_{Ypf} \tag{4.50}$$

$$i_{Y2} = \frac{v_{pf}}{(z_H + w_1)}\left(\frac{z_1}{2z_1 + z_0}\right)\left(\frac{z_X + z_{SY}}{z_X + z_Y + z_{SX} + z_{SY}}\right)$$

$$i_{Y0} = \frac{v_{pf}}{(z_{H0} + w_0)}\left(\frac{z_0}{2z_1 + z_0}\right)\left(\frac{z_{X0} + z_{SX0}}{z_{X0} + z_{Y0} + z_{SX0} + z_{SY0}}\right)$$

4.5.3 Line to Line Fault

For this fault condition, substitute the per-unit version of (4.25) into (4.48), using $z_1 = z_2$, to obtain

$$i_{H1} = -\frac{v_{pf}}{2(z_H + w_1)} + i_{Hpf} \quad , \quad i_{H2} = \frac{v_{pf}}{2(z_H + w_1)} \quad , \quad i_{H0} = 0$$

$$i_{X1} = \frac{v_{pf}}{2(z_H + w_1)}\left(\frac{z_Y + z_{SY}}{z_X + z_Y + z_{SX} + z_{SY}}\right) + i_{Xpf} \tag{4.51}$$

$$i_{X2} = -\frac{v_{pf}}{2(z_H + w_1)}\left(\frac{z_Y + z_{SY}}{z_X + z_Y + z_{SX} + z_{SY}}\right) \quad , \quad i_{X0} = 0$$

4.5.4 Double Line to Ground Fault

For this fault, substitute the per-unit version of (4.30) into (4.48), using $z_1 = z_2$, to get

$$i_{H1} = -\frac{v_{pf}}{(z_H + w_1)}\left(\frac{z_1 + z_0}{z_1 + 2z_0}\right) + i_{Hpf}$$

$$i_{H2} = \frac{v_{pf}}{(z_H + w_1)}\left(\frac{z_0}{z_1 + 2z_0}\right)$$

$$i_{H0} = \frac{v_{pf}}{(z_{H0} + w_0)}\left(\frac{z_0}{z_1 + 2z_0}\right)$$

$$i_{X1} = \frac{v_{pf}}{(z_H + w_1)}\left(\frac{z_1 + z_0}{z_1 + 2z_0}\right)\left(\frac{z_Y + z_{SY}}{z_X + z_Y + z_{SX} + z_{SY}}\right) + i_{Xpf}$$

$$i_{X2} = -\frac{v_{pf}}{(z_H + w_1)}\left(\frac{z_0}{z_1 + 2z_0}\right)\left(\frac{z_Y + z_{SY}}{z_X + z_Y + z_{SX} + z_{SY}}\right)$$

$$i_{X0} = -\frac{v_{pf}}{(z_{H0} + w_0)}\left(\frac{z_0}{z_1 + 2z_0}\right)\left(\frac{z_{Y0} + z_{SY0}}{z_{X0} + z_{Y0} + z_{SX0} + z_{SY0}}\right) \quad (4.52)$$

$$i_{Y1} = \frac{v_{pf}}{(z_H + w_1)}\left(\frac{z_1 + z_0}{z_1 + 2z_0}\right)\left(\frac{z_X + z_{SX}}{z_X + z_Y + z_{SX} + z_{SY}}\right) + i_{Ypf}$$

$$i_{Y2} = -\frac{v_{pf}}{(z_H + w_1)}\left(\frac{z_0}{z_1 + 2z_0}\right)\left(\frac{z_X + z_{SY}}{z_X + z_Y + z_{SX} + z_{SY}}\right)$$

$$i_{Y0} = -\frac{v_{pf}}{(z_{H0} + w_0)}\left(\frac{z_0}{z_1 + 2z_0}\right)\left(\frac{z_{X0} + z_{SX0}}{z_{X0} + z_{Y0} + z_{SX0} + z_{SY0}}\right)$$

4.5.5 Zero Sequence Impedances

Fig. 4.9 lists some examples of zero sequence circuits for 3-terminal transformers. An infinite impedance is represented by a break in the circuit. When substituting into the preceding formulas, a limiting process needs to be used. Although there are many more possibilities than shown in Fig. 4.9, they can serve to illustrate the method for accounting for the different 3-phase connections. The previous formulas apply directly to Fig. 4.8c which represents a transformer with all grounded Y terminal connections.

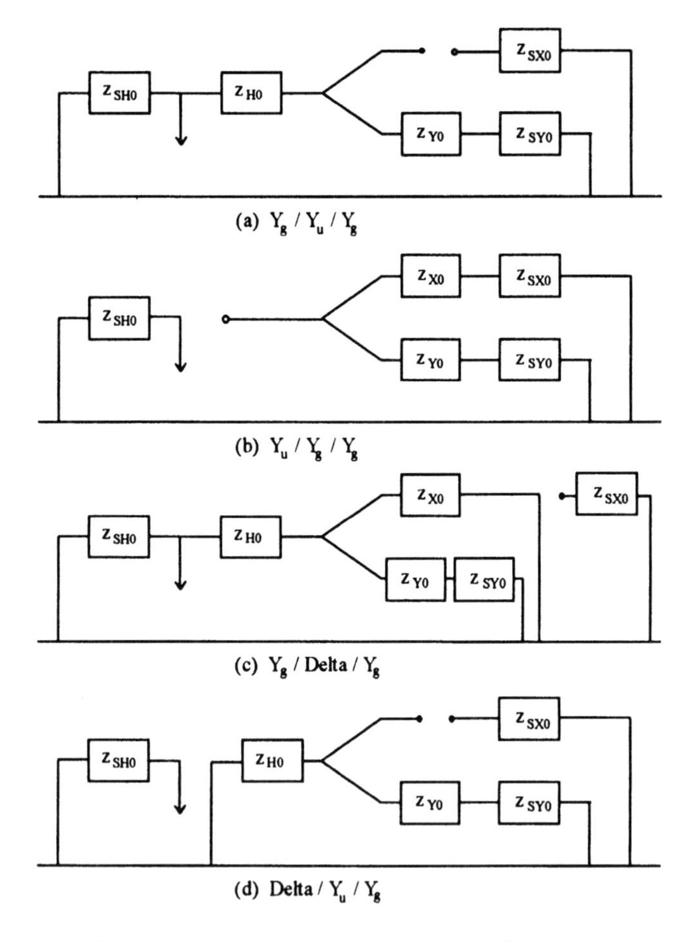

Figure 4.9 Some examples of zero sequence circuit diagrams for 3 terminal transformers. The arrow indicates the fault point. Y_g = grounded Y, Y_u = ungrounded Y

In Fig. 4.9a, we have $z_{X0} \to \infty$, so that $w_0 = z_{Y0} + z_{SY0}$. In the fault current formulas, $(z_{Y0} + z_{SY0})/(z_{X0} + z_{Y0} + z_{SX0} + z_{SY0}) = 0$ and $(z_{X0} + z_{SX0})/(z_{X0} + z_{Y0} + z_{SX0} + z_{SY0}) = 1$. This insures that no zero sequence current flows in the X-terminal.

In Fig. 4.9b, we see that $z_0 = z_{SH0}$. We also have $z_{H0} \to \infty$. This implies that there are no zero sequence fault currents in any of the terminals.

In Fig. 4.9c, we may take $z_{SX0} = 0$ since it is replaced by a short in the circuit as seen by the transformer. This should be substituted into all the formulas.

In Fig. 4.9d, we see that $z_0 = z_{SH0}$. In this case, the only zero sequence current in the fault comes from the high voltage system, none from the transformer.

4.6 ASYMMETRY FACTOR

A factor multiplying the currents calculated above is necessary to account for a transient overshoot when the fault occurs. This factor, called the asymmetry factor, is given by

$$K = \left\{1 + \exp\left[-\left(\phi + \frac{\pi}{2}\right)\frac{r}{x}\right]\sin\phi\right\}\sqrt{2} \tag{4.53}$$

where x is the reactance looking into the terminal, r the resistance and $\phi = \tan^{-1}(x/r)$ in radians. Usually the system impedances are ignored when calculating these quantities, so that for a 2 terminal unit,

$$\frac{x}{r} = \frac{\mathrm{Im}\,(z_{HL})}{\mathrm{Re}\,(z_{HL})} \tag{4.54}$$

while for a 3 terminal unit

$$\begin{aligned} x_H &= \mathrm{Im}(z_H) + \frac{\mathrm{Im}(z_X)\,\mathrm{Im}(z_Y)}{\mathrm{Im}(z_X) + \mathrm{Im}(z_Y)} \\ r_H &= \mathrm{Re}(z_H) + \frac{\mathrm{Re}(z_X)\,\mathrm{Re}(z_Y)}{\mathrm{Re}(z_X) + \mathrm{Re}(z_Y)} \end{aligned} \tag{4.55}$$

with corresponding expressions for x_X , r_X , x_Y , r_Y. When K in (4.53) multiplies the rms short circuit current, it yields the maximum peak short circuit current.

5. RABINS' METHOD FOR CALCULATING LEAKAGE FIELDS, FORCES, AND INDUCTANCES IN TRANSFORMERS

Summary Rabin's method utilizes a simplified transformer geometry, consisting of a core, coils, and yokes of infinite extent, to solve Maxwell's equations. This method works well for calculating the magnetic field near the coils so that quantities such as inductances and forces which depend on this near field are accurately calculated. Since the tank wall, clamping structure, and other details are omitted, this method does not allow one to calculate stray losses in these structures. We use this method to find forces, the two winding leakage inductance, as well as self and mutual inductances between coil sections.

5.1 INTRODUCTION

Modern general purpose computer programs are available for calculating the magnetic field inside the complex geometry of a transformer. These numerical methods generally employ finite elements or boundary elements. Geometric details such as the tank wall and clamping structure can be included. While 3D programs are available, 2D programs using an axisymmetric geometry are adequate for most purposes. Although inputting the geometry, the Ampere-turns in the winding sections, and the boundary conditions can be tedious, parametric procedures are often available for simplifying this task. Along with the magnetic field, associated quantities such as inductances and forces can be calculated by these methods. In addition, eddy currents in structural parts and their accompanying losses can be obtained with the appropriate a.c. solver.

In spite of these modern advances in computational methods, older procedures can often be profitably employed to obtain quantities of interest very quickly and with a minimum of input. One of these is Rabins' method, which assumes an idealized transformer geometry [Rab56]. This simplified geometry permits analytic formulas to be developed for the magnetic field and other useful quantities. The geometry consists of a single leg of a single or possibly 3 phase transformer. The leg consists of a core and surrounding coils which are assumed to be axisymmetric, along with yokes which are assumed to be of infinite extent at the top and bottom of the leg. The entire axisymmetric geometry is of infinite extent radially. Thus there are no tank walls or clamping structures in the geometry. In addition, the core and yokes are assumed to be infinitely permeable.

In spite of these simplifications, Rabins' method does a good job of calculating the magnetic field in the immediate vicinity of the windings. Thus forces and inductances, which depend largely on the field near the windings, are also accurately obtained. This can be shown by direct comparison with a finite element solution applied to a more complex geometry, including tank wall and clamps. Although the finite element procedure can obtain losses in structural parts, Rabins' method is not suited for this. However, because the magnetic field near or inside the windings is obtained accurately, eddy current losses in the windings as well as their spatial distribution can be accurately obtained from formulas based on this leakage field.

In the following, we present Rabins' method and show how it can be used to obtain forces, leakage reactances, and self and mutual inductances between winding sections for use in detailed circuit models of transformers such as are needed in impulse calculations.

5.2 THEORY

We model a cylindrical coil or section of a coil surrounding a core leg with top and bottom yokes as shown in Fig. 5.1. The yokes and core are really boundaries of the geometry which extends infinitely far radially. The coils are assumed to be composed of stranded conductors so that no eddy current effects are modeled. Thus we can assume d.c. conditions. The current density in the coil is assumed to be piecewise constant axially and uniform radially as shown in Fig. 5.2. As indicated, there can be regions within the coil where the current density drops to zero. Since there are no non-linear effects within the geometry modeled, the fields from several coils can be added vectorially so that it is only necessary to model one coil at a time.

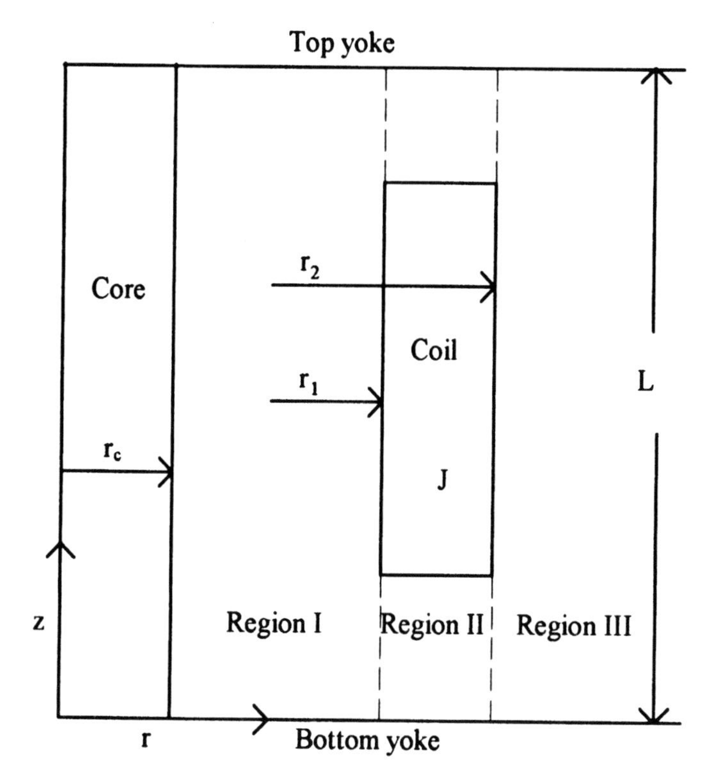

Figure 5.1 Geometry of iron core, yokes, and coil or coil section

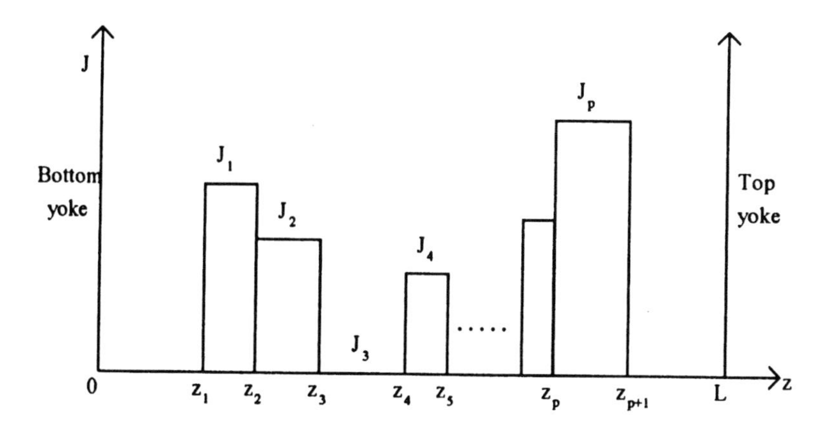

Figure 5.2 Axial distribution of current density in the coil

Maxwell's equations for the magnetic field, applied to the geometry outside the core and yokes and assuming static conditions are, in SI units,

$$\nabla \times \mathbf{H} = \mathbf{J} \tag{5.1}$$

$$\nabla \cdot \mathbf{B} = 0 \tag{5.2}$$

where $\mathbf{H}$ is the magnetic field, $\mathbf{B}$ the induction, and $\mathbf{J}$ the current density. Defining a vector potential $\mathbf{A}$, by

$$\mathbf{B} = \nabla \times \mathbf{A} \tag{5.3}$$

equation (5.2) is automatically satisfied. We also have

$$\mathbf{B} = \mu_o \mathbf{H} \tag{5.4}$$

in the region of interest where μ_o is the permeability of vacuum, oil, or air. Substituting (5.3) and (5.4) into (5.1), we obtain

$$\nabla \times (\nabla \times \mathbf{A}) = \nabla(\nabla \cdot \mathbf{A}) - \nabla^2 \mathbf{A} = \mu_o \mathbf{J} \tag{5.5}$$

The vector potential is not completely defined by (5.3). It contains some arbitrariness which can be removed by setting

$$\nabla \cdot \mathbf{A} = 0 \tag{5.6}$$

Thus (5.5) becomes

$$\nabla^2 \mathbf{A} = -\mu_o \mathbf{J} \tag{5.7}$$

The current density vector is azimuthal so that

$$\mathbf{J} = J_\varphi \mathbf{a}_\varphi \tag{5.8}$$

where $\mathbf{a}_\varphi$ is the unit vector in the azimuthal direction. Because of the axisymmetric geometry, all the field quantities are independent of φ. With these assumptions (5.7) becomes in cylindrical coordinates

$$\frac{\partial^2 A_\varphi}{\partial r^2} + \frac{1}{r}\frac{\partial A_\varphi}{\partial r} - \frac{A_\varphi}{r^2} + \frac{\partial^2 A_\varphi}{\partial z^2} = -\mu_o J_\varphi \tag{5.9}$$

Thus **A** and **J** have only a φ component and we drop this subscript in the following for simplicity.

Let us write the current density as a Fourier series in terms of a fundamental spatial period of length L, the yoke to yoke distance or window height.

$$J = J_0 + \sum_{n=1}^{\infty} J_n \cos\left(\frac{n\pi z}{L}\right) \tag{5.10}$$

where

$$J_0 = \frac{1}{L}\int_0^L J dz \quad , \quad J_n = \frac{2}{L}\int_0^L J \cos\left(\frac{n\pi z}{L}\right) dz \tag{5.11}$$

For the current density described in Fig. 5.2, we have

$$J = \begin{cases} 0 & , \quad 0 \le z \le z_1 \\ J_1 & , \quad z_1 \le z \le z_2 \\ & \quad \vdots \\ J_i & , \quad z_i \le z \le z_{i+1} \\ & \quad \vdots \\ J_p & , \quad z_p \le z \le z_{p+1} \\ 0 & , \quad z_{p+1} \le z \le L \end{cases} \tag{5.12}$$

Using (5.12), the integrals in (5.11) can be evaluated to get

$$\begin{aligned} J_0 &= \frac{1}{L}\sum_{i=1}^{p} J_i \left(z_{i+1} - z_i\right) \\ J_n &= \frac{2}{n\pi}\sum_{i=1}^{p} J_i \left[\sin\left(\frac{n\pi z_{i+1}}{L}\right) - \sin\left(\frac{n\pi z_i}{L}\right)\right] \end{aligned} \tag{5.13}$$

Thus for one section with constant current density J we would have

$$J_0 = J\frac{(z_2 - z_1)}{L}$$

$$J_n = \frac{2J}{n\pi}\left[\sin\left(\frac{n\pi z_2}{L}\right) - \sin\left(\frac{n\pi z_1}{L}\right)\right] \tag{5.14}$$

As shown in Fig. 5.1, we have divided the solution space into 3 regions:

Region I	$r_c \le r \le r_1$,	$0 \le z \le L$
Region II	$r_1 \le r \le r_2$,	$0 \le z \le L$
Region III	$r_2 \le r \le \infty$,	$0 \le z \le L$

In regions I and III, the current density is 0 so equation (5.9) becomes, dropping the subscript φ

$$\frac{\partial^2 A}{\partial r^2} + \frac{1}{r}\frac{\partial A}{\partial r} - \frac{A}{r^2} + \frac{\partial^2 A}{\partial z^2} = 0 \tag{5.15}$$

This is a homogeneous partial differential equation. We look for a solution of the form

$$A(r,z) = R(r)Z(z) \tag{5.16}$$

Substituting this into (5.15) and dividing by RZ, we get

$$\frac{1}{R}\frac{\partial^2 R}{\partial r^2} + \frac{1}{rR}\frac{\partial R}{\partial r} - \frac{1}{r^2} + \frac{1}{Z}\frac{\partial^2 Z}{\partial z^2} = 0 \tag{5.17}$$

This equation contains terms which are only a function of r and terms which are only a function of z whose sum is a constant = 0. Therefore, each set of terms must separately equal a constant whose sum is zero. Let the constant be m^2, a positive number. Then

$$\frac{1}{R}\frac{\partial^2 R}{\partial r^2} + \frac{1}{rR}\frac{\partial R}{\partial r} - \frac{1}{r^2} = m^2 \quad , \quad \frac{1}{Z}\frac{\partial^2 Z}{\partial z^2} = -m^2 \tag{5.18}$$

Rearranging terms, we get

$$r^2\frac{\partial^2 R}{\partial r^2}+r\frac{\partial R}{\partial r}-R\left(m^2r^2+1\right)=0 \quad , \quad \frac{\partial^2 Z}{\partial z^2}+m^2 Z=0 \tag{5.19}$$

We need to consider separately the cases $m = 0$ and $m > 0$.

For $m = 0$, the solution to the z equation which satisfies the boundary conditions at the top and bottom yokes is a constant and the r equation in (5.19) becomes

$$r^2\frac{\partial^2 R}{\partial r^2}+r\frac{\partial R}{\partial r}-R=0 \tag{5.20}$$

The solution to this equation is

$$R_0 = Sr+\frac{T}{r} \tag{5.21}$$

where S and T are constants to be determined by the boundary conditions.

For $m > 0$, the solution to the z equation in (5.19) can be written

$$Z = Z_m\cos\left(mz+\varphi_m\right) \tag{5.22}$$

where Z_m and φ_m are constants to be determined by the boundary conditions. Since we assume that the yoke material has infinite permeability, this requires that the B-field be perpendicular to the yoke surfaces. Using (5.3), we have in cylindrical coordinates

$$\mathbf{B} = -\frac{\partial A}{\partial z}\mathbf{a}_r+\left(\frac{\partial A}{\partial r}+\frac{A}{r}\right)\mathbf{k} \tag{5.23}$$

where $\mathbf{a}_r$ and $\mathbf{k}$ are unit vectors in the r and z directions. For $\mathbf{B}$ to be perpendicular to the upper and lower yokes, we must have

$$\frac{\partial A}{\partial z}=0 \quad \Rightarrow \quad \frac{\partial Z}{\partial z}=0 \quad \text{at} \quad z=0,L \tag{5.24}$$

Using this, (5.22) takes the form

$$Z = Z_n\cos\left(mz\right) \quad , \quad m=\frac{n\pi}{L} \quad , \quad n=1,2,\cdots \tag{5.25}$$

Here we use n rather than m to label the constant. m also depends on n but we omit this reference for simplicity.

For $m > 0$, the radial equation in (5.19) can be written, with the substitution $x = mr$,

$$x^2 \frac{\partial^2 R}{\partial x^2} + x\frac{\partial R}{\partial x} - R\left(x^2 + 1\right) = 0 \tag{5.26}$$

The solution to this equation is

$$R_n = C_n I_1(x) + D_n K_1(x) \tag{5.27}$$

where I_1 and K_1 are modified Bessel functions of the first and second kind respectively and of order 1 [Dwi61]. We have also labeled the constants C and D with the subscript n since the solution depends on n through m which occurs in x.

In general, the solution to (5.15) is expressible as a sum of these individual solutions, each a product of an R and Z term,

$$A = Sr + \frac{T}{r} + \sum_{n=1}^{\infty}\left[C_n I_1(mr) + D_n K_1(mr)\right]\cos(mz) \quad , \quad m = \frac{n\pi}{L} \tag{5.28}$$

where the Z solution constants have been absorbed in the overall constants shown. This solution satisfies the boundary condition at $z = 0$, L. Because we also assume an infinitely permeable core, the B-field must be normal to the core surface. Thus from (5.23), we require in Region I

$$\frac{\partial A}{\partial r} + \frac{A}{r} = 0 \qquad \text{at} \qquad r = r_c \tag{5.29}$$

Substituting (5.28) into this last equation, we get

$$2S + \sum_{n=1}^{\infty}\left[mC_n \frac{\partial I_1}{\partial x} + mD_n \frac{\partial K_1}{\partial x} + m\frac{(C_n I_1 + D_n K_1)}{x}\right]\cos(mz) = 0 \tag{5.30}$$

at $x = mr_c$

Although (5.30) would seem to require that $S = 0$, it will turn out that when all the windings are considered with their Ampere-turns which sum to zero, we can satisfy this boundary condition with $S \neq 0$ for each winding. For the remaining z-dependent terms in (5.30), we require

$$mC_n\left(\frac{\partial I_1}{\partial x}+\frac{I_1}{x}\right)+mD_n\left(\frac{\partial K_1}{\partial x}+\frac{K_1}{x}\right) = mC_n I_0 - mD_n K_0 = 0 \tag{5.31}$$

where we have used modified Bessel function identities [Rab56]. I_0 and K_0 are modified Bessel functions of order 0. Thus we have from (5.31)

$$D_n = C_n \frac{I_0(mr_c)}{K_0(mr_c)} \tag{5.32}$$

Labeling the unknown constants with the region number as superscript, (5.28) becomes

$$A^I = S^I r + \sum_{n=1}^{\infty} C_n^I \left[I_1(mr) + \frac{I_0(mr_c)}{K_0(mr_c)} K_1(mr) \right] \cos(mz) \tag{5.33}$$

We have dropped the T/r term since it approaches infinity as the core radius approaches zero. We also assume implicitly that m depends on n as in (5.28).

In Region III, we require that A be finite as $r \to \infty$. Since $I_1 \to \infty$ as $r \to \infty$ and the Sr term also $\to \infty$, we have from (5.28), using the appropriate region label,

$$A^{III} = \frac{T^{III}}{r} + \sum_{n=1}^{\infty} D_n^{III} K_1(mr) \cos(mz) \tag{5.34}$$

In Region II, we must keep the current density term in (5.9). Substituting the Fourier series (5.10) into (5.9) and dropping the φ subscript, we have in Region II,

$$\frac{\partial^2 A}{\partial r^2} + \frac{1}{r}\frac{\partial A}{\partial r} - \frac{A}{r^2} + \frac{\partial^2 A}{\partial z^2} = -\mu_o \left(J_0 + \sum_{n=1}^{\infty} J_n \cos(mz) \right) \tag{5.35}$$

We look for a solution to this equation in the form of a series expansion,

$$A = \sum_{n=0}^{\infty} R_n(r)\cos(mz) \quad , \quad m = \frac{n\pi}{L} \tag{5.36}$$

Substituting into (5.35), we get

$$\begin{gathered}\sum_{n=0}^{\infty}\left(\frac{\partial^2 R_n}{\partial r^2} + \frac{1}{r}\frac{\partial R_n}{\partial r} - \frac{R_n}{r^2}\right)\cos(mz) - m^2\sum_{n=1}^{\infty} R_n\cos(mz) \\ = -\mu_o\left(J_0 + \sum_{n=1}^{\infty} J_n\cos(mz)\right)\end{gathered} \tag{5.37}$$

Since the cosine functions are orthogonal, we can equate corresponding coefficients on both sides of this equation. We obtain

$$\begin{gathered}\frac{\partial^2 R_0}{\partial r^2} + \frac{1}{r}\frac{\partial R_0}{\partial r} - \frac{R_0}{r^2} = -\mu_0 J_0 \\ \frac{\partial^2 R_n}{\partial r^2} + \frac{1}{r}\frac{\partial R_n}{\partial r} - \frac{R_n}{r^2} - m^2 R_n = -\mu_o J_n\end{gathered} \tag{5.38}$$

The solution to the $n = 0$ equation can be written in terms of a solution of the homogeneous equation plus a particular solution,

$$R_0 = Sr + \frac{T}{r} - \frac{\mu_o J_0 r^2}{3} \tag{5.39}$$

The solution to the $n > 0$ equations in (5.38) consists of a homogeneous solution which was found previously and a particular solution,

$$R_n = C_n I_1(mr) + D_n K_1(mr) - \frac{\pi\mu_o J_n}{2m^2} L_1(mr) \tag{5.40}$$

where L_1 is a modified Struve function of order 1 [Abr72]. Thus the solution (5.36) is given explicitly as

$$\begin{gathered}A^{II} = S^{II} r + \frac{T^{II}}{r} - \frac{\mu_o J_0 r^2}{3} \\ + \sum_{n=1}^{\infty}\left[C_n^{II} I_1(mr) + D_n^{II} K_1(mr) - \frac{\pi\mu_o J_n}{2m^2} L_1(mr)\right]\cos(mz)\end{gathered} \tag{5.41}$$

This equation already satisfies the boundary conditions at $z = 0, L$. The unknown constants must be determined by satisfying the boundary conditions ar $r = r_1$, r_2 . We require that the vector potential be continuous across the interfaces. Otherwise the B-field given by (5.23) would contain infinities. Thus at $r = r_1$, using (5.33) and (5.41),

$$S^I r_1 + \frac{T^I}{r} + \sum_{n=1}^{\infty} C_n^I \left[I_1(mr_1) + \frac{I_0(mr_c)}{K_0(mr_c)} K_1(mr_1) \right] \cos(mz)$$
$$= S^{II} r_1 + \frac{T^{II}}{r_1} - \frac{\mu_o J_0 r_1^2}{3} \tag{5.42}$$
$$+ \sum_{n=1}^{\infty} \left[C_n^{II} I_1(mr_1) + D_n^{II} K_1(mr_1) - \frac{\pi \mu_o J_n}{2m^2} L_1(mr_1) \right] \cos(mz)$$

Since this must be satisfied for all z , we obtain

$$S^I r_1 = S^{II} r_1 + \frac{T^{II}}{r_1} - \frac{\mu_o J_0 r_1^2}{3}$$
$$\tag{5.43}$$
$$C_n^I \left[I_1(x_1) + \frac{I_0(x_c)}{K_0(x_c)} K_1(x_1) \right] = C_n^{II} I_1(x_1) + D_n^{II} K_1(x_1) - \frac{\pi \mu_o J_n}{2m^2} L_1(x_1)$$

where $x_1 = mr_1$, $x_c = mr_c$.

At $r = r_2$, we obtain similarly, using (5.34) and (5.41),

$$\frac{T^{III}}{r_2} = S^{II} r_2 + \frac{T^{II}}{r_2} - \frac{\mu_o J_0 r_2^2}{3}$$
$$\tag{5.44}$$
$$D_n^{II} K_1(x_2) = C_n^{II} I_1(x_2) + D_n^{II} K_1(x_2) - \frac{\pi \mu_o J_n}{2m^2} L_1(x_2)$$

where $x_2 = mr_2$.

In addition to the continuity of A at the interfaces between regions, we also require, according to Maxwell's equations, that the normal component of **B** and the tangential component of **H** be continuous across these interfaces. According to (5.23), the normal **B** components are already continuous across these interfaces since all regional solutions have the same z dependence and because the A's are continuous. Since **B** is proportional to **H** in all regions of interest here, we require that the tangential **B** components be continuous. Thus from (5.23), we require that

$$\frac{\partial A}{\partial r}+\frac{A}{r} = \frac{1}{r}\frac{\partial}{\partial r}(rA) = \frac{m}{x}\frac{\partial}{\partial x}(xA) \tag{5.45}$$

be continuous at $r = r_1$, r_2 $(x = x_1$, $x_2)$. Using this, we obtain the additional conditions on the unknown constants

$$2S^{I} = 2S^{II} - \mu_o J_0 r_1 \quad , \quad 2S^{II} - \mu_o J_0 r_2 = 0$$

$$C_n^{I}\left\{\frac{\partial}{\partial x}\left[xI_1(x)\right]+\frac{I_0(x_c)}{K_0(x_c)}\frac{\partial}{\partial x}\left[xK_1(x)\right]\right\}$$
$$= C_n^{II}\frac{\partial}{\partial x}\left[xI_1(x)\right]+D_n^{II}\frac{\partial}{\partial x}\left[xK_1(x)\right]-\frac{\pi\mu_o J_n}{2m^2}\frac{\partial}{\partial x}\left[xL_1(x)\right] \quad \text{at} \quad x = x_1 = mr_1$$

$$C_n^{II}\frac{\partial}{\partial x}\left[xI_1(x)\right]+D_n^{II}\frac{\partial}{\partial x}\left[xK_1(x)\right]-\frac{\pi\mu_o J_n}{2m^2}\frac{\partial}{\partial x}\left[xL_1(x)\right]$$
$$= D_n^{III}\frac{\partial}{\partial x}\left[xK_1(x)\right] \quad \text{at} \quad x = x_2 = mr_2$$

(5.46)

Using the identities [Rab56, Abr72],

$$\frac{\partial}{\partial x}\left[xI_1(x)\right] = xI_0(x) \quad , \quad \frac{\partial}{\partial x}\left[xK_1(x)\right] = -xK_0(x) \quad , \quad \frac{\partial}{\partial x}\left[xL_1(x)\right] = xL_0(x) \tag{5.47}$$

we obtain for the last two equations in (5.46)

$$C_n^{I}\left[I_0(x_1)-\frac{I_0(x_c)}{K_0(x_c)}K_0(x_1)\right] = C_n^{II}I_0(x_1)-D_n^{II}K_0(x_1)-\frac{\pi\mu_o J_n}{2m^2}L_0(x_1)$$

(5.48)

$$C_n^{II}I_0(x_2)-D_n^{II}K_0(x_2)-\frac{\pi\mu_o J_n}{2m^2}L_0(x_2) = -D_n^{III}K_0(x_2)$$

Solving for the S and T constants from (5.43), (5.44), and (5.46), we obtain

$$S^{I} = \frac{\mu_o J_0 (r_2 - r_1)}{2}, \quad S^{II} = \frac{\mu_o J_0 r_2}{2}$$
$$T^{II} = -\frac{\mu_o J_0 r_1^3}{6}, \quad T^{III} = \frac{\mu_o J_0 (r_2^3 - r_1^3)}{6} \tag{5.49}$$

Solving for the C_n and D_n constants from (5.43), (5.44), and (5.46), we get

$$C_n^{II} = \frac{\pi \mu_o J_n}{2m^2} \left[\frac{K_0(x_2)L_1(x_2) + K_1(x_2)L_0(x_2)}{K_0(x_2)I_1(x_2) + K_1(x_2)I_0(x_2)} \right]$$

$$C_n^{I} = C_n^{II} - \frac{\pi \mu_o J_n}{2m^2} \left[\frac{K_0(x_1)L_1(x_1) + K_1(x_1)L_0(x_1)}{K_0(x_1)I_1(x_1) + K_1(x_1)I_0(x_1)} \right]$$

$$D_n^{II} = \frac{I_0(x_c)}{K_0(x_c)} C_n^{I} + \frac{\pi \mu_o J_n}{2m^2} \left[\frac{I_0(x_1)L_1(x_1) - I_1(x_1)L_0(x_1)}{K_0(x_1)I_1(x_1) + K_1(x_1)I_0(x_1)} \right] \tag{5.50}$$

$$D_n^{III} = D_n^{II} - \frac{\pi \mu_o J_n}{2m^2} \left[\frac{I_0(x_2)L_1(x_2) - I_1(x_2)L_0(x_2)}{K_0(x_2)I_1(x_2) + K_1(x_2)I_0(x_2)} \right]$$

Using the identities [Rab56],

$$I_0(x)K_1(x) + I_1(x)K_0(x) = \frac{1}{x}$$

$$x\left[L_0(x)K_1(x) + L_1(x)K_0(x)\right] = \frac{2}{\pi} \int_0^x t K_1(t) dt \tag{5.51}$$

$$x\left[L_0(x)I_1(x) - L_1(x)I_0(x)\right] = \frac{2}{\pi} \int_0^x t I_1(t) dt$$

we can transform (5.50) into the form

$$C_n^{II} = \frac{\mu_o J_n}{m^2} \int_0^{x_2} tK_1(t)dt$$

$$C_n^{I} = \frac{\mu_o J_n}{m^2} \int_{x_1}^{x_2} tK_1(t)dt$$

$$D_n^{II} = \frac{\mu_o J_n}{m^2} \left[\frac{I_0(x_c)}{K_0(x_c)} \int_{x_1}^{x_2} tK_1(t)dt - \int_0^{x_1} tI_1(t)dt \right] \tag{5.52}$$

$$D_n^{III} = \frac{\mu_o J_n}{m^2} \left[\frac{I_0(x_c)}{K_0(x_c)} \int_{x_1}^{x_2} tK_1(t)dt + \int_{x_1}^{x_2} tI_1(t)dt \right]$$

Summarizing and simplifying the notation slightly, the solutions in the three regions are given by

$$A^{I} = \frac{\mu_o J_0 (r_2 - r_1)}{2} r + \mu_o \sum_{n=1}^{\infty} \frac{J_n}{m^2} \left[C_n I_1(x) + D_n K_1(x) \right] \cos(mz)$$

$$A^{II} = \mu_o J_0 \left(\frac{r_2 r}{2} - \frac{r_1^3}{6r} - \frac{r^2}{3} \right) + \mu_o \sum_{n=1}^{\infty} \frac{J_n}{m^2} \left[E_n I_1(x) + F_n K_1(x) - \frac{\pi}{2} L_1(x) \right] \cos(mz) \tag{5.53}$$

$$A^{III} = \frac{\mu_o J_0 (r_2^3 - r_1^3)}{6r} + \mu_o \sum_{n=1}^{\infty} \frac{J_n}{m^2} G_n K_1(x) \cos(mz)$$

where $m = n\pi/L$, $x = mr$, and

$$C_n = \int_{x_1}^{x_2} tK_1(t)dt \quad , \quad D_n = \frac{I_0(x_c)}{K_0(x_c)} C_n$$

$$E_n = \int_0^{x_2} tK_1(t)dt \quad , \quad F_n = \frac{I_0(x_c)}{K_0(x_c)} \int_{x_1}^{x_2} tK_1(t)dt - \int_0^{x_1} tI_1(t)dt \quad (5.54)$$

$$G_n = \frac{I_0(x_c)}{K_0(x_c)} \int_{x_1}^{x_2} tK_1(t)dt + \int_{x_1}^{x_2} tI_1(t)dt$$

Here $x_1 = mr_1$ and $x_2 = mr_2$.

Note that, using (5.23), the axial component of the induction vector at the core radius is

$$B_z(r_c, z) = \mu_o J_0 (r_2 - r_1) \quad (5.55)$$

This is μ_o times the average current density times the radial build of the winding. Since the axial height L is the same for all the windings, when we add this axial component for all the windings we will get zero, assuming Ampere-turn balance. Thus the flux will enter the core radially as required.

Although the modified Bessel functions are generally available in mathematical computer libraries, the modified Struve functions are not so easily obtained. We therefore indicate here some methods of obtaining these and other quantities of interest in terms of readily available functions or easily evaluated integrals. From reference [Abr72], the modified Struve functions are given in integral form as

$$L_0(x) = \frac{2}{\pi} \int_0^{\pi/2} \sinh(x\cos\theta)d\theta$$

$$(5.56)$$

$$L_1(x) = \frac{2x}{\pi} \int_0^{\pi/2} \sinh(x\cos\theta)\sin^2\theta d\theta$$

It can be seen from (5.56) that these functions become asymptotically large as $x \to \infty$. Such large values of x would occur when evaluating the higher harmonics in (5.53). Ref. [Abr72] also gives integral expressions for the modified Bessel functions. In particular, we have

$$I_0(x)=\frac{1}{\pi}\int_0^{\pi} e^{x\cos\theta}d\theta \quad , \quad I_1(x)=\frac{x}{\pi}\int_0^{\pi} e^{x\cos\theta}\sin^2\theta d\theta \tag{5.57}$$

These also become asymptotically large as x increases. However, by defining the difference functions

$$M_0(x)=I_0(x)-L_0(x) \quad , \quad M_1(x)=I_1(x)-L_1(x) \tag{5.58}$$

we can show that

$$M_0(x)=\frac{2}{\pi}\int_0^{\pi/2} e^{-x\cos\theta}d\theta \quad , \quad M_1(x)=\frac{2}{\pi}\left(1-\int_0^{\pi/2} e^{-x\cos\theta}\cos\theta d\theta\right) \tag{5.59}$$

The integrals in (5.59) are well behaved as x increases and can be evaluated numerically. In subsequent developments, we will also need the integral of M_0. This is readily determined from (5.59) as

$$\int_0^x M_0(t)dt=\frac{2}{\pi}\int_0^{\pi/2}\frac{\left(1-e^{-x\cos\theta}\right)}{\cos\theta}d\theta \tag{5.60}$$

The integrand approaches x as $\theta \to \pi/2$ and so the integral is also well behaved. For high x values, asymptotic series can be found in Ref. [Abr72].

Let us now write other expressions of interest in terms of the M functions and modified Bessel functions. Using (5.51) and (5.58), we can write

$$\int_0^x tI_1(t)dt=\frac{\pi}{2}x\left[M_1(x)I_0(x)-M_0(x)I_1(x)\right]$$

$$\int_0^x tK_1(t)dt=\frac{\pi}{2}\left\{1-x\left[M_1(x)K_0(x)+M_0(x)K_1(x)\right]\right\} \tag{5.61}$$

We will also need a similar integral for the modified Struve function. We need the following identities given in [Rab56].

$$\frac{d}{dt}L_0(t) = \frac{2}{\pi} + L_1(t) \quad , \quad \frac{d}{dt}I_0(t) = I_1(t) \tag{5.62}$$

Multiplying these by t and integrating from 0 to x , we get

$$\begin{aligned} xL_0(x) - \int_0^x L_0(t)dt &= \frac{x^2}{\pi} + \int_0^x tL_1(t)dt \\ xI_0(x) - \int_0^x I_0(t)dt &= \int_0^x tI_1(t)dt \end{aligned} \tag{5.63}$$

Substituting L_0 in terms of the M_0 function and rearranging, we get

$$\int_0^x tL_1(t)dt = -xM_0(x) - \frac{x^2}{\pi} + \int_0^x M_0(t)dt + \int_0^x tI_1(t)dt \tag{5.64}$$

Thus all the functions needed to determine the vector potential are obtainable in terms of the modified Bessel functions and the M functions. We will also need (5.64) in obtaining derived quantities in the following sections.

The determination of the vector potential for the other coils or coil sections proceeds identically to the previous development. Because Maxwell's equations are linear in the fields and potentials in the region outside the core and yokes, we can simply add the potentials or fields from the various coils vectorially at each point to get the net potential or field. We need to be aware of the fact that a given point may not be in the same region number (I, II, or III) for the different coils.

5.3 DETERMINING THE B-FIELD

The B-field or induction vector is given by (5.23). We need to evaluate it in the three regions using (5.53). The radial component is, using the region label as superscript,

$$B_r^I = \mu_o \sum_{n=1}^{\infty} \frac{J_n}{m}\left[C_n I_1(x) + D_n K_1(x)\right]\sin(mz)$$

$$B_r^{II} = \mu_o \sum_{n=1}^{\infty} \frac{J_n}{m}\left[E_n I_1(x) + F_n K_1(x) - \frac{\pi}{2} L_1(x)\right]\sin(mz) \tag{5.65}$$

$$B_r^{III} = \mu_o \sum_{n=1}^{\infty} \frac{J_n}{m} G_n K_1(x)\sin(mz)$$

Using (5.47) and

$$B_z = \frac{1}{r}\frac{\partial}{\partial r}(rA) = \frac{m}{x}\frac{\partial}{\partial x}(xA) \tag{5.66}$$

the axial component is,

$$B_z^I = \mu_o J_0(r_2 - r_1) + \mu_o \sum_{n=1}^{\infty} \frac{J_n}{m}\left[C_n I_0(x) - D_n K_0(x)\right]\cos(mz)$$

$$\begin{aligned} B_z^{II} &= \mu_o J_0(r_2 - r) \\ &\quad + \mu_o \sum_{n=1}^{\infty} \frac{J_n}{m}\left[E_n I_0(x) - F_n K_0(x) - \frac{\pi}{2} L_0(x)\right]\cos(mz) \end{aligned} \tag{5.67}$$

$$B_z^{III} = -\mu_o \sum_{n=1}^{\infty} \frac{J_n}{m} G_n K_0(x)\cos(mz)$$

To find the net B-field associated with a collection of coils, we simply add their components at the point in question, keeping in mind what region the point is in relative to each coil.

5.4 DETERMINING THE WINDING FORCES

The force density vector, $\mathbf{F}$, in SI units (Newtons/m^3), is given by

$$\mathbf{F} = \mathbf{J} \times \mathbf{B} \tag{5.68}$$

Since $\mathbf{J}$ is azimuthal and $\mathbf{B}$ has only r and z components, this reduces to

$$\mathbf{F} = JB_z\mathbf{a}_r - JB_r\mathbf{k} \tag{5.69}$$

where $\mathbf{a}_r$ is the unit vector in the radial direction and $\mathbf{k}$ the unit axial vector. We have omitted the φ subscript on J. Thus the radial forces are due to the axial field component and vice versa.

In (5.69), the B-field values are the resultant from all the coils and J is the current density at the point in question. This force density is non-zero only over those parts of the winding which carry current. In order to obtain net forces over all or part of a winding, it is necessary to integrate (5.69) over the winding or winding part. For this purpose, the winding can be subdivided into as fine a mesh as desired and $\mathbf{F}$ computed at the centroids of these subdivisions and the resulting values times the mesh volume element added.

One force which is useful to know is the compressive force which acts axially at each axial position in the winding. We assume that the windings are constrained at the two ends by pressure rings of some type. We ignore gravity here. Starting from the bottom, we integrate the axial forces upwards along the winding, stopping when the sum of the downward acting forces reach a maximum. This is the net downward force on the bottom pressure ring which is countered by an equal upward force on the winding exerted by the pressure ring. We do the same thing, starting from the top of the winding, integrating the axial forces until the largest upwards acting summed force is reached. This then constitutes the net upward force acting on the top pressure ring which is countered by an equal downward force exerted by the pressure ring. Including the reaction forces of the pressure rings, we then integrate upwards say, starting at the bottom of the winding and at each vertical position, the force calculated will be the compressive force at that position since there will be an equal and opposite force acting from above. This is illustrated in Fig. 5.3a.

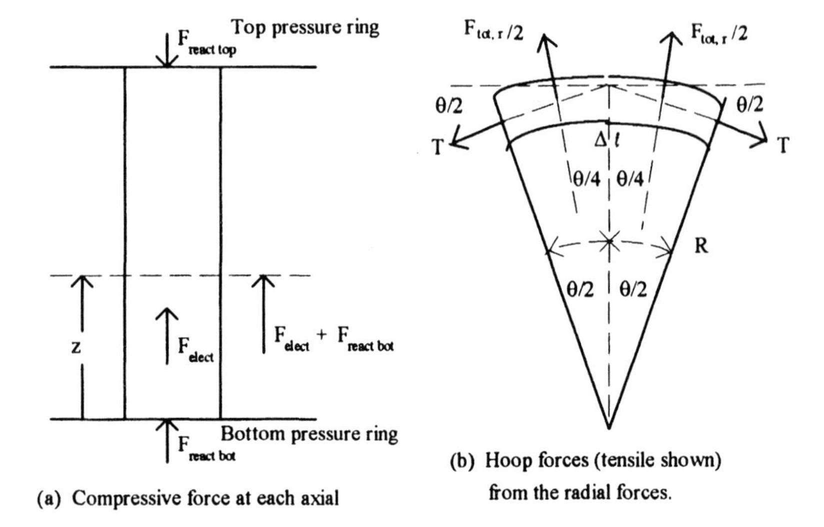

Figure 5.3 Forces on the winding needed for stress analysis

Another force of interest for coil design is the radial force which results in hoop stress, tensile or compressive, on the winding. Because of the cylindrical symmetry, the radial forces vectorially add to zero. We must therefore handle them a bit differently if we are to arrive at a useful resultant for hoop stress calculations. In Fig. 5.3b, we isolate a small portion of the winding and show the forces acting in a horizontal section. We show the radial forces acting outward and the tensile forces applied by the missing part of the winding which are necessary to maintain equilibrium. If the radial forces acted inward, then the rest of the winding would need to apply compressive forces to maintain equilibrium and all the force arrows in the figure would be reversed. The calculation would, however, proceed similarly.

In Fig. 5.3b, θ is assumed to be a very small angle. The force balance in the upward direction in the figure is

$$2\frac{F_{tot,r}}{2}\cos\left(\frac{\theta}{4}\right) = 2T\sin\left(\frac{\theta}{2}\right) \tag{5.70}$$

$F_{tot,\,r}$ is the sum of the radial electromagnetic forces acting on the winding section shown and T is the total tensile force acting on the winding cross-section. For small θ, this reduces to

$$F_{tot,r} = T\theta \tag{5.71}$$

In radians, $\theta = \Delta\ell / R$, where $\Delta\ell$ is the circumferential length of the winding section shown and R its average radius. The average tensile stress is $\sigma = T / A$, where A is the cross-sectional area of the winding section. Noting that the volume of the winding section is $\Delta V = A\,\Delta\ell$. In the limit of small θ, (5.71) becomes

$$\sigma = R \frac{F_{tot,r}}{\Delta V} \tag{5.72}$$

Thus the average tensile stress in the winding is the average radius times the average radial force density, assuming the radial forces act outward. For inward radial forces, σ is compressive. The average radial force density at a particular axial position z can be found by computing the radial component of (5.69) at several radial positions in the winding, keeping z fixed, and taking their average. One should weight these by the volume element which is proportional to r.

Other forces of interest can be found from the known force densities.

5.5 GENERAL METHOD FOR DETERMINING INDUCTANCES AND MUTUAL INDUCTANCES

Self and leakage inductances are usually defined in terms of flux linkages between circuits. However, it is often more convenient to calculate them in terms of magnetic energy. In this section, we show the equivalence of these methods and develop a useful formula for the determination of these inductances in terms of the vector potential. In addition, we show how the vector potential can be employed to calculate mutual inductances.

Consider a set of stationary circuits as shown in Fig. 5.4a. We assume that we slowly increase the currents in them by means of batteries with variable control. In calculating the work done by the batteries, we ignore any I^2R or dissipative losses or, more realistically, we treat these separately. Thus the work we are interested in is the work necessary to establish the magnetic field. Because of the changing flux linking the different circuits, emf's will be induced in them by Faraday's law. We assume that the battery controllers are adjusted so that the battery voltages just balance the induced voltages throughout the process.

Because the circuits can have finite cross sectional areas as shown in the figure, we imagine subdividing them into infinitesimal circuits or tubes carrying an incremental part of the total current, dI, as in Fig. 5.4b.

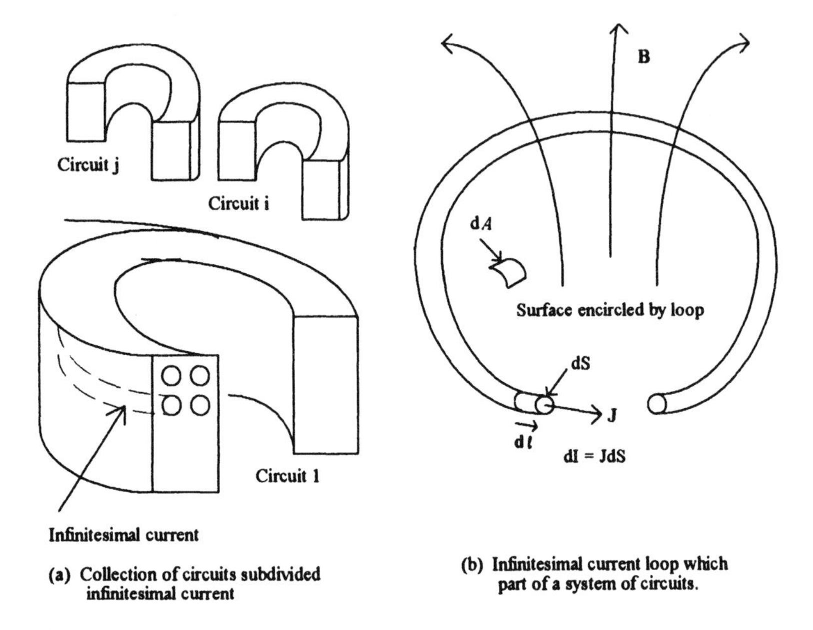

Figure 5.4 Method of circuit subdivision used to calculate self and mutual inductances

The incremental energy, dW , which the batteries supply during a time interval, dt , is

$$dW = dt\int VdI = dt\int_{\text{cross-sectional area of circuits}} VJdS \tag{5.73}$$

Here the integrand is the infinitesimal power flowing through the infinitesimal circuit or loop carrying current dI and along which a voltage V is induced. The integral simply represents the sum of these powers. We have also substituted the current density J times the infinitesimal cross-sectional area dS for the incremental current in the loop so that the last integral is over the cross-sectional area of the circuits. From Faraday's law for a single turn circuit

$$V = \frac{d\Phi}{dt} \tag{5.74}$$

where Φ is the flux linked by the circuit. Substituting into (5.73)

$$dW = \int_{\text{cross-sectional area of circuits}} d\Phi J dS \tag{5.75}$$

By definition

$$\Phi = \int_{\text{area of loop}} \mathbf{B}\cdot\mathbf{n}dA = \int_{\text{area of loop}} (\nabla\times\mathbf{A})\cdot\mathbf{n}dA = \oint_{\text{along loop}} \mathbf{A}\cdot\mathbf{d}\ell \tag{5.76}$$

In (5.76), the surface integral is over the surface encircled by the loop and the line integral is along the loop as shown in Fig. 5.4b. $\mathbf{n}$ is the unit normal to the surface encircled by the loop and $\mathbf{d}\ell$ is the infinitesimal distance vector along the loop. Note that $\mathbf{d}\ell$ points in the same direction as $\mathbf{J}$ considered as a vector. Since the current loops are fixed, we get from (5.76)

$$d\Phi = \oint_{\text{along loop}} \mathbf{dA}\cdot\mathbf{d}\ell \tag{5.77}$$

Substituting this into (5.75), we get

$$dW = \int_{\text{cross-sectional area of circuits}} \left[\oint_{\text{along loop}} \mathbf{dA}\cdot\mathbf{d}\ell\right] J dS \tag{5.78}$$

Since $\mathbf{J}$ points along $\mathbf{d}\ell$ and since the volume element $dV = d\ell dS$, we can rewrite (5.78)

$$dW = \int_{\text{cross-sectional area of circuits}} \left[\oint_{\text{along loop}} \mathbf{dA}\cdot\mathbf{J}d\ell dS\right] = \int_{\text{volume of circuits}} \mathbf{dA}\cdot\mathbf{J}dV \tag{5.79}$$

In changing to a volume integral, we are simply recognizing the fact that the integral over the cross-sectional area of the circuits combined with an integral along the lengths of the infinitesimal circuits amounts to an integration over the volume of the circuits. If $\mathbf{A}$ increases linearly with $\mathbf{J}$, it can be shown that the total work done to establish the final field values is

$$W = \int_{\text{volume of circuits}} dV \int_0^A \mathbf{J} \cdot \mathbf{dA} = \frac{1}{2} \int_{\text{volume of circuits}} \mathbf{A} \cdot \mathbf{J} dV \tag{5.80}$$

It is sometimes more convenient for calculational purposes to express (5.80), or more generally (5.79), differently. First of all the volume integration in both of these equations could be taken over all space since the current density is zero everywhere except within the circuits. (By all space we mean the solution space of the problem of interest, ignoring the rest of the universe.) Substituting (5.1) into (5.79), we get

$$dW = \int_{\text{all space}} \mathbf{dA} \cdot (\nabla \times \mathbf{H}) dV \tag{5.81}$$

Using the vector identity for general vector fields $\mathbf{P}$, $\mathbf{Q}$ [Pug62],

$$\nabla \cdot (\mathbf{P} \times \mathbf{Q}) = \mathbf{Q} \cdot (\nabla \times \mathbf{P}) - \mathbf{P} \cdot (\nabla \times \mathbf{Q}) \tag{5.82}$$

we have, upon substitution into (5.81) with the identification $\mathbf{P} \to \mathbf{H}$, $\mathbf{Q} \to \mathbf{dA}$,

$$\begin{aligned} dW &= \int_{\text{all space}} \mathbf{H} \cdot (\nabla \times \mathbf{dA}) dV + \int_{\text{all space}} \nabla \cdot (\mathbf{H} \times \mathbf{dA}) dV \\ &= \int_{\text{all space}} \mathbf{H} \cdot \mathbf{dB} dV + \int_{\text{boundary surface}} (\mathbf{H} \times \mathbf{dA}) \cdot \mathbf{n} dS \end{aligned} \tag{5.83}$$

In this equation, we have used the definition of A given in (5.3) and the Divergence Theorem [Pug62] to convert the last volume integral into a surface integral. The surface integral generally vanishes since the boundary surface is usually at infinity where the fields drop to zero. In the case of Rabins' solution, it also vanishes on the core and yoke surfaces because of the boundary conditions. Thus we can drop it, to get for the situation where $\mathbf{B}$ increases linearly with $\mathbf{H}$

$$W = \int_{\text{all space}} dV \int_0^B \mathbf{H} \cdot \mathbf{dB} = \frac{1}{2} \int_{\text{all space}} \mathbf{H} \cdot \mathbf{B} dV \tag{5.84}$$

Either (5.80) or (5.84) can be used to calculate the magnetic energy. The second equality in both equations applies to linear systems.

We now show how the magnetic energy can be related to inductances or leakage inductances. The voltages induced in the circuits of Fig. 5.4a can be written in terms of inductances and mutual inductances as

$$V_i = \sum_j M_{ij} \frac{dI_j}{dt} \tag{5.85}$$

where M_{ij} is the mutual inductance between circuits i and j. When j = i, $M_{ii} = L_i$, where L_i is the self inductance of circuit i. The incremental energy for the circuits of Fig. 5.4a can be expressed, using (5.85),

$$dW = \sum_i V_i I_i dt = \sum_{i,j} M_{ij} I_i dI_j = \sum_i L_i I_i dI_i + \sum_{i \neq j} M_{ij} I_i dI_j \tag{5.86}$$

For linear systems, it can be shown that $M_{ij} = M_{ji}$. Thus the second sum in (5.86) can be written

$$\begin{aligned} \sum_{i \neq j} M_{ij} I_i dI_j &= \frac{1}{2}\left(\sum_{i \neq j} M_{ij} I_i dI_j + \sum_{i \neq j} M_{ji} I_j dI_i \right) \\ &= \frac{1}{2} \sum_{i \neq j} M_{ij} \left(I_i dI_j + I_j dI_i \right) \\ &= \sum_{i<j} M_{ij} \left(I_i dI_j + I_j dI_i \right) = \sum_{i<j} M_{ij} d\left(I_i I_j \right) \end{aligned} \tag{5.87}$$

The first equality in this equation is a matter of changing the index labels in the sum and the second equality results from the symmetric nature of M_{ij}. The third equality results from the double counting which happens when i and j are summed independently. Substituting (5.87) into (5.86), we get

$$dW = \frac{1}{2} \sum_i L_i d\left(I_i^2\right) + \sum_{i<j} M_{ij} d\left(I_i I_j\right) \tag{5.88}$$

Integrating this last equation, we obtain

$$W = \frac{1}{2} \sum_i L_i I_i^2 + \sum_{i<j} M_{ij} I_i I_j \tag{5.89}$$

Thus (5.89) is another expression for the magnetic energy in terms of inductances and mutual inductances. As an example, consider a single circuit for which we want to know the self-inductance. From (5.89),

$$L_1 = \frac{2W}{I_1^2} \tag{5.90}$$

where W can be calculated from (5.80) or (5.84) with only the single circuit in the geometry. W is usually available from finite element codes. If we have 2 circuits, then (5.89) becomes

$$L_1 I_1^2 + L_2 I_2^2 + 2M_{12} I_1 I_2 = 2W \tag{5.91}$$

If these two circuits are the high (label 1) and low (label 2) voltage coils of a 2 winding transformer with N_1 and N_2 turns respectively, then we have $I_2 / I_1 = - N_1 / N_2$ so that (5.91) becomes

$$L_1 + \left(\frac{N_1}{N_2}\right)^2 L_2 - 2\left(\frac{N_1}{N_2}\right) M_{12} = \frac{2W}{I_1^2} \tag{5.92}$$

The expression on the left is the leakage inductance referred to the high voltage side [MIT43]. The magnetic energy on the right side can be calculated from (5.80) or (5.84) with the two windings in the geometry and with the Ampere-turns balanced. Multiplying by $\omega = 2\pi f$ produces the leakage impedance.

The mutual inductance between two circuits is needed in detailed circuit models of transformers where the circuits of interest are sections of the same or different coils. We can obtain these quantities in terms of the vector potential. By definition, the mutual inductance between circuits 1 and 2 , M_{12} , is the flux produced by circuit 2 which links circuit 1 divided by the current in circuit 2. 1 and 2 could be interchanged in this definition without changing the value of the mutual inductance for linear systems. We use the infinitesimal loop approach as illustrated in Fig. 5.4 to obtain this quantity. Each infinitesimal loop represents an infinitesimal fraction of a turn. For circuit 1 , an infinitesimal loop carrying current dI_1 represents a fractional turn dI_1 / I_1 . Therefore the flux linkage between this infinitesimal turn due to the flux produced by circuit 2, $d\lambda_{12}$, is

$$d\lambda_{12} = \frac{dI_1}{I_1} \int\limits_{\substack{\text{surface enclosed by} \\ \text{infinitesimal loop in circuit 1}}} \mathbf{B}_2 \cdot \mathbf{n} dA_1 = \frac{dI_1}{I_1} \int\limits_{\substack{\text{surface enclosed by} \\ \text{infinitesimal loop in circuit 1}}} (\nabla \times \mathbf{A}_2) \cdot \mathbf{n} dA_1$$

$$= \frac{dI_1}{I_1} \oint\limits_{\substack{\text{along infinitesimal} \\ \text{loop in circuit 1}}} \mathbf{A}_2 \cdot \mathbf{d}\ell_1 \tag{5.93}$$

Integrating this expression over the cross-sectional area of circuit 1 and using the fact that $dI_1 = J_1 dS_1$ and that $\mathbf{J}_1$ as a vector points along $\mathbf{d}\ell_1$, we can write

$$\lambda_{12} = \frac{1}{I_1} \int\limits_{\text{cross-sectional area}} J_1 dS_1 \oint\limits_{\text{along loop}} \mathbf{A}_2 \cdot \mathbf{d}\ell_1 = \frac{1}{I_1} \int\limits_{\text{volume of circuit}} \mathbf{A}_2 \cdot \mathbf{J}_1 dV_1 \tag{5.94}$$

where we have used the same integration devices as were used previously. Thus the mutual inductance between these circuits is given by

$$M_{12} = \frac{\lambda_{12}}{I_2} = \frac{1}{I_1 I_2} \int\limits_{V_1} \mathbf{A}_2 \cdot \mathbf{J}_1 dV_1 \tag{5.95}$$

Note that this formula reduces to the formula for self-inductance (5.90) when 1 and 2 are the same circuit and letting $M_{11} = L_1$.

5.6 RABINS' FORMULA FOR LEAKAGE REACTANCE

The leakage reactance for a 2 winding transformer as given in (5.92) can be obtained from the total magnetic energy. We assume that the two windings occupy different radial positions as shown in Fig. 5.5. We use (5.80) for the magnetic energy where $\mathbf{A}$ is the total vector potential due to both windings. Since $\mathbf{A}$ and $\mathbf{J}$ are both azimuthally directed, the dot product becomes simply the ordinary product. We also drop the implied φ subscript on J and A.

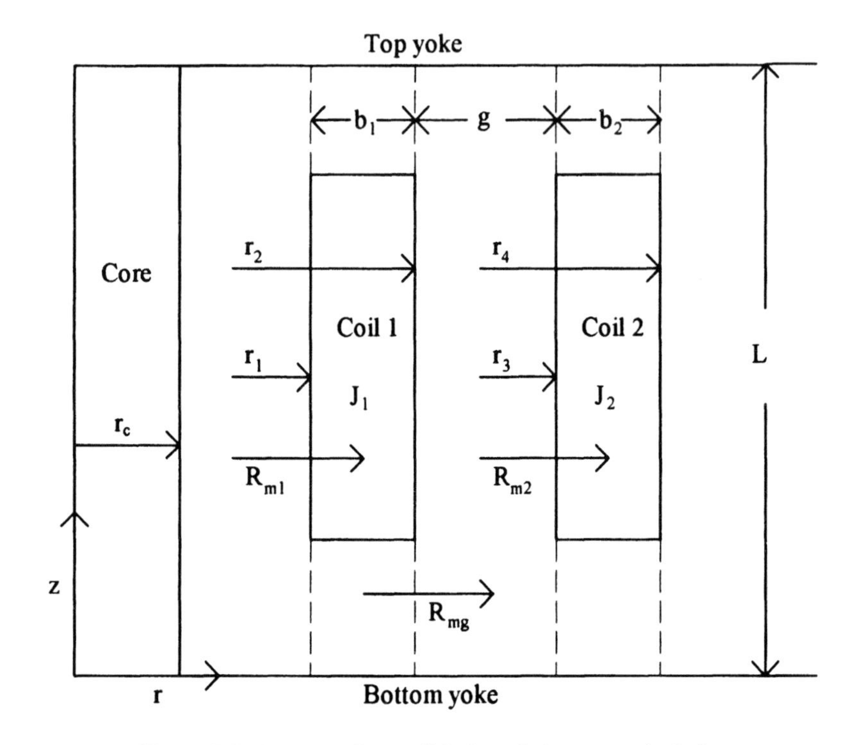

Figure 5.5 Geometry for 2 coil leakage inductance calculation

Thus, writing A_1 , A_2 for the vector potential due to coils 1 and 2 and J_1 , J_2 for their current densities, (5.80) becomes

$$2W = \int_{V_1} (A_1 + A_2) J_1 dV_1 + \int_{V_2} (A_1 + A_2) J_2 dV_2 =$$
$$= \int_{V_1} A_1 J_1 dV_1 + \int_{V_1} A_2 J_1 dV_1 + \int_{V_2} A_1 J_2 dV_2 + \int_{V_2} A_2 J_2 dV_2 \tag{5.96}$$

For the A_1J_1 or A_2J_2 integrals, we must use the Region II solution. Assuming coil 1 is the inner coil, the A_2J_1 integral requires that we use the Region I solution for A_2 while the A_1J_2 integral requires that we use the Region III solution for A_1.

From (5.53) and (5.10), using a second subscript to distinguish the coil 1 quantities from the coil 2 ones,

$$\int_{V_1} A_1 J_1 dV_1$$

$$= 2\pi \int_{r_1}^{r_2} r dr \int_0^L dz \left[J_{0,1} + \sum_{n'=1}^{\infty} J_{n',1} \cos\left(\frac{n'\pi z}{L}\right)\right] \left\{ \mu_o J_{0,1} \left(\frac{r_2 r}{2} - \frac{r_1^3}{6r} - \frac{r^2}{3} \right) \right. \tag{5.97}$$

$$\left. + \mu_o \sum_{n=1}^{\infty} \frac{J_{n,1}}{m^2} \left[E_{n,1} I_1(mr) + F_{n,1} K_1(mr) - \frac{\pi}{2} L_1(mr) \right] \cos\left(\frac{n\pi z}{L}\right) \right\}$$

Performing the z-integral first, we notice that product terms which contain a single cosine term will vanish, i.e.

$$\int_0^L \cos\left(\frac{n\pi z}{L}\right) dz = \frac{L}{n\pi} \sin\left(\frac{n\pi z}{L}\right)\Bigg|_0^L = 0 \tag{5.98}$$

Product terms which contain 2 cosine factors have the value

$$\int_0^L \cos\left(\frac{n\pi z}{L}\right) \cos\left(\frac{n'\pi z}{L}\right) dz = \begin{cases} 0 & , \quad n \neq n' \\ L/2 & , \quad n = n' \end{cases} \tag{5.99}$$

After performing the z-integral, (5.97) becomes

$$\int_{V_1} A_1 J_1 dV_1 = 2\pi \int_{r_1}^{r_2} r dr \left\{ \mu_o L J_{0,1}^2 \left(\frac{r_2 r}{2} - \frac{r_1^3}{6r} - \frac{r^2}{3} \right) + \right. \tag{5.100}$$

$$\left. \frac{\mu_o L}{2} \sum_{n=1}^{\infty} \frac{J_{n,1}^2}{m^2} \left[E_{n,1} I_1(mr) + F_{n,1} K_1(mr) - \frac{\pi}{2} L_1(mr) \right] \right\}$$

Performing the r-integration, we get

$$\int_{V_1} A_1 J_1 dV_1 = 2\pi\mu_o L \left\{ J_{0,1}^2 \left(\frac{r_1^4}{4} - \frac{r_2^3}{12} - \frac{r_2 r_1^3}{3} \right) \right.$$
$$\left. + \frac{1}{2} \sum_{n=1}^{\infty} \frac{J_{n,1}^2}{m^2} \left[E_{n,1} \int_{r_1}^{r_2} r I_1(mr) dr + F_{n,1} \int_{r_1}^{r_2} r K_1(mr) dr - \frac{\pi}{2} \int_{r_1}^{r_2} r L_1(mr) dr \right] \right\}$$
$$= 2\pi\mu_o L \left\{ J_{0,1}^2 \left(\frac{r_1^4}{4} - \frac{r_2^3}{12} - \frac{r_2 r_1^3}{3} \right) \right.$$
$$\left. + \frac{1}{2} \sum_{n=1}^{\infty} \frac{J_{n,1}^2}{m^4} \left[E_{n,1} \int_{x_1}^{x_2} x I_1(x) dx + F_{n,1} \int_{x_1}^{x_2} x K_1(x) dx - \frac{\pi}{2} \int_{x_1}^{x_2} x L_1(x) dx \right] \right\} \tag{5.101}$$

Formulas for the x-integrals above have been given previously in terms of known or easily calculated quantities. The $A_2 J_2$ term in (5.96) is given by (5.101) with a 2 subscript and with the r-integration from r_3 to r_4 (See Fig. 5.5). Thus we have

$$\int_{V_2} A_2 J_2 dV_2 = 2\pi\mu_o L \left\{ J_{0,2}^2 \left(\frac{r_3^4}{4} - \frac{r_4^3}{12} - \frac{r_4 r_3^3}{3} \right) \right.$$
$$\left. + \frac{1}{2} \sum_{n=1}^{\infty} \frac{J_{n,2}^2}{m^4} \left[E_{n,2} \int_{x_3}^{x_4} x I_1(x) dx + F_{n,2} \int_{x_3}^{x_4} x K_1(x) dx - \frac{\pi}{2} \int_{x_3}^{x_4} x L_1(x) dx \right] \right\} \tag{5.102}$$

The $A_2 J_1$ integral in (5.96) is given by

$$\int_{V_1} A_2 J_1 dV_1 = 2\pi \int_{r_1}^{r_2} r dr \int_0^L dz \left[J_{0,1} + \sum_{n'=1}^{\infty} J_{n',1} \cos\left(\frac{n'\pi z}{L} \right) \right]$$
$$\times \left\{ \frac{\mu_o J_{0,2}}{2} (r_4 - r_3) \right.$$
$$\left. + \mu_o \sum_{n=1}^{\infty} \frac{J_{n,2}}{m^2} \left[C_{n,2} I_1(mr) + D_{n,2} K_1(mr) \right] \cos\left(\frac{n\pi z}{L} \right) \right\} \tag{5.103}$$

Using the previous method for carrying out these integrals, we obtain

$$\int_{V_1} A_2 J_1 dV_1 = 2\pi\mu_o L\left\{\frac{J_{0,1}J_{0,2}}{6}(r_4 - r_3)\left(r_2^3 - r_1^3\right)\right.$$
$$\left.+\frac{1}{2}\sum_{n=1}^{\infty}\frac{J_{n,1}J_{n,2}}{m^4}\left[C_{n,2}\int_{x_1}^{x_2} xI_1(x)dx + D_{n,2}\int_{x_1}^{x_2} xK_1(x)dx\right]\right\} \tag{5.104}$$

The $A_1 J_2$ integral in (5.96) is given similarly by

$$\int_{V_2} A_1 J_2 dV_2$$
$$= 2\pi\mu_o L\left\{\frac{J_{0,1}J_{0,2}}{6}(r_4 - r_3)\left(r_2^3 - r_1^3\right) + \frac{1}{2}\sum_{n=1}^{\infty}\frac{J_{n,1}J_{n,2}}{m^4} G_{n,1}\int_{x_3}^{x_4} xK_1(x)dx\right\} \tag{5.105}$$

In spite of their different appearances, (5.104) and (5.105) are the same. Summing up the terms in (5.96), we get after some algebraic manipulation

$$2W = \frac{2\pi\mu_o L}{12}\left\{J_{0,1}^2(r_2 - r_1)^2\left[(r_1 + r_2)^2 + 2r_1^2\right] + J_{0,2}^2(r_4 - r_3)^2\left[(r_3 + r_4)^2 + 2r_3^2\right]\right.$$
$$\left.+4J_{0,1}J_{0,2}(r_2 - r_1)(r_4 - r_3)\left(r_1^2 + r_1 r_2 + r_2^2\right)\right\}$$
$$+\pi\mu_o L\sum_{n=1}^{\infty}\left\{\frac{J_{n,1}^2}{m^4}\left[E_{n,1}\int_{x_1}^{x_2} xI_1(x)dx + F_{n,1}\int_{x_1}^{x_2} xK_1(x)dx - \frac{\pi}{2}\int_{x_1}^{x_2} xL_1(x)dx\right]\right.$$
$$\left.+\frac{J_{n,2}^2}{m^4}\left[E_{n,2}\int_{x_3}^{x_4} xI_1(x)dx + F_{n,2}\int_{x_3}^{x_4} xK_1(x)dx - \frac{\pi}{2}\int_{x_3}^{x_4} xL_1(x)dx\right] + 2\frac{J_{n,1}J_{n,2}}{m^4}G_{n,1}\int_{x_3}^{x_4} xK_1(x)dx\right\}$$
$$\tag{5.106}$$

When the coefficients E_n and F_n etc. have a 1 subscript, then the expressions given by (5.54) apply with x_1, x_2 for the integration limits. However, when the second subscript is a 2, then x_3, x_4 must be substituted for x_1, x_2 in the formulas.

The first set of terms involving the J_0's can be further manipulated by realizing that the average current density in the coil times the area of the coil equals the total Ampere-turns in the coil so that

$$N_1 I_1 = J_{0,1}(r_2 - r_1)L \quad , \quad N_2 I_2 = J_{0,2}(r_4 - r_3)L = -N_1 I_1 \tag{5.107}$$

Using this and the geometric parameters shown in Fig. 5.5, this set of terms can be expressed as

$$\frac{\pi\mu_o(NI)^2}{L}\left(\frac{R_{m1}b_1}{3}+\frac{R_{m2}b_2}{3}+R_{mg}g+\frac{b_1^2}{12}-\frac{b_2^2}{12}\right) \tag{5.108}$$

where R_{m1} , R_{m2} are the mean radii of the coils, b_1 , b_2 their radial thicknesses, R_{mg} is the mean radius of the gap, and g the gap's radial thickness. In this equation NI could refer to either coil since they are equal except for sign.

The leakage inductance, referred to coil 1 as given in (5.92), can be written, using (5.106) and (5.108),

$$\begin{aligned} L_{leak} &= \frac{\pi\mu_o N^2}{L}\left(\frac{R_{m1}b_1}{3}+\frac{R_{m2}b_2}{3}+R_{mg}g+\frac{b_1^2}{12}-\frac{b_2^2}{12}\right) \\ &+\frac{\pi\mu_o L N^2}{(NI)^2}\sum_{n=1}^{\infty}\left\{\frac{J_{n,1}^2}{m^4}\left[E_{n,1}\int_{x_1}^{x_2}xI_1(x)dx+F_{n,1}\int_{x_1}^{x_2}xK_1(x)dx-\frac{\pi}{2}\int_{x_1}^{x_2}xL_1(x)dx\right]\right. \\ &\left.+\frac{J_{n,2}^2}{m^4}\left[E_{n,2}\int_{x_3}^{x_4}xI_1(x)dx+F_{n,2}\int_{x_3}^{x_4}xK_1(x)dx-\frac{\pi}{2}\int_{x_3}^{x_4}xL_1(x)dx\right]+2\frac{J_{n,1}J_{n,2}}{m^4}G_{n,1}\int_{x_3}^{x_4}xK_1(x)dx\right\} \end{aligned} \tag{5.109}$$

The first set of terms in this last equation is the leakage inductance produced by the axial flux from two coils with uniformly distributed currents and no end effects. The second set of terms has been written in terms of NI to emphasize that it, along with N could refer to either coil. To get the leakage impedance multiply L_{leak} by $\omega = 2\pi f$.

There are sometimes transformer designs where two coils are placed on top of each other axially and therefore occupy the same radial position. In this case the coil radial builds are usually the same. Often the coils are duplicates. Rabins' method can also be used to find the leakage inductance for this situation. The current densities for each coil will be non-zero within different axial regions but the Fourier decomposition should reflect this. In this case, only the region II solution is needed for both coils and (5.96) still holds. In this case, $r_3 = r_1$ and $r_4 = r_2$. Evaluating the terms in (5.96), we find that the A_1J_1 and A_2J_2 terms are the same as before except that we must replace r_3 , r_4 , x_3 , x_4 by r_1 , r_2 , x_1 ,x_2.

The A_2J_1 integral is

$$\int_{V_1} A_2 J_1 dV_1$$

$$= 2\pi \int_{r_1}^{r_2} r dr \int_0^L dz \left[J_{0,1} + \sum_{n'=1}^{\infty} J_{n',1} \cos\left(\frac{n'\pi z}{L}\right) \right] \left\{ \mu_o J_{0,2} \left(\frac{r_2 r}{2} - \frac{r_1^3}{6r} - \frac{r^2}{3} \right) \right. \tag{5.110}$$

$$\left. + \mu_o \sum_{n=1}^{\infty} \frac{J_{n,2}}{m^2} \left[E_{n,1} I_1(mr) + F_{n,1} K_1(mr) - \frac{\pi}{2} L_1(mr) \right] \cos\left(\frac{n\pi z}{L}\right) \right\}$$

We can use a 1 subscript on E_n and F_n above since the radial extent of the two windings is the same. However, the 2 subscript is needed on J_n since the current distributions are different. Carrying out the z and r integrations, we get

$$\int_{V_1} A_2 J_1 dV_1 = 2\pi\mu_o L \left\{ J_{0,1} J_{0,2} \left(\frac{r_1^4}{4} - \frac{r_2^3}{12} - \frac{r_2 r_1^3}{3} \right) \right.$$

$$\left. + \frac{1}{2} \sum_{n=1}^{\infty} \frac{J_{n,1} J_{n,2}}{m^4} \left[E_{n,1} \int_{x_1}^{x_2} x I_1(x) dx + F_{n,1} \int_{x_1}^{x_2} x K_1(x) dx - \frac{\pi}{2} \int_{x_1}^{x_2} x L_1(x) dx \right] \right\} \tag{5.111}$$

This expression is also equal to the $A_1 J_2$ term. Hence we get, combining terms

$$2W = \frac{2\pi\mu_o L}{12} \left\{ \left(J_{0,1}^2 + J_{0,2}^2 + 2J_{0,1} J_{0,2} \right) (r_2 - r_1)^2 \left[(r_1 + r_2)^2 + 2r_1^2 \right] \right\}$$

$$+ \pi\mu_o L \sum_{n=1}^{\infty} \left\{ \left(\frac{J_{n,1}^2 + J_{n,2}^2 + 2J_{n,1} J_{n,2}}{m^4} \right) \left[E_{n,1} \int_{x_1}^{x_2} x I_1(x) dx + F_{n,1} \int_{x_1}^{x_2} x K_1(x) dx - \frac{\pi}{2} \int_{x_1}^{x_2} x L_1(x) dx \right] \right\} \tag{5.112}$$

Using (5.107) applied to this case, we see that the first set of terms vanishes. Thus we get for the leakage impedance

$$L_{\substack{\text{leak, axially} \\ \text{displaced coils}}} = \frac{\pi\mu_o L N^2}{(NI)^2} \sum_{n=1}^{\infty} \left\{ \left(\frac{J_{n,1}^2 + J_{n,2}^2 + 2J_{n,1} J_{n,2}}{m^4} \right) \right.$$

$$\left. \times \left[E_{n,1} \int_{x_1}^{x_2} x I_1(x) dx + F_{n,1} \int_{x_1}^{x_2} x K_1(x) dx - \frac{\pi}{2} \int_{x_1}^{x_2} x L_1(x) dx \right] \right\} \tag{5.113}$$

5.7 RABINS' METHOD APPLIED TO CALCULATE SELF AND MUTUAL INDUCTANCES OF COIL SECTIONS

A coil section for our purposes here consists of a part of a coil which has a uniform current density. Then (5.14) holds for the Fourier coefficients. In all other respects it is treated like a full coil. The self-inductance of this section is given by (5.95) with the subscript 2 replaced by 1. In this case, we use the solution for A in Region II since that is where the current density is non-zero. The integral in (5.95) has already been carried out in (5.101) so we have

$$L_1 = \frac{2\pi\mu_o L J_{0,1}^2}{12 I_1^2}(r_2 - r_1)^2\left[(r_2 + r_1)^2 + 2r_1^2\right]$$
$$+ \frac{\pi\mu_o L}{I_1^2}\sum_{n=1}^{\infty}\left\{\frac{J_{n,1}^2}{m^4}\left[E_{n,1}\int_{x_1}^{x_2} xI_1(x)dx + F_{n,1}\int_{x_1}^{x_2} xK_1(x)dx - \frac{\pi}{2}\int_{x_1}^{x_2} xL_1(x)dx\right]\right\} \tag{5.114}$$

Since

$$N_1 I_1 = J_{0,1}(r_2 - r_1)L \tag{5.115}$$

we can rewrite (5.114)

$$L_1 = \frac{\pi\mu_o N_1^2}{6L}\left[(r_2 + r_1)^2 + 2r_1^2\right]$$
$$+ \frac{\pi\mu_o L N_1^2}{(N_1 I_1)^2}\sum_{n=1}^{\infty}\left\{\frac{J_{n,1}^2}{m^4}\left[E_{n,1}\int_{x_1}^{x_2} xI_1(x)dx + F_{n,1}\int_{x_1}^{x_2} xK_1(x)dx - \frac{\pi}{2}\int_{x_1}^{x_2} xL_1(x)dx\right]\right\} \tag{5.116}$$

The mutual induction between two coil sections is given by (5.95). Here again, except for the uniform current density in one coil section, we treat the coil sections as if they were full coils in applying Rabins' method. If the sections are on the same coil or two coils axially displaced, then A_2 is the solution in Region II. The A_2J_1 integral has already been done in this case and is given in (5.111). Thus we have

$$M_{12,\,\substack{\text{same radial}\\ \text{position}}} = \frac{2\pi\mu_o L J_{0,1} J_{0,2}}{12 I_1 I_2}(r_2 - r_1)^2\left[(r_2 + r_1)^2 + 2r_1^2\right]$$
$$+\frac{\pi\mu_o L}{I_1 I_2}\sum_{n=1}^{\infty}\left\{\frac{J_{n,1}J_{n,2}}{m^4}\left[E_{n,1}\int_{x_1}^{x_2} xI_1(x)dx + F_{n,1}\int_{x_1}^{x_2} xK_1(x)dx - \frac{\pi}{2}\int_{x_1}^{x_2} xL_1(x)dx\right]\right\}$$
$$(5.117)$$

Since

$$N_1 I_1 = J_{0,1}(r_2 - r_1)L \quad , \quad N_2 I_2 = J_{0,2}(r_2 - r_1)L \qquad (5.118)$$

we can rewrite (5.117)

$$M_{12,\,\substack{\text{same radial}\\ \text{position}}} = \frac{2\pi\mu_o N_1 N_2}{6L}\left[(r_2 + r)^2 + 2r_1^2\right]$$
$$+\frac{\pi\mu_o L N_1 N_2}{(N_1 I_1)(N_2 I_2)}\sum_{n=1}^{\infty}\left\{\frac{J_{n,1}J_{n,2}}{m^4}\left[E_{n,1}\int_{x_1}^{x_2} xI_1(x)dx + F_{n,1}\int_{x_1}^{x_2} xK_1(x)dx - \frac{\pi}{2}\int_{x_1}^{x_2} xL_1(x)dx\right]\right\}$$
$$(5.119)$$

When the coil sections are part of different coils, radially displaced, then we can assume that 1 is the inner and 2 the outer coil. The result will be independent of this assumption. Thus we need the Region I solution. The A_1J_2 integral has already been done and the result given in (5.104). Thus we have

$$M_{12,\,\substack{\text{different radial}\\ \text{positions}}} = \frac{2\pi\mu_o L J_{0,1} J_{0,2}}{6 I_1 I_2}(r_4 - r_3)(r_2^3 - r_1^3)$$
$$+\frac{\pi\mu_o L}{I_1 I_2}\sum_{n=1}^{\infty}\left\{\frac{J_{n,1}J_{n,2}}{m^4}\left[C_{n,2}\int_{x_1}^{x_2} xI_1(x)dx + D_{n,2}\int_{x_1}^{x_2} xK_1(x)dx\right]\right\} \qquad (5.120)$$

In this case, we have

$$N_1 I_1 = J_{0,1}(r_2 - r_1)L \quad , \quad N_2 I_2 = J_{0,2}(r_4 - r_3)L \qquad (5.121)$$

Using this, we can rewrite (5.120)

$$M_{12,\,\substack{\text{different radial}\\ \text{positions}}} = \frac{\pi\mu_o N_1 N_2}{3L}\left(r_1^2 + r_1 r_2 + r_2^2\right)$$
$$+ \frac{\pi\mu_o N_1 N_2}{(N_1 I_1)(N_2 I_2)} \sum_{n=1}^{\infty} \left\{ \frac{J_{n,1} J_{n,2}}{m^4} \left[C_{n,2} \int_{x_1}^{x_2} x I_1(x) dx + D_{n,2} \int_{x_1}^{x_2} x K_1(x) dx \right] \right\} \tag{5.122}$$

From the formulas for leakage, self, and mutual inductances, we see that all the results can be expressed in terms of the modified Bessel functions of order 0 and 1, the M_0, M_1 functions, and the integral of the M_0 function. The M_0, M_1 and the integral of the M_0 function have convenient expressions in terms of well behaved integrals.

6. MECHANICAL DESIGN

Summary During fault conditions when currents can increase to about 25 times their normal values, transformer windings are subjected to very high forces. Sufficient bracing must be provided so that little movement occurs. In addition, the mechanical design and material properties must be such that the resulting stresses do not lead to permanent deformation, fracture, or buckling of the materials. Although fairly accurate calculations of the forces can be made either analytically or via finite elements, because of the complex structure of the windings, it is difficult to obtain the resulting stresses or strains without resorting to approximations. Nevertheless, with sufficient allowances for factors of safety, design rules to limit the stresses can be reliably established.

6.1 INTRODUCTION

Transformers must be designed to withstand the large forces which occur during fault conditions. Fault currents for the standard fault types such as single line to ground, line to line, double line to ground, and all three lines to ground must be calculated. Since these faults can occur during any part of the ac cycle, the worst case transient overcurrent must be used to determine the forces. This can be calculated and is specified in the standards as an asymmetry factor which multiplies the rms steady state currents. It is given by [IEE93]

$$K = \sqrt{2}\left[1 + e^{-\left(\phi + \frac{\pi}{2}\right)\frac{r}{x}} \sin\phi\right] \tag{6.1}$$

where $\phi = \tan^{-1}(x/r)$ and x/r is the ratio of the effective ac reactance to resistance. They are part of the total impedance which limits the fault current in the transformer when the short circuit occurs. Using this factor, the resulting currents are used to obtain the magnetic field (leakage field) surrounding the coils and, in turn, the resulting forces on the windings. Analytic methods such as Rabins' method [Rab56] as discussed in Chapter 5 or finite element methods can be used to calculate the magnetic field. An example of such a leakage field is shown in Fig. 6.1 which was generated by the finite element program Maxwell® EM2D Field Simulator [Ansoft]. Since this is a 2D program, the figure is cylindrically symmetrical about the core center line and only the bottom half of the windings and core are shown because of assumed symmetry about a horizontal center plane. Although details such as clamps and shields can be included in the calculation using a finite element approach, they are not part of Rabins' analytical approach which assumes a simpler idealized geometry. However, calculations show that the magnetic field in the windings and hence the forces are nearly identical in the two cases.

The force density (force/unit volume), $\mathbf{f}$, generated in the windings by the magnetic induction, $\mathbf{B}$, is given by the Lorentz force law

$$\mathbf{f} = \mathbf{J} \times \mathbf{B} \tag{6.2}$$

where $\mathbf{J}$ is the current density and SI units are used. These force densities can be integrated to get total forces, forces/unit length, or pressures depending on the type of integration performed. The resulting values or the maxima of these values can then be used to obtain stresses or maximum stresses in the winding materials.

The procedure described above is static in that the field and force calculations assume steady-state conditions, even though the currents were multiplied by an asymmetry factor to account for a transient overshoot. Because the fault currents are applied suddenly, the resulting forces or pressures are also suddenly applied. This could result in transient mechanical effects such as the excitation of mechanical resonances which could produce higher forces for a brief period than those obtained by a steady-state calculation. A few studies have been done of these transient mechanical effects in transformers [Hir71, Bos72, Pat80, Tho79]. Although these studies are approximate, where they have indicated a large effect, due primarily to the excitation of a mechanical resonance, we use an appropriate enhancement factor to multiply the steady-state force. In other cases, these studies have indicated little or no transient effects so that no correction is necessary.

Because controlled short-circuit tests are rarely performed to validate a transformer's mechanical design, there are few empirical studies which directly test the theoretical calculations. To some extent, our confidence in these calculations comes from the fact that, in recent years, very few field failures have been directly attributable to mechanical causes.

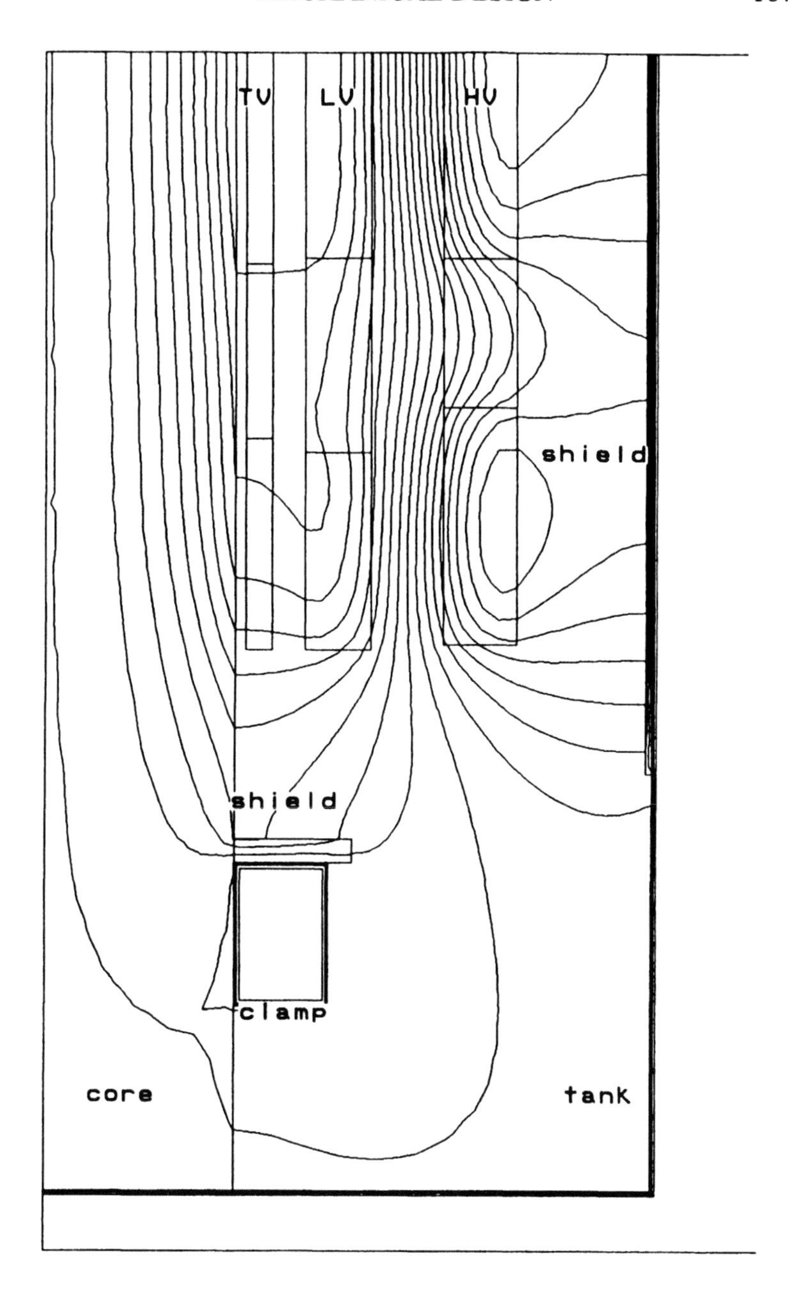

Figure 6.1 Plot of transformer leakage flux. Only the bottom half is shown. The figure is assumed to be cylindrically symmetrical about the core center line.

6.2 FORCE CALCULATIONS

As mentioned previously, the electromagnetic forces acting on the coils are obtained by means of an analytic or finite element magnetic field calculation in conjunction with the Lorentz force law. In our analytic calculation which uses Rabins' method [Rab56], each coil carrying current is subdivided into 4 radial sections and 100 axial sections as shown in Fig. 6.2a. In the case of a finite element method, the coil cross-section is subdivided into an irregular net of triangles as shown in Fig. 6.2b. The principles used to determine the forces or pressures needed in the stress calculation are similar for the two cases so we will focus on the analytic method.

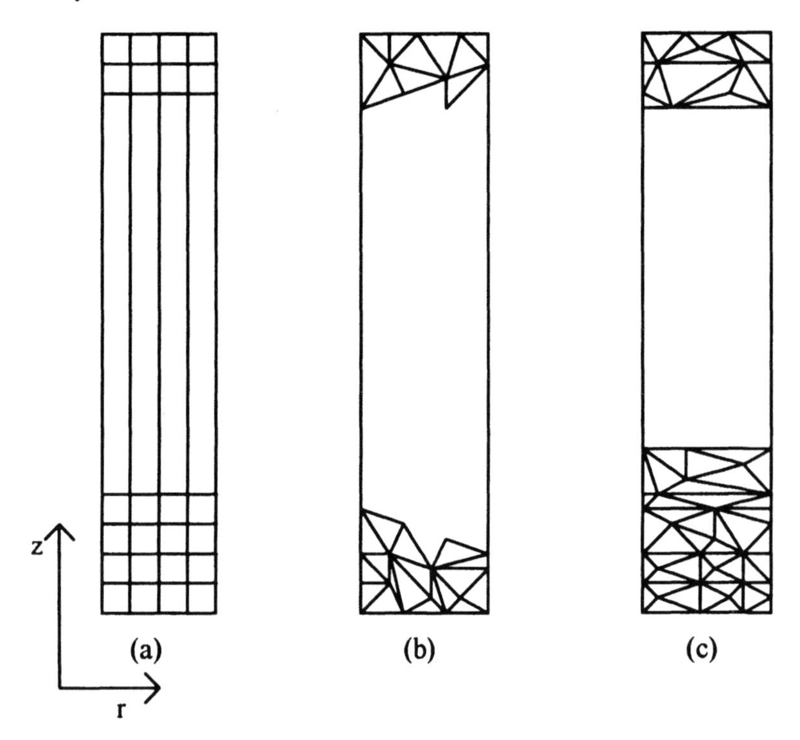

Figure 6.2 Coil subdivisions in (a) analytic and (b) finite element calculations of the magnetic field. In (c), a modified finite element mesh is shown which is more useful for subsequent stress analysis.

The current density is uniform in each block in Fig. 6.2a but it can vary from block to block axially due to tapping out sections of the coil or thinning of sections of the winding to achieve a better amp-turn balance with tapped out sections in adjacent winding. These sections of reduced or zero current density produce radially bulging flux lines as are evident in Fig. 6.1. Form the Lorentz force law, since the current is azimuthally directed, radially directed flux produces axial forces and axially directed flux produces radial forces. There are no azimuthal forces since the geometry is assumed to be cylindrically symmetric.

The radial and axial forces are computed for each block in Fig. 6.2a. Because of the cylindrical symmetry, each block in the figure is really a ring. The radial force can be thought of as a pressure acting inward or outward on the ring depending on its sign. If these forces are summed over the four radial blocks and then divided by the area of the cylindrical surface at the average radius of the winding, we obtain an effective total radial pressure acting on the coil at the axial position of the block. The maximum of these pressures in absolute value for the 100 axial positions is the worst case radial pressure and is used in subsequent stress analysis. Note that these forces are radially directed so that integrating them vectorially over 360 degrees would produce zero. This integration must therefore be done without regard to their vectorial nature.

The axially directed force summed over the four radial blocks is also needed. The maximum of the absolute value of this force for the 100 axial positions is a worst case force used in the stress analysis. These axial forces are also summed, starting at the bottom of the coil. The total of these axial forces (summed over all the blocks of the coil) is a net upward or downward force depending on its sign. This force is countered by an equal and opposite force exerted by the pressure ring. If this net force acts downward, then the bottom pressure ring exerts an upward equal force on the coil. If it acts upward, then the top pressure ring exerts an equal downward force. In the former case, the top ring exerts no force on the coil and in the latter case, the bottom ring exerts no force on the coil We are ignoring the gravitational forces in comparison with the electromagnetic forces here. Starting with the force exerted upward, if any, by the bottom pressure ring, if we keep adding to this the forces produced by the horizontal layers of 4 blocks starting from the bottom row, we will arrive at a maximum net upward force at some vertical position along the winding. This force is called the maximum compressive force and is a worst case force used in the stress analysis.

The total axial forces on the windings are added together if they are upward (positive). Similarly, windings having downward (negative) total axial forces contribute to a total downward force. These total upward or downward forces due to all the windings should be equal in absolute value since the net electromagnetic force acting axially must be zero. In practice there are slight differences in these calculated quantities due to rounding errors. The net upward or downward axial force due to all the windings is called the total end thrust and is used in sizing the pressure rings. If the windings are symmetric about a horizontal center plane, the total axial force on each winding is nearly zero so there is little or no end thrust. However, when one or more windings are offset vertically from the others, even slightly, net axial forces develop on each winding which push some windings up and some down. It is a good practice to include some offset, say 0.635 cm to 1.27 cm (1/4 to 1/2 in), in the calculations to take into account possible misalignment in the transformer's construction.

As can be seen from the way forces needed for the stress analysis were extracted from the block forces in Fig. 6.2a, the finite element mesh shown in Fig. 6.2c would be more useful for obtaining the needed forces. Here the coil is subdivided into a series of axial blocks, which can be of different heights, before the triangular mesh is generated. Then the forces can be summed for all the triangles within each block to correspond to the forces obtained from the 4 radial blocks at the same axial height in Fig. 6.2a.

6.3 STRESS ANALYSIS

We need to relate the forces discussed in the last section to the stresses in the coil in order to determine whether the coil can withstand them without permanently deforming or buckling. The coils have a rather complex structure as indicated in Fig. 6.3 which shows a disk type coil. Helical windings are disk windings with only 1 turn per section and thus are a special case of Fig. 6.3. The sections are separated vertically by means of key spacers made of pressboard. These are spaced around the coil so as to allow cooling oil to flow between them. The coils are supported radially by means of axial sticks and pressboard cylinders or barriers. These brace the coil on the inside against an inner coil or against the core. There are similar support structures on the outside of the coil, except for the outermost coil. Spaces between the sticks allow cooling oil to flow. Because of the dissimilar materials used, pressboard, paper, and copper for the conductors, and because of the many openings for the cooling oil, the stress analysis would be very complicated unless suitable approximations are made.

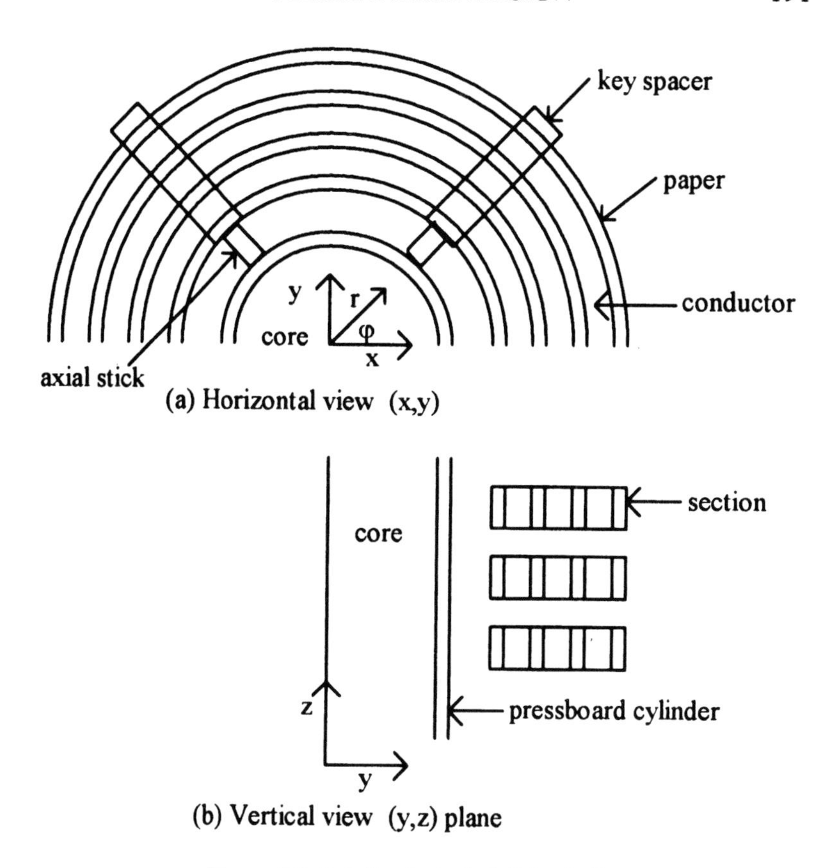

Figure 6.3 Details of a typical disk winding

Although we have shown distinct separate axial sections in Fig. 6.3, in reality the wires must maintain electrical continuity from section to section so that the coil has a helical (spring like) structure. For the stress analysis, it is generally assumed that the helical pitch is small enough that the coil can be regarded as having distinct horizontal sections as shown in the figure. Moreover, these sections are assumed to close on themselves, forming rings.

Another approximation concerns the cable which comprises the winding turns. Fig. 6.4 shows the two types which are commonly used. Magnet wire consists of a single strand of copper surrounded by a paper covering and is treated almost without approximation in the stress analysis. Transposed cable consists of multiple enamel coated copper strands arranged in the nearly rectangular pattern shown. Not shown are how the transpositions are made which rotate the different strands so that they each occupy all the positions shown as one moves along the wire.

The transpositions give some rigidity to the collection of strands. In addition, use is often made of bonded cable in which all the strands are bonded together by means of an epoxy coating over the enamel which is subjected to a heat treatment. In this case the cable can be treated as a rigid structure, although there is some question as to how to evaluate its material properties. With or without bonding, one has to make some approximations as to how to model the cable for stress analysis purposes. Without bonding, we assume for radial force considerations that the cable has a radial thickness equivalent to 2 radial strands. With bonding, we assume a radial thickness equivalent to 80% of the actual radial build.

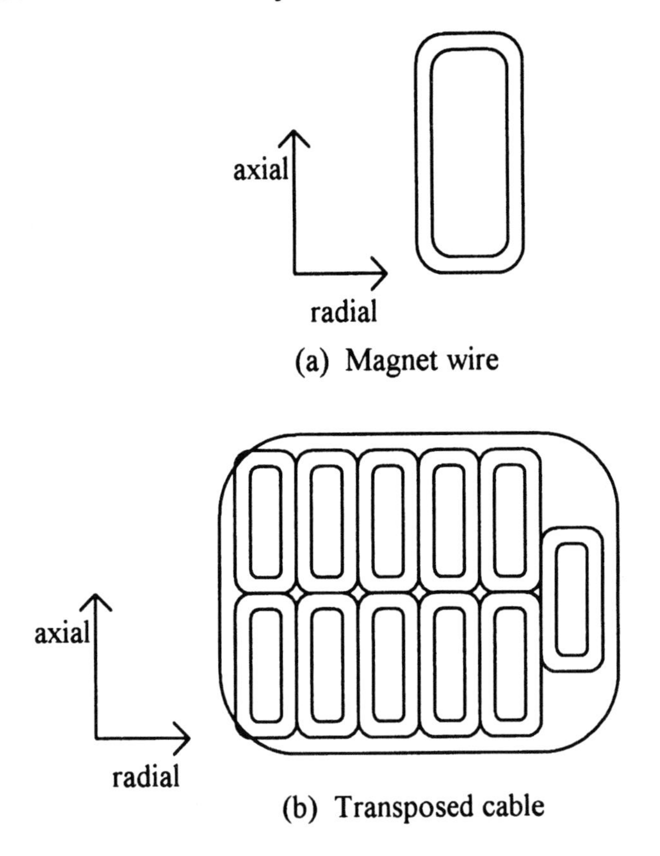

Figure 6.4 Types of wire or cable used in transformer coils

We will begin by looking at the stresses produced by the axial forces. Then we will consider the stresses due to the radial forces. We will need to examine worst case stresses in the copper winding turns, in the key spacers, in the pressure rings and in the tieplates which are attached to the top and bottom clamps and provide the tension which compresses the coils. In the following sections.

6.3.1 Compressive Stress in the Key Spacers

The maximum axial compressive force is used to determine the worst case compressive stress in the key spacers. This force, F_c, obtained for each coil from the magnetic field analysis program, is divided by the area of the key spacers covering one 360 degree disk section to obtain the key spacer compressive stress σ_{ks},

$$\sigma_{ks} = \frac{F_c}{N_{ks} W_{ks} B} \tag{6.3}$$

where N_{ks} is the number of key spacers around one 360 degree section, W_{ks} is the width of a key spacer, and B is the radial build of the coil. We use key spacers made of pre-compressed pressboard which can withstand a maximum compressive stress of 310 MPa (45,000 psi). Therefore, the stress calculated by (6.3) should not exceed this number.

6.3.2 Axial Bending Stress per Strand

The maximum axial force over the 100 vertical subdivisions is computed for each coil. Call this maximum force F_a. In order to compute the bending stress, we need to know the force/unit length acting on an individual strand, since these forces are continuously distributed along the strand. The number of strands in the entire coil, N_s, is given by

$$N_s = N_t N_h N_w N_{st} \tag{6.4}$$

where N_t is the number of turns/leg, N_h is the number of cables/turn high (radially), N_w is the number of cables/turn wide (axially), and N_{st} is the number of strands/cable. We are allowing for the fact that each turn can consist of several cables in parallel, some radially and some axially positioned, each having N_{st} strands. If the coil consists of 2 separate windings stacked axially (center fed) each having N_e electrical turns, then $N_t = 2 N_e$. Since the section with the maximum axial force is only 1/100 of the coil, it only has 1/100 of the above number of strands. Hence, the maximum force/unit length on a single strand, q_{st}, is given by

$$q_{st} = \frac{100F_a}{N_s \pi D_m} \tag{6.5}$$

where D_m is the mean diameter of the coil.

The problem can be analyzed as a uniformly loaded rectangular beam with built in ends as shown in Fig. 6.5a. There are 6 unknowns, the horizontal and vertical components of the reaction forces and the bending moments at the two built in ends, but only 3 equations of statics making this a statically indeterminate problem. The two horizontal reaction forces are equal and opposite and produce a tensile stress in the beam which is small for small deflections and will be ignored. The vertical reaction forces are equal and share the downward load equally. They are therefore given by $R_{1,up} = R_{2,up} = qL/2$, where $q = q_{st}$ is the downward force/unit length along the beam and L the beam's length. By symmetry, the bending moments M_1 and M_2 shown in Fig. 6.5a are equal and produce a constant bending moment along the beam. In addition, a bending moment as a function of position is also present, resulting in a total bending moment at position x of

$$M(x) = \frac{qLx}{2} - \frac{qx^2}{2} - M_1 \tag{6.6}$$

using the sign convention of Ref. [Tim56].

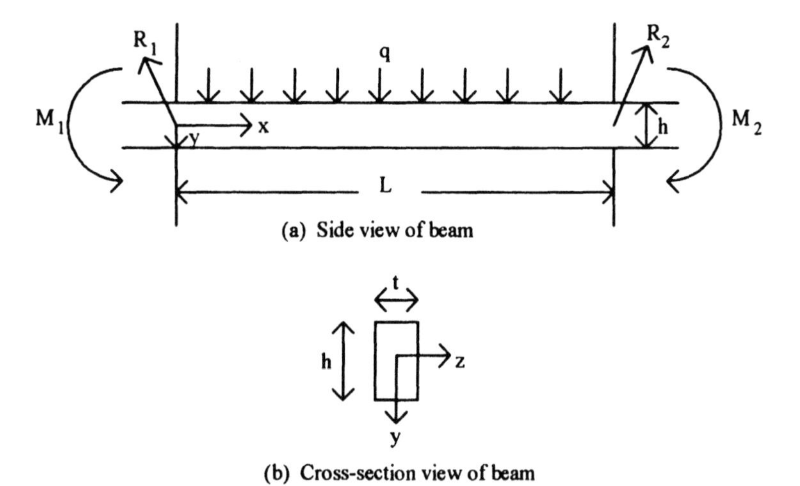

Figure 6.5 Uniformly loaded beam with built in ends

To solve for the unknown M_l , we need to solve the equation of the beam's deflection curve [Tim56]

$$\frac{d^2y}{dx^2} = -\frac{M(x)}{EI_z} \tag{6.7}$$

where E is Young's modulus for the beam material and I_z is the area moment of inertia about the z axis. Substituting (6.6) into (6.7), we have

$$\frac{d^2y}{dx^2} = \frac{1}{EI_z}\left(M_l - \frac{qLx}{2} + \frac{qx^2}{2}\right) \tag{6.8}$$

Integrating once, we get

$$\frac{dy}{dx} = \frac{1}{EI_z}\left(M_l x - \frac{qLx^2}{4} + \frac{qx^3}{6}\right) + C \tag{6.9}$$

where C is a constant of integration. Since the beam is rigidly clamped at the ends, the slope $dy/dx = 0$ at $x = 0$ and $x = L$. Setting $dy/dx = 0$ at $x = 0$ in (6.9), yields $C = 0$. Setting $dy/dx = 0$ at $x = L$, we find that

$$M_l = \frac{qL^2}{12} \tag{6.10}$$

Substituting these values into (6.9), we obtain

$$\frac{dy}{dx} = \frac{q}{EI_z}\left(\frac{L^2x}{12} - \frac{Lx^2}{4} + \frac{x^3}{6}\right) \tag{6.11}$$

Integrating again, we obtain

$$y = \frac{q}{EI_z}\left(\frac{L^2x^2}{24} - \frac{Lx^3}{12} + \frac{x^4}{24}\right) = \frac{qx^2(L-x)^2}{24EI_z} \tag{6.12}$$

where the constant of integration was set to zero since $y = 0$ at $x = 0$ and $x = L$.

Inserting (6.10) into (6.6), the bending moment as a function of position along the beam is

$$M(x) = \frac{q}{2}\left[x(L-x) - \frac{L^2}{6}\right] \tag{6.13}$$

The maximum positive value occurs at $x = L/2$ and is $M_{max} = qL^2/24$. The minimum value occurs at $x = 0$ or $x = L$ and is $M_{min} = -qL^2/12$. Since the minimum bending moment is larger in absolute value, we use it in the formula to obtain the stress due to bending in the beam [Tim56]

$$\sigma_x = \frac{My}{I_z} \tag{6.14}$$

where y is measured downward from the centroid of the beam cross-section as shown in Fig. 6.5. For a given x, σ_x is a maximum or minimum when $y = \pm h/2$, where h is the beam height in the bending direction. If σ_x is positive, the stress is tensile, if negative compressive. Inserting M_{min} into (6.14), taking $y = -h/2$, and using

$$I_z = \frac{t h^3}{12} \tag{6.15}$$

for the area moment of inertia for a rectangular cross section with respect to the z-axis through the centroid (See Fig. 6.5b), we obtain

$$\sigma_{x,max} = \frac{q}{2t}\left(\frac{L}{h}\right)^2 \tag{6.16}$$

Here t is the thickness of the beam perpendicular to the bending plane. This is a tensile stress and occurs at the top of the beam at the supports. There is a compressive stress of equal magnitude at the bottom of the beam at the supports.

Inserting the actual load (6.5) into (6.16), we get for the maximum axial bending stress,

$$\sigma_{max,axial} = \frac{100F_a}{2N_s\pi D_m t}\left(\frac{L}{h}\right)^2 \qquad (6.17)$$

The span length, L , can be determined from the number of key spacers, their width, and the mean circumference,

$$L = \frac{\pi D_m - N_{ks}W_{ks}}{N_{ks}} = \frac{\pi D_m}{N_{ks}} - W_{ks} \qquad (6.18)$$

The strand height h and thickness t apply to a single strand, whether as part of a cable having many strands or as a single strand in a magnet wire. If the cable is bonded, the maximum axial bending stress in (6.17) is divided by 3. This is simply an empirical correction to take into account the greater rigidity of bonded cable.

6.3.3 Tilting Strength

The axial compressive force which is applied to the key spacers can cause the individual strands of the conductors which are pressed between the key spacers to tilt if the force is large enough. Fig. 6.6 shows an idealized geometry of this situation. Depicted is an individual strand which has the form of a closed ring acted on by a uniform axial compressive pressure, P_c . We assume initially that the strand has rounded ends which do not dig into the adjacent key spacers to prevent or oppose the tilting shown. (There could be several layers of strands in the axial direction separated by paper, which plays the same role as the key spacers in the figure.)

Analyzing a small section in the azimuthal direction of length $\Delta\ell$, the applied pressure exerts a torque, τ_c , given by

$$\tau_c = P_c(t\Delta\ell)h\sin\theta \qquad (6.19)$$

where t is the radial thickness of the strand and where $t\,\Delta\ell$ is the area on which the pressure P_c acts. The axial height of the strand is h and θ is the tilting angle from the vertical which is assumed to be small. The tilting causes the material of the ring to stretch above its axial center and to compress below it. This produces stresses in the ring which in turn produce a torque which opposes the torque calculated above. To calculate this opposing torque, let y measure the distance above the axial center of the strand as shown in Fig. 6.6b. The increase in radius at distance y above the center line is $y \tan\theta$. Therefore the strain at position y is

$$\varepsilon = \frac{y \tan\theta}{R} \tag{6.20}$$

where ε is an azimuthal strain. This produces an azimuthal tensile stress (hoop stress) given by

$$\sigma = E\varepsilon = \frac{Ey \tan\theta}{R} \tag{6.21}$$

where E is Young's modulus for the conductor material. This hoop stress results in an inward force, F_r, on the section of strand given by

$$F_r = \sigma(t\,\Delta y)\Delta\varphi = \sigma(t\,\Delta y)\frac{\Delta\ell}{R} \tag{6.22}$$

In (6.22) $t\,\Delta y$ is the area at height y over which the stress σ acts and $\Delta\varphi$ the angle subtended by the azimuthal section of strand of length $\Delta\ell$. Also $\Delta\ell = R\,\Delta\varphi$ has been used.

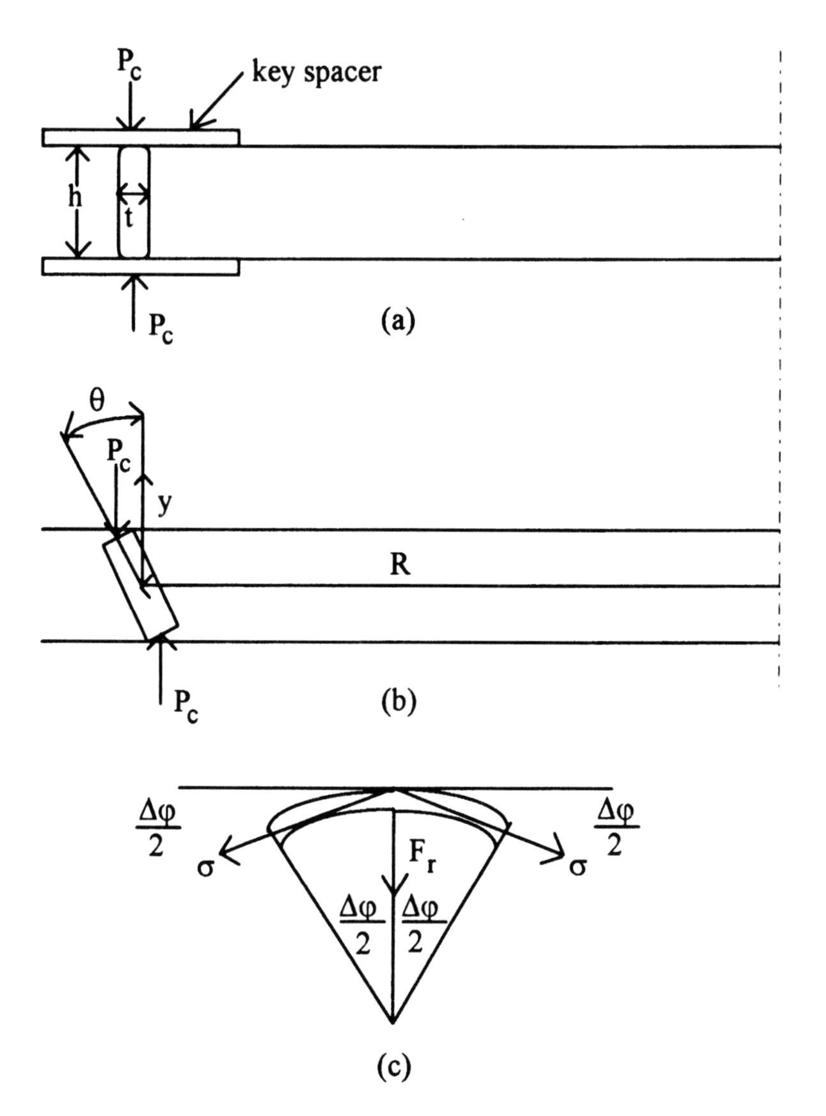

Figure 6.6 Geometry of strand tilting due to the axial compressive force

The inward force F_r produces a counter torque $\Delta\tau$ on the small section of height Δy which for small θ is given by

$$\Delta\tau = F_r y = \frac{\sigma\, t\, \Delta\ell\, y\, \Delta y}{R} \tag{6.23}$$

Using (6.21) this becomes

$$\Delta\tau = \frac{E\,t\,\Delta\ell\,\tan\theta\,y^2\,\Delta y}{R^2} \tag{6.24}$$

Letting Δy become infinitesimal and integrating from $y = 0$ to $h/2$, we get

$$\tau = \frac{E\,t\,\Delta\ell\,\tan\theta\,h^3}{24\,R^2} \tag{6.25}$$

Analyzing the portion of the strand below the center line, the hoop stress is compressive and will result in an outward force on the strand. This will create a torque of the same magnitude and sense as that in (6.25) so to take the whole strand into account we just need to multiply (6.25) by 2. In equilibrium, this resulting torque equals the applied torque given by (6.19). Equating these two expressions and using, for small θ, $\sin\theta \approx \tan\theta \approx \theta$, we get

$$P_c = \frac{E}{12}\left(\frac{h}{R}\right)^2 \tag{6.26}$$

When the conductor strand has squared ends, there is an additional resistance to tilting as a result of the ends digging into the key spacers or paper. This results in an additional resisting torque of magnitude [Ste62, Wat66]

$$C\frac{N_{ks}W_{ks}\,t^3\sin\theta\,\Delta\ell}{6h(2\pi R)} \tag{6.27}$$

Here C is a constant depending on the spacer material. We use a value of $C = 6.21\times10^4$ MPa $(9\times10^6$ psi). This should be added to $2 \times$ equation (6.25), resulting in a tilting pressure of

$$P_c = \frac{E}{12}\left(\frac{h}{R}\right)^2 + C\frac{N_{ks}W_{ks}t^2}{6h^2(2\pi R)} \tag{6.28}$$

If the strand has rounded corners of radius R_c , then t in the above formula is reduced by $2R_c$ so only its flat portion is considered. The resulting critical axial pressure is therefore

$$P_c = \frac{E}{12}\left(\frac{h}{R}\right)^2 + \left(\frac{C}{6}\right)\frac{N_{ks}W_{ks}}{2\pi R}\left(\frac{t-2R_c}{h}\right)^2 \tag{6.29}$$

For comparison with the applied maximum axial compressive force, we multiply (6.29) by the radial surface area of the strands in one horizontal layer, A_{layer} . This is

$$A_{layer} = \pi D_m t N_d N_h \left(\frac{N_{st}-1}{2}\right) \tag{6.30}$$

where N_d is the number of turns in a disk or section and the other symbols have been defined previously. Note that as a result of the strand positioning in a cable as shown in Fig. 6.4b, the expression $(N_{st} - 1)/2$ gives the number of strands radially that are part of the double layer. For magnet wire, the expression in parentheses is taken as 1. Thus the critical axial force is

$$F_{cr} = P_c A_{layer} \tag{6.31}$$

This applies to unbonded cable. For bonded cable, we take $F_{cr} = \infty$ since it is assumed in this case that tilting cannot occur. We compare (6.31) with the maximum applied axial compressive force, F_c , by taking the ratio, F_{cr} / F_c . This ratio must be > 1 for a viable design, i.e. F_c must be $< F_{cr}$.

In the above derivation, the compressive force was assumed to be applied uniformly around the strand ring, whereas in reality it is only applied to the portions of the ring in contact with the key spacers. Since the portions outside the key spacers see no applied pressure, the uniformly applied pressure represents an averaging process over the entire ring and is a reasonable approximation.

6.3.4 Stress in Tie Bars

The tie bars or tieplates are used to join the upper and lower clamping structures which keep the coils under compression. These are generally long rectangular bars of steel which are placed along the sides of the core legs. They are under mild tension during normal transformer operation. During short circuit, the tensile stresses can increase considerably. Also when the transformer is lifted, the tie bars support the entire weight of the coils and core.

The short circuit stress in the tie bars is due to the total end thrust produced by all the coils. This is the sum of the total upward or downward forces acting on the coils and is an output of the force program. Since this output refers to a single leg, the tie bars affected by this force should only be those associated with a single leg. In the case of a 3 phase fault, all the tie bars are affected equally. However, in the case of a single line to ground fault where the forces are much higher on one leg than the other two, the tie bars associated with the leg having the greater force will probably experience the greatest stress. Therefore in this worst case scenario, we are assuming that the legs act independently at least for the short duration of the fault.

The total end thrust is the result of a static force calculation. Because of possible dynamic effects associated with the sudden application of a force to an elastic system, the end thrust could be considerably higher for a short period after the force application. To see what the dynamic force enhancement might be, we analyze an elastic bar subject to a suddenly applied force as illustrated in Fig. 6.7. Let x measure the change in length produced by the force, relative to the unstressed bar of length L. A bar under stress stores elastic energy, U, given by [Tim56]

$$U = \frac{EA}{2L} x^2 \tag{6.32}$$

where E is Young's modulus for the bar material and A is the cross-sectional area of the bar. The applied force causes the bar material to move so that it acquires a kinetic energy. Since each portion of the bar moves with a different velocity - the bar is fixed at one end and moves with maximum velocity at the other end - it is necessary to integrate the kinetic energy of each segment along the bar to get the total. In Fig. 6.7a, a bar segment a distance y from the fixed end of thickness dy is isolated. The parameter u measures the displacement of this bar segment and its velocity is therefore du/dt. But the strain ε is uniform along the bar so we have $\varepsilon = u/y = x/L$. Therefore $u = xy/L$ and $du/dt = (y/L)\, dx/dt$. The segment's kinetic energy is

$$d(KE) = \frac{1}{2}\rho A\, dy\left(\frac{y}{L}\frac{dx}{dt}\right)^2 = \frac{1}{2}\frac{\rho A}{L^2}\left(\frac{dx}{dt}\right)^2 y^2 dy \qquad (6.33)$$

where ρ is the mass density of the bar material. Integrating over the bar, we get

$$KE = \frac{1}{2}\frac{\rho AL}{3}\left(\frac{dx}{dt}\right)^2 \qquad (6.34)$$

This says that effectively 1/3 of the mass of the bar is moving with the end velocity dx/dt .

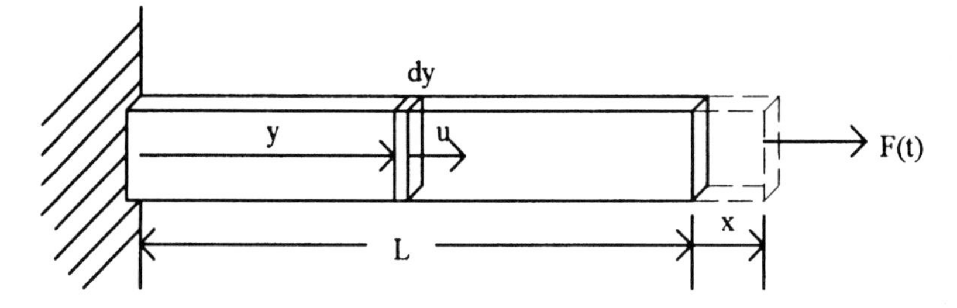

Figure 6.7 Elastic bar fixed at one end and subject to an applied force at the other end

We use Lagrange's method to obtain the equation of motion of the bar. The Lagrangian is $\mathcal{L}$ = KE - U and the equation of motion is

$$\frac{d}{dt}\left[\frac{\partial \mathcal{L}}{\partial(dx/dt)}\right] - \frac{\partial \mathcal{L}}{\partial x} = F \qquad (6.35)$$

Using (6.32) and (6.34), this becomes

$$\frac{d^2x}{dt^2} + \frac{3E}{\rho L^2}x = \left(\frac{3}{\rho AL}\right)F \qquad (6.36)$$

The force applied to the tie bars during a fault is produced by the coils. Although the forces applied to the coils are proportional to the current squared, because of the coil's internal structure, the force transmitted to the tie bars may be modified. However, assuming the coils are well clamped, we expect that the force transmitted to the tie bars is also proportional to the current squared to a good approximation. The fault current has the approximate form [Wat66]

$$I = I_o\left(e^{-at} - \cos\omega t\right)u(t) \tag{6.37}$$

where a is a constant which is a measure of the resistance in the circuit, ω is the angular frequency, and u(t) is the unit step function which is zero for times $t < 0$ and 1 for times $t \geq 0$. The force has the form

$$F = F_o\left(\frac{1}{2} + e^{-2at} - 2e^{-at}\cos\omega t + \frac{1}{2}\cos 2\omega t\right)u(t) \tag{6.38}$$

which results from squaring (6.37) and using a trigonometric identity. This function is sketched in Fig. 6.8a for $a = 22.6$ and $\omega = 2\pi(60)$. To simplify matters and maintain a worst case position, take $a = 0$ so (6.38) reduces to

$$F = F_o\left(\frac{3}{2} - 2\cos\omega t + \frac{1}{2}\cos 2\omega t\right)u(t) \tag{6.39}$$

This is plotted in Fig. 6.8b. It achieves a maximum of $F_{max} = 4\ F_o$ whereas (6.38) reaches a maximum value of about 3.3 F_o for the parameters used. If this maximum force were acting in a steady state manner, the bar's displacement would be, according to Hooke's law,

$$x_{max} = \frac{L}{EA}4F_o = \frac{L}{EA}F_{max} \tag{6.40}$$

This should be compared with the dynamical solution of (6.36) to get the enhancement factor.

To solve (6.36), take Laplace transforms, using the boundary conditions $x(t = 0^-) = 0$ and $dx/dt(t = 0^-) = 0$. We obtain

$$\left(s^2 + \frac{3E}{\rho L^2}\right)\mathcal{L}(x) = \left(\frac{3}{\rho AL}\right)\mathcal{L}(F) \tag{6.41}$$

The Laplace transform of (6.39) is

$$\mathcal{L}(F) = F_o\left[\frac{3}{2}\left(\frac{1}{s}\right) - 2\left(\frac{s}{s^2 + \omega^2}\right) + \frac{1}{2}\left(\frac{s}{s^2 + 4\omega^2}\right)\right] \tag{6.42}$$

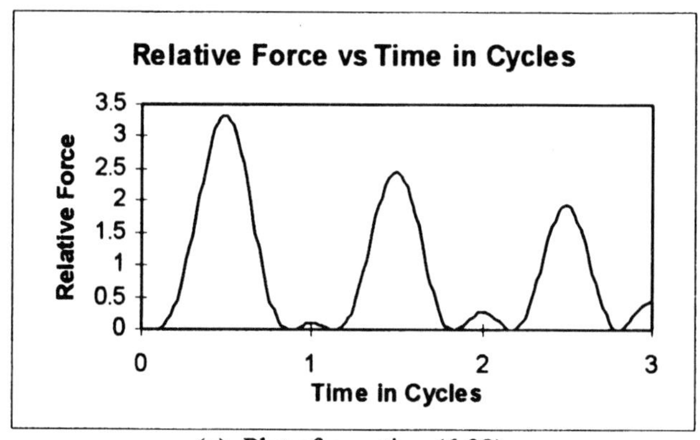

(a) Plot of equation (6.38)

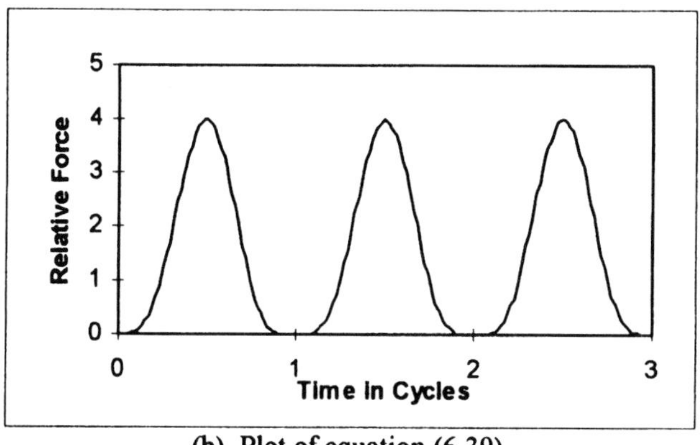

(b) Plot of equation (6.39)

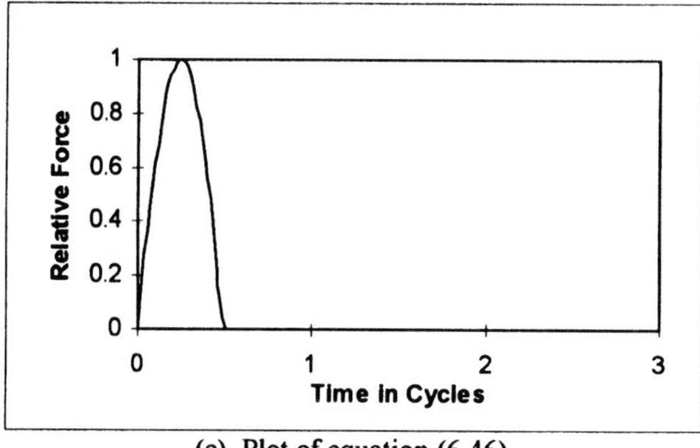

(c) Plot of equation (6.46)

Figure 6.8 Graphs of some tie bar forces versus time in cycles

Substituting into (6.41), we obtain for the Laplace transform of x

$$\ell(x) = \left(\frac{3}{\rho AL}\right)\frac{F_o}{s^2+b^2}\left[\frac{3}{2}\left(\frac{1}{s}\right) - 2\left(\frac{s}{s^2+\omega^2}\right) + \frac{1}{2}\left(\frac{s}{s^2+4\omega^2}\right)\right] \quad (6.43)$$

where $b^2 = 3E/\rho L^2$. Using some algebra to rewrite (6.43) and taking inverse transforms, we obtain

$$x(t) = \left(\frac{L}{EA}\right)F_{max}u(t)$$

$$\times\left\{\frac{1}{4}\left[\frac{3}{2}(1-\cos bt) + \frac{2}{1-(\omega/b)^2}(\cos bt - \cos\omega t) - \frac{1}{2}\left(\frac{1}{1-4(w/b)^2}\right)(\cos bt - \cos 2\omega t)\right]\right\}$$

(6.44)

where the quantity in curly brackets is the enhancement factor.

We need to compare the natural angular frequency b with the applied angular frequencies $\omega = 2\pi f = 377$ Rads/sec, assuming $f = 60$ Hz, and $2\omega = 754$ Rads/sec to see whether a resonance problem might occur. For steel bars, $E = 2.07\times10^5$ MPa (30×10^6 psi), $\rho g = 7.68\times10^4$ N/m^3 (0.283 lb/in^3), where g is the acceleration of gravity = 9.8 m/sec^2 (386 in/sec^2). Thus, we obtain for b,

$$b = \sqrt{\frac{3Eg}{(\rho g)L^2}} = \frac{8.9\times10^3}{L(m)}\text{Rads/sec} \quad (6.45)$$

For a 2.54m (100 in) long tie bar which is typical, this gives $b = 3504$ Rads/sec. Since this is much larger than ω or 2ω, we are far from resonance. Thus $(\omega/b)^2$ can be ignored relative to 1 in (6.44) and it simplifies to $x(t) = (L/EA)\,F$ with F given by (6.39). Hence x_{max} is the same as in the steady-state case and the enhancement factor is 1. Thus, unless the applied or twice the applied frequency is close to the tie bar's natural frequency, there is no dynamic enhancement. By numerically checking over a large grid of times and frequencies, a maximum enhancement of about 1.66 is produced if the time does not exceed one period ($\omega t < 2\pi$). However as time increases, the maximum enhancement factor gradually increases when we are near resonance. This is to be expected since there is no damping in the problem and no cut-off of the force.

Another perhaps more realistic approximation to the applied force is provided by a half wave pulse given by

$$F(t) = F_{max}\left[u(t)\sin\omega t + u\left(t - \frac{\pi}{\omega}\right)\sin\omega\left(t - \frac{\pi}{\omega}\right)\right] \tag{6.46}$$

where u(t) is a unit step at $t = 0$ and $u(t - \pi/\omega)$ is a unit step at $t = \pi/\omega$. This is illustrated in Fig. 6.8c. The Laplace transform of (6.46) is

$$\mathcal{L}(F) = \frac{\omega\left(1 + e^{-\frac{\pi}{\omega}s}\right)}{s^2 + \omega^2} \tag{6.47}$$

Substituting into (6.41), we obtain for the Laplace transform of x,

$$\mathcal{L}(x) = \left(\frac{3}{\rho AL}\right)F_{max}\frac{\omega\left(1 + e^{-\frac{\pi}{\omega}s}\right)}{\left(s^2 + b^2\right)\left(s^2 + \omega^2\right)} \tag{6.48}$$

Taking the inverse transform, we get

$$x = \left(\frac{L}{EA}\right)F_{max}$$

$$\times\left\{\frac{1}{1 - \left(\frac{\omega}{b}\right)^2}\left[u(t)\left(\sin\omega t - \frac{\omega}{b}\sin bt\right) + u\left(t - \frac{\pi}{\omega}\right)\left(\sin\omega\left(t - \frac{\pi}{\omega}\right) - \frac{\omega}{b}\sin b\left(t - \frac{\pi}{\omega}\right)\right)\right]\right\} \tag{6.49}$$

where the quantity in curly brackets is the enhancement factor. For $b \gg \omega$, we get the static response $x = (L/EA)\,F$, with F given by (6.46), i.e. the bar extension is just proportional to the force with no enhancement. By numerically checking over a large grid of times and frequencies, the maximum enhancement factor obtained was 1.77 .

In practice, we use a force of 1.8 times the end thrust provided it is larger than 0.8 times the maximum compressive force over all the windings. Otherwise we use 0.8 times the maximum compressive force over all the windings. However, since the tie bars must support the weight of the core and coils during lifting, we check the stress in the tie bars produced by lifting. During lifting, we only assume the tie bars associated with the outer legs are stressed since this is where the lifting hooks are positioned. Both the short circuit dynamic stresses and the lifting stresses must be below a maximum allowable stress in the tie bar material. We take this maximum allowable stress to be 620 MPa (90,000 psi) if a low carbon steel is used and 414 MPa (60,000 psi) if a stainless steel is used for the tie bar material.

6.3.5 Stress in the Pressure Rings

The pressure ring receives the total end thrust of the windings. In our designs, it is made of pressboard of about 3.8 to 6.35 cm (1.5 to 2.5 in) thickness. The ring covers the radial build of the windings with a little overhang. During a fault, it must support the full dynamic end thrust of the windings, which according to the last section is 1.8 times the total end thrust or 0.8 times the maximum compressive force in all the windings whichever is larger.

The end thrust or force is distributed over the end ring, producing an effective pressure of

$$P_{ring} = \frac{F_{ring}}{A_{ring}} \tag{6.50}$$

where we use ring to label the end force, F_{ring}, and ring area, A_{ring}. This area is given by

$$A_{ring} = \frac{\pi}{4}\left(D_{ring,out}^2 - D_{ring,in}^2\right) \tag{6.51}$$

in terms of the outer and inner ring diameters. The ring is supported on radial blocks with space between for the leads. This produces an unsupported span of a certain length L_u. To a good approximation, the problem is similar to that discussed previously for the axial bending of a strand of wire. Thus we can use formula (6.16) for the maximum stress in the end ring, with $L = L_u$, $t = (1/2)(D_{ring,out} - D_{ring,in})$ the radial build of the ring, $h = h_{ring}$ the ring's thickness, and $q = P_{ring}\, t$ the force/unit length along the unsupported span. We obtain

$$\sigma_{x,max} = \frac{P_{ring}}{2}\left(\frac{L_u}{h_{ring}}\right)^2 = \frac{F_{ring}}{2A_{ring}}\left(\frac{L_u}{h_{ring}}\right)^2 \tag{6.52}$$

For our pressboard rings, the maximum bending stress permissible is σ_{bend} = 103 MPa (15,000 psi). Substituting this value for $\sigma_{x,max}$ and solving for F_{ring} corresponding to this limiting stress, we find

$$F_{ring,max} = \sigma_{bend}\left(\frac{\pi}{2}\right)\left(D^2_{ring,out} - D^2_{ring,in}\right)\left(\frac{h_{ring}}{L_u}\right)^2 \tag{6.53}$$

This is the maximum end force the pressure ring can sustain. It must be greater than the applied maximum end force.

6.3.6 Hoop Stress

The maximum radial pressure acting on the winding as obtained from the force program creates a hoop stress in the winding conductor. The hoop stress is tensile or compressive, depending on whether the pressure acts radially outwards or inwards respectively. In Fig. 6.9, we treat the winding as an ideal cylinder or ring subjected to a radially inward pressure, P_r. Let R_m be the mean radius of the cylinder and H its axial height. In part (b) of the figure, we show 2 compressive reaction forces F in the winding, sustaining the force applied to half the cylinder. The x directed force produced by the pressure P_r cancels out by symmetry and the net y directed force acting downward is given by

$$F_y = P_r H R_m \int_0^{\pi} \sin\varphi \, d\varphi = 2 P_r H R_m \tag{6.54}$$

This is balanced by a force of 2F acting upward so we get $F = P_r H R_m$. Dividing by the cross sectional area A of the material sustaining the force, we get the compressive stress in the material

$$\sigma_{hoop} = \frac{F}{A} = \frac{P_r H R_m}{A} \tag{6.55}$$

For $A = HB$, where B is the radial build of the cylinder, we get

$$\sigma_{hoop} = \frac{P_r R_m}{B} \tag{6.56}$$

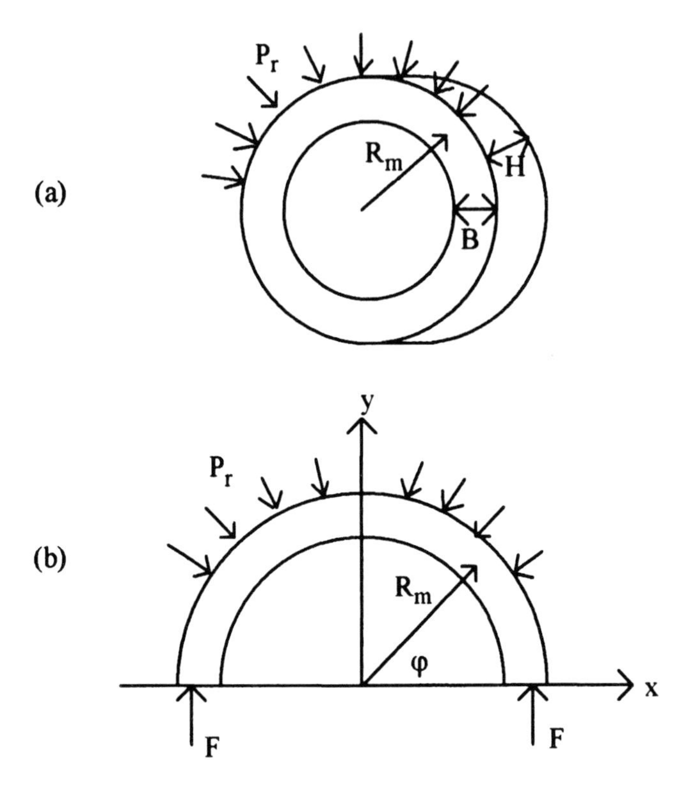

Figure 6.9 Geometry for determining the hoop stress in a cylinder acted on by a radially inward pressure

This last formula assumes the cylinder is made of a homogeneous material. If the cylinder is made of conductors and insulating materials, the conductors primarily support the forces. In this case A should equal the cross-sectional area of all the conductors in the winding, $A = A_t N_t$, where A_t is the cross-sectional area of a turn and N_t is the total number of turns in the winding. If the winding is center fed, i.e. consists of two parallel windings on the same leg, and N_t refers to the total turns/leg or twice the number of electrical turns, then A_t should be 1/2 the turn area. Substituting into (6.55) and using $D_m = 2R_m$, we obtain

$$\sigma_{hoop} = \frac{P_r H D_m}{2 A_t N_t} \tag{6.57}$$

for the hoop stress. This is compressive for P_r acting inward and tensile for P_r acting outward. In either case, this stress should not exceed the proof stress of the winding material.

When the radial pressure acts inwards, the winding is apt to buckle before the proof stress is exceeded. This inward radial buckling is a complex process to analyze. We will present an idealized analysis later. Based on limited experimental results, it has been suggested that this compressive hoop stress not exceed some fraction of the proof stress, the fraction varying from 0.4 to 0.7 depending on the type of cable used and whether it is bonded.

6.3.7 Radial Bending Stress

Windings have inner radial supports such as sticks made of pressboard which are spaced uniformly along their circumference and extend the height of the winding. When an inward radial pressure acts on the winding, the sections of the winding between supports act like a curved beam subjected to a uniform loading. A similar situation occurs in the case of a rotating flywheel with radial spokes. In the flywheel case, the loading (centrifugal force) acts outwards but otherwise the analysis is similar. The flywheel example is analyzed in Timoshenko [Tim56] which we follow here with minor changes.

We need to make use of Castigliano's theorem. This states that if the material of a system follows Hooke's law, i.e. remains within the elastic limit, and if the displacements are small, then the partial derivative of the strain energy with respect to any force equals the displacement corresponding to the force. Here force and displacement have a generalized meaning, i.e. they could refer to torques or moments and angular displacements as well as their usual meanings of force and length displacements. Also the strain energy must be expressed as a quadratic function of the forces. For example, the strain energy associated with tensile or compressive forces N in a beam of length L is

$$U_{tensile} = \int_0^L \frac{N^2}{2EA} dx \tag{6.58}$$

where N , A , and E , can be functions of position along the beam , x. The strain energy associated with a bending moment M in a beam of length L is

$$U_{bending} = \int_0^L \frac{M^2}{2EI} dx \tag{6.59}$$

where M, I, and E can be functions of position, x. Here I is the area moment of inertia.

In Fig. 6.10a, we show a portion of the winding with the inner radial supports spaced an angle 2α apart. There is a normal force X acting radially outwards at the supports which counters the inward pressure which has been converted to a force/unit length q acting on the coil section. The coil section is assumed to form a closed ring of radial build h, axial height t, and mean radius R.

In Fig. 6.10b, we further isolate a portion of the ring which extends between adjacent mid-sections between the supports. The reason for doing this is that there is no radially directed (shearing) force acting on these mid cross-sections. This is because by symmetry, the distributed load between the mid-sections must balance the outward force at the included support. Thus the only reactions at the midsections are an azimuthally directed force N_o and a couple M_o which need to be found.

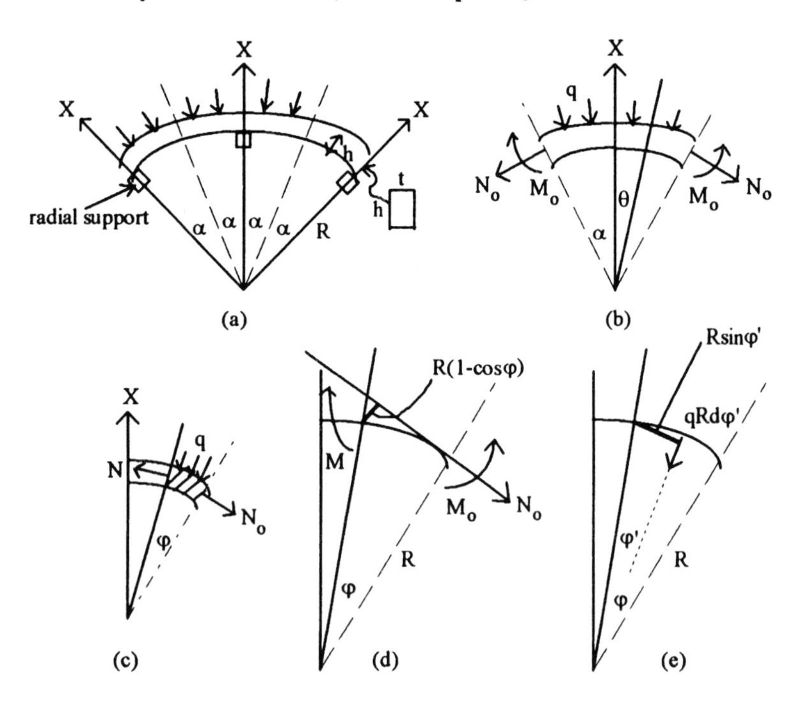

Figure 6.10 Geometry for determining the radial bending stresses

Balancing the vertical forces, we have

$$x - 2N_o \sin\alpha - 2qr\int_0^\alpha \cos\theta d\theta = 0 \qquad (6.60)$$

Performing the integral and solving for N_o, we obtain

$$N_o = \frac{X}{2\sin\alpha} - qR \qquad (6.61)$$

At any cross-section as shown in Fig. 6.10c, measuring angles from the mid-section position with the variable φ, we can obtain the normal force, N, from the static equilibrium requirement

$$N = N_o \cos\varphi - qR\int_0^\varphi \sin\varphi' d\varphi' \qquad (6.62)$$

Integrating and substituting for N_o from (6.61), we get

$$N = \frac{X\cos\varphi}{2\sin\alpha} - qR \qquad (6.63)$$

Similarly, using Fig. 6.10d,e, we can obtain the bending moment at the cross-section an angle φ from the mid-section by balancing the moments

$$M = M_o - N_o R(1 - \cos\varphi) - qR^2\int_0^\varphi \sin\varphi' d\varphi' \qquad (6.64)$$

Performing the integration and substituting for N_o from (6.61), we obtain

$$M = M_o + \frac{XR(\cos\varphi - 1)}{2\sin\alpha} \qquad (6.65)$$

Equations (6.63) and (6.65) express the normal force and bending moments as functions of position along the beam (arc in this case). These can be used in the energy expressions (6.58) and (6.59). Castigliano's theorem can then be used to solve for the unknowns.

However, we are missing the strain energy associated with the supports. The radial supports consist of several different materials as illustrated in Fig. 6.11. We assume they can be treated as a column of uniform cross-sectional area A_{stick}. In general the column consists of winding material (copper), pressboard sticks, and core steel. However, for an innermost winding, the winding material is not present as part of the support column. For such a composite structure, we derive an equivalent Young's modulus, E_{eq}, by making use of the fact that the stress is the same throughout the column. Only the strain differs from material to material. We obtain

$$E_{eq} = \frac{L}{\frac{L_w}{E_w} + \frac{L_s}{E_s} + \frac{L_c}{E_c}} \tag{6.66}$$

where L_w is the length of the winding portion and E_w its Young's modulus. Similarly s refers to the stick and c to the core parameters. $L = L_w + L_s + L_c$ is the total column length.

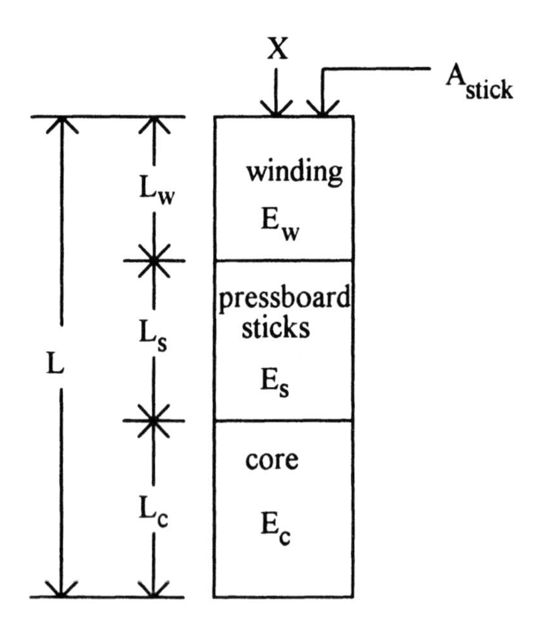

Figure 6.11 Radial support structure

The total strain energy for our system, retaining only the portion shown in Fig. 6.10b, since the entire ring energy is simply a multiple of this, can be written

$$U = 2\int_0^\alpha \frac{N^2 R}{2EA} d\varphi + 2\int_0^\alpha \frac{M^2 R}{2EI} d\varphi + \frac{X^2 L}{2E_{eq} A_{stick}} \tag{6.67}$$

where N and M are given be (6.63) and (6.65), A is the cross-sectional area of the ring, $A = th$, I its bending moment, $I = th^3/12$, and the infinitesimal length along the bar, $Rd\varphi$, is used. The two unknowns are X and M_o. At the fixed end of the support column (center of the core) the displacement is zero, hence by Castigliano's theorem $\partial U/\partial X = 0$. Also the bending moment at the mid-section of the span between the supports produces no angular displacement by symmetry. Hence, by Castigliano's theorem $\partial U/\partial M_o = 0$. Differentiating (6.67), we obtain

$$\frac{\partial U}{\partial X} = \frac{2R}{EA}\int_0^\alpha N \frac{\partial N}{\partial X} d\varphi + \frac{2R}{EI}\int_0^\alpha M \frac{\partial M}{\partial X} d\varphi + \frac{XL}{E_{eq} A_{stick}} = 0 \tag{6.68}$$

and

$$\frac{\partial U}{\partial M_o} = \frac{2R}{EA}\int_0^\alpha N \frac{\partial N}{\partial M_o} d\varphi + \frac{2R}{EI}\int_0^\alpha M \frac{\partial M}{\partial M_o} d\varphi = 0 \tag{6.69}$$

Substituting for N and M from (6.63) and (6.65), we obtain

$$\frac{2R}{EA}\left\{\int_0^\alpha \left(\frac{X\cos\varphi}{2\sin\alpha} - qR\right)\frac{\cos\varphi}{2\sin\alpha} d\varphi + \frac{A}{I}\int_0^\alpha \left[M_o + \frac{XR(\cos\varphi - 1)}{2\sin\alpha}\right]\frac{R(\cos\varphi - 1)}{2\sin\alpha} d\varphi\right\}$$
$$+ \frac{XL}{E_{eq} A_{stick}} = 0 \tag{6.70}$$

and

$$\frac{2R}{EI}\int_0^\alpha \left[M_o + \frac{XR(\cos\varphi - 1)}{2\sin\alpha}\right] d\varphi = 0 \tag{6.71}$$

Integrating the above expressions, we obtain

$$M_o = \frac{XR}{2}\left(\frac{1}{\sin\alpha} - \frac{1}{\alpha}\right) \tag{6.72}$$

and

$$X = \frac{qR}{\frac{1}{4\sin^2\alpha}\left(\alpha + \frac{\sin 2\alpha}{2}\right) + \frac{AR^2}{I}\left[\frac{1}{4\sin^2\alpha}\left(\alpha + \frac{\sin 2\alpha}{2}\right) - \frac{1}{2\alpha}\right] + \frac{EA}{E_{eq}A_{stick}}} \tag{6.73}$$

where we have used the fact that $L = R$. Substituting into (6.63) and (6.65) and defining

$$f_1(\alpha) = \frac{1}{4\sin^2\alpha}\left(\alpha + \frac{\sin 2\alpha}{2}\right)$$

$$f_2(\alpha) = \frac{1}{4\sin^2\alpha}\left(\alpha + \frac{\sin 2\alpha}{2}\right) - \frac{1}{2\alpha} \tag{6.74}$$

we obtain for N and M

$$N = qR\left[\frac{\cos\varphi}{2\sin\alpha}\left(\frac{1}{f_1(\alpha) + \frac{AR^2}{I}f_2(\alpha) + \frac{EA}{E_{eq}A_{stick}}}\right) - 1\right] \tag{6.75}$$

$$M = \frac{qR^2}{2}\left(\frac{\cos\varphi}{\sin\alpha} - \frac{1}{\alpha}\right)\left(\frac{1}{f_1(\alpha) + \frac{AR^2}{I}f_2(\alpha) + \frac{EA}{E_{eq}A_{stick}}}\right) \tag{6.76}$$

We also have

$$\frac{AR^2}{I} = 12\left(\frac{R}{h}\right)^2 \tag{6.77}$$

where h is the radial build of the ring. N gives rise to a normal stress

$$\sigma_N = \frac{N}{A} = \frac{qR}{A}\left[\frac{\cos\varphi}{2}\left(\frac{1}{\sin\alpha}\right)\left(\frac{1}{f_1(\alpha)+12\left(\frac{R}{h}\right)^2 f_2(\alpha)+\frac{EA}{E_{eq}A_{stick}}}\right)-1\right] \tag{6.78}$$

and M gives rise to a bending stress which varies over the cross-section, achieving a maximum tensile or compressive value of (see formula (6.14))

$$\sigma_M = \frac{Mh}{2I} = \frac{qR}{A}\left(\frac{3R}{h}\right)\left(\cos\varphi - \frac{\sin\alpha}{\alpha}\right)\left(\frac{1}{\sin\alpha}\right)\left(\frac{1}{f_1(\alpha)+12\left(\frac{R}{h}\right)^2 f_2(\alpha)+\frac{EA}{E_{eq}A_{stick}}}\right) \tag{6.79}$$

We have factored out the term qR/A since this can be shown to be the hoop stress in the ring. (In formula (6.55), PH corresponds to q and R_m to R in our development here.) Thus σ_{hoop} will be substituted in the following formulas for qR/A, where σ_{hoop} is given by (6.57).

We now need to add σ_N and σ_M in such a way as to produce the worst case stress in the ring. The quantity

$$\frac{1}{\sin\alpha\left(f_1(\alpha)+12\left(\frac{R}{h}\right)^2 f_2(\alpha)+\frac{EA}{E_{eq}A_{stick}}\right)} \tag{6.80}$$

occurs in both stress formulas. We have tabulated $f_1(\alpha)$, $f_2(\alpha)$, and $\sin\alpha$ times these in Table 6.1 for a range of α values.

Table 6.1 Tabulated values for $f_1(\alpha)$ and $f_2(\alpha)$ and $\sin\alpha$ times these quantities

# Sticks /Circle	α (degs)	$f_1(\alpha)$	$f_2(\alpha)$	$\sin\alpha\, f_1(\alpha)$	$\sin\alpha\, f_2(\alpha)$
2	90	0.3927	0.07439	0.3927	0.07439
4	45	0.6427	0.006079	0.4545	0.004299
6	30	0.9566	0.001682	0.4783	0.0008410
8	22.5	1.2739	0.0006931	0.4875	0.0002652
10	18	1.5919	0.0003511	0.4919	0.0001085
12	15	1.9101	0.0002020	0.4944	0.00005228
18	10	2.8648	0.00005942	0.4975	0.00001032
24	7.5	3.8197	0.00002500	0.4986	0.00000326
30	6	4.7747	0.00001279	0.4991	0.00000134

It can be see that for any choice of the other parameters (6.80) is positive. In (6.79), the magnitude of σ_M for fixed α is determined by the term $(\cos\varphi - \sin\alpha/\alpha)$, where φ can range from 0 to α. This achieves a maximum in absolute value at $\varphi = \alpha$. The stress can have either sign depending on whether it is on the inner radial or outer radial surface of the ring. In (6.78), the magnitude of σ_N achieves a maximum at $\varphi = \alpha$ for virtually any choice of the other parameters. It has a negative sign consistent with the compressive nature of the applied force. Thus σ_N and σ_M should be added with each having a negative sign at $\varphi = \alpha$ to get the maximum stress. We obtain

$$\sigma_{max} = \sigma_{hoop}\left[\frac{\cos\alpha}{2\sin\alpha}\left(\frac{1}{f_1(\alpha)+12\left(\frac{R}{h}\right)^2 f_2(\alpha)+\frac{EA}{E_{eq}A_{stick}}}\right)-1\right.$$
$$\left.+\frac{3R}{h}\left(\cos\alpha-\frac{\sin\alpha}{\alpha}\right)\left(\frac{1}{\sin\alpha}\right)\left(\frac{1}{f_1(\alpha)+12\left(\frac{R}{h}\right)^2 f_2(\alpha)+\frac{EA}{E_{eq}A_{stick}}}\right)\right] \tag{6.81}$$

This stress is negative although it is usually quoted as a positive number. It occurs at the support.

We have analyzed a ring subjected to a hoop stress having radial supports. A coil is usually not a monolithic structure but consists of a number of cables radially distributed. The cables could consist of a single strand of conductor as in the case of magnet wire or be multi-stranded. The latter could also be bonded. The average hoop stress in the winding will be nearly the same in all the cables since the paper insulation tends to equalize it. We will examine this in more detail in a later section. The radial thickness, h in the formulas, should refer to a single cable. If it is magnet wire, then its radial thickness should be used. If multi-stranded transposed cable, then something less than its radial thickness should be used since this is not a homogeneous material. If unbonded, we use twice the thickness of an individual strand as its effective radial build. If bonded, we use 80% of its actual radial thickness as its effective radial build.

6.4 RADIAL BUCKLING STRENGTH

Buckling occurs when a sufficiently high force causes a structure to deform its shape to the point where it becomes destabilized and may collapse. The accompanying stresses may in fact be small and well below the proof stress of the material. An example is a slender column subjected to an axial compressive force. At a certain value of the force, a slight lateral bulge in the column could precipitate a collapse. Another example, which we will pursue here, is that of a thin ring subjected to a uniform compressive radial pressure. A slight deformation in the circular shape of the ring could cause a collapse if the radial pressure is high enough. This critical radial pressure produces a hoop stress (see Section 3.6). It is called the critical hoop stress, which could be well below the proof stress of the material. In general, the smaller the ratio of the radial build to the radius, the smaller the critical hoop stress. Thus, buckling is essentially an instability problem and is analyzed by assuming a small distortion in the shape of the system under study and determining under what conditions this leads to collapse.

We wish to examine the possible buckling of a winding subjected to an inward radial pressure. We will treat an individual cable of the winding as a closed ring as was done in the last section, since the cables are not bonded to each other. In free radial buckling, it is assumed that there are no inner supports. Thus the sticks spaced around the inside circumference of a winding are assumed to be absent. It is argued that there is sufficient looseness in this type of support that the onset of buckling occurs as if these supports were absent and once started, the buckling process continues towards collapse or permanent deformation. It could also be argued that even though buckling may begin in the absence of supports, before it progresses very far the supports are engaged and from then on it becomes a different type of buckling, called forced or constrained buckling. The key to the last argument is that even though free buckling has begun, the stresses in the material are quite low, resulting in no permanent deformation and the process is halted before collapse can occur.

We will analyze free buckling of a circular ring in some detail and quote the results for forced buckling. The lowest order shape distortion away from a circle is taken to be an ellipse [Tim56]. This is shown in Fig. 6.12 along with the parameters used to describe the system. With u measuring the displacement radially inward from the circular shape, the differential equation for the deflection of a thin bar (ring) with circular center line of radius R is [Tim56]

$$\frac{d^2u}{ds^2}+\frac{u}{R^2}=-\frac{M}{EI} \tag{6.82}$$

where s is the length along the bar. Since $s = Rd\varphi$ where the angle φ is shown in Fig. 6.12, $ds^2 = R^2 d\varphi^2$, and (6.82) can be written

$$\frac{d^2u}{d\varphi^2}+u=-\frac{R^2M}{EI} \tag{6.83}$$

Here M is the bending moment along the ring, E is Young's modulus, and I the area moment of inertia.

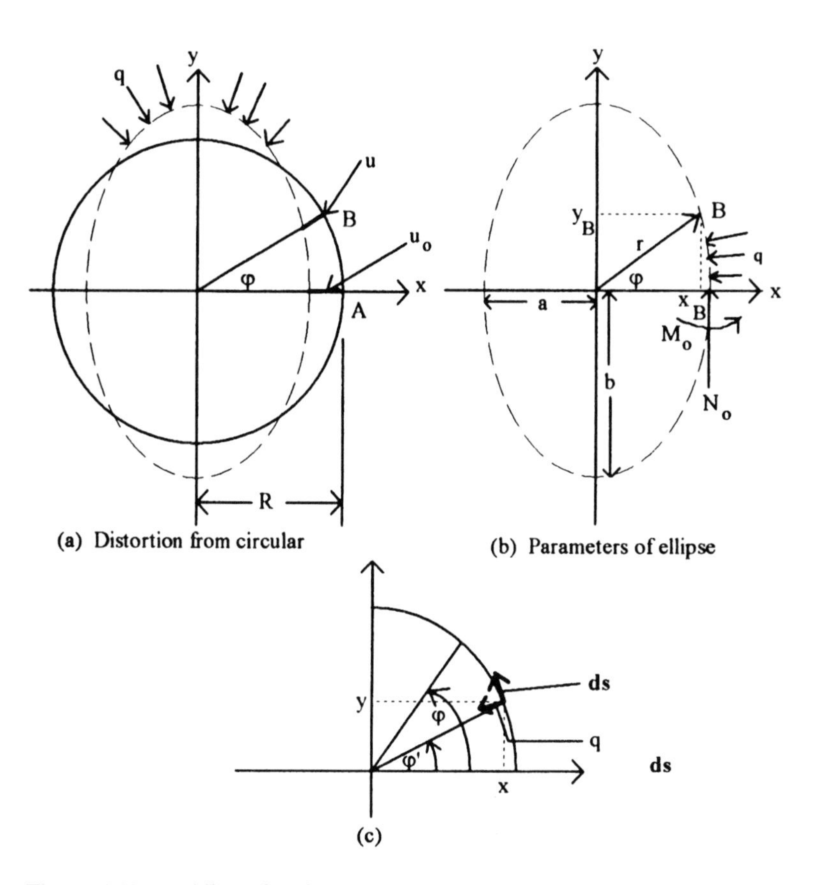

Figure 6.12 Buckling of a circular ring. The distorted elliptical shape and its parametrization are shown.

By symmetry, we need only consider one quadrant of the ring and we have chosen the upper right quadrant. We wish to determine the bending moment at a cross-section B at an angle φ from the x axis as shown in Fig. 6.12. Consider the equilibrium of the section of the ring from A to B. At the cross-section A there is an unknown normal force N_o and an unknown bending moment M_o acting. From symmetry, no shearing force acts at A. The bending moment at B has a contribution from M_o, from N_o, and from the distributed force/unit length q acting on the segment from A to B.

The moment at B due to M_o is just M_o and the moment at B due to N_o is

$$M_{N_o} = N_o(a - x_B) \tag{6.84}$$

where x_B is the x-coordinate of position B as shown in Fig. 6.12b. N_o must balance the net force acting downward on one quadrant of the ellipse. The infinitesimal vectorial length along the ellipse is (Fig. 6.12c)

$$\mathbf{ds} = dx\,\mathbf{i} + dy\,\mathbf{j} = dx\left(\mathbf{i} + \frac{dy}{dx}\mathbf{j}\right) \tag{6.85}$$

The force distribution q acts perpendicular to this length so that the force on an element of length ds is given in magnitude and direction by

$$\mathbf{dq} = q|\mathbf{ds}| = \left[\frac{\frac{dy}{dx}\mathbf{i} - \mathbf{j}}{\sqrt{1 + \left(\frac{dy}{dx}\right)^2}}\right] q|\mathbf{ds}| = \left(\frac{dy}{dx}\mathbf{i} - \mathbf{j}\right) q\,dx \tag{6.86}$$

Thus the downward force acting on the quadrant is given by

$$\mathbf{N}_{down} = \int_{quadrant} (\mathbf{dq})_y = -\mathbf{j}q \int_{quadrant} dx = -\mathbf{j}qa \tag{6.87}$$

Thus N_o acts up with a magnitude of qa so that (6.84) becomes

$$M_{N_o} = qa^2\left(1 - \frac{x_B}{a}\right) \tag{6.88}$$

We need to compute the contribution to the moment at B from the forces **qds** and integrate these from A to B. This is given by

$$\begin{aligned} dM_q &= (\mathbf{dq})_x(y_B - y) + (\mathbf{dq})_y(x - x_B) \\ &= q\,dx\left[\left(\frac{dy}{dx}\right)(y_B - y) + (x_B - x)\right] \end{aligned} \tag{6.89}$$

where (x,y) are the coordinates of **qds** and B labels the coordinates of position B. From the equation of an ellipse,

$$\left(\frac{x}{a}\right)^2+\left(\frac{y}{b}\right)^2=1 \tag{6.90}$$

we obtain

$$y=\frac{b}{a}\sqrt{a^2-x^2} \quad , \quad \frac{dy}{dx}=-\frac{b}{a}\frac{x}{\sqrt{a^2-x^2}} \tag{6.91}$$

Substituting into (6.89), we obtain

$$dM_q=q\,dx\left\{-\frac{b}{a}\frac{x}{\sqrt{a^2-x^2}}y_B+\left[\left(\frac{b}{a}\right)^2-1\right]x+x_B\right\} \tag{6.92}$$

Integrating from $x = x_B$ to $x = a$, we get

$$\begin{aligned} M_q &= q\left\{-\left(\frac{b}{a}\right)^2\left(a^2-x_B^2\right)+\left[\left(\frac{b}{a}\right)^2-1\right]\frac{\left(a^2-x_B^2\right)}{2}+x_B\left(a-x_B\right)\right\} \\ &= -\frac{q}{2}\left\{\left(\frac{b}{a}\right)^2\left(a^2-x_B^2\right)+\left(a-x_B\right)^2\right\} \\ &= -\frac{q}{2}\left(a-x_B\right)\left\{\left(\frac{b}{a}\right)^2\left(a+x_B\right)+\left(a-x_B\right)\right\} \\ &= -\frac{q}{2}\left(1-\frac{x_B}{a}\right)\left\{b^2\left(1+\frac{x_B}{a}\right)+a^2\left(1-\frac{x_B}{a}\right)\right\} \end{aligned} \tag{6.93}$$

Adding the contributions to M at B, M_o, (6.88) and (6.93), we get

$$M=M_o-\frac{q}{2}\left(b^2-a^2\right)\left[1-\left(\frac{x_B}{a}\right)^2\right] \tag{6.94}$$

In terms of u_o in Fig. 6.12, we have for u_o small compared with R

$$a=R-u_o \quad , \quad b=R+u_o \tag{6.95}$$

Therefore

$$b^2-a^2=4Ru_o \tag{6.96}$$

The equation of an ellipse in polar coordinates is

$$r = \frac{b}{\sqrt{1+\left(\frac{b^2-a^2}{a^2}\right)\cos^2\varphi}} \cong \frac{b}{\sqrt{1+\frac{4Ru_o}{a^2}\cos^2\varphi}} \qquad (6.97)$$

where (6.96) has been used to obtain the second approximate equality. For small deviations from a circular shape ($u_o \ll R$), (6.97) becomes approximately

$$r \cong b\left(1-\frac{2Ru_o}{a^2}\cos^2\varphi\right) \qquad (6.98)$$

Using (6.95), this becomes

$$\begin{aligned} r &\cong (R+u_o)\left[1-\frac{\frac{2u_o}{R}}{\left(1-\frac{u_o}{R}\right)^2}\cos^2\varphi\right] \\ &\cong (R+u_o)\left[1-\frac{2u_o}{R}\left(1+\frac{2u_o}{R}\right)\cos^2\varphi\right] \end{aligned} \qquad (6.99)$$

Retaining only terms linear in u_o, we get

$$r \cong R+u_o-2u_o\cos^2\varphi \qquad (6.100)$$

Letting $x_B = a\cos\varphi$ for small u_o, (6.100) becomes

$$r \cong R+u_o-2u_o\left(\frac{x_B}{a}\right)^2 \qquad (6.101)$$

From Fig. 6.12,

$$u = R-r = 2u_o\left(\frac{x_B}{a}\right)^2-u_o = u_o\left[2\left(\frac{x_B}{a}\right)^2-1\right] \qquad (6.102)$$

Therefore,

$$1-\left(\frac{x_B}{a}\right)^2=\frac{1}{2}\left(1-\frac{u}{u_o}\right) \tag{6.103}$$

Substituting (6.96) and (6.103) into (6.94), we obtain

$$M=M_o-qR(u_o-u) \tag{6.104}$$

Substituting (6.104) into (6.83), the differential equation becomes

$$\frac{d^2u}{d\varphi^2}+\left(1+\frac{qR^3}{EI}\right)u=-\frac{R^2}{EI}(M_o-qRu_o) \tag{6.105}$$

The general solution to this equation is

$$u=A\sin(p\varphi)+B\cos(p\varphi)-R^2\left(\frac{M_o-qRu_o}{EI+qR^3}\right) \tag{6.106}$$

where A and B are constants to be determined by the boundary conditions and

$$p=\sqrt{1+\frac{qR^3}{EI}} \tag{6.107}$$

We can see from the shape of the ellipse that a choice of boundary conditions for (6.105) is

$$\left.\frac{du}{d\varphi}\right|_{\varphi=0}=0 \quad \text{and} \quad \left.\frac{du}{d\varphi}\right|_{\varphi=\pi/2}=0 \tag{6.108}$$

These yield

$$A=0 \quad \text{and} \quad pB\sin\left(p\frac{\pi}{2}\right)=0 \tag{6.109}$$

The second equation above can be satisfied if $p=2n$ for $n=1,2,3,\ldots$. For $n=1$, (6.107) becomes

$$2 = \sqrt{1 + \frac{qR^3}{EI}} \tag{6.110}$$

which can be re-expressed as

$$q = \frac{3EI}{R^3} \tag{6.111}$$

This is the critical buckling force/unit length since it is associated with an angular dependence of u which corresponds to an elliptical shape.

We can find the critical hoop stress associated with q in (6.111) by means of the expression $\sigma_{hoop} = qR/A$ developed earlier (see the discussion following formula (6.79)). Thus

$$\sigma_{crit} = \frac{qR}{A} = \frac{3EI}{AR^2} \tag{6.112}$$

For a ring with a radial build of h and an axial height of t (see Fig. 6.10), we have $A = th$ and $I = th^3/12$ so that (6.112) becomes

$$\sigma_{crit} = \frac{1}{4}E\left(\frac{h}{R}\right)^2 \tag{6.113}$$

Thus the critical hoop stress depends geometrically only on the ratio of the radial build to the radius of the ring.

Since buckling occurs after the stress has built up in the ring to the critical value and increases incrementally beyond it, the appropriate modulus to use in (6.113) is the tangential modulus since this is associated with incremental changes. This argument for using the tangential modulus is based on an analogous argument for the buckling of slender columns together with supporting experimental evidence given in Ref. [Tim56]. We will assume it applies to thin rings as well.

The tangential modulus can be obtained graphically from the stress-strain curve for the material as illustrated in Fig. 6.13a. However, for copper which is the material of interest here, the stress-strain curve can be parametrized for copper of different hardnesses according to [Tho79] by

$$\sigma = \frac{E_o \varepsilon}{\left[1 + k\left(\frac{\sigma}{\sigma_o}\right)^m\right]} \tag{6.114}$$

where $k = 3/7$, $m = 11.6$, and $E_o = 1.10 \times 10^5$ MPa (16×10^6 psi). σ_o depends on the copper hardness. The tangential modulus obtained from this is

$$E_t = \frac{d\sigma}{d\varepsilon} = \frac{E_o}{\left[1 + \gamma\left(\frac{\sigma}{\sigma_o}\right)^m\right]} \tag{6.115}$$

where $\gamma = k(m+1) = 5.4$. Substituting E_t from (6.115) for E in (6.113), we obtain a formula for self-consistently determining the critical stress,

$$\sigma_{crit}^{m+1} + \frac{\sigma_o^m}{\gamma}\sigma_{crit} - \frac{\sigma_o^m}{4\gamma}E_o\left(\frac{h}{R}\right)^2 = 0 \tag{6.116}$$

This can be solved by Newton-Raphson iteration.

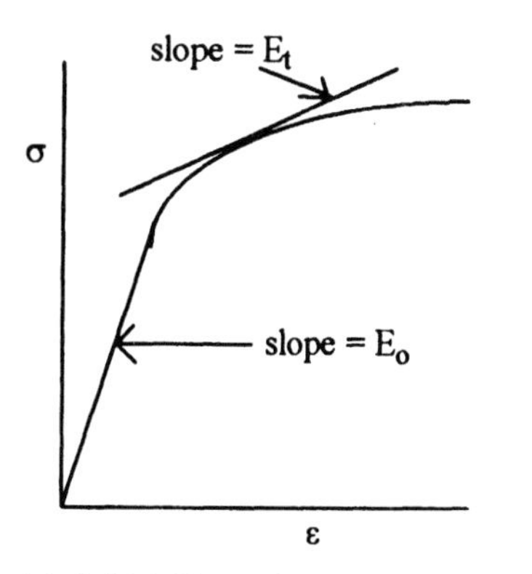

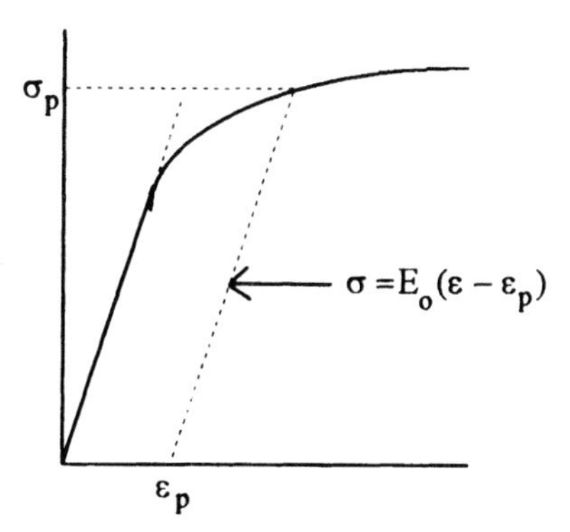

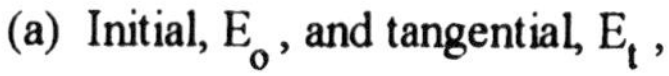
(a) Initial, E_o, and tangential, E_t,

(b) Proof stress illustrated

Figure 6.13 Stress-strain curve and derived quantities

The parameter σ_o is generally not provided by the wire or cable supplier. It could be obtained by fitting a supplied stress-strain curve. Alternatively and more simply, it can be obtained from the proof stress of the material which is generally provided or specified. As one moves along a stress-strain curve and then removes the stress, the material does not move back towards zero stress along the same curve it followed when the stress increased but rather it follows a straight line parallel to the initial slope of the stress-strain curve as illustrated in Fig. 6.13b. This leaves a residual strain in the material, labeled ε_p in the figure, corresponding to the stress σ_p which was the highest stress achieved before it was removed. For $\varepsilon_p = 0.002$ (0.2%), σ_p is called the proof stress. (Some people use $\varepsilon_p = 0.001$ in this definition.)

Thus the proof stress is determined from the intersection of the recoil line

$$\sigma = E_o(\varepsilon - \varepsilon_p) \tag{6.117}$$

with the stress-strain curve given by (6.114). Solving (6.114) and (6.117) simultaneously, we find

$$\sigma_o = \left(\frac{k}{E_o\varepsilon_p}\right)^{1/m} \sigma_p^{(m+1)/m} \tag{6.118}$$

This permits us to find σ_o from a given proof stress σ_p corresponding to the appropriate ε_p.

When the supports (sticks) are engaged in the buckling process, we have forced or constrained buckling. Since there is some looseness in the support due to building tolerances, it can be regarded as a hinged type of attachment for calculational purposes. In this case, the lowest order buckling mode is shown in Fig. 6.14. The corresponding critical force/unit length, q_{crit}, is given by [Tim56]

$$q_{crit} = \frac{EI}{R^3}\left[\left(\frac{2\pi}{\beta}\right)^2 - 1\right] \tag{6.119}$$

where β is the angle between the supports. This corresponds to a critical hoop stress of

$$\sigma_{crit} = \frac{E_t}{12}\left(\frac{h}{R}\right)^2\left[\left(\frac{2\pi}{\beta}\right)^2 - 1\right] \tag{6.120}$$

where h is the radial build of the arch. This will exceed the free buckling critical stress, equation (6.113), when $\beta \le \pi$, i.e. for only 2 diametrical supports. Thus, for most cases where $\beta \ll \pi$, the constrained buckling stress will be much larger than the free buckling stress. Both buckling types depend on the radial build of the ring or arch. We will adopt the same procedure for determining the effective radial build of a cable as was done for the radial bending stress determination at the end of Section 3.7.

Since a loose or hinged support can also be imagined as existing at the center point of the arch in Fig. 6.14, β should be taken as the angle between three consecutive inner supports (sticks). Arched buckling with this value of β appears to provide a more realistic value of buckling strength in practice than totally free unsupported buckling.

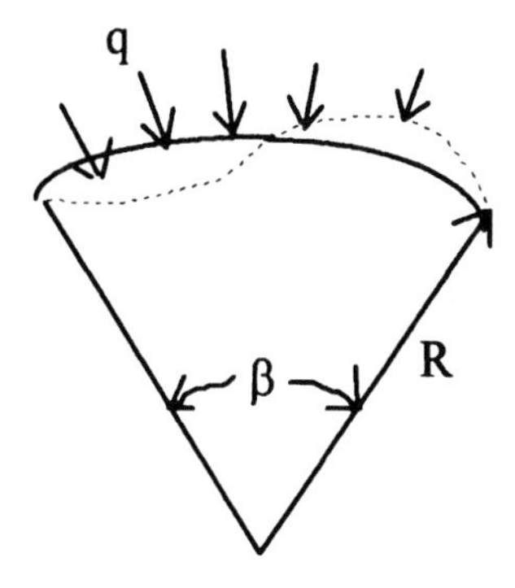

Figure 6.14 Buckling of a circular hinged arch

6.5 STRESS DISTRIBUTION IN A COMPOSITE WIRE-PAPER WINDING SECTION

The hoop stress previously calculated for a winding section or disk (Section 3.6) was an average over the disk. In reality, for the innermost winding, the axial magnetic field varies from nearly zero on the inside of the winding to close to its maximum value at the outer radius of the winding. Since the current density is uniform, the force density also varies in the same fashion as the magnetic field. Thus we might expect higher hoop stresses in the outermost turns as compared with the inner turns. However, because of the layered structure with paper insulation between turns, the stresses tend to be shared more equally by all the turns. This effect will be examined here in order to determine the extent of the stress non-uniformity so that, if necessary, corrective action can be taken.

We analyze an ideal ring geometry as shown in Fig. 6.15. In the figure, r_{ci} denotes the inner radius of the ith conductor layer and r_{pi} the inner radius of the ith paper layer, where $i = 1,...,n$ for the conductors and $i = 1,...,n-1$ for the paper layers. Because of the assumed close contact, the outer radius of the ith conductor layer equals the inner radius of the ith paper layer and the outer radius of the ith paper layer equals the inner radius of the $i+1^{th}$ conductor layer. We do not need to include the innermost or outermost paper layers since they are essentially stress free.

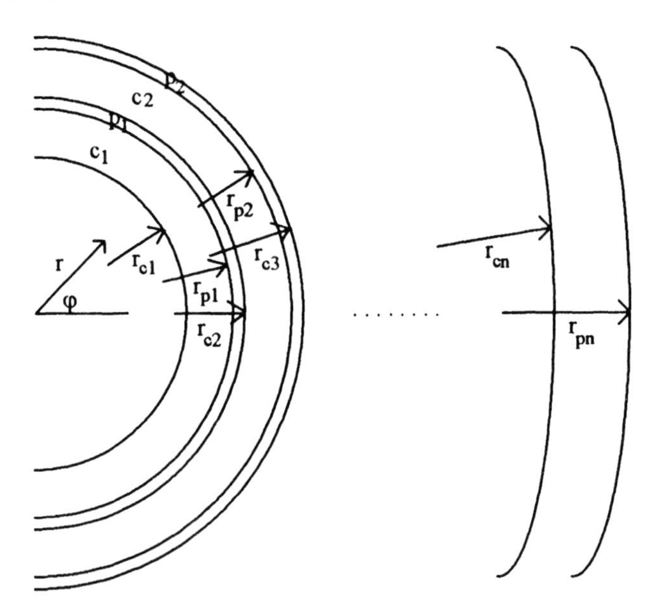

Figure 6.15 Conductor-paper layered ring winding section. ci refers to conductor i and pi to paper layer i.

We will assume that the radial force density varies linearly from the innermost to outermost conductor layers. Thus

$$f_{ci} = \frac{f_o i}{n} \tag{6.121}$$

where f_o is the maximum force density at the outermost conductor. We wish to express this in terms of the average force density, f_{ave}. We have

$$f_{ave} = \frac{1}{n}\sum_{i=1}^{n} f_{ci} = \frac{f_o}{n^2}\sum_{i=1}^{n} i = \frac{f_o(n+1)}{2n} \tag{6.122}$$

Solving for f_o and substituting into (6.121), we get

$$f_{ci} = \frac{2f_{ave}}{n+1}\, i \tag{6.123}$$

It is even more convenient to relate f_{ci} to the average hoop stress in the winding resulting from these radial forces. We related this stress to the radial pressure in equation (6.56). But the average force density is just the average radial pressure divided by the winding radial build. Therefore we find, from (6.56)

$$\sigma_{ave,hoop} = \frac{P_r R_m}{B} = f_{ave} R_m \tag{6.124}$$

where R_m is the mean radius of the winding. Thus (6.123) can be written

$$f_{ci} = \frac{2\sigma_{ave,hoop}}{R_m(n+1)}\, i \tag{6.125}$$

We assume that the winding section can be analyzed as a 2 dimensional stress distribution problem, i.e. stress variations in the axial direction are assumed to be small. The governing equation for this type of problem in polar coordinates when only radial forces are acting and the geometry is cylindrically symmetric is [Tim70]

$$\frac{\partial \sigma_r}{\partial r} + \frac{\sigma_r - \sigma_\varphi}{r} + f_r = 0 \tag{6.126}$$

where σ_r is the radial stress, σ_φ the azimuthal stress, and f_r the radial force density. The stresses are related to the strains for the 2 dimensional plane stress case by

$$\sigma_r = \frac{E}{1-\nu^2}\left(\varepsilon_r + \nu\varepsilon_\varphi\right) \quad , \quad \sigma_\varphi = \frac{E}{1-\nu^2}\left(\varepsilon_\varphi + \nu\varepsilon_r\right) \tag{6.127}$$

where E is Young's modulus and ν is Poisson's ratio ($\nu = 0.25$ for most materials). The radial and azimuthal strains, ε_r and ε_φ, are related to the radial displacement u, in the cylindrically symmetric case, by

$$\varepsilon_r = \frac{du}{dr} \quad , \quad \varepsilon_\varphi = \frac{u}{r} \tag{6.128}$$

Substituting (6.127) and (6.128) into (6.126), we obtain

$$r^2 \frac{d^2u}{dr^2} + r\frac{du}{dr} - u + \frac{f_r\left(1-\nu^2\right)}{E} r^2 = 0 \tag{6.129}$$

with the general solution,

$$u = A\,r + \frac{B}{r} + K\,r^2 \tag{6.130}$$

where A and B are constants to be determined by the boundary conditions and

$$K = -\frac{f_r\left(1-\nu^2\right)}{3E} \tag{6.131}$$

Note that for our problem, f_r is negative (radially inwards) so that K is positive. In the paper layers, $K = 0$ since there is no force density there. Using (6.130) for u in (6.127) and (6.128), we obtain

$$\sigma_r = \frac{E}{\left(1-\nu^2\right)}\left[A(1+\nu) - \frac{B}{r^2}(1-\nu) + K\,r(2+\nu)\right]$$

$$\sigma_\varphi = \frac{E}{\left(1-\nu^2\right)}\left[A(1+\nu) + \frac{B}{r^2}(1-\nu) + K\,r(1+2\nu)\right] \tag{6.132}$$

The solution (6.130) ,(6.131), and (6.132) applies to each layer of conductor or paper. We therefore need to introduce labels to distinguish the layers. Let A_{ci} , B_{ci} apply to conductor layer i and A_{pi} , B_{pi} apply to paper layer i. Let ci and pi also label the displacements, u , and the stresses σ_r , σ_φ for the corresponding layer. At the conductor-paper interface, the displacements must match,

$$u_{ci}\left(r_{pi}\right) = u_{pi}\left(r_{pi}\right) \quad , \quad i = 1,\cdots,n-1$$
$$u_{ci}\left(r_{ci}\right) = u_{p(i-1)}\left(r_{ci}\right) \quad , \quad i = 2,\cdots,n \tag{6.133}$$

There are 2(n - 1) such equations. (See Fig. 6.15 for the labeling.) Also at the interface, the radial stresses must match,

$$\sigma_{r,ci}\left(r_{pi}\right) = \sigma_{r,pi}\left(r_{pi}\right) \quad , \quad i = 1,\cdots,n-1$$
$$\sigma_{r,ci}\left(r_{ci}\right) = \sigma_{r,p(i-1)}\left(r_{ci}\right) \quad , \quad i = 2,\cdots,n \tag{6.134}$$

There are also 2(n - 1) such equations. We also have, at the innermost and outermost radii,

$$\sigma_{r,c1}\left(r_{c1}\right) = 0 \qquad \text{and} \qquad \sigma_{r,cn}\left(r_{pn}\right) = 0 \tag{6.135}$$

This provides 2 more equations. Thus altogether we have 4n - 2 equations. There are 2 unknowns, A_{ci} and B_{ci} , associated with each conductor layer for a total of 2n unknowns and 2 unknowns, A_{pi} and B_{pi} , associated with each paper layer for a total of 2(n - 1) unknowns since there are only n - 1 paper layers. Thus there are altogether 4n - 2 unknowns to solve for and this matches the number of equations. In (6.132) we must use the appropriate material constants for the conductor or paper layer, i.e. $E = E_c$ or E_p is the conductor's or paper's Young's modulus and $\nu = \nu_c$ or ν_p for the conductor's or paper's Poisson's ratio. In addition K needs to be labeled according to the layer, i.e. K_{ci} , and in the case of a paper layer $K_{pi} = 0$.

The resulting set of 4n - 2 equations in 4n - 2 unknowns is a linear system and can be solved by standard methods. Once the solution is obtained, (6.132) can be used to find the stresses. The σ_φ can then be compared with the average hoop stress to see how much deviation there is from a uniform distribution. We show a sample calculation in Table 6.2. The average stresses are calculated for the conductor and paper and then the stresses in the layers are expressed as multipliers of this average stress. These multipliers are averages for the layers since the stresses vary across a layer. The input is the geometric data and the average hoop stress in the conductor which is obtained from the radial pressure via (6.124). The calculated average hoop stress in the conductor does not quite agree with the input, probably because of numerical approximations in the averaging method. The hoop stress in the conductors varies by about 20% from the inner to the outer layers. The other stresses are small in comparison although they show considerable variation across the winding. The stresses are shown as positive, even though they are compressive and therefore negative.

Table 6.2 Sample Stress Distribution in a Composite Conductor-Paper Disk

Input:

Winding inner radius	15 in	Conductor layer radial thickness	0.25 in
Winding outer radius	18 in	Paper layer radial thickness	0.05 in
# Radial turns	10	Average hoop stress	18,000 psi
$E_c = 16 \times 10^6$ psi		$E_p = 0.03 \times 10^6$ psi	$\nu_c = \nu_p = 0.25$

Output:

Average hoop stress in conductor layers	18,500 psi
Average hoop stress in paper layers	156 psi
Average radial stress in conductor layers	440 psi
Average radial stress in paper layers	490 psi

Average Stress Multipliers

	Conductor		Paper	
Layer	Hoop	Radial	Hoop	Radial
1	0.90	0.28	0.60	0.51
2	0.91	0.78	0.89	0.89
3	0.93	1.13	1.10	1.15
4	0.96	1.36	1.23	1.30
5	0.99	1.47	1.27	1.35
6	1.03	1.47	1.24	1.30
7	1.06	1.35	1.12	1.14
8	1.08	1.12	0.92	0.87
9	1.09	0.76	0.63	0.50
10	1.08	0.28		

6.6 ADDITIONAL MECHANICAL CONSIDERATIONS

During a short circuit, the leads or busbars are subjected to an increased force due to the higher fault current they carry interacting with the higher leakage flux from the main windings and from nearby leads. These forces will depend on the detailed positioning of the leads with respect to the main windings and with respect to each other. They will therefore vary considerably from design to design. The leads must be braced properly so that they do not deform or move much during a fault. The leakage flux in the vicinity of the lead can be obtained from a finite element calculation. The flux produced by neighboring leads can be determined from the Biot-Savart law. From these, the forces on the leads can be determined and the adequacy of the bracing checked. In general, the bracing will contain sufficient margin based on past experience so that the above rather laborious analysis will only be necessary for unusual or novel designs.

We have neglected gravitational forces in the preceding sections except for the effect of the core and coil weight on the tie bar stress during lifting. Gravitational forces will affect the compressive force on the key spacers and on the downward end thrust which acts on the bottom pressure ring. Another force which was neglected is the compressive force which is initially placed on the coils by pre-tensioning of the tie bars. This force adds to the compressive force on the key spacers and to the top and bottom thrust on the pressure ring as well as adding some initial tension to the tie bars. Since the compressive forces on the key spacers are involved in conductor tilting, this design criterion will also be affected. These additional forces are present during normal operation and will add to the fault forces when a fault occurs.

We have treated the axial and radial stress calculations independently, whereas in reality axial and radial forces are applied simultaneously, resulting in a biaxial stress condition. Results of such a combined analysis for a circular arched wire segment between supports have been reported and show good agreement with experiment [Ste72]. In this type of analysis, the worst case stress condition is not necessarily associated with the largest axial or radial forces since these would not usually occur at the same position along the winding. It is rather due to some combination of the two which would have to be examined at each position along the winding at which the forces are calculated.

As long as the materials remain linear, i.e. obey Hooke's law, and the displacements are small, the axial and radial analyses can be performed separately. The resulting stresses can then be combined appropriately to get the overall stress state. Various criteria for failure can then be applied to this overall stress state. Our strategy of looking at the worst case stresses produced by the axial and radial forces separately and applying a failure criterion to each is probably a good approximation to that obtained from a combined analysis, especially since the worst case axial and radial forces typically occur at different positions along the winding. The radial forces are produced by axial flux which is high in the middle of the winding whereas the axial forces are produced by radial flux which is high at the ends of the winding.

Dynamical effects have been studied by some authors, particularly the axial response of a winding to a suddenly applied short circuit current [Bos72, Ste72]. They found that the level of pre-stress is important. When the pre-stress was low, ~ 10% of normal, the winding literally bounced against the upper support, resulting in a much higher than expected force. The enhancement factor over the expected non-dynamical maximum force was about 4. However when the pre-stress was normal or above, there was no enhancement over the expected maximum force. The pre-stress also affects the natural frequency of the winding to oscillations in the axial direction. Higher pre-stress tends to shift this frequency towards higher values, away from the frequencies in the applied short circuit forces. Hence little dynamical enhancement is expected under these conditions. Thus, provided sufficient pre-stress is applied to clamp the winding in the axial direction, there should be little or no enhancement of the end thrust over the expected value based on the maximum fault currents. However, as the unit ages, the pre-stress could decrease. With modern pre-compressed pressboard key spacers, this effect should be small. Even so, we allow an enhancement factor of 1.8 in design.

An area of some uncertainty is how to treat transposed cable, with or without bonding, in the stress calculations. It is not exactly a solid homogeneous material, yet it is not simply a loose collection of individual strands. We believe we have taken a conservative approach in our calculations. However, this is one area, among others, where further experimental work would be useful.

7. CAPACITANCE CALCULATIONS

Summary Lumped circuit models of transformers require that the capacitance of coil sections, consisting of one or more disks, as well as coil to coil capacitances be determined. In addition, special treatment of the disks at the high voltage end of a winding, such as the use of static rings to shape the voltage profile, modify the normal disk capacitance. These changed capacitances must also be determined for use in lumped circuit as well as traveling wave models which attempt to simulate transformer behavior under impulse conditions. We calculate these capacitances in this report, using an energy method. We also calculate the capacitance of a pair of winding disks containing wound-in-shields by means of a simple formula and by means of a detailed circuit model. In addition, experiments were performed to check the formula under various conditions such as changing the number of shield turns and attaching the shield to the high voltage terminal at different points or letting it float. Unlike a static capacitance calculation, these capacitances depend on inductive effects.

7.1 INTRODUCTION

Under impulse conditions, very fast voltage pulses are applied to a transformer. These contain high frequency components, eliciting capacitative effects which are absent at normal operating frequencies. Thus, in order to simulate the behavior of a transformer under impulse conditions, capacitances must be determined for use in circuit or traveling wave models [Mik78, Rud40].

Usually, the highest electrical stresses occur at the high voltage end of the winding so that modifications are sometimes made to the first few disks to meet voltage breakdown limits. These modifications commonly take the form of the addition of one or more static rings so that their effect on the disk capacitance must be determined. Other methods such the use of wound-in-shields or interleaving, although effective in increasing the disk capacitance (which is generally favorable), are not treated here.

We employ an energy method to determine the capacitance or, in general, the capacitance matrix. This method is a generalization of the method used by Stein to determine the disk capacitance of a disk embedded in a winding of similar disks [Ste64]. It utilizes a continuum model of a disk so that disks having many turns are contemplated. We also compare capacitances determined in this manner with capacitances determined using a more conventional approach. The conventional method also works for helical windings, i.e. windings having one turn per disk, and so is useful in its own right.

For completeness we also calculate coil-coil, coil-core, and coil-tank capacitances which are qualitatively similar. These are based on a cruder model and assume an infinitely long coil, ignoring end effects. They are really capacitances per unit length.

7.2 THEORY

We try to be as general as possible so that we may apply the results to a variety of situations. Thus we consider a disk or coil section having a series capacitance per unit length of c_s and shunt capacitances per unit length of c_a and c_b to the neighboring objects on either side of the disk or coil section as shown in Fig. 7.1. These neighboring objects are assumed, for generality, to have linear voltage distributions given by

$$\begin{aligned} V_a(x) &= V_{a1} + \frac{x}{L}(V_{a2} - V_{a1}) \\ V_b(x) &= V_{b1} + \frac{x}{L}(V_{b2} - V_{b1}) \end{aligned} \tag{7.1}$$

where L is the length of the disk or coil section and x measures distances from the high voltage end at V_1 to the low voltage end at V_2.

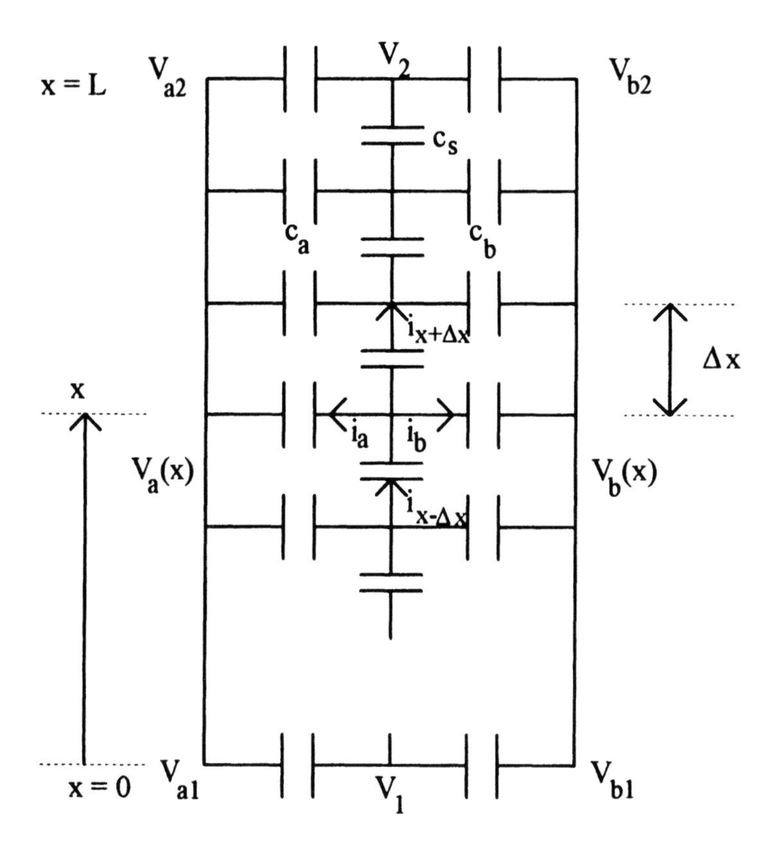

Figure 7.1 Approximately continuous capacitance distribution of coil section

Applying current conservation to the node at x in Fig. 7.1 and assuming the current directions shown, we have

$$i_{x-\Delta x} - i_{x+\Delta x} = i_a + i_b \tag{7.2}$$

Using the current-voltage relationship for a capacitor, $i = C\, dV/dt$, we can rewrite (7.2)

$$\begin{aligned} c_s\Delta x\frac{d}{dt}\left[V(x-\Delta x)-V(x)\right]-c_s\Delta x\frac{d}{dt}\left[V(x)-V(x-\Delta x)\right] \\ = c_a\Delta x\frac{d}{dt}\left[V(x)-V_a(x)\right]+c_b\Delta x\frac{d}{dt}\left[V(x)-V_b(x)\right] \end{aligned} \tag{7.3}$$

Rearranging, we find

$$\frac{d}{dt}\Big\{c_s\big[V(x+\Delta x)-2V(x)+V(x-\Delta x)\big] \\ -c_a\big[V(x)-V_a(x)\big]-c_a\big[V(x)-V_a(x)\big]\Big\} \tag{7.4}$$

Thus the quantity in curly brackets is a constant in time and may for all practical purposes be set to 0. (For an applied pulse, all the voltages are zero after a very long time so the constant must be zero.) Dividing (7.4) by $(\Delta x)^2$, we obtain

$$\frac{V(x+\Delta x)-2V(x)+V(x-\Delta x)}{(\Delta x)^2}-\frac{(c_a+c_b)}{c_s(\Delta x)^2}V(x) \\ =-\frac{\big[c_a V_a(x)+c_b V_b(x)\big]}{c_s(\Delta x)^2} \tag{7.5}$$

The first term in (7.5) is the finite difference approximation to d^2V/dx^2. The combination

$$\frac{(c_a+c_b)}{c_s(\Delta x)^2} \tag{7.6}$$

can be expressed in terms of the total series capacitance C_s and total shunt capacitances C_a and C_b using

$$C_s=\frac{c_s\Delta x}{N} \quad , \quad C_a=c_a\Delta xN \quad , \quad C_b=c_b\Delta xN \tag{7.7}$$

where N is the number of subdivisions of the total length L into units of size Δx, $N=L/\Delta x$ Hence (7.7) becomes

$$C_s=\frac{c_s(\Delta x)^2}{L} \quad , \quad C_a=c_aL \quad , \quad C_b=c_bL \tag{7.8}$$

Substituting into (7.6), we obtain

$$\frac{(c_a+c_b)}{c_s(\Delta x)^2}=\frac{C_a+C_b}{C_sL^2}=\frac{\alpha^2}{L^2} \tag{7.9}$$

where

$$\alpha = \sqrt{\frac{C_a + C_b}{C_s}} \tag{7.10}$$

Similarly,

$$\frac{c_a}{c_s(\Delta x)^2} = \frac{C_a}{C_s L^2} \quad , \quad \frac{c_b}{c_s(\Delta x)^2} = \frac{C_b}{C_s L^2} \tag{7.11}$$

Substituting into (7.5) and taking the limit as $\Delta x \to 0$, results in the differential equation

$$\frac{d^2 V}{dx^2} - \left(\frac{\alpha}{L}\right)^2 V = -\frac{1}{C_s L^2}\left[C_a V_a(x) + C_b V_b(x)\right] \tag{7.12}$$

The solution to the homogeneous part of (7.12) is

$$V = A e^{\frac{\alpha}{L}x} + B e^{-\frac{\alpha}{L}x} \tag{7.13}$$

where A and B are constants to be determined by the boundary conditions. For the inhomogeneous part of (7.12), we try

$$V = F + Gx \tag{7.14}$$

where F and G are determined by substituting into (7.12) and using (7.1)

$$\begin{aligned} &-\left(\frac{\alpha}{L}\right)^2 (F + Gx) \\ &= -\frac{1}{C_s L^2}\left\{C_a V_{a1} + C_b V_{b1} + \frac{x}{L}\left[C_a\left(V_{a2} - V_{a1}\right) + C_b\left(V_{b2} - V_{b1}\right)\right]\right\} \end{aligned} \tag{7.15}$$

Collecting terms, we find

$$F = \frac{1}{\alpha^2 C_s}(C_a V_{a1} + C_b V_{b1})$$
$$G = \frac{1}{\alpha^2 C_s L}\left[C_a(V_{a2} - V_{a1}) + C_b(V_{b2} - V_{b1})\right] \tag{7.16}$$

Thus the general solution to (7.12) is

$$\begin{aligned} V &= Ae^{\frac{\alpha}{L}x} + Be^{-\frac{\alpha}{L}x} \\ &+ \frac{1}{\alpha^2 C_s}(C_a V_{a1} + C_b V_{b1}) + \frac{1}{\alpha^2 C_s}\left(\frac{x}{L}\right)\left[C_a(V_{a2} - V_{a1}) + C_b(V_{b2} - V_{b1})\right] \end{aligned} \tag{7.17}$$

Using the boundary conditions $V = V_1$ at $x = 0$ and $V = V_2$ at $x = L$, we can solve (7.17) for A and B. Performing the algebra, the solution can be cast in the form

$$\begin{aligned} V(x) &= \frac{1}{\sinh\alpha}\left\{\left[V_2 - (\gamma_a V_{a2} + \gamma_b V_{b2})\right]\sinh\left(\alpha\frac{x}{L}\right) + \left[V_1 - (\gamma_a V_{a1} + \gamma_b V_{b1})\right]\sinh\left[\alpha\left(1 - \frac{x}{L}\right)\right]\right\} \\ &+ (\gamma_a V_{a1} + \gamma_b V_{b1}) + \left[\gamma_a(V_{a2} - V_{a1}) + \gamma_b(V_{b2} - V_{b1})\right]\frac{x}{L} \end{aligned} \tag{7.18}$$

where

$$\gamma_a = \frac{C_a}{C_a + C_b} \quad , \quad \gamma_b = \frac{C_b}{C_a + C_b} \tag{7.19}$$

so that $\gamma_a + \gamma_b = 1$. The derivative of this expression is also needed for evaluating the energy and for obtaining turn-turn voltages. It is

$$\begin{aligned} \frac{dV}{dx} &= \frac{\alpha}{L\sinh\alpha}\left\{\left[V_2 - (\gamma_a V_{a2} + \gamma_b V_{b2})\right]\cosh\left(\alpha\frac{x}{L}\right)\right. \\ &\left. - \left[V_1 - (\gamma_a V_{a1} + \gamma_b V_{b1})\right]\cosh\left[\alpha\left(1 - \frac{x}{L}\right)\right]\right\} \\ &+ \frac{1}{L}\left[\gamma_a(V_{a2} - V_{a1}) + \gamma_b(V_{b2} - V_{b1})\right] \end{aligned} \tag{7.20}$$

To determine the capacitance, the stored electrostatic energy must be evaluated. Reverting to the original discrete notation, the energy in the series capacitance is

$$E_{series} = \frac{1}{2}\sum_{1}^{N} c_s \Delta x (\Delta V)^2 \tag{7.21}$$

Using (7.8), this becomes

$$E_{series} = \frac{C_s L}{2}\sum_{1}^{N} \frac{1}{\Delta x}(\Delta V)^2 \tag{7.22}$$

Substituting $\Delta V = (dV/dx)\ \Delta x$ into (7.22) and letting $\Delta x \to 0$, we obtain

$$E_{series} = \frac{C_s L}{2}\int_0^L \left(\frac{dV}{dx}\right)^2 dx \tag{7.23}$$

The energy in the shunt capacitances can be found similarly,

$$E_{shunt} = \frac{1}{2}\sum_{1}^{N} c_a \Delta x \left[V(x) - V_a(x)\right]^2 + \frac{1}{2}\sum_{1}^{N} c_b \Delta x \left[V(x) - V_b(x)\right]^2 \tag{7.24}$$

Again, using (7.8) and taking the limit as $\Delta x \to 0$, we get

$$E_{shunt} = \frac{C_a}{2L}\int_0^L \left[V(x) - V_a(x)\right]^2 dx + \frac{C_b}{2L}\int_0^L \left[V(x) - V_b(x)\right]^2 dx \tag{7.25}$$

Combining (7.23) and (7.25), the total energy is

$$E = \frac{C_s L}{2}\int_0^L \left(\frac{dV}{dx}\right)^2 dx + \frac{C_a}{2L}\int_0^L \left[V(x) - V_a(x)\right]^2 dx + \frac{C_b}{2L}\int_0^L \left[V(x) - V_b(x)\right]^2 dx \tag{7.26}$$

Substituting (7.18) and (7.20) into (7.26) and performing the integrations, we obtain

$$
\begin{aligned}
E = \frac{C_s}{2}\Bigg\{&\left[(V_2-\eta)^2+(V_1-\beta)^2\right]\frac{\alpha}{\tanh\alpha}-2(V_1-\beta)(V_2-\eta)\frac{\alpha}{\sinh\alpha} \\
&-(\eta-\beta)^2-2(\eta-\beta)(V_1-V_2) \\
&+\alpha^2\gamma_a\left[(\beta-V_{a1})^2+(\beta-V_{a1})[(\eta-\beta)-(V_{a2}-V_{a1})]+\frac{1}{3}[(\eta-\beta)-(V_{a2}-V_{a1})]^2\right] \\
&+\alpha^2\gamma_b\left[(\beta-V_{b1})^2+(\beta-V_{b1})[(\eta-\beta)-(V_{b2}-V_{b1})]+\frac{1}{3}[(\eta-\beta)-(V_{b2}-V_{b1})]^2\right]\Bigg\}
\end{aligned}
\tag{7.27}
$$

where

$$\beta = \gamma_a V_{a1} + \gamma_b V_{b1} \quad , \quad \eta = \gamma_a V_{a2} + \gamma_b V_{b2} \tag{7.28}$$

For most of the applications of interest here, the side objects on which the shunt capacitances terminate are at a constant potential. Thus

$$V_{a1} = V_{a2} = V_a \quad , \quad V_{b1} = V_{b2} = V_b \tag{7.29}$$

and

$$\beta = \eta = \gamma_a V_a + \gamma_b V_b \tag{7.30}$$

We have

$$V(x) = \frac{1}{\sinh\alpha}\left\{(V_2-\beta)\sinh\left(\alpha\frac{x}{L}\right)+(V_1-\beta)\sinh\left[\alpha\left(1-\frac{x}{L}\right)\right]\right\}+\beta \tag{7.31}$$

$$\frac{dV}{dx} = \frac{\alpha}{L\sinh\alpha}\left\{(V_2-\beta)\cosh\left(\alpha\frac{x}{L}\right)-(V_1-\beta)\cosh\left[\alpha\left(1-\frac{x}{L}\right)\right]\right\} \tag{7.32}$$

$$
\begin{aligned}
E = \frac{C_s}{2}\Bigg\{&\left[(V_2-\beta)^2+(V_1-\beta)^2\right]\frac{\alpha}{\tanh\alpha}-2(V_1-\beta)(V_2-\beta)\frac{\alpha}{\sinh\alpha} \\
&+\alpha^2\left[\gamma_a(\beta-V_a)^2+\gamma_b(\beta-V_b)^2\right]\Bigg\}
\end{aligned}
\tag{7.33}
$$

7.3 STEIN'S CAPACITANCE FORMULA

As an example, we can apply the above results to the case considered by Stein, consisting of a disk embedded in a coil of similar disks [Ste64]. The situation is shown in Fig. 7.2. Assuming V is the voltage drop across the disk, we can take $V_1 = V$ and $V_2 = 0$. Assuming neighboring disks have the same voltage drop, we can imagine equipotential planes between the disks with the values shown in the figure, namely $V_a = V$ and $V_b = 0$. As we move along the disk, these values are the average of the two potential values on the neighboring disks and can be taken as representing the potential value midway between them. The capacitances to the mid-plane are twice the disk-disk capacitance, $C_a = C_b = 2C_{dd}$. We also have $\beta = V/2$. For these values, the energy, from (7.33), becomes

$$E = \frac{C_s V^2}{2}\left\{\frac{1}{2}\frac{\alpha}{\tanh\alpha} + \frac{1}{2}\frac{\alpha}{\sinh\alpha} + \frac{\alpha^2}{4}\right\} \tag{7.34}$$

The effective disk capacitance is found from $E = 1/2\ CV^2$ and is

$$C_{Stein} = C_s\left\{\frac{1}{2}\frac{\alpha}{\tanh\alpha} + \frac{1}{2}\frac{\alpha}{\sinh\alpha} + \frac{\alpha^2}{4}\right\} \tag{7.35}$$

where C_s is the series capacitance of the turns,

$$\alpha = \sqrt{\frac{C_a + C_b}{C_s}} = \sqrt{\frac{4C_{dd}}{C_s}} \tag{7.36}$$

and C_{dd} is the disk-disk capacitance.

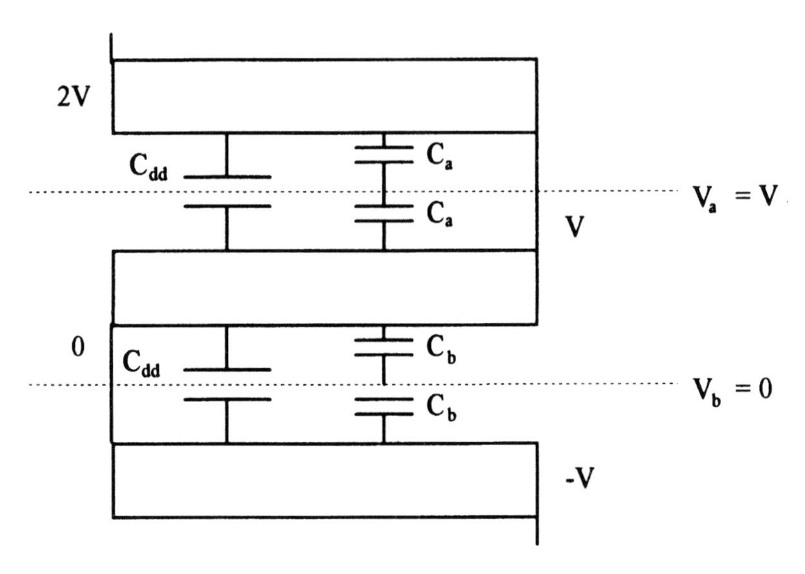

Figure 7.2 Disk embedded in a coil of similar disks with V the voltage drop along a disk

Let us compare this result with the more conventional approach. This assumes that the voltage drop along the disk is linear so that

$$V(x) = V\left(1 - \frac{x}{L}\right) \tag{7.37}$$

The energy in the series turns is, from (7.23)

$$E_{series} = \frac{1}{2} C_s V^2 \tag{7.38}$$

We must consider the shunt energy with respect to the equipotential midplanes as before since
we want the energy associated with a single disk. This is

$$E_{shunt} = \frac{1}{2}\left(\frac{2C_{dd}}{L}\right)\int_0^L \left\{\left[V\left(1 - \frac{x}{L}\right)\right]^2 + \left[V - V\left(1 - \frac{x}{L}\right)\right]^2\right\} dx \tag{7.39}$$

Evaluating this expression, we find

$$E_{shunt} = \frac{1}{2}\left(\frac{4}{3}C_{dd}\right)V^2 \tag{7.40}$$

Combining (7.38) and (7.40) and extracting the effective disk capacitance, we obtain

$$C_{conv} = C_s + \frac{4}{3}C_{dd} \tag{7.41}$$

This expression applies to a helical winding (one turn/disk) with $C_s = 0$.

In order to compare (7.35) with (7.41), let us normalize by dividing by C_s. Thus

$$\frac{C_{Stein}}{C_s} = \frac{1}{2}\frac{\alpha}{\tanh\alpha} + \frac{1}{2}\frac{\alpha}{\sinh\alpha} + \frac{\alpha^2}{4} \tag{7.42}$$

and

$$\frac{C_{conv}}{C_s} = 1 + \frac{4}{3}\frac{C_{dd}}{C_s} \tag{7.43}$$

From (7.36), the right hand side of (7.42) is a function of C_{dd}/C_s as is the right hand side of (7.43). Alternatively both right hand sides are functions of α. For small α, it can be shown that (7.42) approaches (7.43). For larger α, the comparison is shown graphically in Fig. 7.3. The difference becomes noticeable at values of $\alpha > 5$. At $\alpha = 10$, the conventional capacitance is about 15% larger than Stein's.

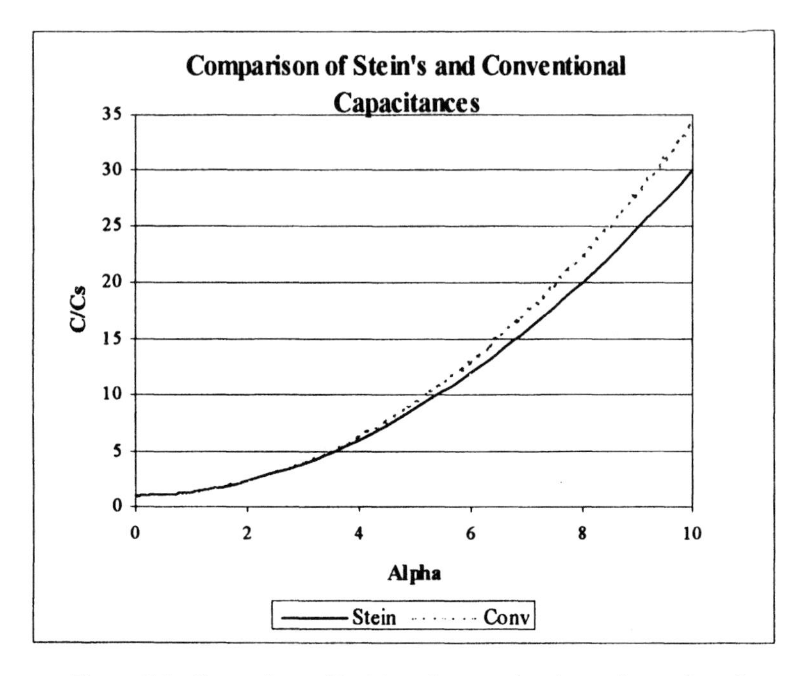

Figure 7.3 Comparison of Stein's and conventional capacitance formulas

The voltage distribution along the disk can be obtained from (7.31), substituting $V_1 = V$, $V_2 = 0$, and $\beta = V/2$,

$$V(x) = \frac{V}{2}\left\{1 + \frac{1}{\sinh\alpha}\left[\sinh\left[\alpha\left(1 - \frac{x}{L}\right)\right] - \sinh\left(\alpha\frac{x}{L}\right)\right]\right\} \tag{7.44}$$

This is plotted in Fig. 7.4 in normalized form. As can be seen, the voltage becomes increasingly less uniform as α increases.

The voltage gradient is obtained from (7.32) with the appropriate substitutions and is

$$\frac{dV}{dx} = -\frac{\alpha V}{2L\sinh\alpha}\left\{\cosh\left[\alpha\left(1 - \frac{x}{L}\right)\right] + \cosh\left(\alpha\frac{x}{L}\right)\right\} \tag{7.45}$$

This is always negative as Fig. 7.4 indicates. Its largest value, in absolute terms, occurs at either end of the disk

$$\left|\frac{dV}{dx}\right| = \frac{\alpha V}{2L\sinh\alpha}(1+\cosh\alpha) = \frac{\alpha V}{2L\tanh(\alpha/2)} \tag{7.46}$$

As $\alpha \to 0$, this approaches the uniform value of V/L. This voltage gradient is equal to the stress (electric field magnitude) in the turn-turn insulation, which must be able to handle it without breakdown.

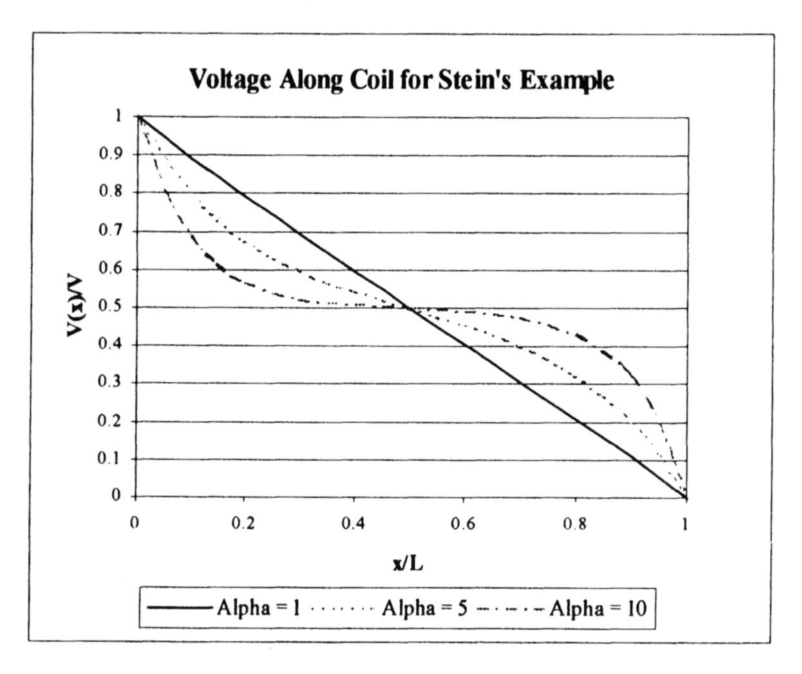

Figure 7.4 Normalized voltage along the disk for various values of alpha in Stein's example.

In this and later applications the series and shunt or disk-disk capacitances can only be approximated for use in the formulas since in reality, we do not have a continuous distribution of capacitances as the model assumes. As shown in Fig. 7.5, a disk consists of N_t turns, usually rectangular in cross-section, with paper thickness τ_p between turns. τ_p is twice the one sided paper thickness of a turn. The disks are separated by means of key spacers of thickness τ_{ks} and width w_{ks} spaced around the circumference.

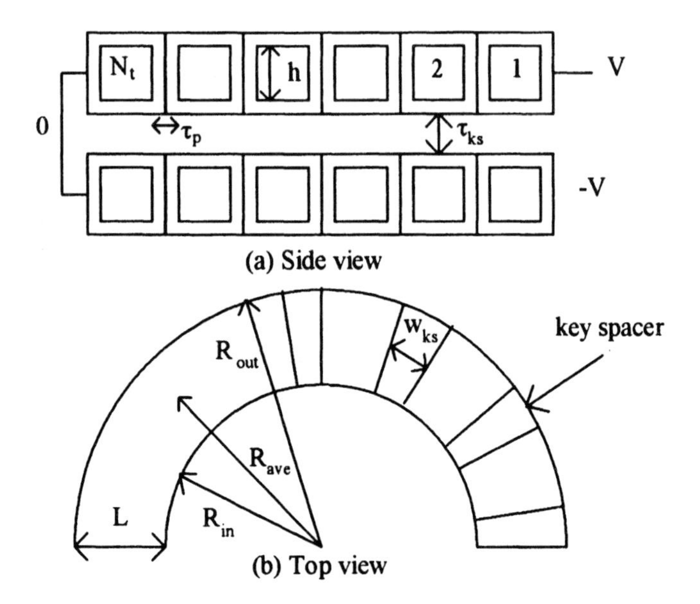

Figure 7.5 Geometry of a practical disk coil

The turn-turn capacitance, C_{tt}, is given approximately by

$$C_{tt} = \varepsilon_o \varepsilon_p 2\pi R_{ave} \frac{(h + 2\tau_p)}{\tau_p} \tag{7.47}$$

where ε_o is the permittivity of vacuum $= 8.854 \times 10^{-12}$ farad/m and ε_p is the relative permittivity of paper ($\cong 3.5$ for oil soaked paper). R_{ave} is the average radius of the disk and h is the bare copper or conductor height. The addition of $2\tau_p$ to h is designed to take fringing effects into account. There are $N_t - 1$ turn-turn capacitances in series which results in a total series capacitance of $C_{tt}/(N_t - 1)$. However they do not see the full disk voltage drop but only the fraction $(N_t - 1)/N_t$. Thus the capacitive energy is

$$E = \frac{1}{2}\left(\frac{C_{tt}}{N_t - 1}\right)\left(\frac{N_t - 1}{N_t}\right)^2 V^2 \tag{7.48}$$

so that based on the full voltage drop V, the equivalent series capacitance is

$$C_s = C_{tt}\frac{(N_t - 1)}{N_t^2} \tag{7.49}$$

This makes sense because for $N_t = 1$, we get $C_s = 0$.

The disk-disk capacitance can be considered to be two capacitances in parallel, namely the capacitance of the portion containing the key spacers and the capacitance of the remainder containing an oil or air thickness instead of key spacers. Let f_{ks} be the key spacer fraction,

$$f_{ks} = \frac{N_{ks} w_{ks}}{2\pi R_{ave}} \tag{7.50}$$

where N_{ks} is the number of key spacers spaced around the circumference and w_{ks} their width. Typically $f_{ks} \approx 1/3$. The key spacer fraction of the disk-disk space is filled with 2 dielectrics, paper and pressboard, the latter being the usual key spacer material. For a planar capacitor containing 2 dielectric layers of permittivity ε_1 and ε_2, it follows from electrostatic theory that the capacitance is

$$C = \frac{A}{\left(\frac{\ell_1}{\varepsilon_1} + \frac{\ell_2}{\varepsilon_2}\right)} \tag{7.51}$$

where A is the area and ℓ_1 and ℓ_2 are the thicknesses of the layers. Applying this to the disk-disk capacitance, we obtain

$$C_{dd} = \varepsilon_o \pi \left(R_{out}^2 - R_{in}^2\right)\left[\frac{f_{ks}}{\left(\frac{\tau_p}{\varepsilon_p} + \frac{\tau_{ks}}{\varepsilon_{ks}}\right)} + \frac{(1 - f_{ks})}{\left(\frac{\tau_p}{\varepsilon_p} + \frac{\tau_{ks}}{\varepsilon_{oil}}\right)}\right] \tag{7.52}$$

Here ε_{ks} is the permittivity of the key spacer material (= 4.5 for oil soaked pressboard) and ε_{oil} is the oil permittivity (= 2.2 for transformer oil). R_{in} and R_{out} are the inner and outer radii of the disk respectively.

7.4 GENERAL DISK CAPACITANCE FORMULA

More generally, if the disk-disk spacings on either side of the main disk are unequal so that $C_a \neq C_b$ as shown in Fig. 7.6, then we have, from (7.33),

$$C_{general} = C_s\left\{\left(\gamma_a^2 + \gamma_b^2\right)\frac{\alpha}{\tanh\alpha} + 2\gamma_a\gamma_b\frac{\alpha}{\sinh\alpha} + \gamma_a\gamma_b\alpha^2\right\} \tag{7.53}$$

where

$$\gamma_a = \frac{C_{dd1}}{C_{dd1} + C_{dd2}}\ ,\ \gamma_a = \frac{C_{dd1}}{C_{dd1} + C_{dd2}}\ ,\ \alpha = \sqrt{\frac{2(C_{dd1} + C_{dd2})}{C_s}} \tag{7.54}$$

with C_{dd1} and C_{dd2} the unequal disk-disk capacitances.

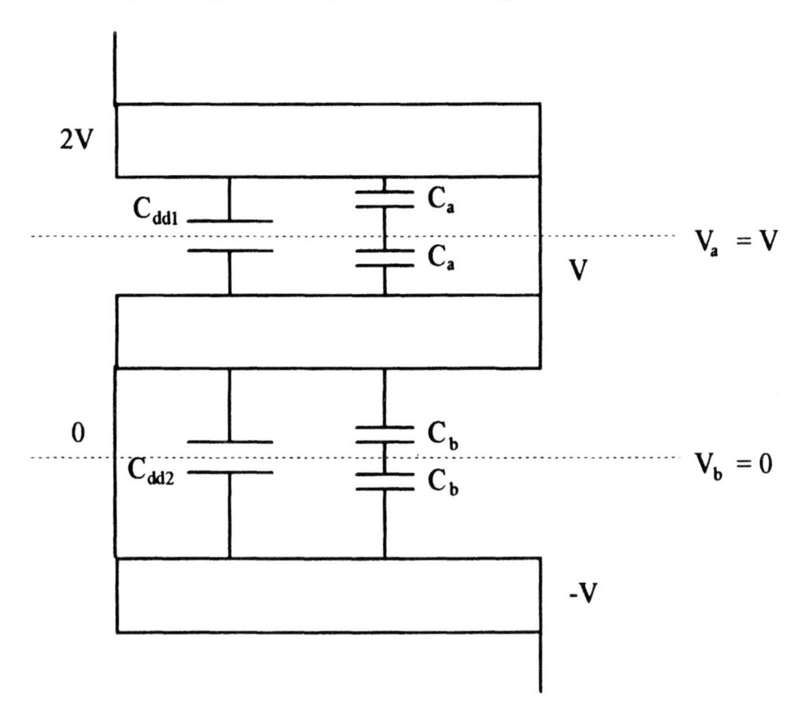

Figure 7.6 More general case of a disk embedded in a coil of similar disks

7.5 COIL GROUNDED AT ONE END WITH GROUNDED CYLINDERS ON EITHER SIDE

An early impulse model for a coil assumed the coil consisted of a uniformly distributed chain of series capacitances connected to ground cylinders on either side by shunt capacitances, i.e. the same model shown in Fig. 7.1 but with $V_{a1} = V_{a2} = V_{b1} = V_{b2} = 0$ [Blu51]. The V_1 terminal was impulsed with a voltage V and the V_2 terminal was grounded. From (7.30) we have $\beta = \eta = 0$ and (7.31) becomes

$$V(x) = V\frac{\sinh\left[\alpha\left(1-\frac{x}{L}\right)\right]}{\sinh\alpha} \tag{7.55}$$

with $\alpha = \sqrt{C_g/C_s}$. Here C_g is the total ground capacitance (both sides) and C_s the series capacitance of the coil. This is shown in normalized form in Fig. 7.7 for several values of α.

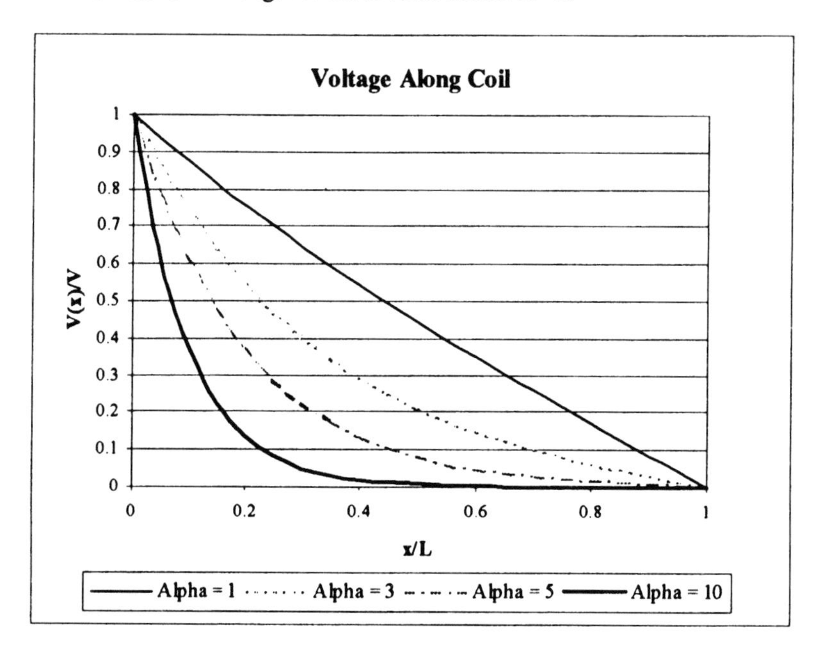

Figure 7.7 Graph of normalized voltage along a coil for several values of α

This voltage distribution is expected to apply immediately after the application of the impulse voltage, before inductive effects come into play. Later, oscillations can cause voltage swings above the values shown in the figure.

The voltage gradient is given by

$$\frac{dV}{dx} = -\frac{\alpha V}{L \sinh \alpha} \cosh\left[\alpha\left(1 - \frac{x}{L}\right)\right] \tag{7.56}$$

The maximum gradient occurs at the line end (x = 0) and is

$$\left|\frac{dV}{dx}\right| = \frac{\alpha V}{L \tanh \alpha} \tag{7.57}$$

Thus the maximum disk-disk voltage immediately after impulse is approximately (7.57) multiplied by the disk-disk spacing.

The coil's total capacitance to ground is, from (7.33),

$$C_{coil} = C_s \frac{\alpha}{\tanh \alpha} \tag{7.58}$$

In this as well as previous formulas in this section, the series capacitance is due to N_d disks in series and if the disk capacitances are obtained by the Stein formula, we have

$$C_s = \frac{C_{Stein}}{N_d} \tag{7.59}$$

For an inner coil, the surfaces of the neighboring coils are usually taken to be the ground cylinders. For the innermost coil, the core determines the ground on one side while for an outermost coil the tank is the ground on one side. In general, the distance to ground is filled with various dielectric materials, including oil or air. One such structure is shown in Fig. 7.8. There are usually multiple pressboard layers, but for convenience there are grouped into a single layer.

The sticks provide spacing for cooling oil or air to flow. This composite structure is similar to that analyzed previously for the disk-disk capacitance. The ground spacing is usually small relative to the coil radius for power transformers so that an approximately planar geometry may be assumed. We obtain for the ground capacitance on one side of the coil, C_{g1}

$$C_{g1} = \varepsilon_o 2\pi R_{gap} H \left[\frac{f_s}{\left(\frac{\tau_{press}}{\varepsilon_{press}} + \frac{\tau_s}{\varepsilon_s} \right)} + \frac{(1-f_s)}{\left(\frac{\tau_{press}}{\varepsilon_{press}} + \frac{\tau_s}{\varepsilon_{oil}} \right)} \right] \tag{7.60}$$

where f_s is the fraction of the space occupied by sticks,

$$f_s = \frac{N_s w_s}{2\pi R_{gap}} \tag{7.61}$$

R_{gap} the mean gap radius, w_s the stick width, N_s the number of sticks around the circumference, H the coil height, τ_{press} the pressboard thickness, ε_{press} the pressboard permittivity, and τ_s, ε_s corresponding quantities for the sticks. The ground capacitance of both gaps, C_{g1} and C_{g2} would be added to obtain the total ground capacitance, C_g.

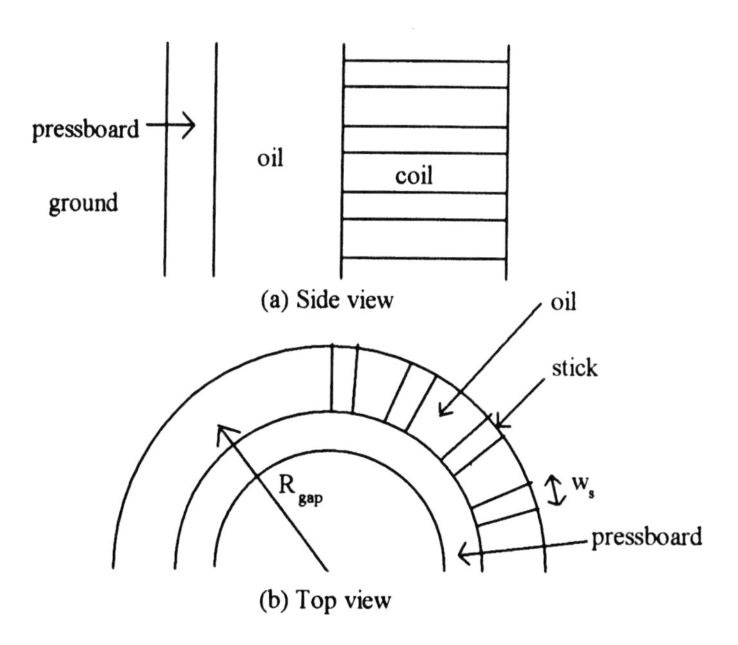

Figure 7.8 Ground capacitance geometry

7.6 STATIC RING ON ONE SIDE OF DISK

If a static ring is present on one side of a disk and connected to the terminal voltage as shown in Fig. 7.9, then we have a situation similar to that considered in the general disk capacitance section. The only difference is that C_a is the disk-static ring capacitance since the static ring is an equipotential surface. Thus (7.53) applies with

$$\gamma_a = \frac{C_a}{C_a + 2C_{dd}} \quad , \quad \gamma_b = \frac{2C_{dd}}{C_a + 2C_{dd}} \quad , \quad \alpha = \sqrt{\frac{C_a + 2C_{dd}}{C_s}} \tag{7.62}$$

with C_{dd} the disk-disk capacitance to the lower disk.

This case would be identical to Stein's if the static ring were spaced at 1/2 the normal disk-disk spacing or whatever is required to achieve $C_a = 2C_{dd} = C_b$. Then $\gamma_a = \gamma_b = 1/2$, $\alpha = \sqrt{4C_{dd}/C_s}$ and (7.53) would reduce to (7.35). Thus the end disk would have the same capacitance as any other disk.

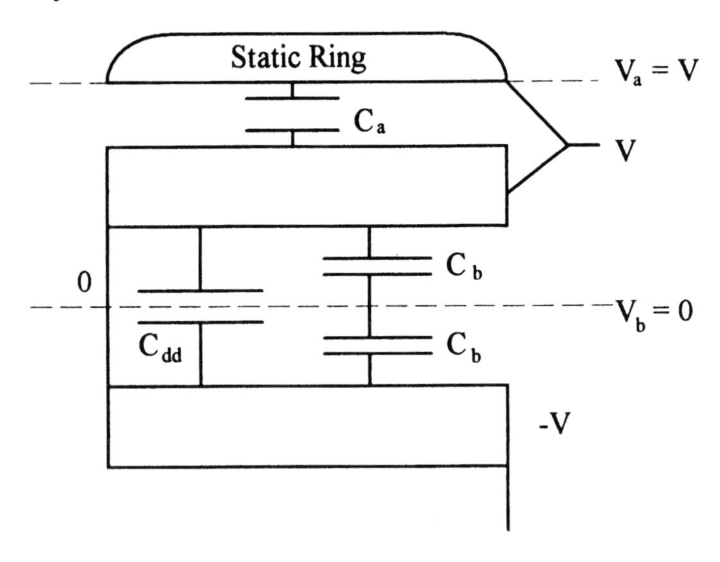

Figure 7.9 Static ring on one side of a disk at the end of a coil

7.7 TERMINAL DISK WITHOUT A STATIC RING

In case the end disk does not have an adjacent static ring, we assume that the shunt capacitance on the end side is essentially 0. Then we have the situation shown in Fig. 7.10a. We have $C_a = 0$ so that $\gamma_a = 0$, $\gamma_b = 1$ and from (7.53)

$$C_{end} = C_s \frac{\alpha}{\tanh \alpha} \tag{7.63}$$

with

$$\alpha = \sqrt{\frac{2C_{dd}}{C_s}} \tag{7.64}$$

This situation also applies to a center fed winding without static ring. In this case, as Fig. 7.10b shows, there is no capacitative energy between the two center disks so that effectively $C_a = 0$. This result would also follow if both center disks were considered as a unit and the energy divided equally between them.

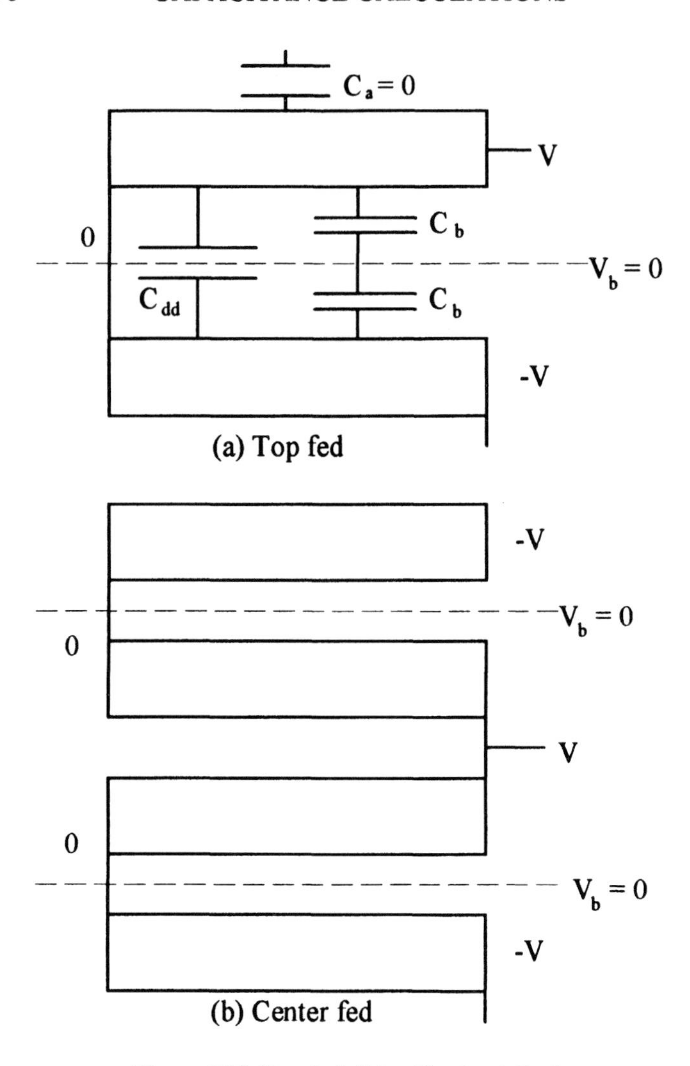

Figure 7.10 Terminal disk without a static ring

7.8 CAPACITANCE MATRIX

Before proceeding to other cases of interest, we need to introduce the capacitance matrix. For a system of conductors having voltages V_i and total electrostatic energy E, it follows from the general theory of linear capacitors that [Smy68]

$$\frac{\partial E}{\partial V_1} = Q_1 = C_{11}V_1 + C_{12}V_2 + \cdots$$
$$\frac{\partial E}{\partial V_2} = Q_2 = C_{21}V_1 + C_{22}V_2 + \cdots \quad (7.65)$$
$$\vdots$$

where Q_i is the charge on conductor i, C_{ii} is the self capacitance of conductor i and C_{ij} the mutual capacitance between conductors i and j. The C's can be grouped into a capacitance matrix which is symmetric, $C_{ij} = C_{ji}$. The diagonal terms are positive while the off-diagonal terms are negative. This follows because if V_k is a positive voltage while all other voltages are 0, i.e. the other conductors are grounded, then the charge of conductor k must be positive, $Q_k = C_{kk}V_k > 0 \Rightarrow C_{kk} > 0$. The charges induced on the other conductors must be negative so $Q_j = C_{jk}V_k < 0 \Rightarrow C_{jk} < 0$. By charge conservation, again assuming $V_k > 0$ and all other V's = 0,

$$\sum Q_i = \sum_{i,j} C_{ij}V_j = \sum_j C_{kj}V_k = 0 \quad (7.66)$$

which implies

$$C_{kk} = -\sum_{j \neq k} C_{kj} \quad (7.67)$$

i.e. the negative of the sum of the off-diagonal terms equals the diagonal term.

Let's apply this to the general energy expression (7.27). Consider the V_1 voltage node,

$$\frac{\partial E}{\partial V_1} = Q_1 = \frac{C_s}{2}\left\{-2(\eta - \beta) + 2(V_1 - \beta)\frac{\alpha}{\tanh\alpha} - 2(V_2 - \eta)\frac{\alpha}{\sinh\alpha}\right\}$$
$$= C_s\left\{V_1\frac{\alpha}{\tanh\alpha} - V_2\frac{\alpha}{\sinh\alpha} - \gamma_a\left(\frac{\alpha}{\tanh\alpha} - 1\right)V_{a1}\right. \quad (7.68)$$
$$\left. - \gamma_a\left(1 - \frac{\alpha}{\sinh\alpha}\right)V_{a2} - \gamma_b\left(\frac{\alpha}{\tanh\alpha} - 1\right)V_{b1} - \gamma_b\left(1 - \frac{\alpha}{\sinh\alpha}\right)V_{b2}\right\}$$

Using the labeling scheme $1,2,3,4,5,6 \leftrightarrow 1,2,a1,a2,b1,b2$, the off-diagonal mutual capacitances are, from (7.68)

$$
\begin{aligned}
&C_{1,2} = -C_s \frac{\alpha}{\sinh\alpha} \\
&C_{1,a1} = -C_s\gamma_a\left(\frac{\alpha}{\tanh\alpha} - 1\right) \;,\quad C_{1,a2} = -C_s\gamma_a\left(1 - \frac{\alpha}{\sinh\alpha}\right) \\
&C_{1,b1} = -C_s\gamma_b\left(\frac{\alpha}{\tanh\alpha} - 1\right) \;,\quad C_{1,b2} = -C_s\gamma_b\left(1 - \frac{\alpha}{\sinh\alpha}\right)
\end{aligned}
\tag{7.69}
$$

These are all negative and the negative of their sum is C_{11} which is

$$
C_{11} = C_s \frac{\alpha}{\tanh\alpha} \tag{7.70}
$$

This is an example of (7.67).

The capacitance diagram corresponding to this situation is shown in Fig. 7.11, which shows only the capacitances attached to voltage node V_1. The other mutual capacitances can be filled in by a similar procedure. If the side voltages are constant so that $V_{a1} = V_{a2} = V_a$ and $V_{b1} = V_{b2} = V_b$, then there is only one mutual capacitance connecting 1 to a and it is given by

$$
C_{1a} = C_{1,a1} + C_{1,a2} = -C_s\left(\frac{\alpha}{\tanh\alpha} - \frac{\alpha}{\sinh\alpha}\right) \tag{7.71}
$$

and similarly for C_{1b}. For small α, this approaches

$$
C_{1a} \rightarrow -\frac{C_a}{2} \tag{7.72}
$$

so that on the capacitance diagram, 1/2 the shunt capacitance is attached to the V_1 node. If we carried through the analysis, we would find that 1/2 of the shunt capacitance would also be attached to the V_2 node, producing a π capacitance diagram.

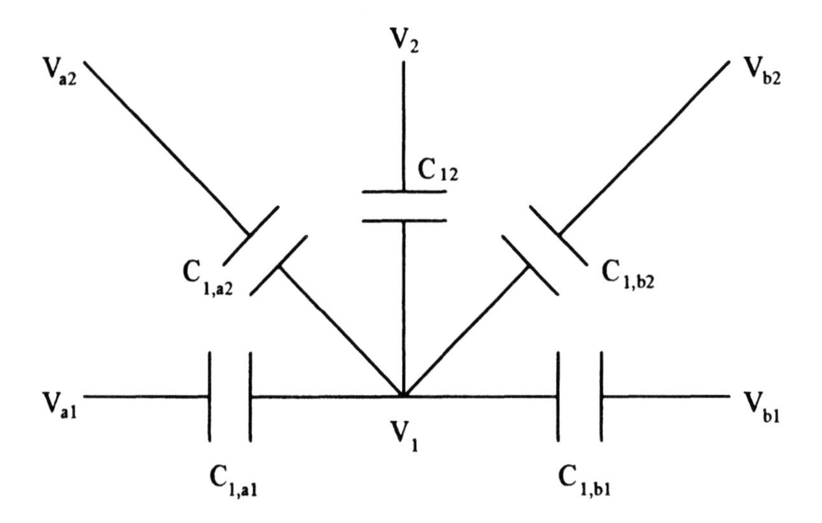

Figure 7.11 Lumped capacitance model of a general lattice capacitor network. The C's are taken to be positive.

7.9 TWO END STATIC RINGS

When two static rings are positioned at the end of a coil, they are situated as shown in Fig. 7.12. Both are attached to the terminal voltage V_1 . This situation also applies to a center fed coil with three static rings since adjacent pairs are configured similarly with respect to the top or bottom coil. It is necessary to analyze more than one disk at a time since their electrostatic energies are coupled via the static rings. We are allowing for the possibility of different spacings between the static rings and adjacent disks and between disks by letting the disk-static ring and disk-disk capacitances be different.

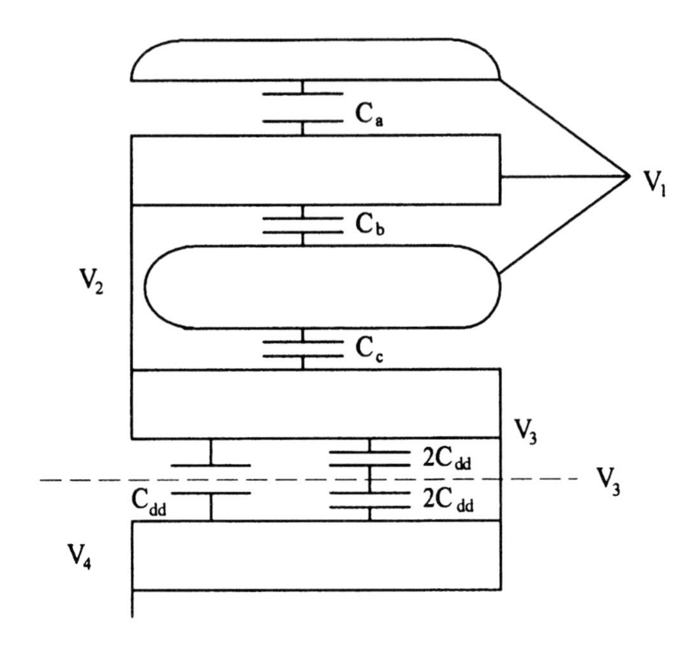

Figure 7.12 Two static rings at the end of a coil

The energy associated with the first or top disk is found, using (7.33) with $V_a = V_b = V_1$ so that $\beta = V_1$,

$$D_{1st\ Disk} = \frac{C_s}{2}(V_2 - V_1)^2 \frac{\alpha}{\tanh\alpha} \tag{7.73}$$

where $\alpha = \sqrt{(C_a + C_b)/C_s}$, C_a and C_b being the disk-static ring capacitances. The energy of the second disk, assuming the same series capacitance C_s, is obtained from (7.33) with the substitutions $V_a = V_1$, $V_b = V_3$, $V_1 = V_2$, $V_2 = V_3$ so that $\beta = \gamma_a V_1 + \gamma_b V_3$. We find

$$\begin{aligned} E_{2nd\ Disk} \\ &= \frac{C_s}{2}\Bigg\{\left[(V_3 - \gamma_a V_1 - \gamma_b V_3)^2 + (V_2 - \gamma_a V_1 - \gamma_b V_3)^2\right]\frac{\alpha_1}{\tanh\alpha_1} \\ &- 2(V_2 - \gamma_a V_1 - \gamma_b V_3)(V_3 - \gamma_a V_1 - \gamma_b V_3)\frac{\alpha_1}{\sinh\alpha_1} \\ &+ \alpha_1^2\left[\gamma_a (V_1 - \gamma_a V_1 - \gamma_b V_3)^2 + \gamma_b (V_3 - \gamma_a V_1 - \gamma_b V_3)^2\right]\Bigg\} \end{aligned} \tag{7.74}$$

with $\alpha_1 = \sqrt{(C_c + 2C_{dd})/C_s}$, $\gamma_a = C_c/(C_c + 2C_{dd})$, $\gamma_b = 2C_{dd}/(C_c + 2C_{dd})$. Using $\gamma_a + \gamma_b = 1$, this can be simplified

$$E_{2nd\ Disk} = \frac{C_s}{2}\left\{\left[\gamma_a^2(V_3 - V_1)^2 + (V_2 - \gamma_a V_1 - \gamma_b V_3)^2\right]\frac{\alpha_1}{\tanh\alpha_1}\right.$$
$$\left. +2\gamma_a(V_2 - \gamma_a V_1 - \gamma_b V_3)(V_1 - V_3)\frac{\alpha_1}{\sinh\alpha_1} + \alpha_1^2\gamma_a\gamma_b(V_1 - V_3)^2\right\} \quad (7.75)$$

Thus the total energy in the first two disks with static rings is

$$E_{Both\ Disks}$$
$$= \frac{C_s}{2}\left\{(V_1 - V_2)^2\frac{\alpha}{\tanh\alpha} + \left[\gamma_a^2(V_3 - V_1)^2 + (V_2 - \gamma_a V_1 - \gamma_b V_3)^2\right]\frac{\alpha_1}{\tanh\alpha_1}\right. \quad (7.76)$$
$$\left. +2\gamma_a(V_2 - \gamma_a V_1 - \gamma_b V_3)(V_1 - V_3)\frac{\alpha_1}{\sinh\alpha_1} + \alpha_1^2\gamma_a\gamma_b(V_1 - V_3)^2\right\}$$

The lumped capacitance network associated with this configuration can be obtained by the procedure described in the previous section. Thus

$$\frac{\partial E_{Both\ Disks}}{\partial V_1} = C_s\left\{(V_1 - V_2)\frac{\alpha}{\tanh\alpha} + \left[\gamma_a^2(V_1 - V_3) - \gamma_a(V_2 - \gamma_a V_1 - \gamma_b V_3)\right]\frac{\alpha_1}{\tanh\alpha_1}\right.$$
$$\left. +\gamma_a\left[(V_2 - \gamma_a V_1 - \gamma_b V_3) - \gamma_a(V_1 - V_3)\right]\frac{\alpha_1}{\sinh\alpha_1} + \alpha_1^2\gamma_a\gamma_b(V_1 - V_3)\right\}$$
$$= C_s\left\{\left[\frac{\alpha}{\tanh\alpha} + 2\gamma_a^2\left(\frac{\alpha_1}{\tanh\alpha_1} - \frac{\alpha_1}{\sinh\alpha_1}\right) + \alpha_1^2\gamma_a\gamma_b\right]V_1 - \left[\frac{\alpha}{\tanh\alpha} + \gamma_a\left(\frac{\alpha_1}{\tanh\alpha_1} - \frac{\alpha_1}{\sinh\alpha_1}\right)\right]V_2\right.$$
$$\left. -\left[\gamma_a(2\gamma_a - 1)\left(\frac{\alpha_1}{\tanh\alpha_1} - \frac{\alpha_1}{\sinh\alpha_1}\right) + \alpha_1^2\gamma_a\gamma_b\right]V_3\right\}$$
$$(7.77)$$

Reading off the mutual capacitances from (7.77), we find

$$C_{12} = -C_s\left[\frac{\alpha}{\tanh\alpha} + \gamma_a\left(\frac{\alpha_1}{\tanh\alpha_1} - \frac{\alpha_1}{\sinh\alpha_1}\right)\right]$$
$$C_{13} = -C_s\left[\gamma_a(2\gamma_a - 1)\left(\frac{\alpha_1}{\tanh\alpha_1} - \frac{\alpha_1}{\sinh\alpha_1}\right) + \alpha_1^2\gamma_a\gamma_b\right] \quad (7.78)$$

These are negative and minus their sum is the self capacitance C_{11}.

We also need to find C_{23}. This is obtained by differentiating (7.76) with respect to V_2

$$\frac{\partial E_{\text{Both Disks}}}{\partial V_2}$$

$$= C_s\left\{-(V_1 - V_2)\frac{\alpha}{\tanh\alpha} + (V_2 - \gamma_a V_1 - \gamma_b V_3)\frac{\alpha_1}{\tanh\alpha_1} + \gamma_a(V_1 - V_3)\frac{\alpha_1}{\sinh\alpha_1}\right\}$$

$$= C_s\left\{-\left[\frac{\alpha}{\tanh\alpha} + \gamma_a\left(\frac{\alpha_1}{\tanh\alpha_1} - \frac{\alpha_1}{\sinh\alpha_1}\right)\right]V_1 + \left[\frac{\alpha}{\tanh\alpha} + \frac{\alpha_1}{\tanh\alpha_1}\right]V_2\right.$$

$$\left. - \left[\gamma_b\frac{\alpha_1}{\tanh\alpha_1} + \gamma_a\frac{\alpha_1}{\sinh\alpha_1}\right]V_3\right\} \tag{7.79}$$

Extracting the mutual capacitances, we find

$$\begin{aligned} C_{21} &= -C_s\left[\frac{\alpha}{\tanh\alpha} + \gamma_a\left(\frac{\alpha_1}{\tanh\alpha_1} - \frac{\alpha_1}{\sinh\alpha_1}\right)\right] \\ C_{23} &= -C_s\left[\gamma_b\frac{\alpha_1}{\tanh\alpha_1} + \gamma_a\frac{\alpha_1}{\sinh\alpha_1}\right] \end{aligned} \tag{7.80}$$

These are negative and the negative of their sum is C_{22}. Moreover, we see that $C_{21} = C_{12}$ as expected since the capacitance matrix is symmetric. Differentiating (7.76) with respect to V_3 would give us no new information.

Thus we can draw the lumped capacitance diagram for this configuration as shown in Fig. 7.13. We assume the capacitances shown are positive (the negative of the mutual capacitances). Hence the total capacitance between the V_1 and V_3 terminals is given by

$$C_{\text{Both Disks}} = C_{13} + \frac{C_{12}C_{23}}{(C_{12} + C_{23})} \tag{7.81}$$

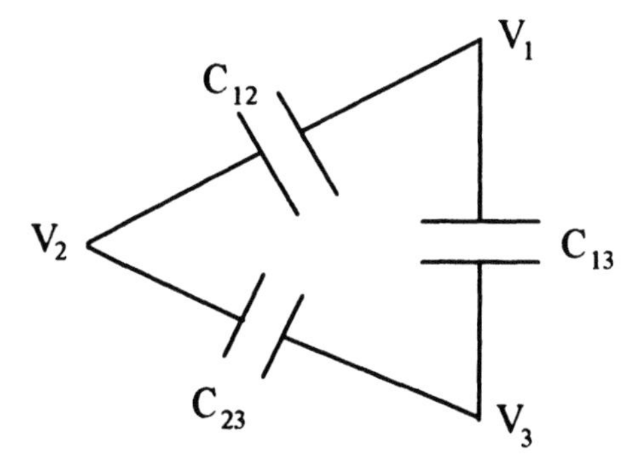

Figure 7.13 Lumped capcitance diagram for the 2 static ring configuration. The C's are taken to be positive.

In model impulse voltage calculations, one could treat the first two disks as a unit having the capacitance given by (7.81). Then the voltages across each disk could be obtained from the overall voltage difference $(V_1 - V_3)$ via

$$(V_1 - V_2) = \frac{C_{23}}{(C_{12} + C_{23})}(V_1 - V_3)$$

$$(V_2 - V_3) = \frac{C_{12}}{(C_{12} + C_{23})}(V_1 - V_3) \tag{7.82}$$

7.10 STATIC RING BETWEEN THE FIRST TWO DISKS

Sometimes a static ring is placed between the first two disks. This is usually only considered for center fed windings so that there are two symmetrically spaced static rings, one for each of the two stacked coils. This case is very similar to the previous case. The only difference is that the energy in the first disk is given by

$$E_{1st\ Disk} = \frac{C_s}{2}(V_1 - V_2)^2 \frac{\alpha}{\tanh\alpha} \tag{7.83}$$

with $\alpha = \sqrt{C_b / C_s}$ since $C_a = 0$. Thus the formulas of the last section apply with this value of α.

7.11 WINDING DISK CAPACITANCES WITH WOUND-IN-SHIELDS

This section is essentially a reprint of our published paper [Del98].

In order to improve the voltage distribution along a transformer coil, i.e. to reduce the maximum disk-disk voltage gradient, it is necessary to make the distribution constant, α, as small as possible, where $\alpha = \sqrt{C_g/C_s}$ with C_g the ground capacitance and C_s the series capacitance of the coil. One way of accomplishing this is to increase C_s. Common methods for increasing the series capacitance include interleaving [Nuy78] and the use of wound-in-shields [For69]. Both of these techniques rely on inductive effects. Geometric methods for increasing C_s such as decreasing turn-turn or disk-disk clearances are generally ruled out by voltage withstand or cooling considerations.

Interleaving can produce large increases in C_s which may be necessary in very high voltage applications. However, it can be very labor intensive and in practice tends to be limited to magnet wire applications. Wound-in-shields tend to produce more modest increases in C_s compared with interleaving. However, they require less labor and are suitable for use with transposed cable. In addition, they can easily provide a tapered capacitance profile to match the voltage stress profile of the winding.

Here we present a simple analytic formula for calculating the disk capacitance with a variable number of wound-in-shield turns. Since this formula rests on certain assumptions, a detailed circuit model is developed to test these assumptions. Finally, experiments are carried out to check the formula under a wide variety of conditions.

7.11.1 Analytic Formula

Fig. 7.14 shows the geometry of a pair of disks containing a wound-in-shield. Also shown is the method of labeling turns of the coil and shield. Since the shield spans two disks, it is necessary to calculate the capacitance of the pair. Each disk has N turns and n wound-in-shield turns, where $n \leq N\text{-}1$. The voltage across the pair of disks is V and we assume the rightmost turn of the top disk, $i = 1$, is at voltage V and the rightmost turn of the bottom disk, $j = 1$, is at 0 volts, so that the coil is wound in a positive sense from outer to inner turn on the bottom disk and from inner to outer turn on the top disk. The shield turns are placed between the coil turns and are wound in the same sense as the coil. However, their cross-over is at the outermost turn rather than the innermost one as is the case for the coil. This means that the positive voltage sense for the shield is from the leftmost turn, $i = n$, on the top coil to turn $i = 1$ on the top disk and then from turn $j = 1$ to $j = n$ on the bottom disk.

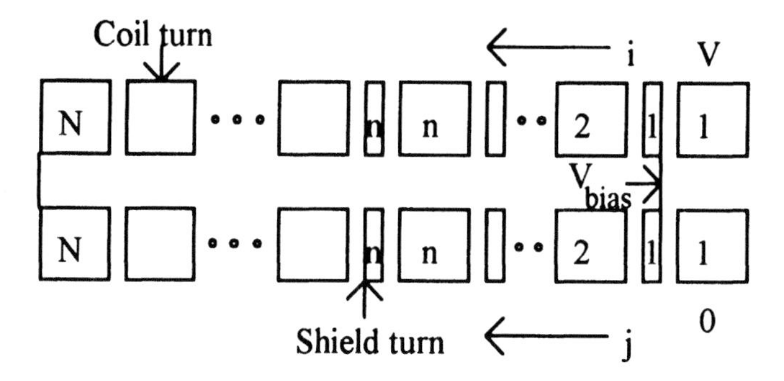

Figure 7.14 Disk pair with wound-in-shields with labelling and other parameters indicated

We assume the voltage rise per turn is ΔV where

$$\Delta V = \frac{V}{2N} \tag{7.84}$$

We assume this same volts per turn applies to the shield as well. For definiteness, we assume the voltage at the shield cross-over point is V_{bias} as shown in Fig. 7.14. Taking the voltage at the midpoint of a turn, we have for the top disk

$$V_c(i) = V - (i - 0.5)\Delta V \tag{7.85}$$

where $V_c(i)$ is the coil voltage for turn i, $i = 1, \ldots, N$. Also for the top disk

$$V_w(i) = V_{bias} - (i - 0.5)\Delta V \tag{7.86}$$

where $V_w(i)$ is the shield voltage for turn i, $i = 1, \ldots, n$. For the bottom disk, we have

$$V_c(j) = V + (j - 2N - 0.5)\Delta V \tag{7.87}$$

for $j = 1,\ldots,N$ and

$$V_w(j) = V_{bias} + (j - 0.5)\Delta V \tag{7.88}$$

for $j = 1,\ldots,n$.

Letting c_w be the capacitance between a coil turn and adjacent shield turn, the energy stored in the capacitance between shield turn i and its adjacent coil turns is

$$\frac{1}{2}c_w\left\{\left[V_c(i) - V_w(i)\right]^2 + \left[V_c(i+1) - V_w(i)\right]^2\right\} \tag{7.89}$$

The two terms reflect the fact that there are two adjacent coil turns for every shield turn. Using the previous expressions for the V's, (7.89) becomes

$$\frac{1}{2}c_w\left\{\left[V - V_{bias}\right]^2 + \left[V - V_{bias} - \Delta V\right]^2\right\} \tag{7.90}$$

This does not depend on i and so is the same for all n shield turns on the top disk. For the bottom disk, (7.89) applies with j replacing i so the capacitative energy between shield turn j and its surrounding coil turns is

$$\frac{1}{2}c_w\left\{\left[V - V_{bias} - 2N\Delta V\right]^2 + \left[V - V_{bias} - (2N-1)\Delta V\right]^2\right\} \tag{7.91}$$

This again does not depend on j and so is the same for all n shield turns on the bottom disk.

To the above energies, we must add the capacitative energy of the turns without wound-in-shields between them. Letting c_t be the turn-turn capacitance, this energy is simply

$$\frac{1}{2}c_t(\Delta V)^2 \tag{7.92}$$

There are $2(N - n - 1)$ such terms in the energy. In addition, there is energy stored in the disk-disk capacitance, c_d. Since we are assuming the voltage varies linearly along the disks, this capacitative energy is given by

$$\frac{1}{2}\left(\frac{c_d}{3}\right)V^2 \tag{7.93}$$

We are ignoring capacitative coupling to other disk pairs at this stage since our experimental setup consisted of an isolated disk pair. However, if this disk pair were embedded in a larger coil of similar disks, then (7.93) would need to be doubled before adding to the energy.

The total capacitative energy of the disk pair is found by adding the above contributions. To simplify the formula, we define

$$\beta = \frac{V - V_{bias}}{\Delta V} \tag{7.94}$$

In terms of this parameter, the total capacitative energy, E_{tot}, is given by

$$E_{tot} = \frac{1}{2}V^2\left\{c_w \frac{n}{4N^2}\left[\beta^2 + (\beta-1)^2 + (\beta-2N)^2 + (\beta-2N+1)^2\right] + c_t \frac{(N-n-1)}{2N^2} + \frac{c_d}{3}\right\} \tag{7.95}$$

where (7.84) has been used. Extracting the equivalent or total capacitance from (7.95), we get

$$C_{tot} = c_w \frac{n}{4N^2}\left[\beta^2 + (\beta-1)^2 + (\beta-2N)^2 + (\beta-2N+1)^2\right] + c_t \frac{(N-n-1)}{2N^2} + \frac{c_d}{3} \tag{7.96}$$

At this point, V_{bias} and hence β is unspecified. This will depend on whether the shield is floating or whether it is attached at some point to a coil voltage. If the shield is floating, we expect $V_{bias} = V/2$. If the shield is attached at the cross-over to the high voltage terminal, then $V_{bias} = V$. If the leftmost or end shield turn on the top disk (i = n) is attached to the high voltage terminal, then $V_{bias} = V + (n-0.5)\Delta V$. In terms of β,

Shield floating	$\beta = N$	
Shield attached at cross-over to V	$\beta = 0$	(7.97)
Shield attached at top end to V	$\beta = -(n-0.5)$	

Other situations can be considered as well.

For the turn-turn capacitance, we used the expression (in SI units)

$$c_t = \frac{2\pi\varepsilon_o\varepsilon_p R_{ave}\left(h+2\tau_c\right)}{\tau_c} \tag{7.98}$$

where R_{ave} is the average radius of the coil, h the bare copper height of a coil turn in the axial direction, τ_c the 2-sided paper thickness of a coil turn, $\varepsilon_o = 8.85\times10^{-12}$ F/m, and ε_p the relative permittivity of paper. The addition of $2\tau_c$ to h is designed to take fringing effects into account. Also the use of R_{ave} is an approximation which is reasonably accurate for coils with radial builds small compared with their radii. For the capacitance c_w the same expression was used but with τ_c replaced by $0.5(\tau_c + \tau_w)$ where τ_w is the 2-sided paper thickness of a shield turn. Since there are key spacers separating the disks and gaps between them, the disk-disk capacitance is given by

$$c_d = \pi\varepsilon_o\left(R_o^2 - R_i^2\right)\left\{\frac{f}{\left(\tau_c/\varepsilon_p + \tau_k/\varepsilon_k\right)} + \frac{(1-f)}{\left(\tau_c/\varepsilon_p + \tau_k/\varepsilon_a\right)}\right\} \tag{7.99}$$

where R_i and R_o are the inner and outer radii of the disk respectively, f is the fraction of the disk-disk space occupied by key spacers, τ_k is the key spacer thickness, ε_k the key spacer relative permittivity, and ε_a the relative permittivity of air (= 1) since the coils were tested in air.

7.11.2 Circuit Model

Since quite a few assumptions went into deriving the capacitance formula in the last section, we decided to model the disk pair by means of a circuit model, including capacitative, inductive, and resistive effects as sketched in Fig. 7.15. We include all mutual couplings between the turns of the coil and wound-in-shield. The circuit is assumed to be excited by a current source which is nearly a step function. The circuit equations are

$$\begin{aligned} C\frac{d\mathbf{V}}{dt} &= A\mathbf{I} \\ M\frac{d\mathbf{I}}{dt} &= B\mathbf{V} - R\mathbf{I} \end{aligned} \tag{7.100}$$

where C is a capacitance matrix, M an inductance matrix, R a diagonal resistance matrix and A , B matrices of ±1's and 0's. The voltage and current vectors, **V** and **I** , include the coil and shield turn voltages and currents. These equations are solved by means of a Runge-Kutta solver.

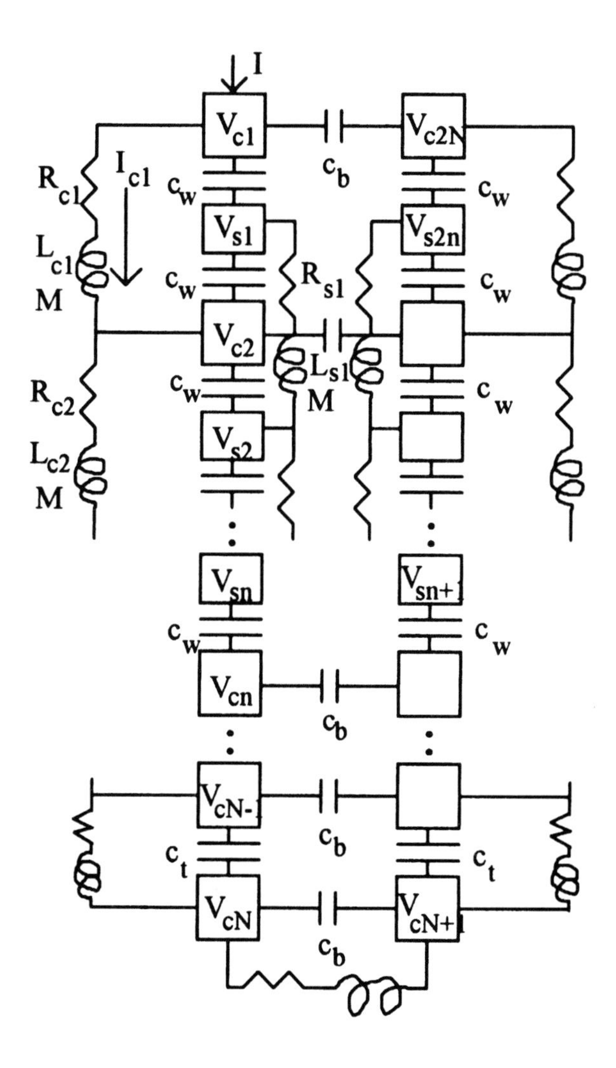

Figure 7.15 Circuit model for a disk pair with wound-in-shields. The labelling scheme and circuit parameters are indicated.

Since the experimental setup was in air, we used air core inductance and mutual inductance expressions. The mutual inductance between two thin wire coaxial loops of radii r_1 and r_2 spaced a distance d apart is given by [Smy68] in MKS units,

$$M = \frac{2\mu_o}{k}\sqrt{r_1 r_2}\left[\left(1-\frac{k^2}{2}\right)K(k)-E(k)\right]$$
$$\text{with}\quad k^2 = \frac{4r_1 r_2}{\left[\left(r_1+r_2\right)^2+d^2\right]} \qquad (7.101)$$

K(k) and E(k) are complete elliptic integrals of the first and second kinds respectively and $\mu_o = 4\pi\times10^{-7}$ H/m. For rectangular cross-section coils, Lyle's method in conjunction with (7.101) could be used for a more accurate determination of the mutual inductance [Gro73]. However, for the turn-turn mutual inductances in our experimental coils, treating the turns as thin circular loops was nearly as accurate as Lyle's method. The self inductance of a single turn circular coil of square cross section with an average radius of a and square side length c is given by [Gro73] in MKS units

$$L = \mu_o a\left\{\frac{1}{2}\left[1+\frac{1}{6}\left(\frac{c}{2a}\right)^2\right]\ln\left[\frac{8}{(c/2a)^2}\right]+0.2041\left(\frac{c}{2a}\right)^2-0.84834\right\} \qquad (7.102)$$

This applies for $c/2a \le 0.2$. When the cross-section is not square, it can be subdivided into a number of squares and (7.102) together with (7.101) can be applied to compute the self inductance more accurately. In our experimental coil, the turn dimensions were such that the simple formula with c taken as the square root of the turn area agreed well with the more accurate calculation.

The turn-turn and turn-shield capacitances were the same as given in the last section. The capacitance c_b in Fig. 7.15 was taken as c_d/N. We did not include the capacitance between shield turns on neighboring disks in our final calculations. Their inclusion had little effect on the total capacitance.

The resistances used in the circuit model were based on the wire dimensions but were multiplied by a factor to account for high frequency losses. This factor may be estimated by examining the frequency dependence of the two main contributors to the coil loss, namely the Joule or I^2R loss and the eddy current loss due to stray flux carried by the conductor strands. Based on a formula in Ref. [Smy68] for cylindrical conductors, the Joule loss is nearly independent of frequency at low frequencies which includes 60 Hz for our conductor dimensions. At high frequencies, the loss divided by the dc or 60 Hz loss is given by,

$$\frac{W_{Joule}(f)}{W_{Joule}(f=0)} = \frac{r_{cond}}{2}\sqrt{\pi\mu_o\sigma f} \tag{7.103}$$

where r_{cond} is the radius of the conductor or in our case an effective radius based on the wire dimensions, σ is the wire's conductivity, and f in this context is the frequency in Hz. Based on our cable dimensions and for a typical frequency encountered in our calculations and experiment of ~ 0.15 MHz, we estimate that $W_{joule}(f)$ / $W_{Joule}(f=0) \sim$ 15.9 The eddy current frequency dependence due to stray flux is given in Ref. [Lam66]. At low frequencies, this loss varies as f^2 whereas at high frequencies, it varies as $f^{0.5}$. Taking ratios of the high to low frequency eddy current loss, we obtain

$$\frac{W_{eddy}(f)}{W_{eddy}(60Hz)} = \frac{6\sqrt{f}}{b^3(\pi\mu_o\sigma)^{1.5}(60)^2} \tag{7.104}$$

where b is the thickness of the lamination or strand in a direction perpendicular to the stray magnetic field. In our case, the cable was made of rectangular strands with dimensions 0.055 in. × 0.16 in. (1.4 mm × 4.06 mm). For the small dimension, (7.104) gives a ratio of 68,086 while for the large dimension, we get a ratio of 2,766. A detailed field mapping is necessary to obtain the eddy current loss contribution at low frequencies but typically this amounts to about 10% of the dc loss, depending on the cable construction. If we assume that the small and large dimension eddy losses are equal at low frequencies, i.e. each is 5% of the dc loss, then we find that the ratio of high to low frequency total loss based on the above ratios is given by $W_{tot}(f)$ / $W_{tot}(60Hz)$ = 3557. This is only a crude estimate. Our data show that this ratio is about 1500 for our cable. Since the capacitance obtained from the simulation is nearly independent of the resistance used, we did not try to match the model's resistance with the experimental values. A sample output is shown in Fig. 7.16.

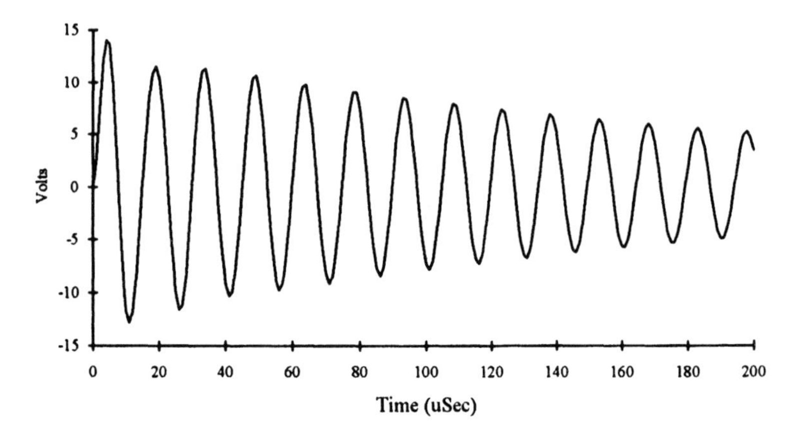

Figure 7.16 Voltage across the coil pair versus time from the detailed circuit model calcualtions.

Two methods were used to extract the equivalent total capacitance from the circuit model. Since the voltages of each coil and shield turn were determined at each time step, the capacitive energy was simply summed and the total capacitance determined via

$$E_{tot} = \frac{1}{2} C_{tot} V^2 \tag{7.105}$$

where V is the voltage of turn $i = 1$ at the particular time step. At the end of the total time duration of about 200 time steps, the average and median total capacitances were calculated. These two values generally agreed fairly closely. The other method consisted of extracting the total capacitance from a simplified equivalent circuit as shown in Fig. 7.17. This latter method was also used to obtain the total capacitance experimentally.

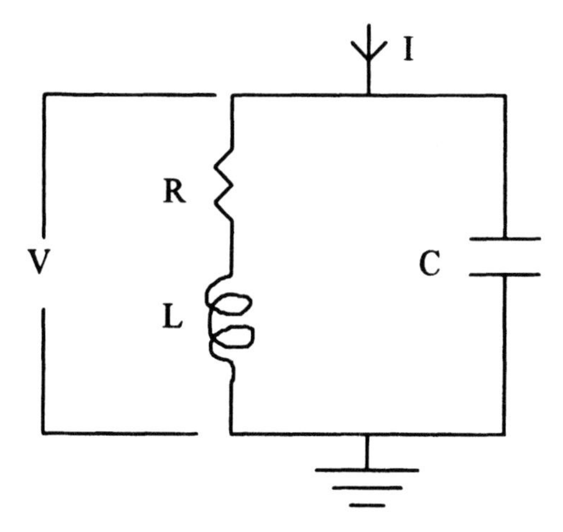

Figure 7.17 Simplified circuit model for capacitance determination

Using Laplace transforms, the circuit of Fig. 7.17 can be solved analytically, assuming a step function current input of magnitude I. We obtain

$$V(t) = IR\left\{1 - e^{-\frac{R}{2L}t}\left[\cos\omega_0 t + \frac{1}{2}\left(\frac{R}{2L}\right)\left(\frac{1}{\omega_0} - \frac{\omega_0}{(R/2L)^2}\right)\sin\omega_0 t\right]\right\} \tag{7.106}$$

for $t \geq 0$. Here

$$\omega_0 = \sqrt{\frac{1}{LC} - \left(\frac{R}{2L}\right)^2} \tag{7.107}$$

In the limiting case as $R \to 0$, (7.106) becomes

$$V(t) = I\sqrt{\frac{L}{C}}\sin\omega_0 t \tag{7.108}$$

In all the cases examined computationally or experimentally, the term $(R/2L)^2$ was much smaller than $1/LC$ so that it can be ignored in the expression for ω_0. Thus by measuring the oscillation frequency, we determine the combination LC. To obtain C, an additional capacitance,

C_1 ,was placed in parallel with the coil and a new oscillation frequency, ω_1 , determined. Since this added capacitance doesn't change L, we have, taking ratios and squaring (7.107),

$$\frac{C+C_1}{C}=\left(\frac{\omega_o}{\omega_1}\right)^2 \tag{7.109}$$

Since ω_o and ω_1 can be measured or determined from the output of the circuit model and C_1 is known, C can be obtained. We found good agreement between the two methods of determining C.

7.11.3 Experimental Methods

A coil containing two disk sections was made of transposed cable. There were 10 turns per disk. The cable turns were 0.462 in. radial build by 0.362 in. axial height, including a 30 mil (2-sided) paper cover (1.17 cm × 0.919 cm with 0.76 mm paper). The inner radius was 9.81 in. (0.249 m). The outer radius depended on the number of shield turns but was approximately 15.4 in. (0.392 m). The wound-in-shield turns consisted of 0.14 in. radial build by 0.365 in. axial height magnet wire, including a 20 mil (2-sided) paper cover (0.355 cm × 0.927 cm with 0.51 mm paper). The disks were separated by means of 18 key spacers equally spaced around the circumference. The key spacers were 1.75 in wide by 0.165 in. thick (4.45 cm × 0.419 cm).

The coil was excited by means of a current source which produced a near step function current. A sample of the current input and coil voltage output is shown in Fig. 7.18. This voltage vs time plot was Fourier analyzed to extract the resonant frequency. Frequencies were measured with and without an external capacitor of 10 nF across the coil in order to obtain the coil capacitance by the ratio method described previously.

Shield turns/disk of $n = 3, 5, 7, 9$ were tested as well as the no shield case, $n = 0$. In addition, for each n value, tests were performed with the shield floating, attached to the high voltage terminal at the cross-over, and attached to the high voltage terminal at the end shield turn on the top disk. Other methods of attachment were also made for the $n = 9$ case to test for expected symmetries.

Because the disk pair was not contained in a tightly wound coil, there tended to be some looseness in the winding. This could be determined by squeezing the disk turns tightly together and noting how much the radial build decreased. From this we could determine how much looseness there was in the insulation and correct for it. This amounted to about a 5% correction.

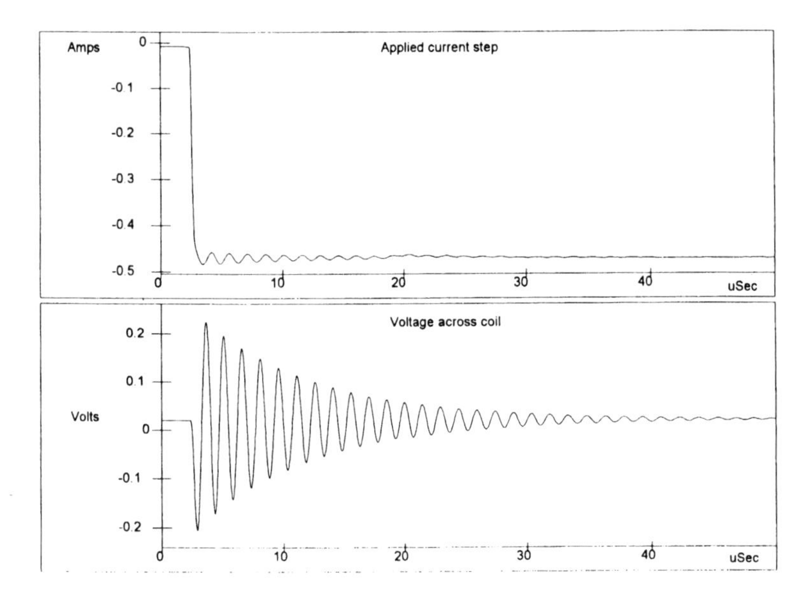

Figure 7.18 Experimental output from one of the test runs on the coil showing the input current and voltage across the coil vs time.

7.11.4 Results

There are some uncertainties associated with the values of the relative permittivities to use in the capacitance calculations, particularly that of paper in air. For the pressboard key spacers, Ref. [Mos87] gives a method for determining its permittivity in terms of the board density, the permittivity of the fibers, and the permittivity of the substance filling the voids. For our key spacers in air, we obtain $\varepsilon_k = 4.0$ using this method. (For pressboard in oil, the value is $\varepsilon_k = 4.5$) For paper, Ref. [Cla62] presents a graph of the dielectric constant vs density in air and oil. Unfortunately paper wrapping on cable in not as homogeneous a substance as pressboard. Its effective density would depend on how the wraps overlap and how loosely or tautly it is wound. In addition, for transposed cable, the paper-copper interface is not a clean rectilinear one. This is because of the extra unpaired strand on one side of the strand bundle. We found that a value of $\varepsilon_p = 1.5$ was needed to get good overall agreement with the test data. This would correspond to an effective paper density of 0.5 gm/cm^3 according to the graph in Ref. [Cla62]. At that density, paper in oil would have $\varepsilon_p = 3.0$ according to the same graph. For tightly wound paper in oil, values of $\varepsilon_p = 3.5 - 4.0$ are typically used. Thus a paper permittivity of 1.5 for paper in air is not unreasonable.

Table 7.1 shows the test results along with the calculated values of the capacitance for different numbers of shield turns and different shield biasing. The overall agreement is good. Certainly the trends are well reproduced. Even for the floating shield case which has the lowest capacitative enhancement, there is a capacitance increase of a factor of 5 for 3 shield turns/disk and a factor of about 13 for 9 shield turns/disk over the unshielded case. By biasing the shield in different ways, even greater increases are achieved.

Table 7.1 Capacitance of coil with 2 disks of 10 turns/disk and variable number of wound-in-shield turns/disk.

		C_{tot} (nF)		
Shield turns n	Calc method	Floating	Attached to V at cross-over	Attached to V at top end
	Measured	0.24		
0	Analytic	0.264		
	Circuit model	0.287		
	Measured	1.20	2.33	3.21
3	Analytic	1.31	2.44	3.08
	Circuit model	1.41	2.66	3.31
	Measured	1.78	3.53	6.07
5	Analytic	2.03	3.94	6.04
	Circuit model	2.21	4.34	6.73
	Measured	2.80	5.73	11.8
7	Analytic	2.75	5.45	10.1
	Circuit model	2.99	6.01	11.4
	Measured	3.17	6.82	16.1
9	Analytic	3.49	6.99	15.5
	Circuit model	3.83	7.61	17.9

In Table 7.2, the coil turn and shield turn voltages as calculated by the circuit model at time $t = 30$ μsec are shown for the $n = 9$ case with the shields floating. As can be seen, the coil volts/turn is about the same as the shield volts/turn, verifying the assumption made in the analytic model. In addition, $V_{bias} \approx V/2$ as was also assumed in the simple model. Also the voltage differences between the coil and shield turns fall into two groups of either ~ 12.8 or 11.5 V in this case. This also corresponds to the simple model prediction of either $N\Delta V = V/2$ or $(N-1)\,\Delta V = V/2 - \Delta V$ volts. Thus, in the case where the shields are floating, the maximum turn-shield voltage is V/2. Based on the simple model, this holds regardless of the number of shield turns.

Table 7.2 Turn voltages at t = 30 μsec for a coil of 2 disks with 10 turns/disk and 9 shield turns/disk with the shield floating.

Coil Turns			Adjacent Shield Turn			$\lvert\Delta V\rvert$
#	Volts	ΔV	#	Volts	ΔV	Shld-adj coil
1	24.27		1	11.43		12.84, 11.49
2	22.92	1.35	2	10.12	1.31	12.80, 11.39
3	21.51	1.45	3	8.73	1.39	12.78, 11.35
4	20.08	1.43	4	7.30	1.43	12.78, 11.35
5	18.65	1.43	5	5.87	1.43	12.78, 11.46
6	17.24	1.41	6	4.45	1.42	12.79, 11.44
7	15.89	1.35	7	3.09	1.36	12.80, 11.53
8	14.62	1.27	8	1.79	1.30	12.83, 11.63
9	13.42	1.20	9	0.52	1.27	12.90, 11.77
10	12.29	1.13				
11	11.49	0.80				
12	10.50	0.99	10	23.43		12.93, 11.94
13	9.45	1.05	11	22.27	1.16	12.82, 11.77
14	8.28	1.17	12	21.08	1.19	12.80, 11.63
15	6.98	1.30	13	19.78	1.30	12.80, 11.50
16	5.64	1.34	14	18.39	1.39	12.75, 11.41
17	4.25	1.39	15	16.99	1.40	12.74, 11.35
18	2.83	1.42	16	15.56	1.43	12.73, 11.31
19	1.40	1.43	17	14.14	1.42	12.74, 11.31
20	0	1.40	18	12.75	1.39	12.75, 11.35

$V_{bias} = (V_{shield\ turn\ 1} + V_{shield\ turn\ 18})/2 = 12.09 \approx V/2 = 12.14$

In Table 7.3, we show the corresponding output from the circuit model at t = 150 μsec for the case where the shield is attached at the cross-over to the high voltage terminal. (It was actually attached to shield turn i = 1 in the model.) We see again that the coil volts/turn ≈ shield volts/turn. According to the simple model prediction (β = 0), the voltage differences between shield and adjacent turn on the top disk are 0 and ΔV as is also nearly the case for the circuit model. Along the bottom disk, the simple model gives voltage differences of $2N\Delta V = V$ and $(2N-1)\,\Delta V = V - \Delta V$ and this is also nearly the case for the circuit model. Thus the maximum coil turn to shield turn voltage is V. According to the simple model, this holds regardless of the number of shield turns.

Table 7.3 Turn voltages at t =150 μsec for a coil of 2 disks with 10 turns/disk and 9 shield turns/disk with the shield attached at the cross-over to the high voltage terminal.

Coil Turns			Adjacent Shield Turn			$\lvert\Delta V\rvert$
#	Volts	ΔV	#	Volts	ΔV	Shld-adj coil
1	5.13		1	5.13		0, 0.29
2	4.84	0.29	2	4.87	0.26	0.03, 0.33
3	4.54	0.30	3	4.58	0.29	0.04, 0.35
4	4.23	0.31	4	4.28	0.30	0.05, 0.35
5	3.93	0.30	5	3.99	0.29	0.06, 0.36
6	3.63	0.30	6	3.69	0.30	0.06, 0.34
7	3.35	0.28	7	3.41	0.28	0.06, 0.33
8	3.08	0.27	8	3.14	0.27	0.06, 0.30
9	2.84	0.24	9	2.90	0.24	0.06, 0.28
10	2.62	0.22				
11	2.42	0.20				
12	2.24	0.18	10	7.52		5.28, 5.10
13	2.02	0.22	11	7.30	0.22	5.28, 5.06
14	1.77	0.25	12	7.05	0.25	5.28, 5.03
15	1.50	0.27	13	6.79	0.26	5.29, 5.02
16	1.22	0.28	14	6.51	0.28	5.29, 5.01
17	0.92	0.30	15	6.23	0.28	5.31, 5.01
18	0.62	0.30	16	5.94	0.29	5.32, 5.02
19	0.31	0.31	17	5.66	0.28	5.35, 5.04
20	0	0.31	18	5.38	0.28	5.38, 5.07

The case where the end turn of the shield on the top disk is attached to the high voltage terminal could be analyzed similarly although this configuration is harder to achieve in practice. This configuration also has a higher coil turn to shield turn voltage difference than the other methods of shield attachment. As expected, however, the price to pay for the higher capacitances are higher coil turn to shield turn voltage differences.

Several symmetric situations were noted in the experimental data. These symmetries were only checked for the $n = 9$ case but should apply to all n values according to the simple model. We found that attaching the high voltage terminal to the end shield turn on the top disk ($i = n$) produced the same capacitance as attaching the low voltage terminal, at 0 volts here, to the end turn on the bottom disk ($j = n$). This symmetry can be seen in the simple formula (7.96) by using the appropriate values for β. Attaching the top end shield turn to the low voltage terminal produced the same capacitance as attaching the bottom end shield turn to the high voltage terminal. This can also be shown by means of the simple formula. Also the same capacitance was produced whether the shield cross-over was attached to the high or low voltage terminal. This also follows from the simple model.

For consistency, the inductance of our coil was extracted from the experimental data as well as from the circuit model output. We found L(data) = 280 μH and L(circuit model) = 320 μH. Using an algorithm in Grover [Gro73], we obtained L(calc) = 340 μH. These are all within reasonable agreement. The decay constant $R/2L$ appearing in (7.106) could be extracted from the data by analyzing the envelope of the damped sinusoid (Fig. 7.18) and we found that it could be dropped in the formula for ω_o (7.107). We also observed a significant resistance change when the frequency was changed by adding the external capacitance.

7.12 MULTI-START WINDING CAPACITANCE

Multi-start windings are a simple example of an interleaved type of winding. They are commonly used as tap windings. In these windings, adjacent turns have voltage differences which can differ from the usual turn-turn voltage drop along typical disk or helical windings. Although multi-start windings are helical windings, they can be thought of as a collection of superposed series connected helical windings as shown in Fig. 7.19.

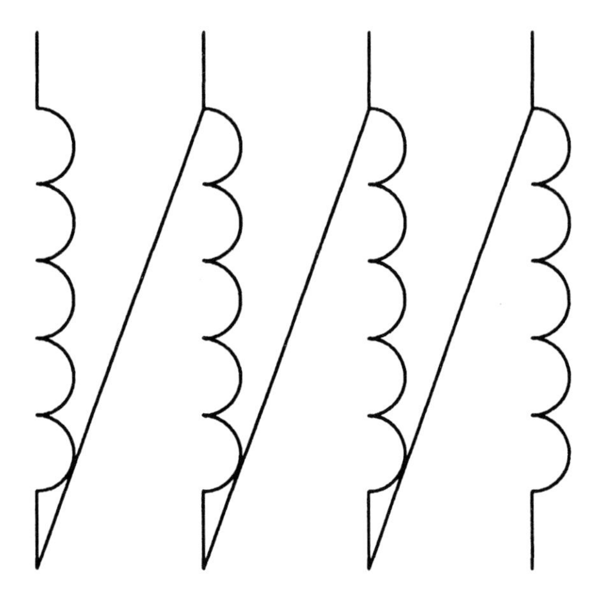

Figure 7.19 Schematic illustration of a multi-start winding

These winding essentially start over again and again which motivates the name. Each start represents a constitutent winding having a certain number of turns called turns per start. The number of starts is the same as the number of constituent windings. By connecting taps between the starts, the turns per start become the number of tap turns. Their advantage as tap windings compared with the standard type is that they allow a more balanced force distribution regardless of the tap setting and do not require thinning of adjacent windings.

Although shown as side by side windings in the figure for explanatory purposes, the windings are superposed into one helical type of winding. The botton to top connections are made external to this winding. The constituent windings are meshed in such a way that the voltage difference between adjacent turns in the composite winding is kept to one or two times the voltage drop along a constituent winding which is really the best that can be done. Letting 1, 2, 3, ... label turns from the different constituent windings, acceptable meshing schemes are shown in Fig. 7.20 for various numbers of starts, N_s ,

$N_s = 4$		$N_s = 5$		$N_s = 5$	
1	2	1	2	1	2
3	1	3	2	3	2
4	2	5	1	5	1
2	1	4	2	6	2
1	2	2	1	4	2
3	1	1	2	2	1
4	2	3	2	1	2
2	1	5	1	3	2
1		4	2	5	1
⋮		2	1	6	2
		1		4	2
		⋮		2	1
				1	
				⋮	

Figure 7.20 Winding schemes for multi-start windings. The numbers beside the turns are the voltage differences between turns in units of the voltage drop between starts.

The turns are arranged along the winding according to Fig. 7.20. The voltage differences between adjacent turns, measured in units of the voltage drop between starts, are also indicated in the figure. A pattern can be seen in the organization: Start with turn 1 from start coil 1. Put turn 2 from start coil 2 at the end of the group. Put turn 3 from start coil 3 below turn 1. Put turn 4 from start coil 4 above turn 2. Put turn 5 from start coil 5 below turn 3. Put turn 6 from start coil 6 above turn 4. Etc. until you run out of start coils. Then repeat the pattern until you run out of turns in the start coils.

The capacitance is obtained by summing up the capacitative energy associated with the winding configuration. Letting c_{tt} be the turn to turn capacitance and ΔV_{tt} the voltage difference between turns, the energy associated with a pair of adjacent turns is given by Energy(turn-turn) = $1/2\ c_{tt}(\Delta V_{tt})^2$. Letting the voltage drop between starts be ΔV_s , the total energy is obtained by summing all the turn-turn energies. As can be seen, each group of turns has two 1's except for the last group which has only 1. If there are n turns/start, there are n such groups. Hence the energy is 2n-1 times the energy associated with voltage ΔV_s across capacitance c_{tt}. The remaining turn-turn voltages in the group have a voltage of $2 \times \Delta V_s$ across them. There are $N_s - 2$ such turns in the group. Since there are n such groups, the energy associated with these is $n \times (N_s - 2)$ times the energy associated with a voltage of $2\Delta V_s$ across capacitance c_{tt}. Combining these energies, we get at total energy of

$$\begin{aligned} E_{total} &= (2n-1)\frac{1}{2}c_{tt}(\Delta V_s)^2 + n(N_s - 2)\frac{1}{2}c_{tt}(2\Delta V_s)^2 \\ &= \frac{1}{2}c_{tt}(\Delta V_s)^2(4nN_s - 6n - 1) \end{aligned} \tag{7.110}$$

Since $\Delta V_s = V/N_s$, where V is the total energy across the entire winding, by substituting this into (7.110), we can extract a total capacitance for the winding, C_{m-s},

$$C_{m-s} = c_{tt}\frac{(4nN_s - 6n - 1)}{N_s^2} \tag{7.111}$$

c_{tt} will depend on the insulation structure of the winding, i.e. whether the turns are touching, paper to paper, in the manner of a layer winding or whether there is an oil gap separating them. In any event, the capacitance is much higher than that across a comparable helical winding where the turn-turn voltages are much smaller.

8. HIGH VOLTAGE INSULATION DESIGN

Summary Transformer insulation must not only be designed to withstand the normal operating voltages but to survive the effects of lightning strikes and other possible disturbances such as switching operations which may occur on the electrical system. In order to assess the adequacy of the insulation, it is necessary to have some understanding of the breakdown process, especially in liquids and solids which are normally used in combination in large power transformers. Although our understanding in this area is very incomplete, some trends or correlations have been deduced from test data. Using these and much accumulated experience, design rules have been formulated whose main justification is that they work in practice at least in most cases. It is also necessary to have some means of calculating the voltages and electric stresses which occur in a transformer under normal and especially abnormal conditions such as lighting strikes. From such calculations, voltages, voltage differences, and electrical stresses can be obtained and compared with the breakdown limits. Although a standardized waveform has been developed to represent a typical lightning strike which reaches a transformer, simple step function approximations are often used in approximate analytical calculations. The test of design adequacy comes when the unit is built and is subjected to a variety of electrical tests to simulate the abnormal conditions. The ultimate test comes from the unit's survival in service for long periods of time.

8.1 INTRODUCTION

A transformer's insulation system must be designed to withstand not only the a.c. operating voltages, with some allowance for an ~ 15% overvoltage, but also the much higher voltages produced by lightning strikes or switching operations. These latter voltages can be limited by protective devices such as lightning or surge arrestors but these devices are usually set to protect at levels well above the normal a.c. operating voltage. Fortunately the transformer's insulation can withstand higher voltages for the shorter periods of time characteristic of lightning or switching disturbances. Thus, insulation designed to be adequate at the operating voltage can also be sufficient for the short duration higher voltages which may be encountered.

Insulation design is generally an iterative process. A particular winding type is chosen such as disk, helix, or layer, for each of the transformer's windings. They must have the right number of turns to produce the desired voltage and must satisfy thermal, mechanical, and impedance requirements. The voltage distribution is then calculated throughout the windings, using a suitable electrical model together with the appropriate input such as a lightning impulse excitation. Voltage differences and/or electric fields are then calculated to determine if they are high enough to cause breakdown, according to some breakdown criterion, across the assumed insulation structure. More elaborate path integrals are sometimes used to determine breakdown. If the breakdown criterion is exceeded at some location, the insulation is redesigned and the process repeated until the breakdown criteria are met. Insulation redesign can consist of adding more paper insulation to the wire or cable, increasing the size of the oil or air ducts, or resorting to interleaving the winding conductors or adding wound-in-shields or other types of shields.

Although voltages and electric fields can be calculated to almost any desired accuracy, assuming the material properties are well known, the same cannot be said for breakdown fields in solids or liquids. The theory of breakdown in gases is reasonably well established, but the solid or liquid theory of breakdown is somewhat rudimentary. Nevertheless, design rules have evolved based on experience. With suitable margins, these rules generally produce successful designs. Success is usually judged by whether a transformer passes a series of dielectric tests using standard impulse waveforms or a.c. power frequency voltages for specified time periods without breakdown or excessive corona. These tests have been developed over the years in an effort to simulate a typical lightning or switching waveshape.

8.2 PRINCIPLES OF VOLTAGE BREAKDOWN

We briefly discuss some of the proposed mechanisms of voltage breakdown in solids, liquids, and gases with primary emphasis on transformer oil. This is because in oil filled transformers, due to the higher dielectric constant of the solid insulation, the highest electric stress tends to occur in the oil. In addition, the breakdown stress of the oil is generally much lower than that of the solid insulation. The same situation occurs in dry type transformers, however the breakdown mechanism in the gas is much better understood.

In gases, breakdown is thought to occur by electron avalanche, also called the Townsend mechanism [Kuf88]. In this process, the electric field imparts sufficient energy to the electrons between collisions with the atoms or molecules of the gas that they release or ionize additional electrons upon subsequent collisions. These additional electrons, in turn, acquire sufficient energy between collisions to release more electrons in an avalanche process. The process depends in detail on the collision cross-sections for the specific gas. These cross-sections can lead to elastic scattering, ionization, as well as absorption and are highly energy dependent. They have been measured for a variety of gases. In principle, breakdown can be calculated from a knowledge of these collision cross-sections, together with corrections due to the influence of the positive ions, photo-excitation, etc. In practice, the theory has served to illuminate the parametric dependence of the breakdown process and is even in reasonable quantitative agreement for specific gases.

One of the major results of the theory of gaseous breakdown is that the breakdown voltage across a uniform gap depends on the product of pressure and gap thickness or, more generally, on the product of gas density and gap thickness. This relationship is called the Paschen curve. A sketch of such a curve is shown in Fig. 8.1. A fairly common feature of such curves is the existence of a minimum. Thus for a given gap distance, as the pressure is lowered, assuming we are to the right of the minimum, the breakdown voltage will drop and rise again as the pressure is lowered past the minimum. Care must be taken that the gap voltage is below the minimum in such a process to avoid breakdown. For a given gas pressure or more accurately density, the breakdown field depends only on the gap thickness.

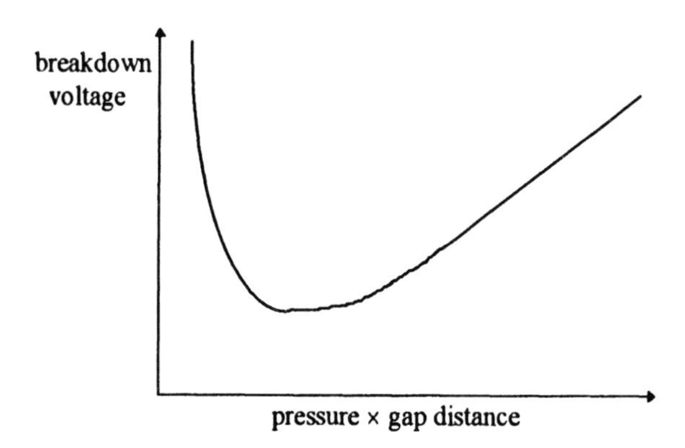

Figure 8.1 Schematic Paschen curve

During the avalanche process, because the negatively charged electrons move much more rapidly than the positively charged ions and because they are pulled towards opposite electrodes, a charge separation occurs in the gas. When the excess charge is large enough, as can occur in a well developed avalanche for large gap distances or high pressure × gap values, the electric field produced by the excess charge approaches the applied field. When this occurs, the Townsend mechanism of breakdown gives way to a streamer type of breakdown. In this process, secondary breakdown paths or plasma channels form at the front of the avalanche, leading to a more rapid breakdown than can be accounted for by the Townsend mechanism alone. Theoretical calculations, based on idealized charge configurations, can account approximately for this type of breakdown.

In solids and liquids where the distance between electron collisions is much shorter than in gases so that the electrons have a harder time acquiring enough energy to produce an avalanche, the Townsend mechanism is not considered to be operative except possibly for extremely pure liquids. A streamer type of mechanism is considered to be much more likely but the theory is not as developed. Moreover, especially in liquids, there are usually many types of impurities whose presence even in small concentrations can lower the breakdown stress considerably. This is well established experimentally where further and further purifications lead to higher breakdown stress to the point where the so called intrinsic breakdown stress of a pure liquid has rarely been measured.

In solids and liquids, the breakdown stress does not appear to be strictly a function of the gap thickness but rather appears to depend on the area of the electrodes or the volume of the material under stress. This would argue against a strictly Townsend mechanism of breakdown according to which, with the nearly constant density of most solids and liquids, the breakdown should depend on the gap distance only. It should be noted, however, that the experimental evidence is fragmentary and sometimes contradictory.

An electrode area or volume dependence of breakdown is usually explained by means of a weak link theory. According to this theory, there is some weak spot, imperfection, or mechanism based on the presence of imperfections which causes the failure. Thus, as the size of the specimen grows, weaker spots or more and greater imperfections are uncovered, resulting in failure at lower electric stress. Some support for this type of failure mechanism comes from studying the statistics of breakdown. There is much experimental evidence to show that breakdown probabilities follow an extreme value distribution, in particular the Weibull distribution [Gum58]. This type of distribution is consistent with a weak link mechanism. In fact, Weibull invented it to account for failure statistics in fracture mechanics which can be associated with material flaws. In general form, the distribution function giving the probability of failure for a voltage $\leq V$, $P(V)$, is

$$P(V) = \begin{cases} 1 - e^{-\left(\frac{V - V_o}{a}\right)^m} & , \quad V \geq V_o \\ 0 & , \quad V < V_o \end{cases} \tag{8.1}$$

where V_o, a, m are parameters > 0. The density function, which is the derivative of the distribution function, when multiplied by ΔV gives the probability of failure in an interval ΔV, assumed small, about V. It is

$$p(V) = \frac{dP(V)}{dV} = \frac{m}{a}\left(\frac{V - V_o}{a}\right)^{m-1} e^{-\left(\frac{V - V_o}{a}\right)^m} \tag{8.2}$$

The density function is, in general, asymmetric about the mode or most probable value and this asymmetry is usually taken as evidence that one is dealing with an extreme value distribution, in contrast to a Gaussian density function which is symmetric about the mode or mean in this case.

According to advocates of an electrode area dependence of the breakdown stress, the weak link can be a protrusion on the electrode surface where the field will be enhanced or it can be an area of greater electron emissivity on the surface. Advocates of the volume dependence of breakdown emphasize impurities in the material which increase with volume. According to Kok [Kok61], impurities in liquids such as transformer oil tend to have a higher dielectric constant than the oil, particularly if they have absorbed some water. Thus they are attracted to regions of higher electric field by the presence of gradients in the field. They will tend to acquire an induced dipole moment so that other dipoles will attach to them in a chain-like fashion. Such a chain can lead to a relatively high conductivity link between the electrodes, due to the presence of water and possibly dissolved ions, leading to breakdown. The probability of such a chain forming increases with the amount of impurity present and hence with the volume of liquid.

We know from numerous experiments that the breakdown electric stress in transformer oil is lowered by the presence of moisture, by particles such as cellulose fibers shed by paper or pressboard insulation, and also by the presence of dissolved gas. It was even shown in one experiment that the breakdown stress between two electrodes depended on whether the electrodes were horizontal or vertical. Presumably the dissolved gas and its tendency to form bubbles when the electric stress was applied was influencing the results, since in one orientation the bubbles would be trapped while in the other they could float away. In other experiments, causing the oil to flow between the electrodes increased the breakdown stress compared with stationary oil. The above influences make it difficult to compare breakdown results from different investigators. However, for a given investigator, using a standardized liquid or solid preparation and testing procedure, observed trends in the breakdown voltage or stress with other variables are probably valid.

Breakdown studies not concerned with time as a variable are generally done under impulse or a.c. power frequency conditions. In the latter case, the time duration is usually 1 minute. In impulse studies, the waveform is a unidirectional pulse having a rise time of t_{rise} = 1 - 1.5 μs and a fall time to 50% of the peak value of t_{fall} = 40 - 50 μs as sketched in Fig. 8.2. In most such studies, the breakdown occurs on the tail of the pulse but the breakdown voltage level is taken as the peak voltage. However, in front of wave breakdown studies, breakdown occurs on the rising part of the pulse and the breakdown voltage is taken as the voltage reached when breakdown occurs. Although the variability in times for the rise and fall of the pulse are not considered too significant when breakdown occurs on the tail of the pulse, front of wave breakdown voltages are generally higher that those occurring on the tail. The polarity of the impulse can also differ between studies, although the standard is negative polarity since that is the polarity of the usual lightning strike. There is also a fair degree of variability in the experimental conditions for a.c. power frequency breakdown studies. The frequency used can be 50 or 60 Hz, depending on the power frequency in the country where the study was performed. Holding times at voltage can be between 1 - 3 minutes or the breakdown occurs on a rising voltage ramp where the volts/sec rise can differ from study to study. These differences can lead to differences in the breakdown levels reported but they should not amount to more than a few percent. The ratio between the full wave impulse breakdown voltage and the a.c. rms breakdown voltage is called the impulse ratio and is found to be in the range of 2 - 3. This ratio applies to a combination of solid (pressboard) insulation and transformer oil, although it is not too different for either considered separately.

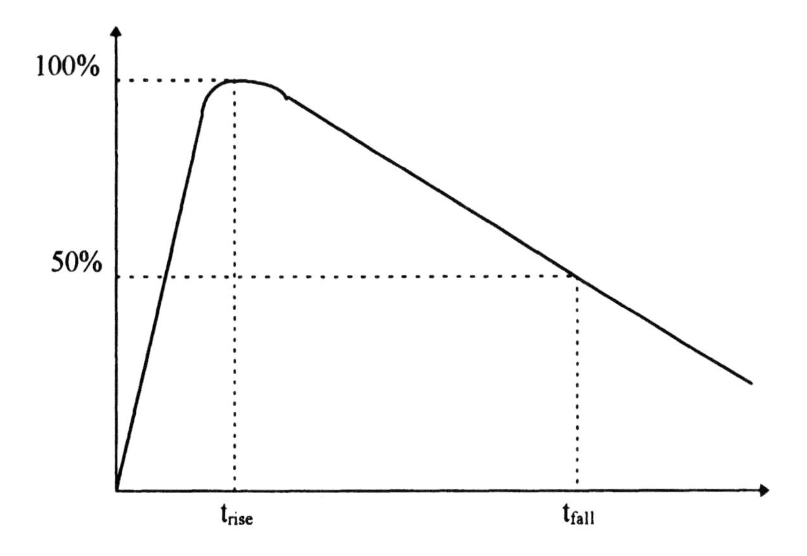

Figure 8.2 Impulse waveshape

There is some controversy concerning the breakdown mechanism in the different time regimes for transformer oil. Endicott and Weber [End57] found the same asymmetric probability density distributions for impulse and a.c. breakdown voltages, indicating that extreme value statistics are operative. They also found that both types of breakdown depended on the electrode area. On the other hand, Bell [Bel77] found that the impulse (front of wave) breakdown voltages had a symmetrical Gaussian probability density whereas the a.c. breakdowns had an asymmetric probability density. They found, nevertheless, that the impulse breakdown levels depended on the stressed oil volume. According to this finding, the volume effect under impulse is not linked to extreme value statistics. One would have difficulty imagining a chain of dipoles aligning in the short times available during a front of wave impulse test. Palmer and Sharpley [Pal69] found that both impulse and a.c. breakdown voltages depended on the volume of stressed oil. However, they reported that both impulse and a.c. breakdown statistics followed a Gaussian distribution. It would appear from these and similar studies that the breakdown mechanism in transformer oil is not understood enough to conclude that different mechanisms are operative in the different time regimes.

One of the consequences of the theoretical uncertainty in transformer oil breakdown and to a certain extent in breakdown in solids also is that there is no accepted way to parametrize the data. Thus some authors present graphs of breakdown voltage or stress vs electrode area, others show breakdown vs stressed oil volume, and others breakdown vs gap thickness. An attractive compromise appears in the work of Danikas [Dan90]. In this work and others [Bel77], breakdown is studied as a function of both electrode area and gap thickness. This also allows for the possibility of a volume effect should the dependence be a function of area × gap thickness. In fact, in [Dan90] the area effect appears to saturate at large areas, i.e. the breakdown level is unchanged as the area increases beyond a given value. Thus, breakdown becomes purely a function of gap spacing at large enough areas. Higaki, et. al. [Hig75] found that the breakdown electric stress becomes constant for large gap distances as well as for large volumes. This would suggest caution in extrapolating experimental results either in the direction of larger or smaller parameter values from those covered by the experiment.

At this point, we present some of the breakdown data and parameter dependencies reported in the literature. When graphs are presented, we have converted the best fit into an equation. Also voltage values are converted to electric field values when the voltage is applied across a uniform gap. For consistency, we use kV/mm units for breakdown stress, mm for gap thickness, and mm^2 or mm^3 for areas or volumes in the formulas. This will allow us to compare different results not only with respect to parameter dependencies but also with respect to numerical values. Generally the breakdown voltages are those for which the probability is 50% to have a breakdown. Thus some margin below these levels is needed in actual design.

We begin with breakdown in solid insulation, namely oil saturated paper and pressboard. Samples are prepared by drying and vacuum impregnation and tested under oil. For paper at 25 °C, Blume, et. al. [Blu51] report breakdown voltage stress $E_{b,ac}$ vs thickness d for a.c. 60 Hz voltages,

$$E_{b,ac}\left(\frac{kV_{rms}}{mm}\right)=\frac{17.1}{d^{0.33}}\left(0.85+\frac{0.15}{t^{1/4}}\right) \tag{8.3}$$

with d in mm and t, the duration of the voltage application, in minutes. Palmer and Sharpley [Pal69] report the impulse breakdown in paper, $E_{b,imp}$, vs thickness d in mm at 90 °C as

$$E_{b,imp}\left(\frac{kV_{peak}}{mm}\right)=\frac{79.43}{d^{0.275}} \tag{8.4}$$

In going from 90 °C to 20 °C, the impulse breakdown stress increases by about 10 % according to [Pal69]. Clark [Cla62] reports for Kraft paper at room temperature and a.c. test conditions

$$E_{b,ac}\left(\frac{kV_{rms}}{mm}\right)=\frac{32.8}{d^{0.33}} \tag{8.5}$$

Results from different investigators are difficult to compare because the thickness buildup is achieved by stacking thin layers of paper. The stacking processes could differ. Some could use lapping with different amounts of overlap as well as different thicknesses of the individual layers. Other possible differences could include the shape and size of the electrodes. Nevertheless the exponent of the thickness dependence is nearly the same in different studies. Ref. [Cla62] reports an area effect but it is difficult to quantify. The impulse ratio for paper, corrected for temperature, based on (8.4) and (8.5) is ~ 2.7.

For pressboard in oil at 25 °C, Ref. [Blu51] reports

$$E_{b,ac}\left(\frac{kV_{rms}}{mm}\right)=\frac{25.7}{d^{0.33}}\left(\frac{1.75}{f^{0.137}}\right)\left(0.5+\frac{0.5}{t^{1/4}}\right) \tag{8.6}$$

where the frequency f in Hz and time duration t in minutes are taken into account. The frequency dependence was only tested in the range of 25 - 420 Hz but is expected to hold for even higher frequencies. For pressboard at room temperature, using 25 mm sphere electrodes, Moser [Mos79] reports

$$E_{b,ac}\left(\frac{kV_{rms}}{mm}\right)=\frac{33.1}{d^{0.32}}$$

$$E_{b,imp}\left(\frac{kV_{peak}}{mm}\right)=\frac{94.6}{d^{0.22}} \tag{8.7}$$

Thus the impulse ratio for pressboard obtained from (8.7) is ~ 3.0. Ref. [Pal69] reports for pressboard in oil at 90 °C

$$E_{b,ac}\left(\frac{kV_{rms}}{mm}\right)=\frac{27.5}{d^{0.26}}$$

$$E_{b,imp}\left(\frac{kV_{peak}}{mm}\right)=\frac{91.2}{d^{0.26}} \tag{8.8}$$

The impulse ratio for pressboard, based on (8.8) is ~ 3.3. The trend in the data between Refs. [Mos79] and [Pal69] is in the right direction since pressboard breakdown strength decreases with increasing temperature.

Based on the above data, there does not appear to be much difference between the breakdown strength of oil soaked paper or pressboard insulation either under a.c. or impulse test conditions. Even the thickness dependencies are similar. Although Cygan and Laghari [Cyg87] find an area and thickness dependence on the dielectric strength of polypropylene films, it is not known how applicable this is to paper or pressboard insulation.

For transformer oil at 90 °C, Ref. [Pal69] reports breakdown electric fields which depend on volume Λ in mm^3 according to

$$E_{b,ac}\left(\frac{kV_{rms}}{mm}\right)=34.9-1.74\ln\Lambda$$

$$E_{b,imp}\left(\frac{kV_{peak}}{mm}\right)=82.5-3.69\ln\Lambda \tag{8.9}$$

This would imply an impulse ratio for oil of ~ 2.5. In contrast to paper insulation, the breakdown strength of transformer oil increases slightly with temperature in the range of -5 to 100 °C [Blu51]. Nelson [Nel89] summarizes earlier work on the volume effect for breakdown in oil by means of the formula

$$E_{b,ac}\left(\frac{kV_{rms}}{mm}\right)=\frac{46.1}{\Lambda^{0.137}} \tag{8.10}$$

with Λ in mm^3. We should note that the logarithmic dependence on volume given in (8.9) cannot be valid as volume increases indefinitely since the breakdown stress would eventually become negative. In (8.10), the breakdown stress becomes unrealistically close to zero as volume increases.

Refs. [End57, Web56] present breakdown strength in oil vs electrode area, A, in the form

$$(E_1 - E_2)_{b,ac}\left(\frac{kV_{rms}}{mm}\right) = 1.74\ln\left(\frac{A_2}{A_1}\right)$$

$$(E_1 - E_2)_{b,imp}\left(\frac{kV_{peak}}{mm}\right) = 4.92\ln\left(\frac{A_2}{A_1}\right) \tag{8.11}$$

The gap spacing in both of these studies was 1.9 mm. Although the impulse conditions in this study were front of wave, the voltage ramp was kept slow enough in an attempt to approximate full wave conditions. Thus, we can reasonably obtain an impulse ratio from (8.11) of ~ 2.8.

Ref. [Mos79] gives the a.c. (50 Hz, 1 min) partial discharge inception electric stress, $E_{pd,ac}$, for oil as a function of the gap thickness only. While not strictly the breakdown strength, partial discharges maintained over a long enough time period can lead to breakdown. Separate curves are given for gas saturated or degassed oil and for insulated and non-insulated electrodes. Fitting the curves, we obtain

$$E_{pd,ac}\left(\frac{kV_{rms}}{mm}\right) = \frac{14.2}{d^{0.36}} \quad \text{(gas saturated oil, non insulated electrodes)}$$

$$E_{pd,ac}\left(\frac{kV_{rms}}{mm}\right) = \frac{17.8}{d^{0.36}} \quad \text{(degassed oil, non insulated electrodes)}$$

$$E_{pd,ac}\left(\frac{kV_{rms}}{mm}\right) = \frac{19.0}{d^{0.38}} \quad \text{(gas saturated oil, insulated electrodes)} \tag{8.12}$$

$$E_{pd,ac}\left(\frac{kV_{rms}}{mm}\right) = \frac{21.2}{d^{0.36}} \quad \text{(degassed oil, insulated electrodes)}$$

with d in mm.

Trinh, et. al. [Tri82] analyzed transformer oil for dielectric strength dependence on both area and volume. They conclude that both area and volume effects can be present, with the area effect becoming more important for ultra clean oil and the volume effect for oil having higher particle content. Although the data shows much scatter, they present curves for oils of different purities. For their technical grade transformer oil, the middle grade of the three analyzed, the following formulas are approximate fits to their curves.

$$E_{b,ac}\left(\frac{kV_{rms}}{mm}\right) = 5\left(1+\frac{10}{A^{0.2}}\right)$$

$$E_{b,imp}\left(\frac{kV_{peak}}{mm}\right) = 10\left(1+\frac{23.7}{A^{0.25}}\right) \tag{8.13}$$

$$E_{b,ac}\left(\frac{kV_{rms}}{mm}\right) = 5\left(1+\frac{7.9}{\Lambda^{0.14}}\right)$$

$$E_{b,imp}\left(\frac{kV_{peak}}{mm}\right) = 15\left(1+\frac{9.7}{\Lambda^{0.18}}\right) \tag{8.14}$$

with A in mm^2 and Λ in mm^3. They seem to be saying that it is immaterial whether one describes breakdown in terms of an area or volume effect. However, the consequences of these two approaches are quite different. For areas and volumes in the range of ~ 10^3 - 10^7 mm^2 or mm^3, the impulse ratio implied by the above formulas is in the range of ~ 2 - 3. In these formulas, as is also evident in the work of refs. [Dan90, Hig75], the dielectric strength approaches a constant value as the area or volume become very large. This is a reasonable expectation. On the other hand, dielectric strengths approach infinity as areas, volumes, and gap distances approach zero in all the above formulas. This is surely inaccurate, although the formulas seem to hold for quite small values of these quantities.

It can be seen, by putting in typical values for d, A, and Λ as found in transformers in the above formulas, that the dielectric strength of paper or pressboard is approximately twice that of oil. However, the electric stress which occurs in the oil is typically greater than that which occurs in the paper or pressboard. For this reason, breakdown generally occurs in the oil gaps first. However, once the oil gaps break down, the solid insulation will see a higher stress so that it could in turn break down. Even if the solid insulation can withstand the higher stress, the destructive effects of the oil breakdown such as arcing or corona could eventually puncture the solid insulation. Thus it would seem inappropriate to design a solid-oil insulation system so that the solid by itself could withstand the full voltage applied across the gap as has sometimes been the practice in the past, unless the insulation is all solid. In fact excess solid insulation, because of its higher dielectric constant than oil, increases the stress in the oil above that of a design more sparing of the solid insulation.

Because of the distance and volume dependence of oil breakdown strength, it is a common practice to subdivide large oil gaps, as occur for example between the transformer windings, by means of one or more thin pressboard cylinders. Thus a gap having a large distance or volume is reduced to several smaller gaps, each having a higher breakdown strength. Hence, current practice favors a distance or volume effect over a pure area effect since gap subdivision would not be of benefit for an area effect. A combination gap-area or gap-volume effect would also be consistent with current practice.

The breakdown data referred to above applies to uniform gaps or as reasonably uniform as practical. This situation is rarely achieved in design so the question arises as to how to apply these results in practice. In volume dependent breakdown, it is suggested that only the oil volume encompassing electric field values between the maximum and 90 % of the maximum be used [Pal69, Wil53]. Thus for concentric cylinder electrodes, for example, only the volume between the inner cylinder and a cylinder at some fraction of the radial distance to the outer cylinder would be used in the volume dependent breakdown formulas. For more complicated geometries, numerical methods such as finite elements could be used to determine this effective volume. For gap distance dependent breakdown, the suggestion is to subdivide a possible breakdown path into equal length subdivisions and to calculate the average electric field over each subdivision. The maximum of these average fields is then compared with the breakdown value corresponding to a gap length equal to the subdivision length. If it exceeds the breakdown value, breakdown along the entire path length is assumed to occur. This procedure is repeated for coarser and coarser subdivisions until a single subdivision consisting or the entire path is reached. Other possible breakdown paths are then chosen and the procedure repeated [Nel89, Franc].

Another type of breakdown which can occur in insulation structures consisting of solids and liquids or solids and gases is creep breakdown. This occurs along a solid surface in contact with a liquid or gas. These potential breakdown surfaces are nearly unavoidable in insulation design. For example, the oil gaps present in the region between windings are kept uniform by means of sticks placed around the circumference. The surfaces of these sticks bridge the gap, providing a possible surface breakdown path. Since breakdown along such surfaces generally occurs at a lower stress than breakdown in the oil or air through the gap itself, surface breakdown is often design limiting. Ref. [Pal69] parametrizes the surface creep breakdown stress along pressboard surfaces in oil at power frequency, $E_{cb,ac}$, in terms of the creep area, A_c, in mm^2 according to

$$E_{cb,ac}\left(\frac{kV_{rms}}{mm}\right) = 16.0 - 1.09\ln A_c \tag{8.15}$$

On the other hand, Ref. [Mos79] describes creep breakdown along pressboard surfaces in oil in terms of the creep distance along the surface, d_c, in mm according to

$$E_{cb,ac}\left(\frac{kV_{rms}}{mm}\right) = \frac{16.6}{d_c^{0.46}} \tag{8.16}$$

For non-uniform field situations, the same procedure of path subdivision and comparison with the creep breakdown strength calculated by (8.16) is followed as was described earlier for gap breakdown.

8.3 INSULATION COORDINATION

Insulation coordination concerns matching the insulation design to the protective devices used to limit the voltages applied to the terminals of a transformer by lightning strikes or switching surges and possibly other potentially hazardous events. Since the insulation must withstand the normal operating voltages which are present continuously as well as lightning or switching events which are of short duration, how breakdown depends on the time duration of the applied voltage is of major importance in insulation coordination. We have seen previously that the impulse ratio is ~ 2 - 3 for oil filled transformer insulation. This means that the breakdown strength of a short duration impulse voltage lasting ~ 5 - 10 μs is much higher than the breakdown strength of a long duration (~ 1 min) a.c. voltage. It appears that, in general, breakdown voltages or stresses decrease with the time duration of the voltage application for transformer insulation. The exact form of this breakdown vs time dependence is still somewhat uncertain, probably because of variations in material purity and/or experimental methods among different investigators.

Refs. [Blu51, Cla62] show a time dependent behavior of relative strength of oil or pressboard which is schematically illustrated in Fig. 8.3. Three regions are evident on the curve, labeled A, B, C. For short durations, < 10 μs, there is a rapid fall-off in the strength with increasing time. This is followed by a flat portion, B, which extends to about 1000 μs for oil and to about 20,000 μs for pressboard. In region A there is a more gradual fall-off approaching an asymptotic value at long times. The combination of an oil pressboard gap also produces a curve similar to Fig. 8.3.

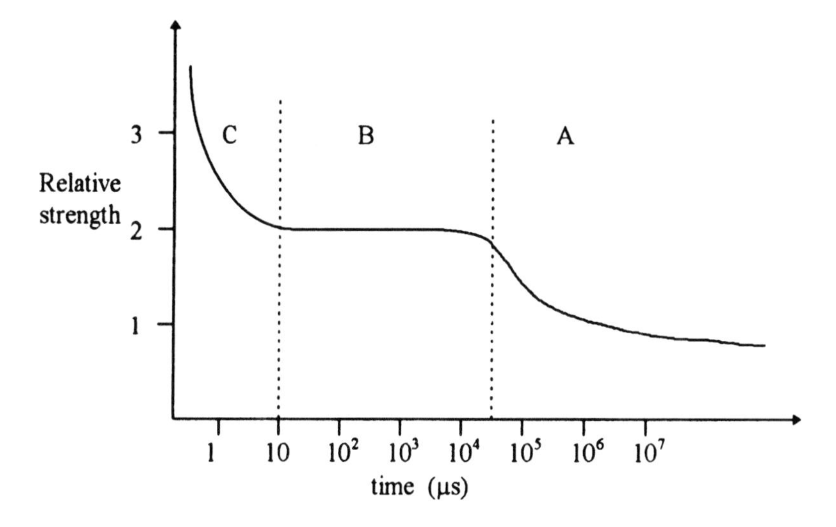

Figure 8.3 Oil or pressboard breakdown relative strength vs time - schematic

Using statistical arguments, i.e. equating the probability of breakdown for times less than t for a fixed voltage across an oil gap under corona free conditions which is $1 - \exp(-t/t_o)$ where t_o is the average time to breakdown, to equation (8.1) with V_o taken to be 0, Ref. [Kau68] obtains the breakdown voltage vs time relationship for an oil gap in the form

$$\frac{t_2}{t_1} = \left(\frac{V_1}{V_2}\right)^m \qquad (8.17)$$

Experimentally they find that for times from a few seconds to a few weeks, m is between 15 and 30. This relationship cannot, however, hold for very long times since it predicts zero breakdown at infinite time. To correct this, one could use the same statistical argument to show that

$$\frac{t_2}{t_1} = \left(\frac{V_1 - V_o}{V_2 - V_o}\right)^m \tag{8.18}$$

where V_o is the infinite time breakdown voltage. This can be rearranged to the form

$$V_2 = V_o + \frac{(V_1 - V_o)t_1^{1/m}}{t_2^{1/m}} \tag{8.19}$$

Picking a V_1 , t_1 pair, this can be written for a general V , t pair,

$$V = V_o + \frac{K'}{t^{1/m}} = V_o\left(1 + \frac{K}{t^{1/m}}\right) \tag{8.20}$$

Dividing by the gap thickness, this can be written in terms of the breakdown stress, E_b , as

$$E_b = E_{b,o}\left(1 + \frac{K}{t^{1/m}}\right) \tag{8.21}$$

where $E_{b,o}$ is the infinite time breakdown stress. This last equation is very similar to (8.6) in its time dependence. As shown in Fig. 8.3, the time dependent behavior of transformer oil breakdown is more complicated over a time span from μs to years than the above formulas would suggest. Presumably these expressions would apply to a.c. breakdown stress covering region A in the curve of Fig. 8.3.

At short times in the μs region, the breakdown voltage or stress vs time characteristic changes rapidly with time. Impulse waveshapes as shown in Fig. 8.2 are somewhat complicated functions of time. In order to extract a single voltage - time duration pair from this waveshape for comparison with the breakdown curve, the common practice is to take the peak voltage and the time during which the voltage exceeds 90 % of its peak value. Thus a standard impulse wave which rises to its peak value in 1.2 μs and decays to 50% of its peak value in 50 μs spends about 10 μs above its 90 % voltage level. A switching surge test waveshape is similar to that of Fig. 8.2 but the rise time to peak value is ~ 100 μs and the fall time to the 50 % level is ~ 500 μs. The time spent above the 90 % of peak voltage level is ~ 200 μs. Therefore we would expect breakdown to occur on a switching surge test at a lower voltage than for a full wave impulse test.

Two other types of impulse test are the chopped wave test and front of wave test, although the latter is considered unnecessary in view of modern methods of protection. A standard chopped wave, as shown in Fig. 8.4, rises to its peak in ~ 1.2 μs and abruptly falls to zero with a slight undershoot at the chop time of ~ 3 μs. It is above its 90 % of peak voltage for ~ 3 μs. The front of wave is chopped on the rising part of the wave and has a duration above 90 % of its maximum value of ~ 0.5 μs. The generally accepted breakdown levels corresponding to these different times, normalized to the full wave breakdown level, are given in Table 8.1.

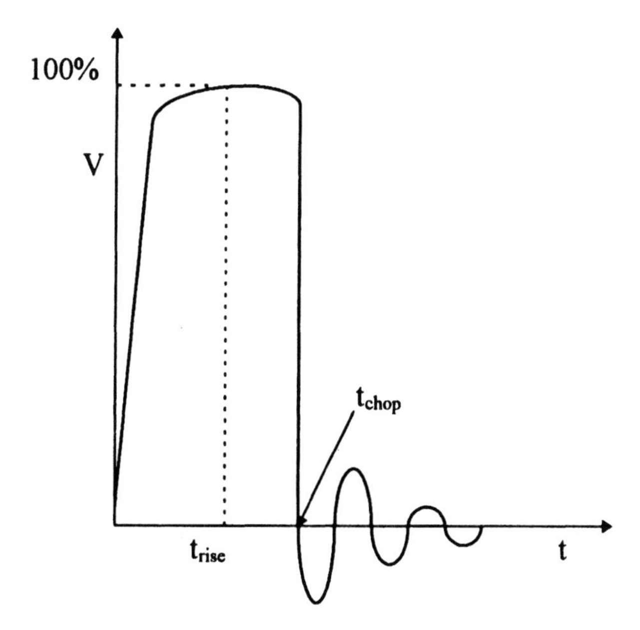

Figure 8.4 Chopped wave impulse waveshape

Table 8.1 Breakdown voltages normalized to the full wave impulse level

Type	Duration (μs)	Breakdown level
Front of wave	0.5	1.3
Chopped wave	3	1.1
Full wave	10	1.0
Switching surge	200	0.83
1 minute a.c.	17.2×10^6	0.5 peak
Nominal a.c.	∞	0.25 peak

We have included in Table 8.1 two other points, the 1 minute a.c. test point and the essentially infinite time nominal voltage point. For the 1 min test point, we have used an impulse ratio of 2.8. Thus the normalized breakdown peak voltage level is $1/2.8 \times \sqrt{2} = 0.5$ and the duration above 90 % of the peak voltage is 0.287×60 sec = 17.2 sec. We have taken the nominal system voltage to be half the 1 min test voltage. Other values for these could have been chosen with equal justification. For instance, an impulse ratio of 2.4 is commonly assumed in setting test values and the nominal voltage can be a factor of 2.5 - 3.0 below the 1 min test level. Note that the nominal system voltage used here is from terminal to ground.

We can produce a reasonable fit to the above tabular values with an equation of the form (8.20). Letting $V_{b,rel}$ be the breakdown levels relative to the full wave level, we obtain

$$V_{b,rel} = 0.25\left(1 + \frac{3.8}{t^{1/12}}\right) \tag{8.22}$$

with t in μs. Table 8.1 is based on practical experience in testing transformers. The numbers can only be regarded as approximate and different choices are often made within a reasonable range about the values shown. This would influence the fit given in (8.22) or possibly require a different parametrization. This parametrization may have no fundamental significance since it does not apply to any carefully controlled experiment. However its form, which agrees with the more fundamentally based equation (8.20), is noteworthy.

In practical applications, a table like Table 8.1 provides a way of linking the various test voltages to the full wave impulse voltage which is also called the basic impulse level (BIL). Since the required impulse test levels are usually linked to the lightning or surge arrestor protection level whereas the a.c. test levels or nominal voltage level are not, they can be specified independently. For example, a 500 kV_{rms} line to line transformer (= 288 kV_{rms} line to ground = 408.2 kV_{peak} line to ground) would correspond to a full wave impulse level of 1633 kV according to Table 8.1. However, the user may have sufficient protection that only a 1300 kV impulse test is required. Nevertheless, the insulation would have to be designed to withstand a full wave 1633 kV impulse test, even though not performed, since that level of protection is required to guarantee satisfactory operation at the nominal 500 kV_{rms} line to line voltage. As another example, one may have breakdown stress vs distance, area, or volume curves for 1 min a.c. test conditions. In order to compare these to the stress levels generated in a simulated impulse test calculation, one needs to know the relative breakdown level factors given in Table 8.1.

8.4 CONTINUUM MODEL OF WINDING USED TO OBTAIN THE IMPULSE VOLTAGE DISTRIBUTION

For very short times after the application of a voltage to a winding terminal, the voltage distribution along the winding is governed primarily by capacitive coupling. This is because the winding inductance limits the flow of current initially. In this approximation, a winding can be modeled by the capacitive ladder diagram shown in Fig. 8.5a. Although discrete capacitances are shown, this is meant to be a continuum model. The calculation outlined here is similar to that given in Chapter 7. Some of the steps are repeated in order to more easily compare the results with the travelling wave theory. Thus, as shown in Fig. 8.5b, the series capacitors, c_s , are separated by a distance Δx , where Δx will eventually approach zero. The series capacitors are determined by the winding structure, including the number of disks, their spacing, and the number of turns/disk. The ground capacitors, c_g , are determined by the distance to neighboring windings or to the tank walls or core. These are all assumed to be at ground potential relative to the impulsed winding, which should be true initially.

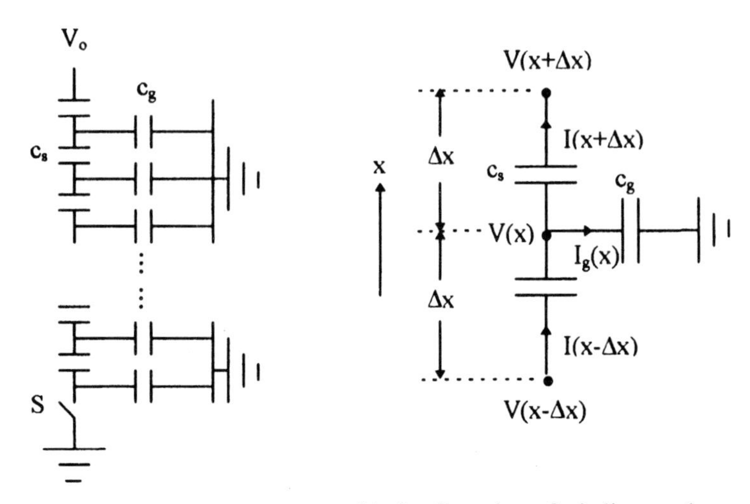

(a) Capacitive model of winding

(b) Small section of winding used for analytical purposes

Figure 8.5 Initial voltage distribution winding model

Analyzing a small portion of the winding at a distance x from the bottom of the winding (at the end opposite to the impulsed terminal), we can write Kirchoff's current law at the center node shown in Fig. 8.5b as

$$I(x-\Delta x)-I(x+\Delta x)=I_g(x) \tag{8.23}$$

However, using the voltage - current relationship for capacitors, we can express this as

$$c_s\frac{d}{dt}[V(x-\Delta x)-V(x)]-c_s\frac{d}{dt}[V(x)-V(x+\Delta x)]=c_g\frac{dV(x)}{dt} \tag{8.24}$$

Rearranging, we get

$$\frac{d}{dt}\{c_s[V(x+\Delta x)-2V(x)+V(x-\Delta x)]-c_g V(x)\}=0 \tag{8.25}$$

The time derivative of the expression in curly brackets is zero so this expression is a constant in time. The constant may be taken as zero since the winding starts out at zero potential. Dividing by $(\Delta x)^2$, (8.25) can be written

$$\frac{V(x+\Delta x)-2V(x)+V(x-\Delta x)}{(\Delta x)^2}=\frac{c_g}{c_s(\Delta x)^2}V(x) \tag{8.26}$$

The term on the left hand side can be recognized as the second order difference operator which approaches the second derivative as $\Delta x \rightarrow 0$. If L is the length of the winding, there are $N = L/\Delta x$ series capacitors, c_s , and ground capacitors, c_g. In terms of the total series and ground capacitances, C_s and C_g , we have

$$c_s = C_s N = C_s \frac{L}{\Delta x} \quad , \quad c_g = \frac{C_g}{N} = C_g \frac{\Delta x}{L} \tag{8.27}$$

Substituting into (8.26) and letting $\Delta x \rightarrow 0$, we obtain

$$\frac{d^2V}{dx^2} = \frac{C_g}{C_s L^2} V = \left(\frac{\alpha}{L}\right)^2 V \tag{8.28}$$

where the distribution constant α has been defined as

$$\alpha = \sqrt{\frac{C_g}{C_s}} \tag{8.29}$$

If the bottom of the winding is grounded (switch S closed in Fig. 8.5a) and a voltage V_o is applied at the line end, the solution to (8.28) is

$$V(x) = V_o \frac{\sinh\left(\frac{\alpha}{L}x\right)}{\sinh \alpha} \tag{8.30}$$

When the switch S is open so that the bottom of the winding is floating when the voltage V_0 is applied to the top of the winding, the solution to (8.28) is

$$V(x) = V_o \frac{\cosh\left(\frac{\alpha}{L}x\right)}{\cosh \alpha} \tag{8.31}$$

Equation (8.30) is plotted in Fig. 8.6 for several values of α. As can be seen, the slope of the curve becomes very steep at the line end as α increases. This implies that the disk-disk voltage, ΔV_d, increases with increasing α since this is given approximately as

$$\Delta V_d = \frac{dV}{dx} w = V_o \alpha \left(\frac{w}{L}\right) \frac{\cosh\left(\frac{\alpha}{L} x\right)}{\sinh \alpha} \tag{8.32}$$

where w is the disk-disk spacing, assumed small compared with L. This has its maximum value at the line end where $x = L$,

$$\Delta V_{d,line} = V_o \frac{\alpha}{\tanh \alpha} \left(\frac{w}{L}\right) \tag{8.33}$$

For the ungrounded winding, the maximum stress again occurs at the line end and, using (8.31), is

$$\Delta V_{d,line} = V_o \alpha \tanh \alpha \left(\frac{w}{L}\right) \tag{8.34}$$

which is somewhat less than (8.33) since $|\tanh \alpha| \leq 1$. If the voltage distribution is uniform or α small, $\Delta V_{d,line} = V_o(w/L)$ for the grounded end case.

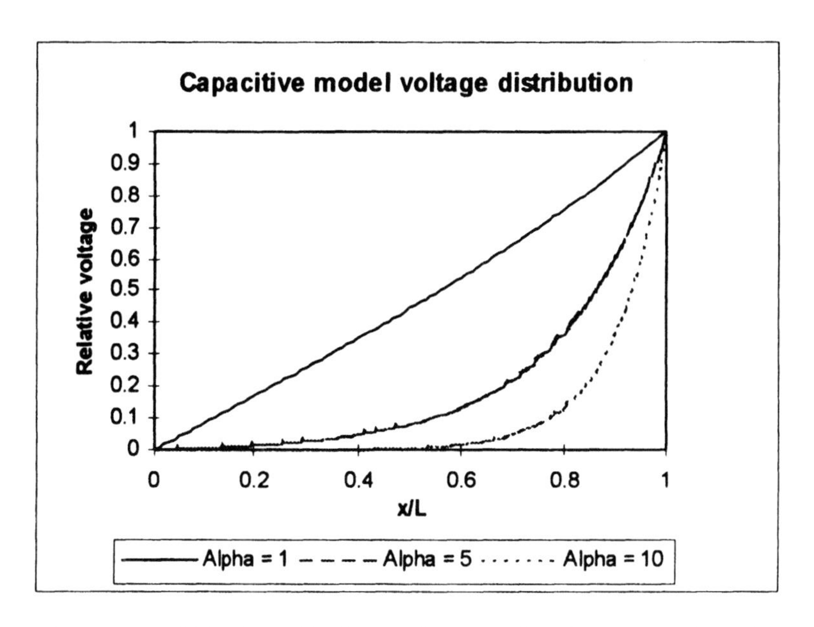

Figure 8.6 Initial voltage distribution along a winding for various values of α.

Since the disk-disk stress increases with α and by implication the turn-turn stress also, (8.29) suggests that to reduce this we need to increase C_s or decrease C_g or both. This has led to winding schemes such as interleaving which can dramatically increase the series capacitance. Interleaving schemes can be quite complicated and need not be applied to the whole winding. In addition, the scheme may vary within a given winding. A multi-start winding is an example of a simple interleaving scheme for increasing the series capacitance of a tap winding. Another method of increasing the series capacitance is by means of wound-in-shields which can be applied to a section of the winding near the line end where they are most needed. Decreasing C_g is not as promising since this would involve, at least in a simple approach, increasing the winding to winding or winding to tank distance for an outer winding or decreasing the winding height for a given radius. Such changes could conflict with impedance or cooling requirements.

An improvement over the static capacitance model is a model which includes the winding inductance in an approximate way. This is the traveling wave theory as developed by Rudenberg [Rud68, Rud40]. This is also a continuum model with the circuit parameters defined on a per unit length basis. Thus, using primes to indicate per unit length, the capacitances per unit length become, in terms of our previous notation with w the disk-disk distance,

$$c'_s = \frac{C_s}{w} \qquad , \qquad c'_g = \frac{C_g}{w} \tag{8.35}$$

and a new variable ℓ' which is the inductance/unit length. Thus the total series and ground capacitances, expressed in terms of the primed quantities are, using (8.27),

$$C_s = \frac{c'_s w^2}{L} \qquad , \qquad C_g = c'_g L \tag{8.36}$$

since $w = \Delta x$ in this context. Hence α can be written,

$$\alpha = \frac{L}{w}\sqrt{\frac{c'_g}{c'_s}} \tag{8.37}$$

The use of an inductance per unit length can only be regarded as approximate since it ignores the mutual inductance between the different sections of the winding. This approximation can be justified to some extent mathematically and by comparison with experiment.

Without going into all the details, the resulting differential equation can be solved by a superposition of traveling waves of the form

$$V = V_o e^{j\omega\left(t - \frac{x}{v}\right)} \tag{8.38}$$

with v the wave velocity and ω its angular frequency related by

$$v = \sqrt{\frac{1}{c'_g \ell'} - \frac{c'_s w^2 \omega^2}{c'_g}} \tag{8.39}$$

At low frequencies, $\omega \to 0$,

$$v \to v_o = \frac{1}{\sqrt{c'_g \ell'}} \tag{8.40}$$

At higher frequencies the wave velocity v eventually becomes zero at the critical frequency, ω_o, given by

$$\omega_o = \frac{1}{w\sqrt{c'_s \ell'}} \qquad (8.41)$$

Thus input waves or Fourier components of input waves of this frequency or higher cannot travel into the winding. However, at higher frequencies, v becomes imaginary and the solution (8.38) takes the form

$$V = V_o e^{j\omega t - \beta x} \qquad (8.42)$$

with β given by

$$\beta = \sqrt{\frac{c'_g \ell' \omega^2}{c'_s \ell' w^2 \omega^2 - 1}} \qquad (8.43)$$

Thus the higher frequencies exponentially decay with distance into the winding. (Here x measures the distance from the line end.) At high ω, this becomes asymptotically

$$\beta \to \beta_\infty = \frac{1}{w}\sqrt{\frac{c'_g}{c'_s}} = \frac{\alpha}{L} \qquad (8.44)$$

where the last equality follows from (8.37).

Rudenberg [Rud68, Rud40] considers the special case of a step function input to the line end of the winding as shown in Fig. 8.7. At the line terminal where $x = 0$, this wave can be written as a Fourier integral,

$$V(t) = \frac{V_o}{2} + \frac{V_o}{\pi}\int_0^\infty \frac{\sin(\omega t)}{\omega} d\omega \qquad (8.45)$$

where V_o is its amplitude. As the integral shows, the frequency component amplitudes decrease with increasing frequency. However, as we have seen previously, only frequency components below the critical frequency ω_o can travel into the winding. Higher frequency components give rise to an exponentially decaying standing wave. Thus, the wave penetrating into the winding has the form

$$V'(t) = \frac{V_o}{2} + \frac{V_o}{\pi}\int_0^{\omega_o} \frac{\sin(\omega t)}{\omega} d\omega \tag{8.46}$$

Because of the truncation of the integral, V' no longer has a step function shape but penetrates into the winding with the shape shown in Fig. 8.8. Thus the steep front has been flattened.

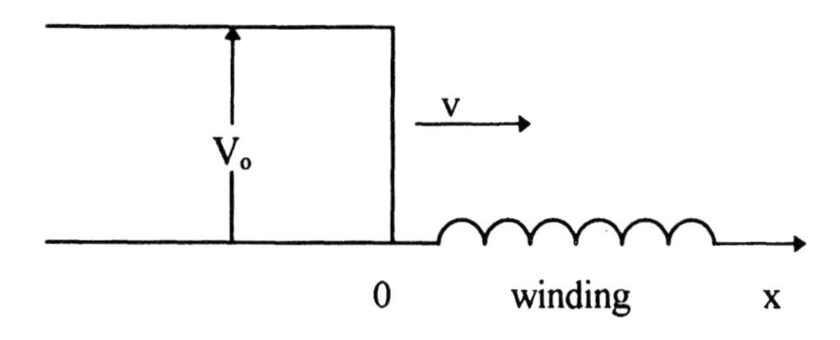

Figure 8.7 Step function voltage wave impinging on winding terminal

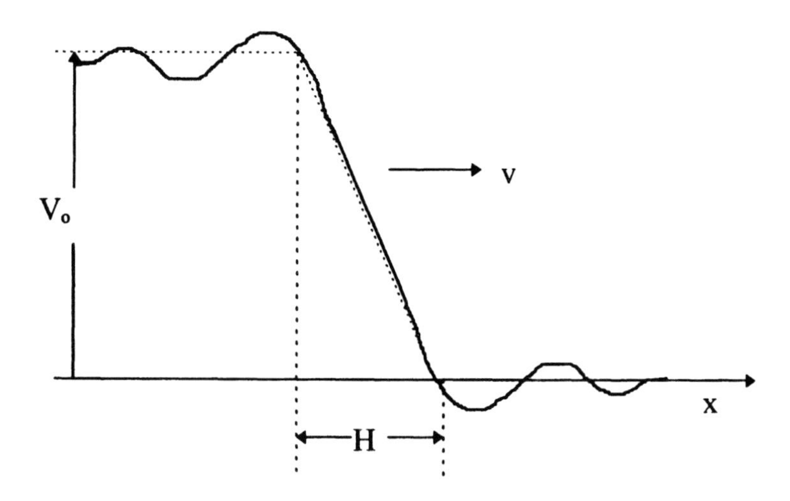

Figure 8.8 Flattened wave which enters the winding starting from the rectangular wave shown in Fig. 8.7

The flattened wave can be approximated by the dotted lines shown in Fig. 8.8. This consists of a sloped straight line front connecting the horizontal straight line asymptotic values. The time duration of this front, τ , is determined by the properties of the integral in (8.46) and is given by

$$\tau = \frac{\pi}{\omega_o} = \pi w \sqrt{c'_s \ell'} \tag{8.47}$$

using (8.41). For the lower frequencies comprising this pulse, we can approximate their velocity by v_o as given in (8.40). Thus the length of the sloped portion of the pulse in Fig. 8.8, H, is given approximately by

$$H \approx H_o = v_o \tau = \pi w \sqrt{\frac{c'_s}{c'_g}} \tag{8.48}$$

Using (8.44), this can be written

$$H_o = \frac{\pi}{\beta_\infty} = \frac{\pi L}{\alpha} \tag{8.49}$$

This sloped front causes a voltage difference across the winding disks which is given by

$$\Delta V'_d = \frac{w}{H_o} V_o = \frac{\alpha}{\pi}\left(\frac{w}{L}\right) V_o \tag{8.50}$$

For the decaying exponential part of the pulse, β can be taken to a good approximation as β_∞. Its amplitude, according to Rudenberg, is $V_o/2$. It is therefore given by

$$V'' = \frac{V_o}{2} e^{-\beta_\infty x} \tag{8.51}$$

Thus the disk-disk voltage drop due to this voltage is given in magnitude by

$$\Delta V''_d = \left|\frac{dV''}{dx}\right| w = \frac{\beta_\infty V_o w}{2} e^{-\beta_\infty x} = \frac{\alpha}{2}\left(\frac{w}{L}\right) V_o e^{-\beta_\infty x} \tag{8.52}$$

Combining (8.50) and (8.52), the overall disk-disk voltage drop is given by

$$\Delta V_d = \Delta V'_d + \Delta V''_d = \alpha\left(\frac{w}{L}\right)V_o\left(\frac{1}{2}e^{-\beta_\infty x} + \frac{1}{\pi}\right) \tag{8.53}$$

At $x = 0$, this becomes

$$\Delta V_{d,line} = 0.818\alpha\left(\frac{w}{L}\right)V_o \tag{8.54}$$

For $\alpha > 3$ which is somewhat typical, $\tanh\alpha \approx 1$, so that comparing this last equation with its counterpart for the purely capacitive case, (8.33) or (8.34), we see that the voltage drop in the traveling wave case is ~ 18 % less than for the capacitive case. However, because of the slight over and under shoots of the actual wave as shown in Fig. 8.8, this difference is not that great.

Norris [Nor48] has proposed an improved version of (8.53) which has the form

$$\Delta V_d = \alpha\left(\frac{w}{L}\right)V_o\left(0.6e^{-\beta_\infty x} + 0.4K_g\right) \tag{8.55}$$

where $K_g = H_o/H_x$, where H_x is the length of the sloped front of the traveling wave a distance x into the winding. The wave spreads out because the different frequency components travel with different velocities and thus $K_g \le 1$. K_g can be found as a solution of the equation [Rud68, Rud40]

$$\frac{\beta_\infty}{\pi}x = \begin{cases} \frac{1}{K_g^2}\left(1+\sqrt{1-K_g^2}\right) & , \quad x > \frac{\pi}{\beta_\infty} \\ 1 & , \quad x < \frac{\pi}{\beta_\infty} \end{cases} \tag{8.56}$$

This reduces to a quartic equation which can be solved analytically. At $x = 0$ and for $\alpha > 3$, (8.55) is equal to disk-disk voltage drop in the capacitive case as can bee seen by comparison with (8.33). Norris has also suggested corrections for incoming waves having finite rise and fall times and gives procedures for handling non-uniform windings.

8.5 LUMPED PARAMETER MODEL FOR TRANSIENT VOLTAGE DISTRIBUTION

In this section, we develop a circuit model of the transformer which includes capacitive, inductive, and resistive elements. The inductive elements include mutual inductances between elements in the same winding and in different windings and the effects of the iron core. The coils can be subdivided into as fine or coarse a manner as is consistent with the desired accuracy. Moreover, the subdivisions can be unequal so that accuracy in certain parts of the coil can be increased relative to that in other parts. The approach taken is similar to that of Miki, et. al. [Mik78] except that it includes winding resistance and the effects of the iron core. Also the differential equations describing the circuit are organized in such a way that circuit symmetries can be exploited and other circuit elements such as non-linear varistors may be included.

8.5.1 Circuit Description

A transformer is approximated as a collection of lumped circuit elements as shown in Fig. 8.9. Although not shown, mutual inductances between all the inductors are assumed to be present. The subdivisions may correspond to distinctly different sections of the coil, having different insulation thicknesses for example, or may simply be present for increased accuracy. The core and tank are assumed to be at ground potential. The presence of the tank only affects the capacitance to ground of the outer coil but not the inductance calculation. Other elements such as grounding resistors, capacitors, inductors, or non-linear elements may be added. Terminals or nodes may be interconnected, shorted to ground, or connected to ground via a resistor, reactor, capacitor, or varistor.

In order to analyze the circuit of Fig. 8.9, we isolate a representative portion as shown in Fig. 8.10. We adopt a node numbering scheme starting from the bottom of the innermost coil and proceeding upward. Then continue from the bottom of the next coil, etc. Similarly, a section or subdivision numbering scheme is adopted, starting from the bottom of the innermost coil, proceeding upward, etc. These numbering schemes are related. In the figure, we have simply labeled the nodes with i (p's and q's for adjacent nodes) and the sections with j. Our circuit unknown are the nodal voltages V_i and section currents I_j directed upward as shown.

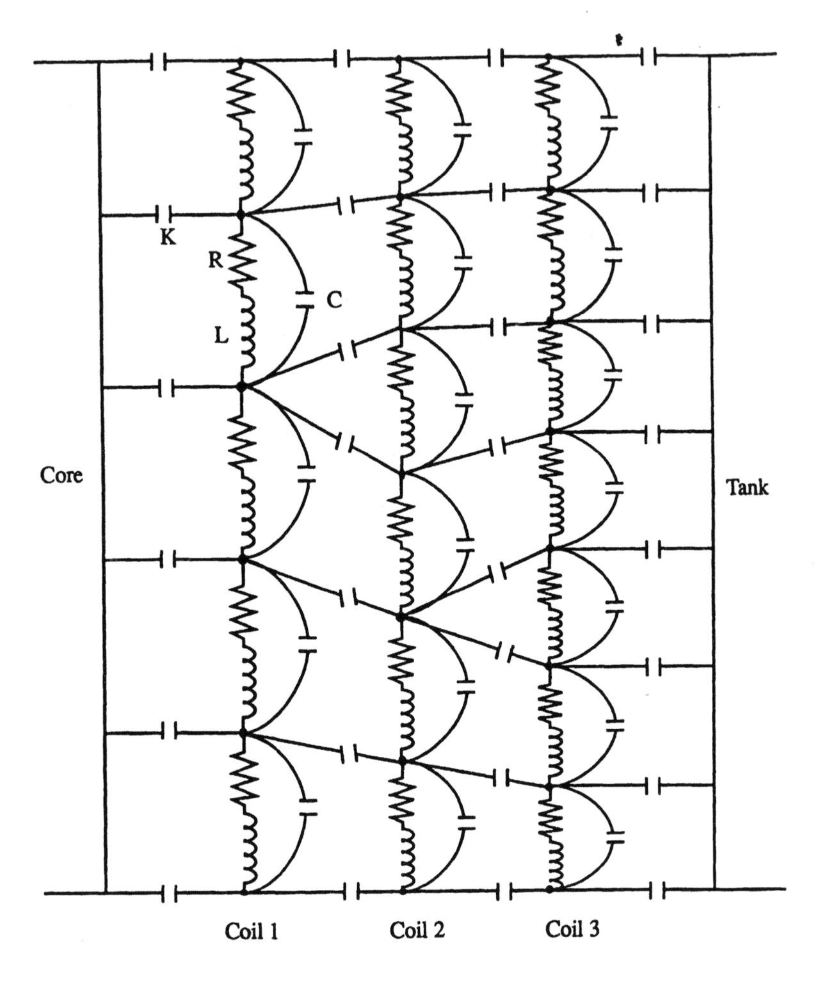

Figure 8.9 Circuit model of transformer. The number of coils and subdivisions within a coil are arbitrary.

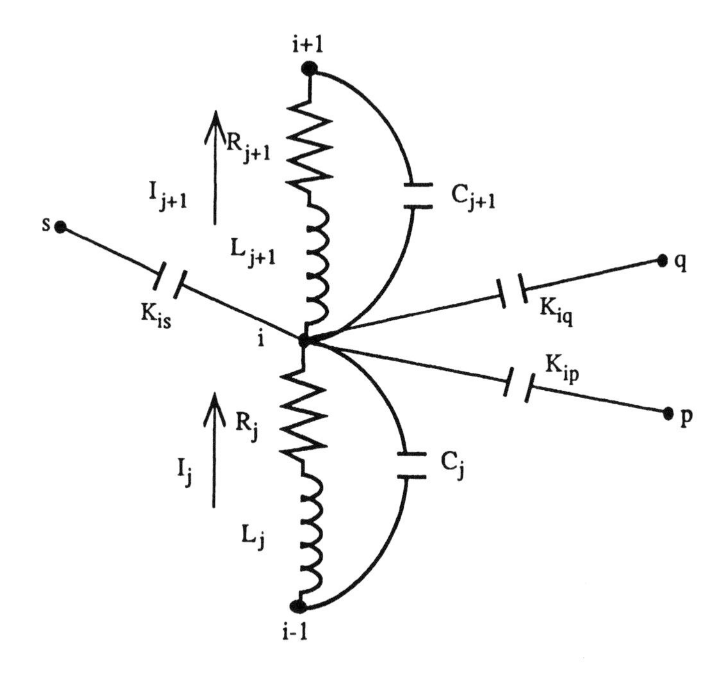

Figure 8.10 Representative portion of Fig. 8.9.

Considering the voltage drop from node i-1 to node i , we can write

$$L_j \frac{dI_j}{dt} + \sum_{k \neq j} M_{jk} \frac{dI_k}{dt} + R_j I_j = V_{i-1} - V_i \tag{8.57}$$

where M_{jk} is the mutual inductance between sections j and k. Defining a section current vector **I**, a nodal voltage vector **V**, an inductance matrix M, where $M_{jj} = L_j$, and a diagonal resistance matrix R, we can compress the above formula as

$$M \frac{d\mathbf{I}}{dt} = B\mathbf{V} - R\mathbf{I} \tag{8.58}$$

where B is a rectangular matrix whose rows correspond to sections and columns to nodes such that $B_{ji-1} = 1$ and $B_{ji} = -1$, where nodes i-1 and i bracket section j. The inductance matrix M is a symmetric positive definite matrix. Equation (8.58) holds independently of how the terminals are interconnected or what additional circuit elements are present.

Applying Kirchoff's current law to node i, we obtain

$$\left(C_j + C_{j+1} + \sum_p K_{ip}\right)\frac{dV_i}{dt} - C_j\frac{dV_{i-1}}{dt} - C_{j+1}\frac{dV_{i+1}}{dt} - \sum_p K_{ip}\frac{dV_p}{dt} = I_j - I_{j+1} \tag{8.59}$$

where the p-sum is over all nodes connected to node i via shunt capacitances K_{ip}. C_j is set to zero when node i is at the bottom of the coil and $C_{j+1} = 0$ when node i is at the top of the coil, although a small value for capacitive coupling to the yokes could be used. $dV_p/dt = 0$ when node p is at ground potential such as for the core or tank. Defining a capacitance matrix C, (8.59) can be rewritten as

$$C\frac{d\mathbf{V}}{dt} = A\mathbf{I} \tag{8.60}$$

where A is a rectangular matrix whose rows correspond to nodes and columns to sections and where $A_{ij} = 1$ and $A_{ij+1} = -1$, with sections j and j+1 on either side of node i. One of these terms is 0 when i is at the end of a coil.

When a nodal voltage V_i is specified, for example at the impulsed terminal, as $V_i = V_s$, equation (8.59) is replaced by

$$\frac{dV_i}{dt} = \frac{dV_s}{dt} \tag{8.61}$$

If $V_s = 0$ as for a grounded terminal, then (8.61) becomes $dV_i/dt = 0$. When node i is shorted to ground via a resistor R_s, a term V_i/R_s is added to the left hand side of (8.59). If node i is grounded by means of a capacitor C_s, a term $C_s dV_i/dt$ is added to the left hand side of (8.59). For a shorting inductor L_s uncoupled from all the other inductors, a term $\int V_i dt/L_s$ is added to the left hand side of (8.59). Other situations such as a resistor or varistor joining two nodes can be easily accommodated.

When several nodes are joined together, say nodes i, r, s, ..., their Kirchoff's current law equations are simply added. The resulting equation replaces the node i equation. Then the node r, s, ... equations are replaced by

$$\frac{dV_r}{dt} - \frac{dV_i}{dt} = 0 \quad , \quad \frac{dV_s}{dt} - \frac{dV_i}{dt} = 0 \quad , \quad \cdots \tag{8.62}$$

The net result of all these circuit modifications is that the capacitance matrix in (8.60) may be altered and the right hand side may acquire additional terms, e.g. terms involving V for a resistor, $\int V dt$ for an inductor, and the impulsed voltage V_s. Thus (8.60) is replaced by

$$C' \frac{d\mathbf{V}}{dt} = A\mathbf{I} + f(\mathbf{V}, V_s) \tag{8.63}$$

where C' is the new capacitance matrix and f depends on the added elements. The above procedure lends itself to straightforward computer implementation.

Equations (8.58) and (8.63) can now be solve simultaneously by means of for example a Runge-Kutta algorithm starting from a given initial state. Linear equations must be solved at each time step to determine dI/dt and dV/dt. Since M is a symmetric positive definite matrix, the Cholesky algorithm may be used to solve (8.58) while (8.63) may be solved by Gaussian elimination. LL^T or LU factorization is first performed on the respective matrices and the factors are used subsequently to solve the linear equations at each time step, saving much computation time.

8.5.2 Mutual and Self Inductance Calculations

The transformer core and coil geometry including the iron yokes is assumed to have cylindrical symmetry. The iron is assumed to be infinitely permeable. A typical coil section is shown in Fig. 8.11. It is assumed to be rectangular in cross-section and carry a uniform current density J, azimuthally directed. The yokes extend outward to infinity and we ignore the tank walls. This geometry corresponds to that used in the Rabins' inductance and mutual inductance calculations of Chapter 5. However, we will use a slightly different method to obtain the self and mutual inductances. For two coil sections, labeled p and q, where p = q indicates a self inductance calculation, the mutual (or self) inductance is given by

$$M_{pq} = \frac{\int J_q A_p dV_q}{I_p I_q} \tag{8.64}$$

where A_p is the vector potential generated by section p, I_p and I_q the total currents in the two coil sections, and the volume integral is only over the coil section q and not the entire window height L since this is where all the current would be located if we used an infinite number of terms in the Fourier expansion of the current density. This produces the same result as before but we use slightly different notation here.

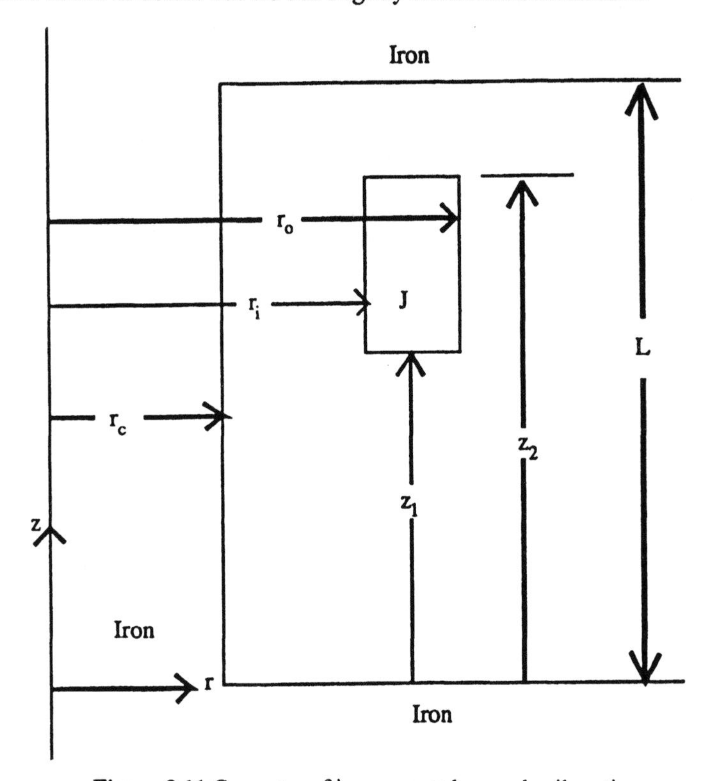

Figure 8.11 Geometry of iron core, yokes, and coil section.

Using the vector potential solution given in Chapter 5 but labeling the inner and outer coil section radii by r_{ip} , r_{op} , r_{iq} , r_{oq} for sections p and q respectively, we find for coil sections on different radially displaced coils,

$$M_{pq} = \frac{\pi\mu_o L}{I_p I_q}\left\{\frac{J_{o,p}J_{o,q}}{3}\left(r_{op}^3 - r_{ip}^3\right)\left(r_{oq} - r_{iq}\right) + \sum_{n=1}^{\infty} G_{n,p}\frac{J_{n,p}J_{n,q}}{m^4}\int_{mr_{iq}}^{mr_{oq}} tK_1(t)dt\right\} \tag{8.65}$$

where $m = n\pi/L$ as before. For coil sections on the same coil, we obtain

$$M_{pq} = \frac{\pi\mu_o L}{I_p I_q}\left\{\frac{J_{o,p}J_{o,q}}{3}\left[r_{op}\left(r_{oq}^3 - r_{iq}^3\right) - r_{ip}^3\left(r_{oq} - r_{iq}\right) - \frac{\left(r_{oq}^4 - r_{iq}^4\right)}{2}\right]\right.$$
$$\left. + \sum_{n=1}^{\infty}\frac{J_{n,q}J_{n,q}}{m^4}\left[E_{n,p}\int_{mr_{iq}}^{mr_{oq}} tI_1(t)dt + F_{n,p}\int_{mr_{iq}}^{mr_{oq}} tK_1(t)dt - \frac{\pi}{2}\int_{mr_{iq}}^{mr_{oq}} tL_1(t)dt\right]\right\} \tag{8.66}$$

The coefficients $G_{n,p}$, $E_{n,p}$, $F_{n,p}$ are as defined in Chapter 5. The modified Bessel and Struve function integrals can be evaluated by the numerical techniques given there.

8.5.3 Capacitance Calculations

The series capacitances must take into account the type of coil, whether helix, disk, multi-start, or other. Series and shunt capacitances incorporate details of the paper and pressboard insulation, the placement of key spacers and sticks, and the oil duct geometry. Except when wound-in-shields are present, the formula developed by Stein as discussed in Chapter 7 is used for the series disk capacitances. This formula was found to produce the best results in the work reported by Miki et. al. [Mik78]. Simple standard formulas are used for helical and multi-start windings as well as for shunt capacitances as discussed in Chapter 7. Basically a shunt capacitance per unit length is calculated, assuming the coils are infinitely long, and this is multiplied by the length of the section to get the section to section or section to ground shunt capacitances.

The series capacitance for a section containing several disks or turns in the case of a helical winding is obtained by adding the single disk capacitances in series, i.e. dividing the single disk capacitance by the number of disks in the section. Since the subdivisions need not contain an integral number of disks, fractional disks are allowed in this calculation. Other subdivision schemes could be adopted which restrict the sections to contain an integral number of disks. When wound-in-shields are present, only an integral number of disk pairs are allowed in the subdivision. The finer the subdivisions or the greater their number, the greater the expected accuracy. However, too fine a subdivision, say less than the height of a disk, could be counterproductive. Fine subdivisions can also lead to convergence problems for the Fourier sums in the mutual inductance calculation.

8.5.4 Impulse Voltage Calculations and Experimental Comparisons

The standard impulse voltage waveform can be mathematically represented as

$$V_s(t) = V_o\left(e^{-\kappa_1 t} - e^{-\kappa_2 t}\right) \tag{8.67}$$

where κ_1, κ_2, and V_o are adjusted so that V_s rises to its maximum value in 1.2 μs and decays to half its value in 50 μs. It is desirable to have an analytic formula which is smoothly differentiable such as (8.67) since the derivative is used in the solution process. Other waveforms such as a chopped wave may also be used. Procedures may be developed for extracting the parameters used in (8.67) from the desired rise and fall times and peak voltage. In our experimental RSO tests discussed below, the actual rise and fall times were 3 and 44 μs respectively and the parameters in (8.67) were adjusted accordingly.

We tested a 45 MVA autotransformer having 4 windings as shown in Fig. 8.12. The high voltage (HV or series) winding consists of 2 coils in parallel with two ends joined at the center where the impulse is applied. The other two ends are joined at the autopoint with the low voltage (LV or common) and tap windings. The HV and LV windings are disk windings. The tap winding is a multi-start winding and is grounded for the impulse test. The tertiary voltage winding (TV) is a helical winding and is grounded at both ends for the impulse test. These windings represent one phase of a three phase transformer but are essentially isolated from each other for the impulse test.

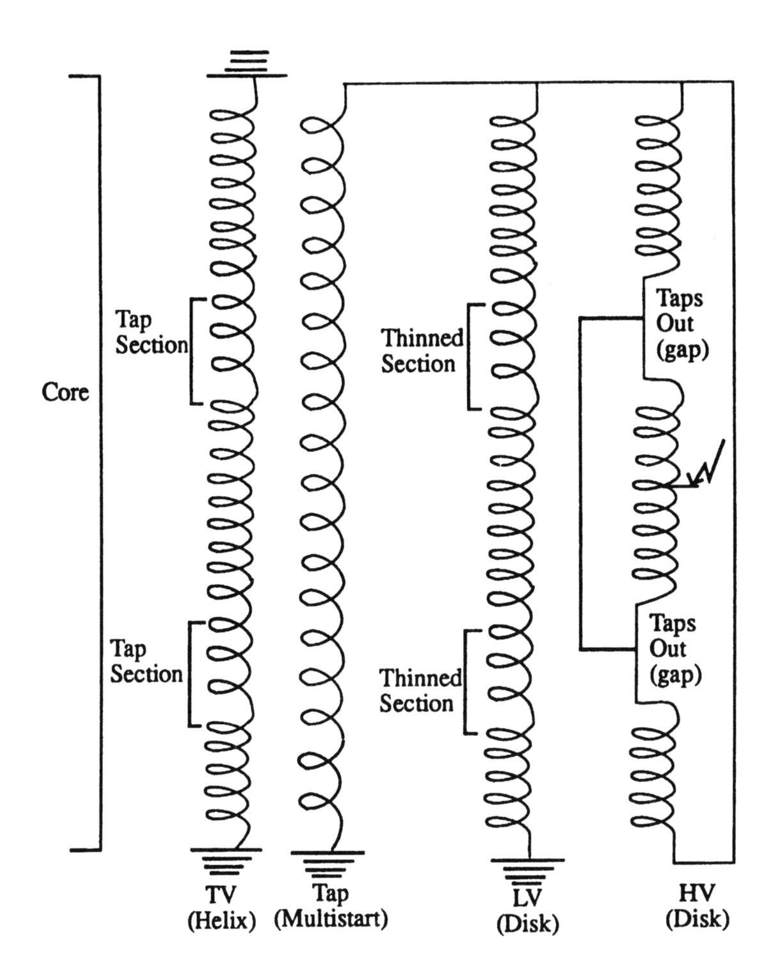

Figure 8.12 Schematic diagram of impulsed autotransformer

Both the HV and TV windings contain tapped sections along their lengths. The tap turns are out on the HV winding and in on the TV winding during the impulse test. In the calculations a direct short is placed across the tapped out sections. In the vicinity of these tap sections, the turn density is lowered in the LV winding (thinning) for better short circuit strength. In the calculations, these thinned areas are specified as one or more separate subdivisions with their own properties which differ from those of the subdivisions in the rest of the winding.

Although nodal voltages and section currents are calculated throughout the transformer as a function of time, voltage differences can easily be obtained. In fact disk-disk voltages are calculated throughout all the windings and the maximum occurring over the time duration of the pulse is printed out for each winding. Similarly, the maximum voltage difference occurring between adjacent windings for the duration of the pulse is also printed out. Secondary quantities such as maximum electric fields in the oil can likewise be obtained.

Experimentally only voltages along the high voltage coil and at the tap positions on the tertiary winding were easily accessible in the tests conducted. In a recurrent surge oscillograph (RSO) test, a repetitive series of pulses having the shape described by (8.67) are applied to, in our case, the HV terminal. This repetitive input results in persistent oscilloscope displays of the output voltages for easy recording. The peak applied voltage is usually low, typically several hundred volts. This test simulates an impulse test done at much higher voltages to the extent that the system is linear. This is likely to be a good assumption provided the core doesn't saturate and provided non-linear circuit elements such as varistors do not come into play. In our RSO test, the transformer was outside the tank. Thus the dielectric constant of air was used in the capacitance formulas. In addition the tank distance, which affects the ground capacitance of the outer coil, was taken as very large.

For the tests conducted here, the computer outputs of interest were the voltages to ground at the experimentally measured points. We present a sufficient number of these in the following figures to indicate the level of agreement between calculation and experiment. We chose to normalize the input to 100 at the peak of the impulse waveform. The units can therefore be interpreted as a percent of the BIL.

It quickly became apparent in comparing the simulations with experiment that the section resistances which were obtained from the wire geometry and the d.c. or power frequency resistivity was not adequate to account for the damping observed in the output waveforms. In fact, we found it necessary to increase this resistivity by a factor of about 3000 to account for the damping. Such a factor is not unreasonable in view of the fact that the currents induced by the impulse waveform contain high frequency components which induce much higher losses than occur at power frequency. A factor of this magnitude was estimated for the experiment involving wound-in-shields in Chapter 7. A better approach which we subsequently adopted is to Fourier analyze the current waveforms and calculate the effective resistivity using the formula given in Chapter 7. This involves running the calculation at least twice, once to obtain the waveforms using an assumed resistivity, Fourier analyzing the waveforms to obtain a better estimate of the effective resistivity, and then re-running the calculation with the recalculated resistivity.

Each of the two HV coils had 52 disks having 12 radial turns each. The impulse was applied to the center of the leg where the two coils are joined. Output voltages are recorded relative to this center point. Fig. 8.13 shows the experimental and calculated voltage waveform 4 disks below the impulsed terminal. The impulse voltage is also shown for comparison. Fig. 8.14 shows the voltage 12 disks below the impulsed terminal. There is good agreement in the major oscillations but some differences in the lower amplitude higher frequency oscillations. Fig. 8.15 shows the experimental and theoretical voltages at the tap position which is about 30 disks below the impulsed terminal. Fig. 8.16 shows the voltage transferred to the upper tap position on the TV winding. There is good agreement between theory and experiment in overall magnitude and major oscillations but the higher frequency ripple is not as well predicted. Fig. 8.17 indicates another way to present the calculations. This shows the voltage as a function of relative coil position along the top HV winding at various instants of time. Note the flat portion of the curves at a relative position of ~ 0.75 which corresponds to the tap section with the taps out. Note also that the high capacitance of the multi-start tap winding effectively grounds the autopoint. Maximum disk-disk voltages could be obtained by examining such curves although this is more easily done by programming.

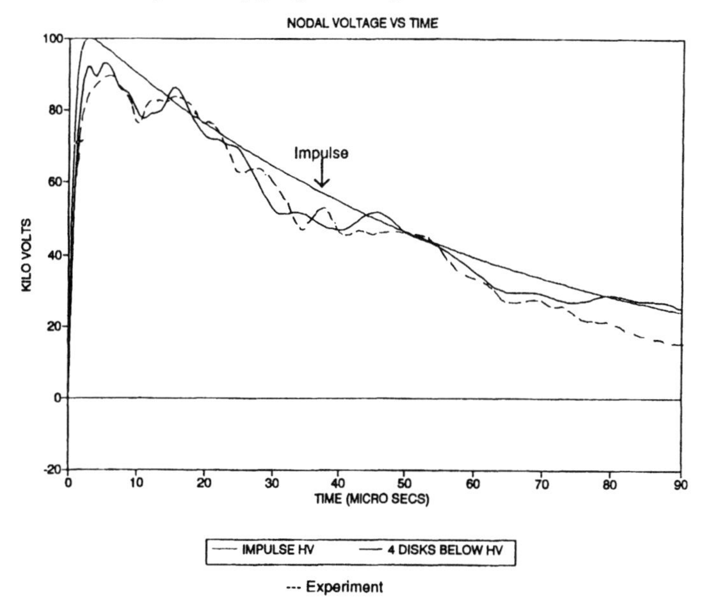

Figure 8.13 Experimental and calculated voltage to ground 4 disks below the HV impulsed terminal. The impulse voltage is also shown..

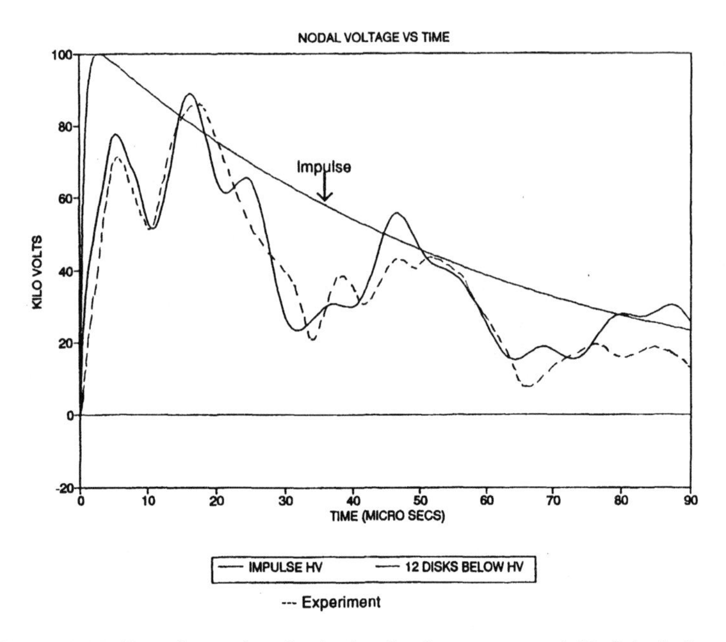

Figure 8.14 Experimental and calculated voltage to ground 12 disks below the HV impulsed terminal

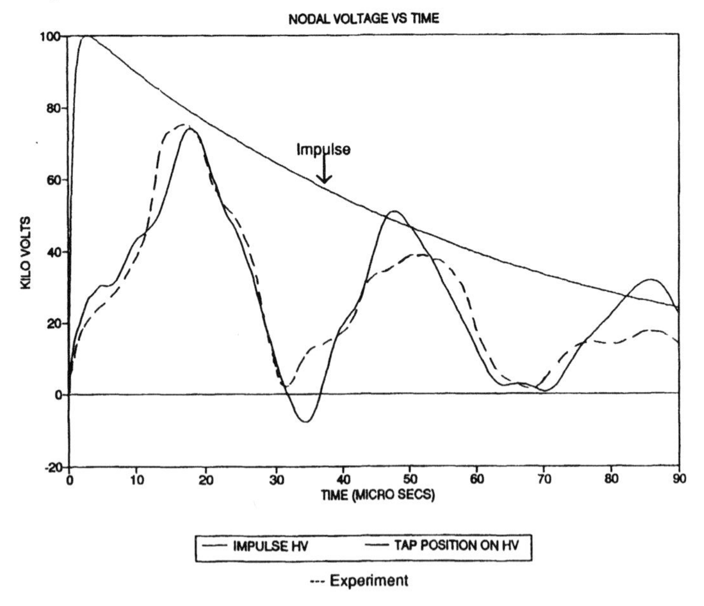

Figure 8.15 Experimental and theoretical voltages to ground at the tap position on the HV winding, about 30 disks below the impulsed terminal

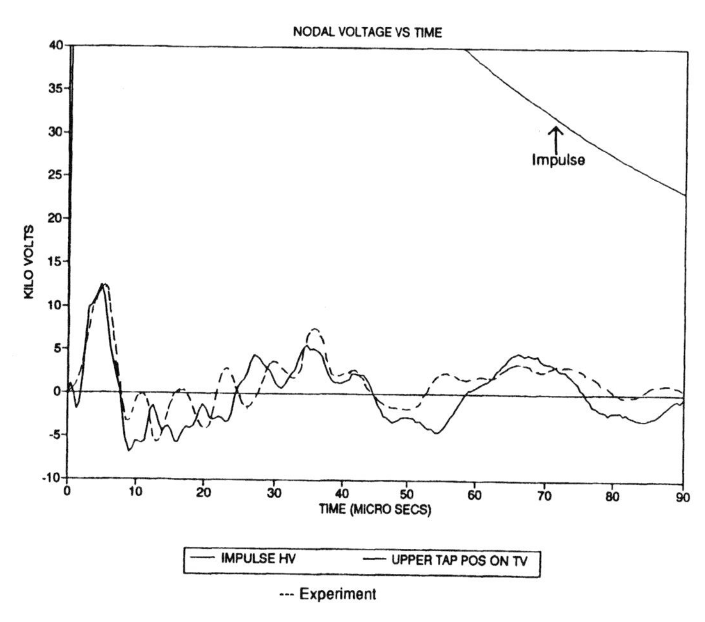

Figure 8.16 Experimental and theoretic voltages to ground at the center of the upper tap position on the TV winding.

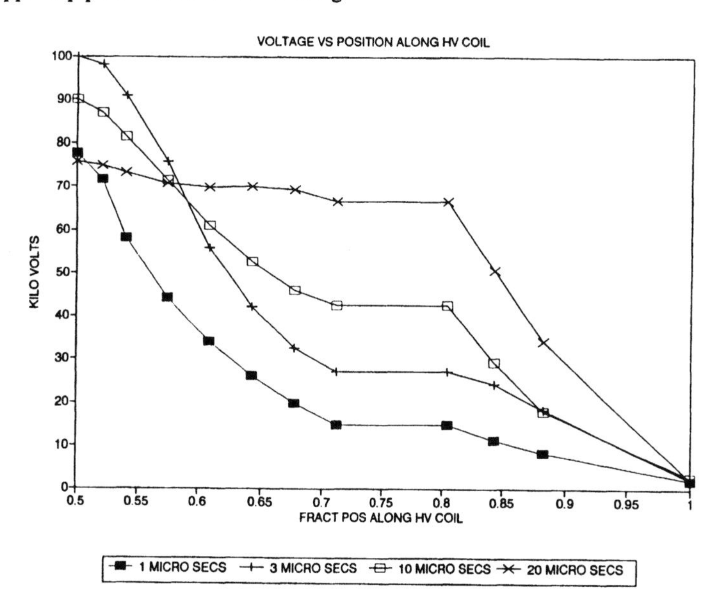

Figure 8.17 Calculated voltage profiles at various instants of time along the upper HV coil. Relative position 0.5 is the impulsed terminal and 1 the top of the winding.

8.5.5 Sensitivity Studies

We explore the sensitivity of the calculations to two of the inputs to which some uncertainty is attached, namely the effective resistivity of the copper and the number of subdivisions along the coils. Figs. 8.18 - 8.20 show the effect of increasing the resistivity to 3000 times its d.c. value relative to the d.c. calculation at several locations. The increased damping has little effect on the waveforms at short times but progressively lowers the amplitude of the longer time oscillations.

The effect of varying the number of coil subdivisions is shown in Figs. 8.21, 8.22. Generally each coil has a minimum number of subdivisions dictated by the number of physically different coil sections. In our case this was about 6 - 8 per coil. The multi-start coil, however, has only one section. Because of its construction, it cannot be meaningfully subdivided into axial sections as can the other coils. In fact the capacitances to neighboring coils must be coupled equally to both the top and bottom node because its voltage is not a unique function of position. We see from the figures that once the number of subdivisions is ~ 12/coil, the results are fairly insensitive to any further refinement. Thus one doesn't have to be overly concerned about the number of subdivisions used, provided they are above some reasonable minimum which in this case is ~ 12/coil.

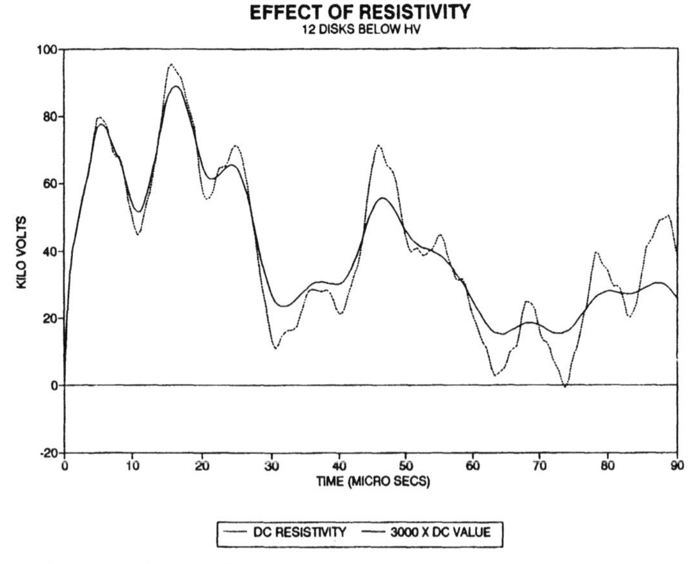

Figure 8.18 Sensitivity of the calculations to the resistivity of the nodal voltage 12 disks below the impulsed terminal

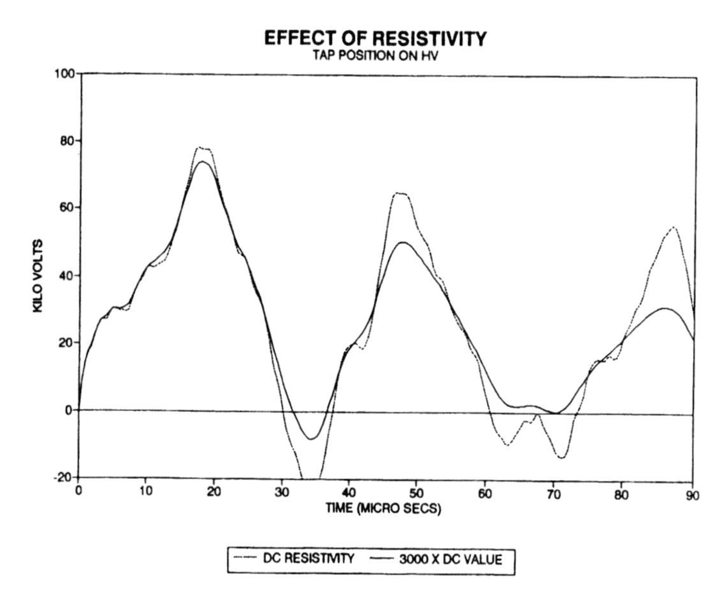

Figure 8.19 Sensitivity of the calculations to the resistivity for the HV tap position voltage

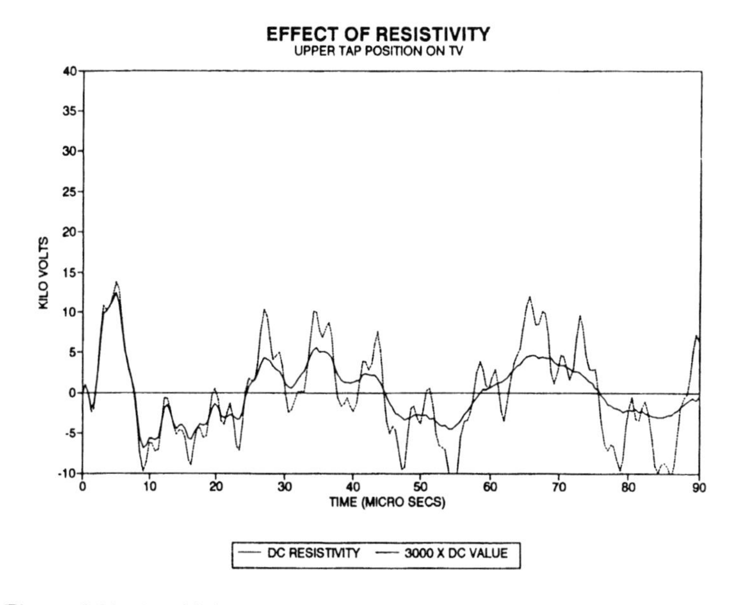

Figure 8.20 Sensitivity of the calculations to the resistivity for the upper tap position on the TV winding

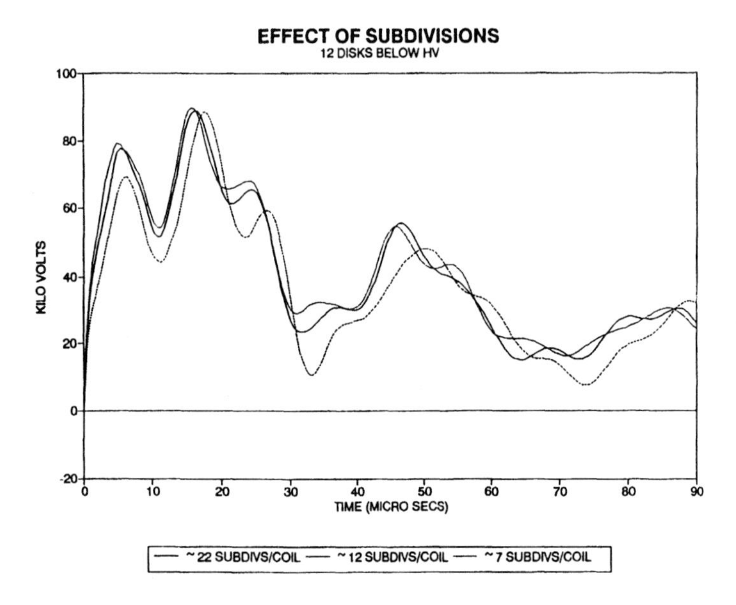

Figure 8.21 Effect of the number of coil subdivisions on the voltage for a point 12 disks below the impulsed terminal

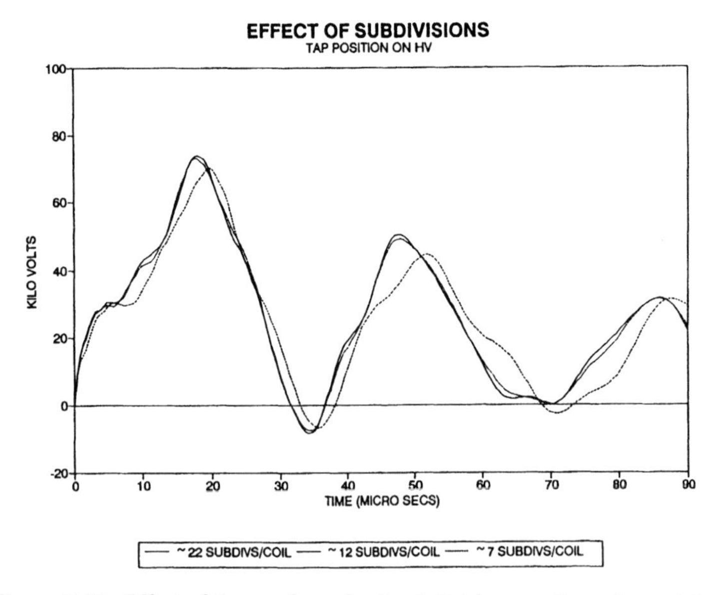

Figure 8.22 Effect of the number of coil subdivisions on the voltage at the tap position on the HV winding

9. ELECTRIC FIELD CALCULATIONS

Summary The electric field is calculated for some idealized insulation structures which can often be useful approximations to actual insulation systems. This can provide insight into the parameter dependence of such fields and can suggest ways of reducing such fields if necessary. The electric fields in the oil gaps between pairs of disks in a transformer winding are strongest at the corners. These corner fields therefore determine the gap spacing and paper insulation required to avoid breakdown. We determine these fields by means of a conformal mapping technique for a 2-D geometry consisting of 2 conductors at different potentials separated by a gap (disk - disk spacing). Both conductors are separated by another gap from a ground plane. This latter gap could be a winding - winding gap or a winding - core or tank gap. The analytic solution does not include the effect of the paper insulation. This leads to an enhancement of the field in the oil over the situation without insulation. A method is proposed to account for the insulation based on a comparison with a finite element solution. Finite element solution methods are discussed for complex geometries.

9.1 SIMPLE GEOMETRIES

It is often possible to obtain a good estimation of the electric field in a certain region of a transformer by idealizing the geometry to such an extent that the field can be calculated analytically. This has the advantage of exhibiting the field as a function of several parameters so that the effect of changing these and how this affects the field can be appreciated. Such insight is often worth the price of the slight inaccuracy which may exist in the numerical value of the field.

As a first example, we consider a layered insulation structure having a planar geometry as shown in Fig. 9.1. This could represent the major insulation structure between two cylindrical windings having large radii. We are further approximating the disk structure as a smooth surface so that the resulting field calculation would be representative of the field away from the corner of the disks. We treat this corner field in the next section.

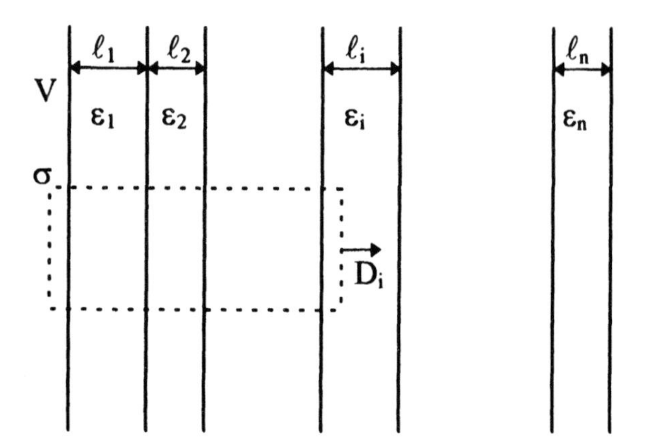

Figure 9.1 Geometry of a planar layered insulation structure

We use one of Maxwell's equations in integral form to solve this,

$$\oint_S \mathbf{D} \cdot \mathbf{dA} = q \tag{9.1}$$

where $\mathbf{D}$ is the displacement vector, $\mathbf{dA}$ a vectorial surface area with an outward normal, and q the charge enclosed by the closed surface S. Because of the assumed ideal planar geometry, the surface charge density on the electrode at potential V is uniform and is designated σ in the figure. An opposite surface charge of $-\sigma$ exists on the ground electrode. Note that both electrode potentials could be raised by an equal amount without changing the results. Only the potential difference matters. We will also assume that the materials have linear electrical characteristics so that

$$\mathbf{D} = \varepsilon \mathbf{E} \tag{9.2}$$

holds within each material where ε, the permittivity, can differ within the various layers as shown in the figure.

Because of the planar geometry, the D and E fields are directed perpendicular to the planes of the electrodes and layers. Thus if we take our closed surface to be the dotted rectangle shown in the figure which has some depth into the figure so that the two vertical sides represent surfaces of area A, then the only contrifution to the integral in (9.1) which is non-zero is the part over the right vertical surface where the displacement vector has the uniform value labelled D_i in the figure. Thus, for this closed surface (9.1) becomes

$$D_i A = q = \sigma A \quad \Rightarrow \quad D_i = \sigma \tag{9.3}$$

Using (9.2) applied to layer i, (9.3) becomes

$$E_i = \frac{\sigma}{\varepsilon_i} \tag{9.4}$$

In terms of the potential V, we can write by definition

$$V = -\int \mathbf{E} \cdot d\boldsymbol{\ell} \tag{9.5}$$

where the line integral starts at the 0 potential electrode and ends on the V potential electrode. In terms of the E fields in the different materials and their thicknesses ℓ_i, (9.5) becomes

$$V = E_1 \ell_1 + E_2 \ell_2 + \cdots = \sum_{j=1}^{n} E_j \ell_j \tag{9.6}$$

Using (9.4), this can be written

$$V = \sigma \sum_{j=1}^{n} \frac{\ell_j}{\varepsilon_j} \tag{9.7}$$

Solving for σ and substituting into (9.4), we get

$$E_i = \frac{V}{\varepsilon_i \left(\sum_{j=1}^{n} \frac{\ell_j}{\varepsilon_j} \right)} \tag{9.8}$$

Letting ℓ be the total distance between the electrodes so that $\ell = \ell_1 + \ell_2 + \cdots$ and defining the fractional lengths, $f_i = \ell_i/\ell$, (9.8) can be expressed as

$$E_i = \frac{(V/\ell)}{\varepsilon_i \left(\sum_{j=1}^{n} \frac{f_j}{\varepsilon_j} \right)} = \frac{E_o}{\varepsilon_i \left(\sum_{j=1}^{n} \frac{f_j}{\varepsilon_j} \right)} \tag{9.9}$$

where E_o is the electric field between the electrodes if there was only one layer of uniform material between them.

Let's apply these results to an oil-pressboard insulation system. Even if there are many layers of pressboard used to subdivide the oil gap, only the total fractional thickness, f_{press} , matters in the calculation. Similarly, the subdivided oil gap's total fractional thickness, $f_{oil} = 1 - f_{press}$, is all that is needed to perform the calculation. For this situation, (9.9) becomes

$$E_{oil} = \frac{E_o}{\varepsilon_{oil} \left(\frac{f_{oil}}{\varepsilon_{oil}} + \frac{f_{press}}{\varepsilon_{press}} \right)} \quad , \quad E_{press} = \frac{E_o}{\varepsilon_{press} \left(\frac{f_{oil}}{\varepsilon_{oil}} + \frac{f_{press}}{\varepsilon_{press}} \right)} \tag{9.10}$$

Since the relative permittivities of pressboard and oil are $\varepsilon_{press} \cong 4.4$, $\varepsilon_{oil} \cong 2.2$, the electric field in the oil is about twice as high as the electric field in the pressboard for a given oil-pressboard combination. Thus the oil's electric field is usually the most important to know for purposes of breakdown estimation. We plot E_{oil}/E_o vs f_{press} in Fig. 9.2. We see from the figure that for a given oil gap, the lowest field results when there is no pressboard. As more pressboard displaces the oil, the field in the oil increases, approaching a value of twice its all oil value when the gap is nearly filled with pressboard.

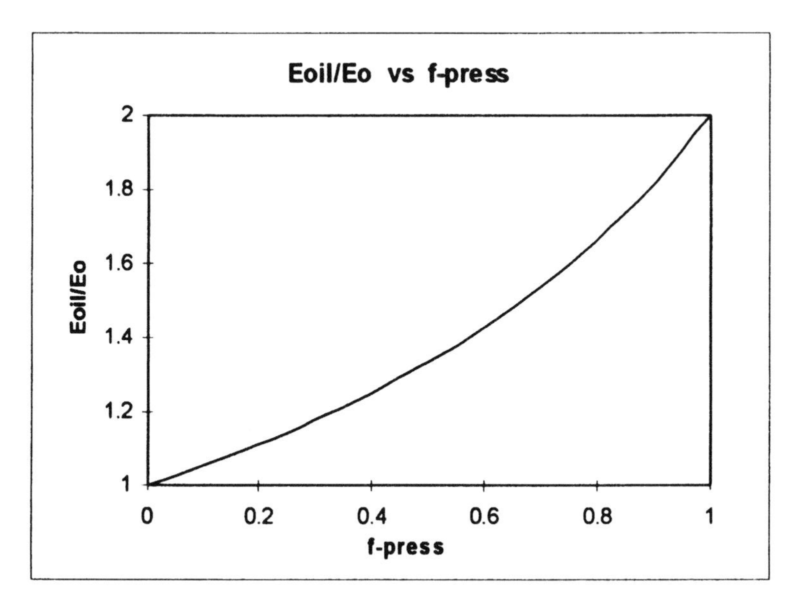

Figure 9.2 Relative electric field in the oil in a planar oil gap as a function of the fractional amount of pressboard.

A somewhat better estimate can be made of the electric field in the major insulation structure between two coils if we consider approximating the geometry as an ideal cylindrical geometry. This geometry is also useful for approximating the field around a long cable of circular cross section. We consider the general case of a multi-layer concentric cylindrical insulation structure as shown in Fig. 9.3. The inner most cylinder is at potential V and the outer most at zero potential, although it is only their potential difference which matters.

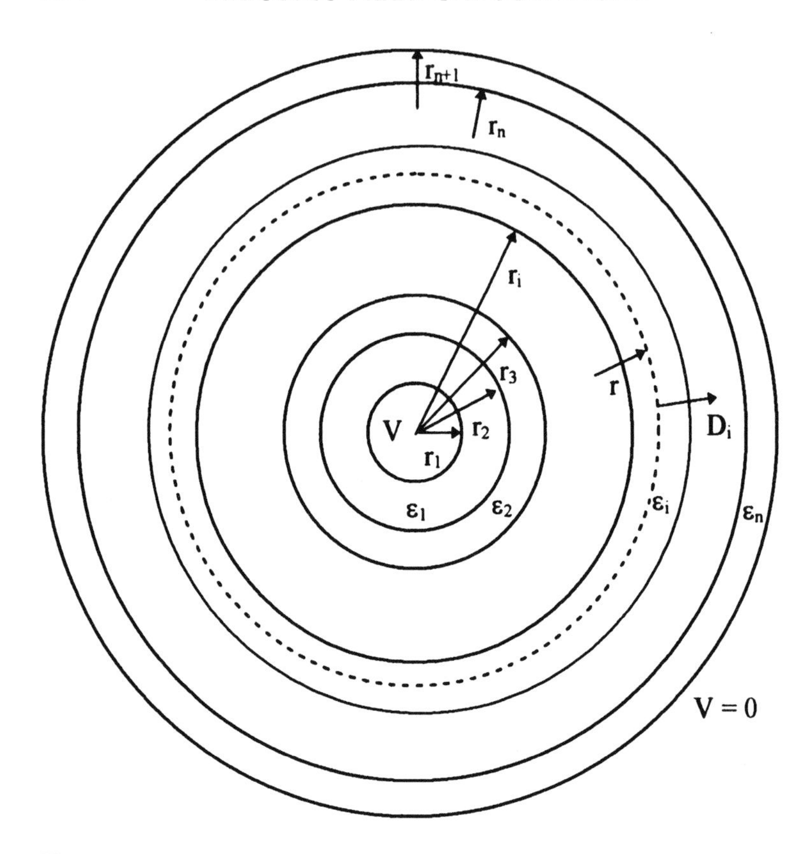

Figure 9.3 Ideal layered cylindrical insulation structure. This same drawing can be reinterpreted to refer to a spherical geometry.

From symmetry, we see that the **D** and **E** fields are directed radially. We assume there is a surface charge per unit length along the inner cylinder of λ. We apply (9.1) to the dashed line cylindrical surface in Fig. 9.3 which is assumed to extend a distance L along the axis with disk like surfaces on either end. The only contribution to the integral is along the lateral sides of the cylinder and we find

$$2\pi r L D_i = \lambda L \quad \Rightarrow \quad D_i = \frac{\lambda}{2\pi r} \tag{9.11}$$

Assuming linear materials, we get for layer i,

$$E_i = \frac{\lambda}{2\pi\varepsilon_i r} \tag{9.12}$$

Using the definition (9.5) and starting the line integral from the outer zero potential electrode, we have, using (9.12),

$$V = -\frac{\lambda}{2\pi}\left\{\frac{1}{\varepsilon_n}\int_{r_{n+1}}^{r_n}\frac{dr}{r}+\cdots+\frac{1}{\varepsilon_i}\int_{r_{i+1}}^{r_i}\frac{dr}{r}+\cdots+\frac{1}{\varepsilon_1}\int_{r_2}^{r_1}\frac{dr}{r}\right\}$$
$$= \frac{\lambda}{2\pi}\sum_{j=1}^{n}\frac{1}{\varepsilon_j}\ln\left(\frac{r_{j+1}}{r_j}\right) \tag{9.13}$$

Solving for λ and substituting into (9.12), we get

$$E_i = \frac{V}{\varepsilon_i r\left[\sum_{j=1}^{n}\frac{1}{\varepsilon_j}\ln\left(\frac{r_{j+1}}{r_j}\right)\right]} \tag{9.14}$$

For any given layer, the maximum field, $E_{i,\,max}$, occurs at its inner radius, so we have

$$E_{i,max} = \frac{V}{\varepsilon_i r_i\left[\sum_{j=1}^{n}\frac{1}{\varepsilon_j}\ln\left(\frac{r_{j+1}}{r_j}\right)\right]} \tag{9.15}$$

We see from the last two equations that, for a given layer, the electric field is inversely proportional to the permittivity. Thus an oil layer at a given position will see about twice the eclectric field of a pressboard layer at the same position, assuming the quantity in square brackets is the same or nearly so. Since the maximum field in a layer is also inversely proportional to the radius of the layer to first order, we see that this field can be reduced by increasing the layer's radius. Thus for a cable surrounded by solid insulation such as paper and immersed in oil, the cirtical field will probably occur in the oil at the outer surface of the paper. To reduce this, one could increase its radius by adding more paper or start with a larger radius cylindrical conductor to begin with.

Another geometry of some interest is the spherical geometry. The general case of a multi-layered syperical insulation structure is shown in Fig. 9.3 by interpreting it as a cross-section through a sphercial system of insulators. In this case, the D and E fields are again directed radially by symmetry and the charge on the inner conductor at potential V is taken as q. The dashed circle in the figure now defines a spherical surface and (9.1) applied to this results in

$$E_i = \frac{q}{4\pi\varepsilon_i r^2} \tag{9.16}$$

From (9.5) and (9.16), we obtain

$$\begin{aligned} V &= -\frac{q}{4\pi}\left\{\frac{1}{\varepsilon_n}\int_{r_{n+1}}^{r_n}\frac{dr}{r^2}+\cdots+\frac{1}{\varepsilon_i}\int_{r_{i+1}}^{r_i}\frac{dr}{r^2}+\cdots+\frac{1}{\varepsilon_1}\int_{r_2}^{r_1}\frac{dr}{r^2}\right\} \\ &= \frac{q}{4\pi}\sum_{j=1}^{n}\frac{1}{\varepsilon_j}\left(\frac{1}{r_j}-\frac{1}{r_{j+1}}\right) \end{aligned} \tag{9.17}$$

Solving for q and substituting into (9.16), we obtain

$$E_i = \frac{V}{\varepsilon_i r^2\left[\sum_{j=1}^{n}\frac{1}{\varepsilon_j r_j}\left(1-\frac{r_j}{r_{j+1}}\right)\right]} \tag{9.18}$$

The maximum field in layer i, $E_{i,\,max}$, occurs at the inner radius and is given by

$$E_{i,max} = \frac{V}{\varepsilon_i r_i^2\left[\sum_{j=1}^{n}\frac{1}{\varepsilon_j r_j}\left(1-\frac{r_j}{r_{j+1}}\right)\right]} \tag{9.19}$$

This expression can be used to approximate the field near a sharp bend in a cable immersed in oil with the various radii defined appropriately. We see that as the radius of the layer increases, the field in it decreases. Thus the field in the oil can be reduced by adding more insulation or decreasing the sharpness of the bend.

9.2 ELECTRIC FIELD CALCULATIONS USING CONFORMAL MAPPING

9.2.1 Physical Basis

In a region of space without charge, Maxwell's electrostatic equation is

$$\nabla \cdot \mathbf{D} = 0 \tag{9.20}$$

where $\mathbf{D}$ is the electric displacement. If the region has a uniform permeability ε, then $\mathbf{D} = \varepsilon\mathbf{E}$, where $\mathbf{E}$ is the electric field. Hence (9.20) becomes

$$\nabla \cdot \mathbf{E} = 0 \tag{9.21}$$

Introducing a potential function V, where

$$\mathbf{E} = -\nabla V \tag{9.22}$$

we obtain from (9.21)

$$\nabla^2 V = 0 \tag{9.23}$$

In 2-dimensions, using Cartesian coordinates, this last equation reads

$$\frac{\partial^2 V}{\partial x^2} + \frac{\partial^2 V}{\partial y^2} = 0 \tag{9.24}$$

The solution of (9.24), including boundary conditions, can then be used to determine the electric field via (9.22).

Boundary conditions are generally of two types, Dirichlet or Neumann. A Dirichlet boundary condition specifies the voltage along a boundary. This voltage is usually a constant as would be appropriate for a metallic surface. A Neumann boundary condition specifies the normal derivative along a boundary. The normal derivative is usually taken to be 0, which says that the equipotential surfaces (surfaces of constant V) or lines in 2-D intersect the boundary at right angles. This type of boundary is often used to enforce a symmetry condition.

Functions satisfying equation (9.24) are called harmonic functions. In the theory of functions of a complex variable, analytic functions play a special role. These are functions which are continuous and differentiable in some region of the complex plane. It turns out that the real and imaginary parts of analytic functions are harmonic functions. Further, analytic mappings from one complex plane to another have properties which allow a solution of (9.24) in a relatively simple geometry in one complex plane to be transformed to a solution of this equation in a more complicated geometry in another complex plane. We briefly describe some of the important properties of these functions which are needed in the present application. See reference [Chu60] for further details.

9.2.2 Conformal Mapping

Let $z = x + iy$ denote the complex variable where $i = \sqrt{-1}$ is the unit imaginary. A function $f(z)$ can be written in terms of its real and imaginary parts as

$$f(z) = u(x,y) + iv(x,y) \tag{9.25}$$

where u and v are real functions of 2 variables. If f is analytic, it is differentiable at points z in its domain of definition. Since we are in the z-plane , the derivative can be taken in many directions about a given point and the value must be independent of direction. Taking this derivative in the x direction and then in the iy direction and equating the results, we obtain the Cauchy-Riemann equations

$$\frac{\partial u}{\partial x} = \frac{\partial v}{\partial y} \quad , \quad \frac{\partial u}{\partial y} = -\frac{\partial v}{\partial x} \tag{9.26}$$

Differentiating the first of these equations with respect to x and the second with respect to y we have

$$\frac{\partial^2 u}{\partial x^2} = \frac{\partial^2 v}{\partial x \partial y} \quad , \quad \frac{\partial^2 u}{\partial y^2} = -\frac{\partial^2 v}{\partial y \partial x} \tag{9.27}$$

Since the mixed partial derivatives on the right hand sides of these equations are equal for differentiable functions, when we add these equations we obtain

$$\frac{\partial^2 u}{\partial x^2} + \frac{\partial^2 u}{\partial y^2} = 0 \tag{9.28}$$

Thus u is a harmonic function. By differentiating the first of equations (9.26) with respect to y, the second with respect to x , and adding, we similarly find that v is a harmonic function.

The solution of the potential problem, equation (9.24) with boundary conditions, is often needed in a rather complicated region geometrically. The idea behind using complex variable theory is to formulate the problem in a simpler geometric region where the solution is easy and then use an analytic function to map the easy solution onto the more complicated geometry of interest. The possibility of doing this derives from several additional properties of analytic functions.

First of all, an analytic function of an analytic function is also analytic. Since the real and imaginary parts of the original and composite functions are harmonic, this says that harmonic functions are transformed into harmonic functions by means of analytic transformations. In terms of formulas, if f(z) is an analytic function of z and z = g(w) expresses z in terms of an analytic mapping from the w-plane , where $w = u + iv$, then f(g(w)) is an analytic function of w. The real and imaginary parts of f are transformed into harmonic functions of the new variables u and v.

Given an analytic mapping from the z to w-plane, w = f(z) , the inverse mapping
z = F(w) is analytic at points where $f'(z) = df/dz \neq 0$. Moreover at such points

$$F'(w) = \frac{1}{f'(z)} \qquad \text{or} \qquad \frac{dz}{dw} = \frac{1}{dw/dz} \tag{9.29}$$

This result will be useful in later applications to the electrostatic problem.

Perhaps the most important characteristic of analytic mappings in the present context is that they are conformal mappings. This means that if two curves intersect at an angle α in the z-plane, their images in the w-plane under an analytic mapping w = f(z) intersect at the same angle α at the transformed point. To see this, we use the fact that an analytic function can be expanded about a point z_o using a Taylor's series expansion

$$\begin{aligned} w = f(z) &= f(z_o) + f'(z_o)(z - z_o) + \cdots \\ &= w_o + f'(z_o)(z - z_o) + \cdots \end{aligned} \tag{9.30}$$

where we assume that $f'(z_o) \neq 0$. Writing $\Delta w = w - w_o$ and $\Delta z = z - z_o$, this last equation becomes

$$\Delta w = f'(z_o)\Delta z \tag{9.31}$$

If Δz is an incremental distance along a curve in the z-plane, Δw is the corresponding incremental distance along the transformed curve in the w-plane (See Fig. 9.4). Since we can write any complex number in polar form,

$$z = |z|e^{i\phi} \tag{9.32}$$

in terms of its magnitude $|z|$ and argument ϕ, (9.31) can be expressed as

$$|\Delta w|e^{i\beta} = |f'(z)||\Delta z|e^{i(\psi_o + \alpha)} \tag{9.33}$$

where ψ_o = argument ($f'(z)$) and α and β are shown in Fig. 9.4. Thus we see from (9.33) that

$$\beta = \psi_o + \alpha \tag{9.34}$$

As Δz and Δw approach zero, their directions approach that of the tangent to their respective curves. Thus the angle which the transformed curve makes with the horizontal axis β is equal to the angle which the original curve made with its horizontal axis α rotated by the amount ψ_o. Since ψ_o is characteristic of the derivative $f'(z_o)$, which is independent of the curve passing through z_o, this says that any curve through z_o will be rotated by the same angle ψ_o. Thus, since any two intersecting curves are rotated by the same angle under the transformation, the angle between the curves will be preserved.

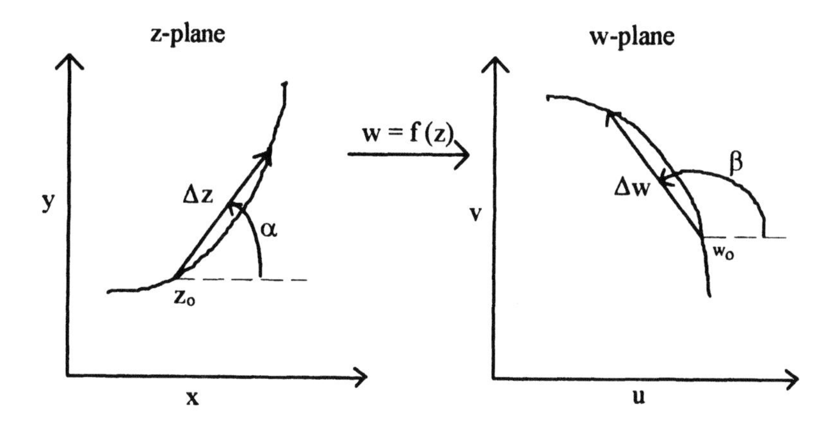

Figure 9.4 Analytic mapping of a curve from the z to the w-plane

In terms of the electrostatic problem, this last mapping characteristic says that a set of equipotential (non-intersecting) curves in one geometry will remain non-intersecting in the transformed geometry obtained from the first by means of an analytic map. Similarly, the orthogonal relationship between the equipotentials and the electric field lines will be preserved in the new geometry.

Finally we need to look at the boundary conditions. If H is a harmonic function which is constant along some curve or boundary,

$$H(x,y) = C \tag{9.35}$$

then, changing variables by means of $z = f(w)$, this becomes, in the new variables,

$$H(x(u,v),y(u,v)) = C \tag{9.36}$$

ie. a transformed curve along which H has the same constant value. Thus Dirichlet boundary conditions are transformed into Dirichlet boundary conditions with the same boundary value along the transformed curve.

A Neumann boundary condition means that the normal derivative of H along the boundary vanishes. Since the normal derivative is the scalar product of the gradient and unit normal vector, the vanishing of this derivative means that the gradient vector points along the boundary, i.e. is tangential to it. However, the gradient vector is perpendicular to curves along which H is constant. Therefore these curves of constant H

are perpendicular to the boundary curve. Under a conformal mapping, this perpendicularity is preserved so that the transformed boundary curve is normal to the transformed curves of constant H. Therefore the gradient of these transformed curves is parallel to the transformed boundary so that their normal derivative vanishes in the transformed geometry. Thus Neumann boundary conditions are preserved under analytic transformations.

Since a solution of (9.24) which satisfies Dirichlet or Neumann boundary conditions is unique, under an analytic mapping, the transformed solution subject to the transformed boundary conditions will also be unique in the new geometry.

9.2.3 Schwarz-Christoffel Transformation

This transformation is an analytic mapping (except for a few isolated points) from the upper half plane to the interior of a closed polygon. The closed polygon can be degenerate in the sense that some of its vertices may be at infinity. This type of polygon includes the type of interest here.

Let us consider the general case as illustrated in Fig. 9.5. Part of the x-axis from x_1 to $x_n = \infty$ is mapped onto the boundary of a closed polygon in the w-plane. We have also drawn unit tangent vectors **s** and **t** along corresponding boundary curves in the z and w-planes. We showed earlier that the angle which the transformed curve makes with the horizontal axis at a point is given by the angle which the original curve makes with its horizontal plus the argument of the derivative of the mapping. In this case, the original curve is along the x-axis in the positive sense and so makes zero angle with this axis. Therefore the transformed curve makes an angle with its axis given by $\arg(f'(z))$ where arg = argument of. Thus if the mapping has a constant argument between two consecutive points along the x-axis, the transformed boundary curve will have a constant argument also and therefore be a straight line. However, as the w value moves along the boundary through a point w_i where the polygon transitions from one side to another, the argument of the tangent abruptly changes value. At these points, the mapping cannot be analytic (or conformal). However, there are only n such points for an n sided polygon.

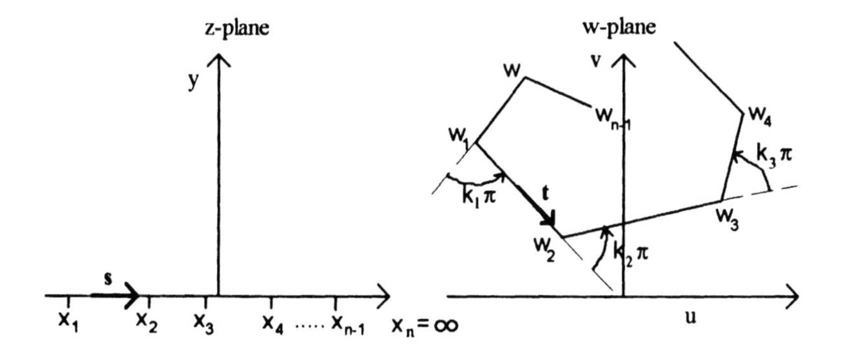

Figure 9.5 Schwarz-Christoffel mapping geometry

A mapping which has the above characteristics is given, in terms of its derivative, by

$$f'(z) = A(z - x_1)^{-k_1}(z - x_2)^{-k_2} \cdots (z - x_{n-1})^{-k_{n-1}} \tag{9.37}$$

where $x_1 < x_2 < \cdots < x_{n-1}$. Since

$$(z - x_i)^{-k_i} = \left(|z - x_i| e^{\arg(z - x_i)}\right)^{-k_i} = |z - x_i|^{-k_i} e^{-k_i \arg(z - x_i)} \tag{9.38}$$

and since the argument of a product of terms is the sum of their arguments,

$$\begin{aligned} &\arg(f'(z)) \\ &= \arg A - k_1 \arg(z - x_1) - k_2 \arg(z - x_2) \cdots k_{n-1} \arg(z - x_{n-1}) \end{aligned} \tag{9.39}$$

When $z = x < x_1$,

$$\arg(x - x_1) = \arg(x - x_2) = \cdots = \arg(x - x_{n-1}) = \pi \tag{9.40}$$

However, when $z = x$ moves to the right of x_1, $\arg(x - x_1) = 0$ but $\arg(x - x_i) = \pi$ for $i > 2$. Thus the argument of $f(z)$ abruptly changes by $k_1\pi$ as $z = x$ moves to the right of x_1. This is shown in Fig. 9.5. Similarly, when $z = x$ passes through x_2, $\arg(x - x_1) = \arg(x - x_2) = 0$ and all the rest equal π, so that $\arg(f'(z))$ jumps by $k_2\pi$. These jumps

are the exterior angles of the polygon traced in the w-plane. As such, they can be restricted to
$-\pi \le k_i\pi \le \pi$ so that

$$-1 \le k_i \le 1 \tag{9.41}$$

Since the sum of the exterior angles of a closed polygon equals 2π, we have for the point at infinity

$$k_n\pi = 2\pi - (k_1 + k_2 + \cdots + k_{n-1})\pi \tag{9.42}$$

so that

$$k_1 + k_2 + \cdots + k_n = 2 \tag{9.43}$$

Note that the point x_n could be a finite point, in which case it must be included in equation (9.37). However, the transformation is simplified if it is at infinity.

To obtain the Schwarz-Christoffel transformation, we must integrate (9.37) to get

$$w = f(z) = A\int_{z_0}^{z} (\zeta - x_1)^{-k_1}(\zeta - x_2)^{-k_2}\cdots(\zeta - x_{n-1})^{-k_{n-1}}\, d\zeta + B \tag{9.44}$$

The complex constants A and B and the x_i values can be chosen to achieve the desired map. There is some arbitrariness in the choice of these values and this freedom should be used to simplify the calculations. With this brief background, we now proceed to determine the mapping of interest here. Further details and proofs can be obtained by consulting Ref. [Chu60] or other standard books on complex variables.

9.2.4 Conformal Map for the Electrostatic Field Problem

Fig. 9.6 shows the w-plane geometry of interest for the electrostatic field problem and the corresponding z-plane boundary points. Note that some of the image points, w_2, w_3, and w_5 are at infinity, albeit in different directions in the complex plane. In assigning values to x_1, x_2, x_3, and x_4, we have taken advantage of the symmetry in the w-plane geometry. In fact the coordinate systems were chosen to exploit this symmetry. Note that, in Fig. 9.5 the positive direction for the exterior angles was chosen so that if we are moving along a side towards the next vertex, the angle increases if we make a left turn and decreases if we turn towards the right. In Fig. 9.6, as we move through the vertex w_1 from w_5, we turn towards the right by $90°$, hence $k_1 = -1/2$. Moving through w_2, we see that we make a $180°$ turn to the left, so $k_2 = 1$. Passing through w_3, we again make a $180°$ left turn so $k_3 = 1$. We turn right by $90°$ in going through w_4, so $k_4 = -1/2$. The angular change through w_5 is $180°$ to the left so $k_5 = 1$. Thus $k_1 + k_2 + k_3 + k_4 + k_5 = -1/2 + 1 + 1 - 1/2 + 1 = 2$ as required.

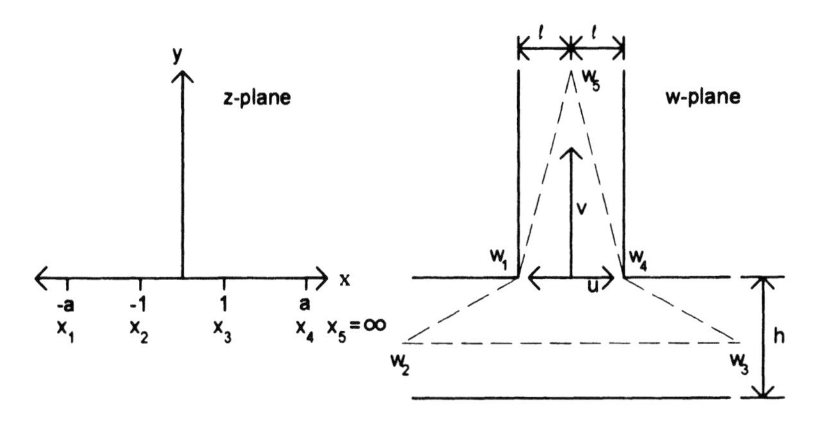

Figure 9.6 Schwarz-Christoffel Transformation for the electrostatic problem

The derivative of the transformation is, from (9.37),

$$f'(z) = A\frac{(z+a)^{1/2}(z-a)^{1/2}}{(z+1)(z-1)} = A\frac{\left(z^2-a^2\right)^{1/2}}{\left(z^2-1\right)} \tag{9.45}$$

where $a > 1$. The integral can be carried out by writing

$$\frac{1}{z^2-1} = \frac{1}{2}\left(\frac{1}{z-1} - \frac{1}{z+1}\right) \tag{9.46}$$

Substituting into (9.45), we get

$$f'(z) = \frac{A}{2}\left[\frac{\left(z^2-a^2\right)^{1/2}}{z-1} - \frac{\left(z^2-a^2\right)^{1/2}}{z+1}\right] \tag{9.47}$$

Using tables of integrals and the definitions of the complex version of the standard functions (See Ref. [Dwi61]), we can integrate this to obtain

$$w = f(z) = A\left\{\ln\left(2\sqrt{z^2 - a^2} + 2z\right) - \frac{\sqrt{a^2 - 1}}{2}\left[\sin^{-1}\left(\frac{z - a^2}{a(z-1)}\right) + \sin^{-1}\left(\frac{z + a^2}{a(z+1)}\right)\right]\right\} + B \tag{9.48}$$

where ln is the natural logarithm. This solution can be verified by taking its derivative.

In order to fix the constants, we must match the image points w_1, w_2, etc. to their corresponding x-axis points. Thus, we must get $w_1 = -\ell$, when $x_1 = -a$. Substituting into (9.48), we find

$$-\ell = A\left\{\ln(-2a) - \frac{\sqrt{a^2 - 1}}{2}\left[\sin^{-1}(1) + \sin^{-1}(-1)\right]\right\} + B \tag{9.49}$$

Since

$$\ln z = \ln|z| + i \arg(z) \tag{9.50}$$

(9.49) becomes

$$-\ell = A\left[\ln(2a) + i\pi\right] + B \tag{9.51}$$

since $\arg(-2a) = \pi$. Similarly, we must have $w_4 = \ell$ correspond to $x_4 = a$. Substituting into (9.48), we find

$$\begin{aligned} \ell &= A\left\{\ln(2a) - \frac{\sqrt{a^2 - 1}}{2}\left[\sin^{-1}(-1) + \sin^{-1}(1)\right]\right\} + B \\ &= A\ln(2a) + B \end{aligned} \tag{9.52}$$

Subtracting (9.51) from (9.52), we obtain $2\ell = -i\pi A$, so that

$$A = i\frac{2\ell}{\pi} \tag{9.53}$$

Adding (9.51) and (9.52), we obtain $0 = 2A \ln(2a) + i\pi A + 2B$. Using (9.53) and solving for B, we get

$$B = \ell - i\frac{2\ell}{\pi}\ln(2a) \tag{9.54}$$

Substituting A and B into (9.48), we obtain

$$w = i\frac{2\ell}{\pi}\left\{\ln\left(2\sqrt{z^2 - a^2} + 2z\right) - \frac{\sqrt{a^2-1}}{2}\left[\sin^{-1}\left(\frac{z-a^2}{a(z-1)}\right) + \sin^{-1}\left(\frac{z+a^2}{a(z+1)}\right)\right]\right\}$$
$$+ \ell - i\frac{2\ell}{\pi}\ln(2a) \tag{9.55}$$

The log terms can be combined, resulting in

$$w = i\frac{2\ell}{\pi}\left\{\ln\left(\frac{\sqrt{z^2-a^2}+z}{a}\right) - \frac{\sqrt{a^2-1}}{2}\left[\sin^{-1}\left(\frac{z-a^2}{a(z-1)}\right) + \sin^{-1}\left(\frac{z+a^2}{a(z+1)}\right)\right]\right\} + \ell \tag{9.56}$$

Further manipulation of the log term leads to

$$\ln\left(\frac{\sqrt{z^2-a^2}+z}{a}\right) = \ln\left(\sqrt{\left(\frac{z}{a}\right)^2 - 1} + \left(\frac{z}{a}\right)\right) = \ln\left(i\sqrt{1-\left(\frac{z}{a}\right)^2} + \left(\frac{z}{a}\right)\right)$$
$$= \ln\left[i\left(\sqrt{1-\left(\frac{z}{a}\right)^2} - i\left(\frac{z}{a}\right)\right)\right] = \ln(i) + \ln\left[\sqrt{1-\left(\frac{z}{a}\right)^2} - i\left(\frac{z}{a}\right)\right] \tag{9.57}$$

Now, substituting $\ln(i) = \ln(1) + i \arg(i) = i\,\pi/2$ and using the identity from Ref. [Dwi61],

$$\sin^{-1}(C) = i\ln\left(\pm\sqrt{1-C^2} - iC\right) \tag{9.58}$$

where C is a complex number, (9.57) becomes,

$$\ln\left(\frac{\sqrt{z^2-a^2}+z}{a}\right)=i\frac{\pi}{2}-i\sin^{-1}\left(\frac{z}{a}\right) \tag{9.59}$$

Substituting into (9.56), we get

$$w=\frac{2\ell}{\pi}\sin^{-1}\left(\frac{z}{a}\right)-i\frac{\ell\sqrt{a^2-1}}{\pi}\left[\sin^{-1}\left(\frac{z-a^2}{a(z-1)}\right)+\sin^{-1}\left(\frac{z+a^2}{a(z+1)}\right)\right] \tag{9.60}$$

To determine the constant a, we must use the correspondence between another set of points, such as x_2, w_2 or x_3, w_3. Using the x_3, w_3 pair, we note that as x approaches
$x_3 = 1$ from below, w must approach ∞ - ih. In (9.60) therefore let $z = x \to 1$ from below. Here we use another expression for the complex inverse sin function [Dwi61],

$$\sin^{-1}(x\pm iy)=\sin^{-1}\left(\frac{2x}{p+q}\right)\pm i\cosh^{-1}\left(\frac{p+q}{2}\right) \tag{9.61}$$

where $p=\sqrt{(1+x)^2+y^2}$, $q=\sqrt{(1-x)^2+y^2}$, and $y \geq 0$.

If $y = 0$ and $x > 1$, $p = 1 + x$, $q = x - 1$, and $p + q = 2x$. If $y = 0$ and $x < -1$, $p = -(1-x)$, $q = 1 - x$, and $p + q = -2x$. If $y = 0$ and $|x| < 1$, $p = 1 + x$, $q = 1 - x$, and $p + q = 2$. This latter case corresponds to the real sine function. Substituting $z \to 1$ into (9.60), we get

$$\infty-ih=\frac{2\ell}{\pi}\sin^{-1}\left(\frac{1}{a}\right)-i\frac{\ell\sqrt{a^2-1}}{\pi}\left[\sin^{-1}(\infty)+\sin^{-1}\left(\frac{1+a^2}{2a}\right)\right] \tag{9.62}$$

where the infinities must be interpreted in a limiting sense. Keeping this in mind and using (9.61), we find

$$\sin^{-1}(\infty) = \sin^{-1}(1)+i\cosh^{-1}\left(\frac{1+a^2}{2a}\right) = \frac{\pi}{2}+i\cosh^{-1}\left(\frac{1+a^2}{2a}\right) \tag{9.63}$$

Since

$$\frac{1+a^2}{2a} = \frac{2a+(a-1)^2}{2a} = 1+\frac{(a-1)^2}{2a} > 1 \tag{9.64}$$

we have

$$\sin^{-1}\left(\frac{1+a^2}{2a}\right) = \sin^{-1}(1)+i\cosh^{-1}\left(\frac{1+a^2}{2a}\right) = \frac{\pi}{2}+i\cosh^{-1}\left(\frac{1+a^2}{2a}\right) \tag{9.65}$$

Substituting into (9.62), and neglecting finite real terms compared with ∞, we get

$$\infty - i\,h = \infty - i\,\ell\sqrt{a^2-1} \tag{9.66}$$

Therefore $\sqrt{a^2-1} = h/\ell$ and solving for a,

$$a = \sqrt{1+\left(\frac{h}{\ell}\right)^2} \tag{9.67}$$

Thus (9.60) becomes

$$w = \frac{2\ell}{\pi}\sin^{-1}\left(\frac{z}{a}\right) - i\frac{h}{\pi}\left[\sin^{-1}\left(\frac{z-a^2}{a(z-1)}\right)+\sin^{-1}\left(\frac{z+a^2}{a(z+1)}\right)\right] \tag{9.68}$$

with a given by (9.67). Other points of correspondence can be checked for consistency by similar procedures.

9.2.4.1 Electric Potential and Field Values

In the w-plane which is the plane of interest for the electrostatic problem, the boundary values of the potential are shown in Fig. 9.7. Two conductors are at potentials V_1 and V_2 relative to a plane at zero potential. This can be done without loss of generality, since if the plane were not at zero potential, its value could be subtracted from all the potential values without altering the values of the electric field. The corresponding boundary values in the z-plane are also shown in Fig. 9.7.

We wish to solve the problem in the simpler z-plane geometry and then use the conformal map given by (9.68) to transfer the solution to the w-plane. A method of doing this is suggested by Fig. 9.8. The angles θ_1 and θ_2 are angles between the vectors from points 1 and -1 to z. In terms of these angles, a potential which satisfies the boundary conditions in the z-plane is given by

$$V = V_1 - \frac{V_1}{\pi}\theta_1 + \frac{V_2}{\pi}\theta_2 \tag{9.69}$$

When $z = x > 1$, θ_1 and θ_2 are zero, so $V = V_1$. When $-1 < z = x < 1$, $\theta_1 = \pi$ and $\theta_2 = 0$, so $V = 0$. When $z = x < -1$, $\theta_1 = \theta_2 = \pi$, so $V = V_2$. Thus the boundary conditions are satisfied.

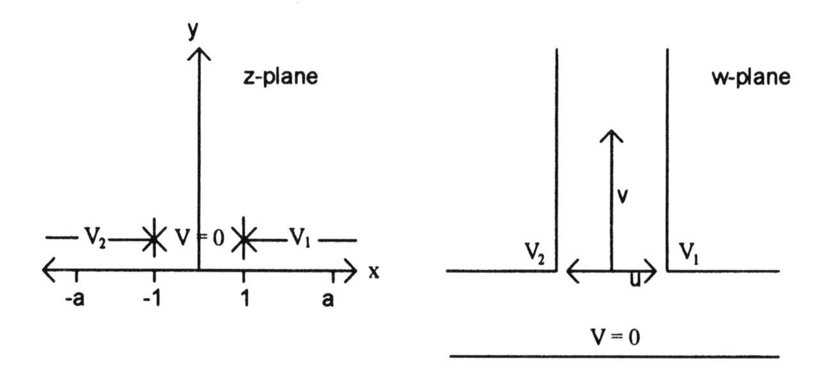

Figure 9.7 Correspondence between potential boundary values for the z and w-planes

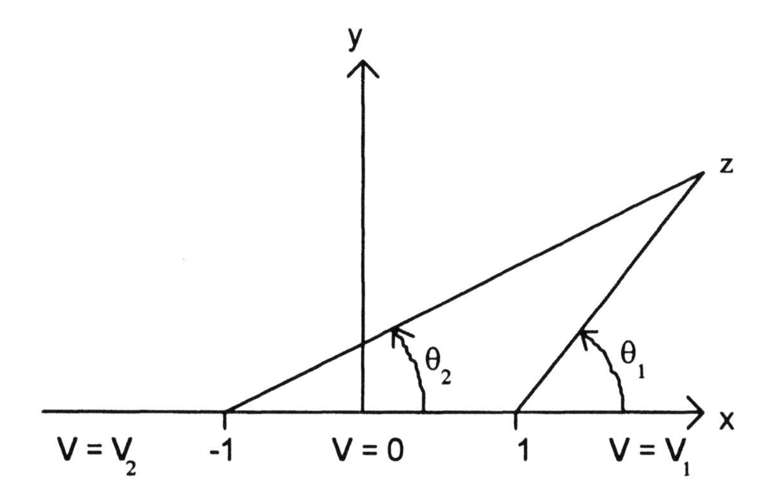

Figure 9.8 The potential function in the z-plane

We must now show that V is a harmonic function. Note that

$$\theta_1 = \arg(z-1) \quad , \quad \theta_2 = \arg(z+1) \tag{9.70}$$

Note also that

$$\ln(z \pm 1) = \ln|z \pm 1| + i\arg(z \pm 1) \tag{9.71}$$

Since the log function is analytic, its real and imaginary components are harmonic functions as was shown previously for analytic functions. Thus θ_1 and θ_2 are harmonic and so is V since sums of harmonic functions are also harmonic.

In terms of x and y where $z = x + iy$, (9.69) can be written

$$V = V_1 - \frac{V_1}{\pi}\tan^{-1}\left(\frac{y}{x-1}\right) + \frac{V_2}{\pi}\tan^{-1}\left(\frac{y}{x+1}\right) \tag{9.72}$$

Since the transformation (9.68) is analytic except at a few points, the inverse transformation is defined and analytic at all points where $w' = dw/dz \neq 0$. Thus V can be considered a function of u , v through the dependence of x and y on these variables via the inverse transformation. This will enable us to obtain expressions for the electric field in the w-plane. We have,

$$E_u = -\frac{\partial V}{\partial u} \quad , \quad E_v = -\frac{\partial V}{\partial v} \tag{9.73}$$

Referring to(9.72), note that

$$\frac{\partial}{\partial u}\tan^{-1}\left(\frac{y}{x-c}\right) = \frac{(x-c)\frac{\partial y}{\partial u} - y\frac{\partial x}{\partial u}}{(x-c)^2 + y^2} \tag{9.74}$$

Substituting $c = \pm 1$, corresponding to the two terms in (9.72), and applying a similar formula for the v derivative, we get for the electric field components,

$$E_u = \frac{V_1}{\pi}\left[\frac{(x-1)\frac{\partial y}{\partial u} - y\frac{\partial x}{\partial u}}{(x-1)^2 + y^2}\right] - \frac{V_2}{\pi}\left[\frac{(x+1)\frac{\partial y}{\partial u} - y\frac{\partial x}{\partial u}}{(x+1)^2 + y^2}\right]$$

$$E_v = \frac{V_1}{\pi}\left[\frac{(x-1)\frac{\partial y}{\partial v} - y\frac{\partial x}{\partial v}}{(x-1)^2 + y^2}\right] - \frac{V_2}{\pi}\left[\frac{(x+1)\frac{\partial y}{\partial v} - y\frac{\partial x}{\partial v}}{(x+1)^2 + y^2}\right] \tag{9.75}$$

The derivatives in these formulas can be determined by means of (9.29), (9.45), and (9.53). Thus

$$\frac{dz}{dw} = \frac{1}{dw/dz} = -i\frac{\pi}{2\ell}\frac{\left(z^2-1\right)}{\left(z^2-a^2\right)^{1/2}} \tag{9.76}$$

Expressing this in terms of x and y and separating into real and imaginary parts, we get

$$\frac{dz}{dw}=\frac{\pi}{2\ell}\left\{\frac{\left(x^2-y^2-1\right)^2+4x^2y^2}{\left[\left(x^2-y^2-a^2\right)^2+4x^2y^2\right]^{\frac{1}{2}}}\right\}^{\frac{1}{2}}$$

$$\times\left\{\sin\left[\tan^{-1}\left(\frac{2xy}{x^2-y^2-1}\right)-\frac{1}{2}\tan^{-1}\left(\frac{2xy}{x^2-y^2-a^2}\right)\right]\right. \tag{9.77}$$

$$\left.-i\cos\left[\tan^{-1}\left(\frac{2xy}{x^2-y^2-1}\right)-\frac{1}{2}\tan^{-1}\left(\frac{2xy}{x^2-y^2-a^2}\right)\right]\right\}$$

Now, using the uniqueness of the derivative when taken in the u or iv directions which led to the Cauchy-Riemann equations mentioned earlier, we can write

$$\frac{dz}{dw}=\frac{\partial x}{\partial u}+i\frac{\partial y}{\partial u}=\frac{\partial y}{\partial v}-i\frac{\partial x}{\partial v} \tag{9.78}$$

This, together with (9.77) can be used to extract the appropriate derivatives for use in (9.75). Thus

$$\frac{\partial x}{\partial u}=\frac{\partial y}{\partial v}=\frac{\pi}{2\ell}\left\{\frac{\left(x^2-y^2-1\right)^2+4x^2y^2}{\left[\left(x^2-y^2-a^2\right)^2+4x^2y^2\right]^{\frac{1}{2}}}\right\}^{\frac{1}{2}}$$

$$\times\sin\left[\tan^{-1}\left(\frac{2xy}{x^2-y^2-1}\right)-\frac{1}{2}\tan^{-1}\left(\frac{2xy}{x^2-y^2-a^2}\right)\right]$$

(9.79)

$$\frac{\partial y}{\partial u}=-\frac{\partial x}{\partial v}=-\frac{\pi}{2\ell}\left\{\frac{\left(x^2-y^2-1\right)^2+4x^2y^2}{\left[\left(x^2-y^2-a^2\right)^2+4x^2y^2\right]^{\frac{1}{2}}}\right\}^{\frac{1}{2}}$$

$$\times\cos\left[\tan^{-1}\left(\frac{2xy}{x^2-y^2-1}\right)-\frac{1}{2}\tan^{-1}\left(\frac{2xy}{x^2-y^2-a^2}\right)\right]$$

Thus the electric field in the w-plane can be expressed completely in terms of x and y via (9.75) and (9.79). However, we must invert equation (9.68) to do this. The first step is to express (9.68) in terms of x and y and in terms of its real and imaginary parts. Thus we write the first $\sin^{-1}$ term, using (9.61),

$$\sin^{-1}\left(\frac{z}{a}\right) = \sin^{-1}\left(\frac{x}{a}+i\frac{y}{a}\right) = \sin^{-1}\left[\frac{2\left(\frac{x}{a}\right)}{p+q}\right]+i\cosh^{-1}\left(\frac{p+q}{2}\right) \tag{9.80}$$

where

$$p=\sqrt{\left(1+\frac{x}{a}\right)^2+\left(\frac{y}{a}\right)^2}\ ,\quad q=\sqrt{\left(1-\frac{x}{a}\right)^2+\left(\frac{y}{a}\right)^2}$$

The second $\sin^{-1}$ term in (9.68) can be written similarly,

$$\begin{aligned}\sin^{-1}\left(\frac{z-a^2}{a(z-1)}\right) &= \sin^{-1}\left\{\frac{\left(x-a^2\right)+iy}{a\left[(x-1)+iy\right]}\right\}\\ &= \sin^{-1}\left\{\frac{\left(x-a^2\right)(x-1)+y^2+iy\left(a^2-1\right)}{a\left[(x-1)^2+y^2\right]}\right\}\\ &= \sin^{-1}\left\{\frac{2\left[\left(x-a^2\right)(x-1)+y^2\right]}{a\left[(x-1)^2+y^2\right](p_1+q_1)}\right\}+i\cosh^{-1}\left(\frac{p_1+q_1}{2}\right)\end{aligned} \tag{9.81}$$

where

$$\left.\begin{matrix}p_1\\ q_1\end{matrix}\right\}=\sqrt{\left[1\pm\frac{\left(x-a^2\right)(x-1)+y^2}{a\left[(x-1)^2+y^2\right]}\right]^2+\left[\frac{y\left(a^2-1\right)}{a\left[(x-1)^2+y^2\right]}\right]^2}$$

with the upper sign referring to p_1 and the lower sign to q_1. Treating the third $\sin^{-1}$ term in (9.68) similarly and separating into real and imaginary parts, we get

$$u = \frac{\ell}{\pi}\left\{2\sin^{-1}\left(\frac{2\left(\frac{x}{a}\right)}{p+q}\right) + \frac{h}{\ell}\left[\cosh^{-1}\left(\frac{p_1+q_1}{2}\right) - \cosh^{-1}\left(\frac{p_2+q_2}{2}\right)\right]\right\}$$

$$v = \frac{\ell}{\pi}\left\{2\cosh^{-1}\left(\frac{p+q}{2}\right)\right.$$

$$\left. - \frac{h}{\ell}\left[\sin^{-1}\left(\frac{2\left[\left(x-a^2\right)(x-1)+y^2\right]}{a\left[(x-1)^2+y^2\right](p_1+q_1)}\right) + \sin^{-1}\left(\frac{2\left[\left(x+a^2\right)(x+1)+y^2\right]}{a\left[(x+1)^2+y^2\right](p_2+q_2)}\right)\right]\right\} \tag{9.82}$$

where

$$a = \sqrt{1+\left(\frac{h}{\ell}\right)^2}$$

$$\left.\begin{matrix} p \\ q \end{matrix}\right\} = \sqrt{\left(1 \pm \frac{x}{a}\right)^2 + \left(\frac{y}{a}\right)^2}$$

$$\left.\begin{matrix} p_1 \\ q_1 \end{matrix}\right\} = \sqrt{\left[1 \pm \frac{\left(x-a^2\right)(x-1)+y^2}{a\left[(x-1)^2+y^2\right]}\right]^2 + \left[\frac{y\left(a^2-1\right)}{a\left[(x-1)^2+y^2\right]}\right]^2}$$

$$\left.\begin{matrix} p_2 \\ q_2 \end{matrix}\right\} = \sqrt{\left[1 \pm \frac{\left(x+a^2\right)(x+1)+y^2}{a\left[(x+1)^2+y^2\right]}\right]^2 + \left[\frac{y\left(a^2-1\right)}{a\left[(x+1)^2+y^2\right]}\right]^2}$$

The upper and lower signs refer to the p's and q's respectively.

Given u and v, (9.82) can be inverted by means of a Newton-Raphson procedure. Using this method, the problem reduces to one of solving the following equations,

$$f_1(x,y) = u(x,y) - u_o = 0$$
$$f_2(x,y) = v(x,y) - v_o = 0 \tag{9.83}$$

where u_o and v_o are the desired coordinates at which to evaluate the field in the w-plane. Since $\partial f_1/\partial x = \partial u/\partial x$, etc., the Newton-Raphson equations for the increments Δx, Δy are

$$\begin{pmatrix}\Delta x\\ \Delta y\end{pmatrix} = \frac{-1}{\left[\left(\frac{\partial u}{\partial x}\right)^2 + \left(\frac{\partial u}{\partial y}\right)^2\right]}\begin{pmatrix}\frac{\partial u}{\partial x} & -\frac{\partial u}{\partial y}\\ \frac{\partial u}{\partial y} & \frac{\partial u}{\partial x}\end{pmatrix}\begin{pmatrix}f_1(x,y)\\ f_2(x,y)\end{pmatrix} \tag{9.84}$$

where we have used the Cauchy-Riemann equations (9.26). At each iteration, we let $x_{new} = x_{old} + \Delta x$ and $y_{new} = y_{old} + \Delta y$ and stop when Δx and Δy are sufficiently small. The derivatives in (9.84) can be obtained by a procedure similar to that used to derive (9.79).

9.2.4.2 Calculations and Comparison with a Finite Element Solution

A computer program was written to implement the above procedure for calculating the electric field. According to the above formulas, the field is infinite at the corners of the conductors, however, because real corners are not perfectly sharp, the field remains finite in practice. In transformer applications, the conductors are usually covered with an insulating layer of paper and the remaining space is filled with transformer oil. Because the paper has a much higher breakdown stress than the oil, the field in the oil is usually critical for design purposes. The highest oil fields will occur at least a paper's thickness away from the corner of the highest potential conductor which we assume is at potential V_1. We have accordingly calculated the field at the three points shown in Fig. 9.9 which are a paper's thickness, d, away from the corner of the V_1 potential conductor.

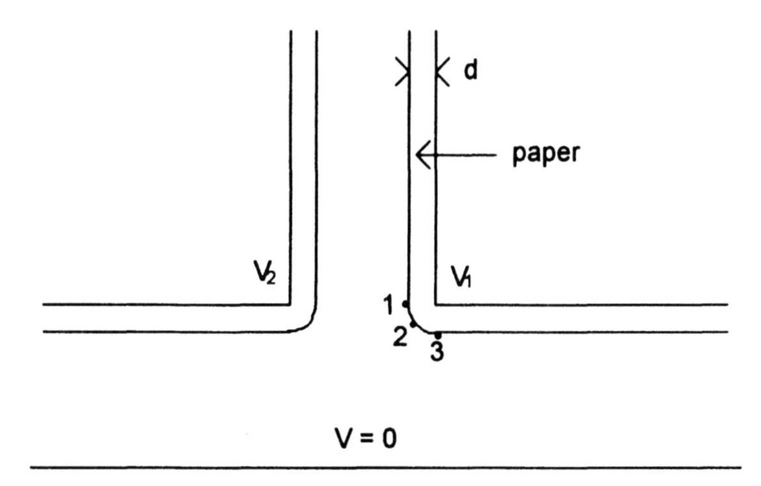

Figure 9.9 Points near the corner of the highest potential conductor where the electric field is calculated

It should be noted that the presence of the paper, which has a different dielectric constant than oil, will modify the electric field. In general, it will increase it in the oil and decrease it in the paper. We will attempt to estimate this oil enhancement factor later. Here, we wish to compare the analytic results as given by the above formulas with results from a finite element calculation without the presence of paper but with the fields calculated at the points shown in Fig. 9.9, a paper's distance away from the corner. The magnitudes of the fields, E, are compared, where

$$E = \sqrt{E_u^2 + E_v^2} \tag{9.85}$$

The results are shown in Table 9.1 for several different potential combinations and conductor separations. The agreement is very good, especially with the sharp corner finite element results. Also shown for comparison are finite element results for a 0.02" radius on the corners. These are also reasonably close to the analytic results and show that the sharpness or smoothness of the corners is washed out at distances greater than or equal to a paper's distance away. Although not shown in the table, the field deep in the gap between the V_1 and V_2 conductors was also calculated with the analytic formulas and produced the expected result, $E = (V_2 - V_1)/\ell$.

Table 9.1 Comparison of analytic and finite element electric field magnitudes at points 1 , 2 , 3 of Fig. 9.9. Enhancement factors are shown in parentheses.

Lengths - inches	Potentials - kV	Fields - kV / in	
	Conformal Mapping	Finite Element (sharp corner)	Finite Element (rounded corner) #
(a) $V_1 = 825$	$V_2 = 725$ $\ell = 0.26$	$h = 2.5$	$d = 0.025$
E_1	703	704 (1.08)	701 (1.12)
E_2	708	716 (1.22)	726 (1.14)
E_3	720	724 (1.08)	721 (1.14)
(b) $V_1 = 850$	$V_2 = 750$ $\ell = 0.2$	$h = 2.5$	$d = 0.02$
E_1	830	831 (1.06)	823 (1.14)
E_2	824 , 728*	737* (1.29)	836 (1.15)
E_3	828	828 (1.05)	828 (1.13)
(c) $V_1 = 800$	$V_2 = 800$ $\ell = 0.2$	$h = 2.5$	$d = 0.02$
E_1	288	290 (0.90)	312 (1.04)
E_2	338 , 309*	315* (1.17)	343 (1.16)
E_3	380	386 (1.04)	370 (1.09)
(d) $V_1 = 100$	$V_2 = -50$ $\ell = 0.5$	$h = 2.5$	$d = 0.02$
E_1	451	447 (1.07)	437 (1.14)
E_2	430 , 382*	384* (1.33)	429 (1.14)
E_3	410	406 (1.03)	406 (1.13)
* $d = 0.028$	# The rounded corner conductor radius = 0.02"		

Also shown in Table 9.1 are the enhancement factors in parentheses. These are obtained by performing the finite element calculations with paper having its normal dielectric constant $\varepsilon = 4.0$ and then repeating it with the paper layer given the dielectric constant of oil $\varepsilon = 2.2$ and taking the ratio of the field magnitudes in the two cases. This is normally greater than one, however in one case shown in the table it is 0.90. This is probably due to discretization inaccuracies in the finite element calculation or possibly in pin-pointing the exact location of where to evaluate the field in the two cases since this was done with the cursor. Note that there seems to be a tendency for greater enhancement factors at the corner point 2 than on either side of it for the sharp corner case.

The conformal mapping technique does not apply to the paper-oil situation so other approximate approaches must be employed to account for the oil enhancement. The equipotential line plot for a finite element calculation with paper present is shown in Fig. 9.10, with a blow-up of the region of interest shown in Fig. 9.11. This is for the conditions given in Table 9.1a with rounded corners. Fig. 9.12 is a similar blow-up for the conditions in Table 9.1a with sharp metal corners.

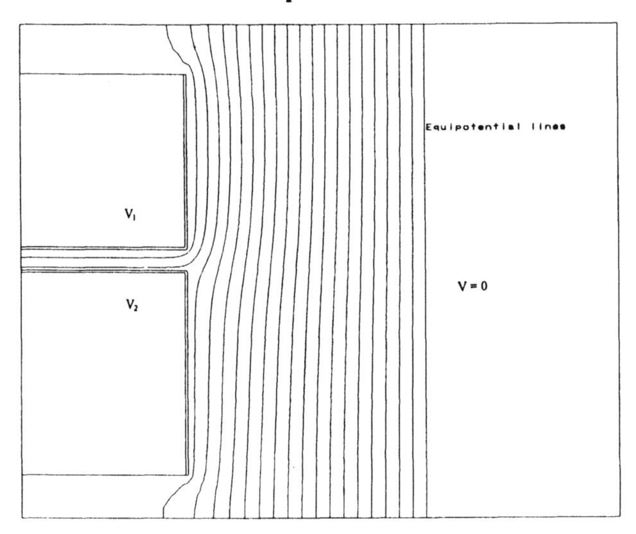

Figure 9.10 Equipotential line plot from a finite element calculation showing the full geometry for the conditions given in Table 9.1a with rounded corners

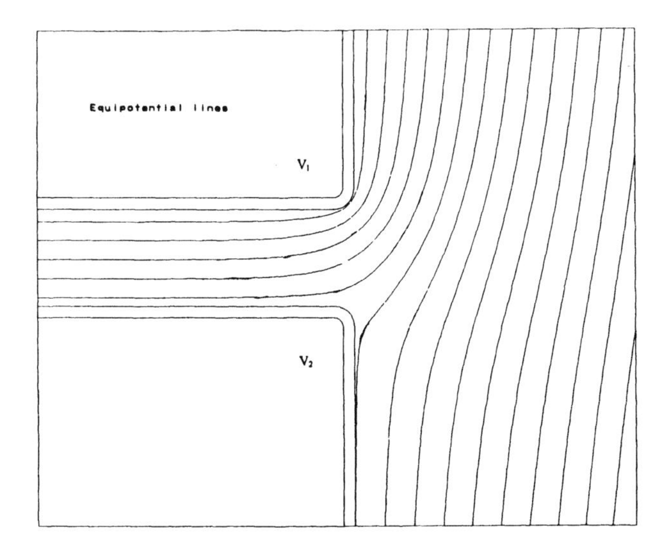

Figure 9.11 Equipotential line plot from a finite element calculation showing a blow-up of the region near the conductor's corners for the same conditions as Fig. 9.10

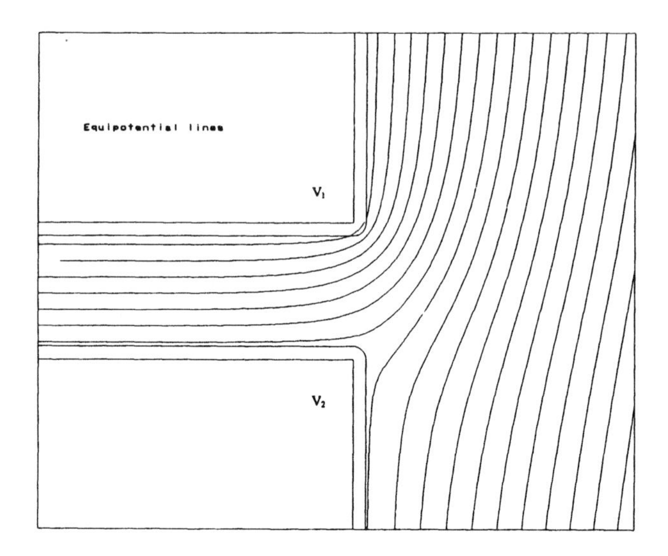

Figure 9.12 Equipotential line plot from a finite element calculation for the same conditions as Fig. 9.11 but with the conductors having sharp corners

9.2.4.3 **Estimating Enhancement Factors**

For a paper-oil layering in a planar geometry as shown in Fig. 9.13a, all the paper can be lumped into one layer for calculational purposes. Letting 1 refer to the paper layer and 2 to the oil layer, the enhancement factor $\eta = E / E_o$, ie. the ratio of the fields with and without a paper layer, is

$$\eta = \frac{1}{\varepsilon_2\left(\frac{f_1}{\varepsilon_1} + \frac{f_2}{\varepsilon_2}\right)} \tag{9.86}$$

where $f_1 = \ell_1/\ell$, $f_2 = \ell_2/\ell$ with $\ell = \ell_1 + \ell_2$ are the fractional lengths of materials 1 and 2 respectively. It is assumed that when the paper is absent, it is replaced by oil, keeping the total distance ℓ between the metal surfaces the same.

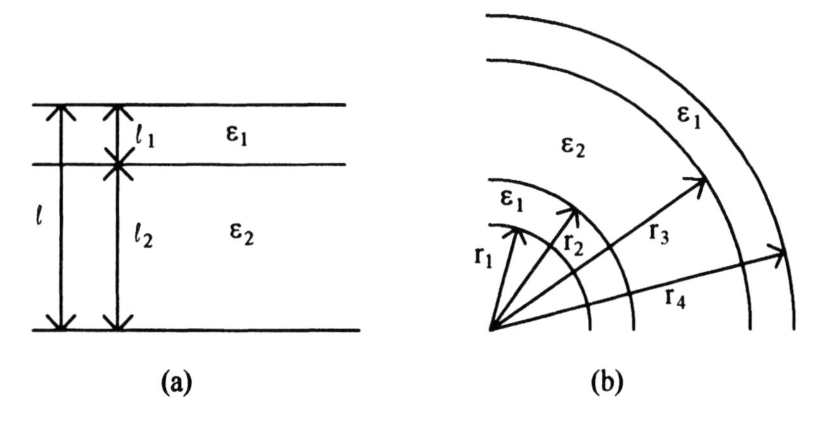

Figure 9.13 Paper-oil configurations for (a) planar and (b) cylindrical geometries resulting in enhanced fields in the oil compared with the all oil case

In the cylindrical case, shown in Fig. 9.13b, the position of the paper layer or layers is important in calculating the enhancement factors. We will assume that one paper layer is next to the inner conductor and one next to the outer conductor. When additional layers are present, such as pressboard barriers, their position must be known and they can be included in the calculation. The oil enhancement factor for the situation shown in Fig. 9.13b is

$$\eta = \frac{\ln\left(\frac{r_2}{r_1}\right) + \ln\left(\frac{r_3}{r_2}\right) + \ln\left(\frac{r_4}{r_3}\right)}{\varepsilon_2\left[\frac{1}{\varepsilon_1}\ln\left(\frac{r_2}{r_1}\right) + \frac{1}{\varepsilon_2}\ln\left(\frac{r_3}{r_2}\right) + \frac{1}{\varepsilon_1}\ln\left(\frac{r_4}{r_3}\right)\right]} \tag{9.87}$$

The cylindrical enhancement factor is larger than the planar enhancement factor for the same paper and oil layer thicknesses. It reduces to the planar case when r_1 becomes large.

The above enhancement factors refer to ideal geometries. For the geometry of interest as shown in Fig. 9.9, the enhancement factors reduce to the planar case for field points away from the corner and deep into the $V_1 - V_2$ gap or the $V_1 - 0$ or $V_2 - 0$ gaps. This is borne out by the finite element calculations. However, near the corner, the geometry is closer to the cylindrical case.

For a sharp corner, corresponding to $r_1 = 0$, (9.87) shows that $\eta \rightarrow \varepsilon_1 / \varepsilon_2 = 1.82$ for the oil-paper case. This is an upper limit on the enhancement factor and is more than enough to account for the sharp corner enhancement factors shown in Table 9.1. For our purposes, the rounded corner enhancement factors are the most relevant ones. Table 9.1 shows that these are nearly the same for the 3 corner points. Treating the rounded corner case as a cylindrical geometry, we can take $r_1 = 0.02"$. It is less clear what radius to use for the outer conductor. In Table 9.2, we calculate the planar and cylindrical oil enhancement factors for the two different gaps, where the outer conductor radius is taken to be the inner conductor radius plus the gap length which is ℓ or h for the two gaps (See Fig. 9.6). Thus, referring to Fig. 9.13b,

$$r_4(V_1 - V_2 \text{ gap}) = r_4' = r_1 + \ell$$
$$r_4(V_1 - 0 \text{ gap}) = r_4'' = r_1 + h \tag{9.88}$$

Note that the V_1 - 0 gap has only one layer of paper. We also show a weighted cylindrical enhancement factor, which uses a weighted average of the above two radii. The weighting is taken to be inversely proportional to the radius so that the outer conductor closest to the corner is given the highest weighting. This results in an effective outer conductor radius of

$$r_4 = \frac{2r_4' r_4''}{r_4' + r_4''} \tag{9.89}$$

These weighted enhancements come closest to the corner enhancements determined by the finite element calculations in most cases. We will therefore use a cylindrical enhancement factor with this weighted outer conductor radius in our design calculations if it is above the V_1 - V_2 gap planar enhancement factor. Otherwise the V_1 - V_2 gap planar enhancement factor will be used.

Table 9.2 Calculated oil enhancement factors for the cases in Table 9.1.

Case	Planar		Cylindrical		Weighted Cylindrical	Table 9.1 Ave Rounded
	V_1 -V_2 Gap	V_1 - 0 Gap	V_1 -V_2 Gap	V_1 - 0 Gap		
(a)	1.095	1.004	1.182	1.081	1.137	1.133
(b)	1.099	1.004	1.174	1.069	1.125	1.140
(c)	1.099	1.004	1.174	1.069	1.125	1.096
(d)	1.037	1.004	1.113	1.069	1.094	1.137

The conformal mapping calculations of the electric field near the corner of one of a pair of conductors at different potentials relative to a neighboring ground plane agree well with calculations obtained by means of a finite element program. Since the electric field is infinite at the corner of a perfectly sharp conductor, the calculations were compared a small distance from the corner, taken to be the thickness of a paper layer in an actual transformer winding. The finite element calculations were made with sharp and rounded corners and the resulting fields were nearly the same a paper's thickness away from the corner. Since the oil breakdown fields are the most critical in transformer design and these occur beyond the paper's thickness, this result shows that a conformal mapping calculation is appropriate for determining these fields.

The only problem with the conformal mapping approach is that it does not take into account the different dielectric constants of oil and paper. We have proposed a method to take these approximately into account by means of an oil enhancement factor. By comparing with a finite element calculation, a formula was developed to obtain this enhancement factor.

9.3 FINITE ELEMENT ELECTRIC FIELD CALCULATIONS

Finite element methods permit the calculation of electric potentials and fields for complicated geometries. Modern commercial finite element codes generally provide a set of drawing tools which allow the user to input the geometry in as much detail as desired. More sophisticated versions even allow parametric input so that changes in one or more geometric parameters such as the distance between electrodes can be easily accomplished without redoing the entire geometry. Both 2 and 3 dimensional versions are available although the input to the 3D versions is, of course, much more complicated. For many problems, a 2D geometry can be an adequate approximation to the real configuration. 2D versions usually allow an axisymmetric geometry by inputting a cross-section of it. It this sense, it is really solving a 3D problem which happens to have cylindrical symmetry. The x-y 2D geometry is really modelling an infinitely long object having the specified 2D cross-section.

The basic geometry which the user inputs is then subdivided into a triangular mesh. Smaller triangles are used in regions where the potential is expected to change most rapidly. Larger triangles can adequately describe more slowly varying potential regions. Some programs automatically perform the triangular meshing and, through an iterative process, even refine the mesh in critical regions until the desired solution accuracy is achieved. When linear triangles are used, the potential is solved for only at the triangle nodes and a linear interpolation scheme is used to approximate it inside the triangle. For higher order triangles, additional nodes are added per triangle and higher order polynomial approximations are used to find the potential inside the triangles. Some programs use only second order triangles since these provides sufficient accuracy for reasonable computer memory and execution times.

Some art is required even for the geometric input. Very often, complete detail is unnecessary to a determination of the fields in critical regions. Thus the user must know when it is reasonable to ignore certain geometric details which are irrelevent to the problem. This not only saves on the labor involved in inputting the geometry but it can also considerably reduce required computer memory and solution times.

Finite element programs require that the user input sources and the appropriate boundary conditions for the problem at hand. In the case of electric potential calculations, the sources are electric charges and the boundary conditions include specifying the voltage at one or more electrode surfaces. These are often the sufaces of metallic objects and thus have a constant potential throughout. It is therefore unnecessary to model their interiors. Typically, the program will allow the user to declare such metallic objects nonexistant so that the solution is not solved for over their interiors. Their surface, however, is still included as an equipotential surface. Sometimes the equipotential surface is a boundary surface so it already has no interior. A metallic object can also be allowed to float so that its potential is part of the problem solution.

On external boundaries where no specification is made, the assumption is that these have natural or Neumann boundary conditions. This means that the normal derivative of the potential vanishes along them. This implies that the potential lines (in 2D) or surfaces (in 3D) enter the boundary at right angles. These types of boundary are usually used to express some symmetry condition. For instance a long conducting cylinder centered inside a rectangular grounded box can by modelled by means of a circle inside the box. However, by symmetry, only a quarter of the geometry, centered on the circle need be modelled and natural boundary conditions imposed on the new boundaries created by isolating this region. This is depicted in Fig. 9.14

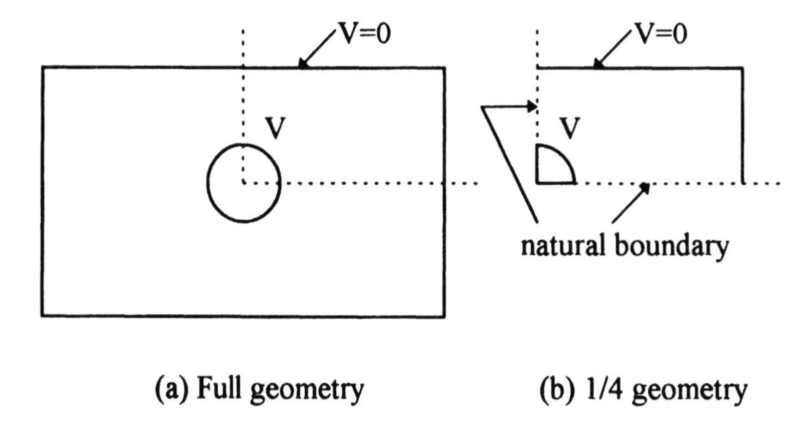

Figure 9.14 Using symmetry to simplify the finite element problem

The method works here because the potential lines are concentric circles about the conducting cylinder and therefore enter the dotted line boundaries in the 1/4 geometry at right angles. Although the problem depicted in Fig. 9.14 is relatively simple to solve in the full geometry so that the use of symmetry does not save much in input or solution times, this technique can save much effort when more complicated geometries are involved.

Another boundary condition which some programs allow is the balloon boundary. This type of boundary is like specifying that the boundary doesn't exist and the solution continues beyond it as if there were empty space out to infinity in that direction. There is a practical necessity for this type of boundary condtition since the finite element technique requires that the entire solution space of interest be subdivided into triangles or elements. When this solution space extends infinitely far or far enough that the geometric region of interest would be dwarfed relative to the whole space, it is convenient to specify such balloon boundaries rather that model vast regions of empty space.

We now give some examples of soving electrostatic problems with a finite element program. We use Ansoft's Maxwell® 2D software [Ansoft]. The first is a varistor stack assembly shown in Fig. 9.15. This is an axisymmetric geometry. The varistors themselves are modelled as two continuous cylinders separated by a metallic region in the center which is floating. The ends consist of shaped metal electrodes to help reduce the end fields. There are pressboard disks which mechanically hold the assembly together inside a pressboard cylinder. There is another pressboard cylinder outside followed by the ground cylinder which defines the outermost boundary. The top and bottom boundaries are balloon boundaries. Although the geometry of the varistor stack itself ir really cylindrical, the ground may not, in fact, be a concentric cylinder as modelled. For instance, it may be a tank wall. However, it is far enough away that a cylindrical approximation is reasonable.

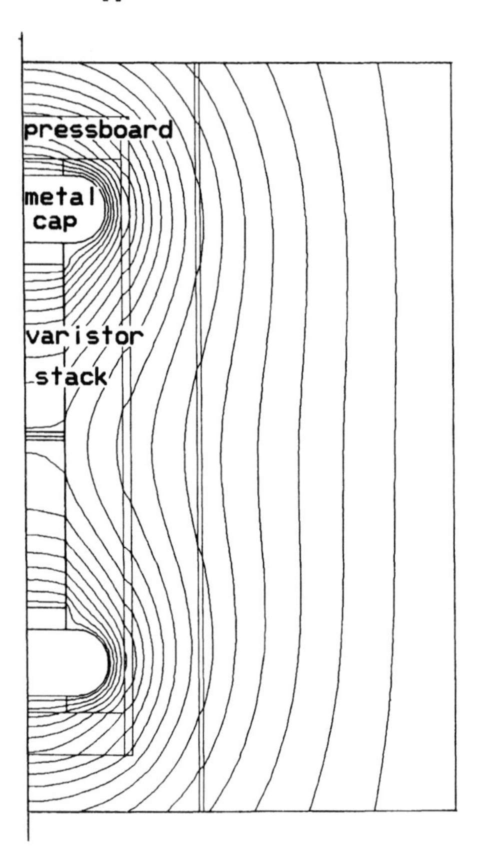

Figure 9.15 Axisymmetric model of a varistor stack with equipotential lines

The varistor cylinders, the pressboard elements, and the oil were given appropriate permittivities. The metallic end caps and center region were declared nonexistant. Different voltages were specified on the two end caps and the outer boundary cylinder was given a zero voltage. The equipotential lines obtained by solving this problem are shown. Since the electric field is the gradient of the potential, where the lines are closely spaced, the field is highest. The field itself could be calculated and displayed vectorially, if desired. The field can also be obtained numerically at specified points along with the maximum field value.

The second example is of two long parallel cylindrical cables. Using symmetry, only the top half of one of these cables actually needs to be modelled as shown in Fig. 9.16. This is an x-y geometry so that it is assumed to be infinitely long in the direction into the page. The metallic portion of the cable is surrounded by an insulating layer having a certain permittivity. The outside space is filled with oil, having a different permittivity. The metallic portion of the cylinder is declared nonexistant and its boundary is given half the potential difference between the two cylinders. This allows one to make the left boundary line a zero potential line since it is assumed to bisect the distance between the two cylinders. The Top and right boundary lines are balloon boundaries. The bottom boundary line is unspecified (natural boundary conditions assumed) since we are taking advantage of the symmetry. The paper layer around the cable is actually subdivided into two layers separated by a thin aluminum layer which is floating.

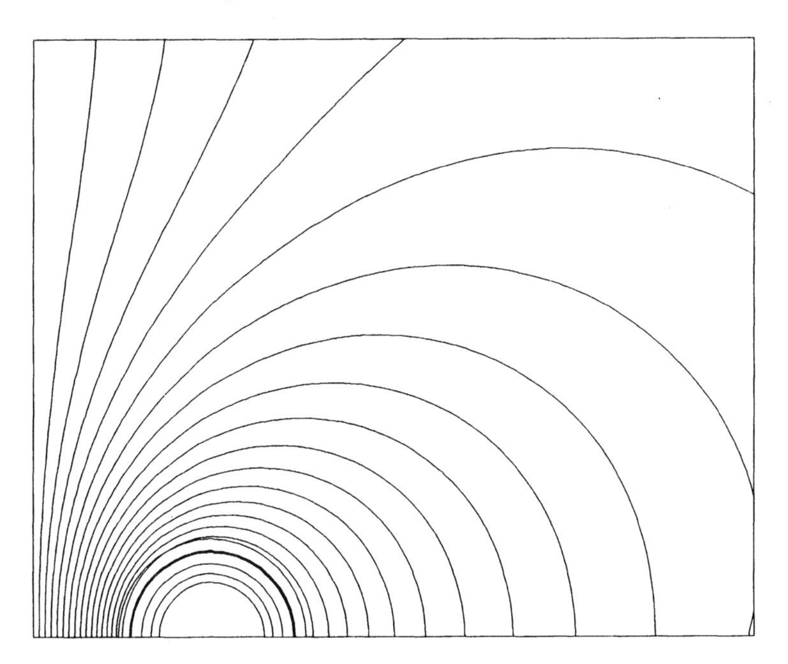

Figure 9.16 Model of two parallel cylindrical cables, using symmetry to simplify it. The equipotential lines are shown.

The equipotential solution lines are shown. They are closest together along a line connecting the center lines of the two cylinders as might be anticipated so that the field is highest here. A closer inspection shows that the highest field occurs in the oil at the surface of the paper along this line. As expected, the potential lines enter the bottom boundary line at right angles.

10. LOSSES

Summary Transformer losses comprise a small percentage of the power throughput in a transformer. Yet these losses can produce localized heating which can compromise its operation. It is important to be able to calculate these losses at the design stage so that adequate cooling can be provided. In addition, such calculations and their parameter dependencies can suggest ways of reducing these losses should that be necessary based on cost considerations or design feasibility. There are two main categories of losses, no-load and load losses. No-load losses are basically core losses associated with energizing the transformer and driving flux through the core. Load losses are further subdivided into I^2R losses and stray losses. The I^2R losses are resistive losses in the windings and leads caused by the main current flow. The stray losses are the result of the stray flux from the windings or leads impinging on metal parts such as the tank walls, the clamps, and even the windings themselves, resulting in induced eddy currents. We present formulas or methods for obtaining these losses in this report.

10.1 INTRODUCTION

Transformer losses are broadly classified as no-load and load losses. No-load losses occur when the transformer is energized with its rated voltage at one set of terminals but the other sets of terminals are open circuited so that no through or load current flows. In this case, full flux is present in the core and only the necessary exciting current flows in the windings. The losses are predominately core losses due to hysteresis and eddy currents produced by the time varying flux in the core steel. Load losses occur when the output is connected to a load so that current flows through the transformer from input to output terminals. Although core losses also occur in this case, they are not considered part of the load losses. When measuring load losses, the output terminals are shorted to ground and only a small impedance related voltage is necessary to produce the desired full load current. In this case, the core losses are small because of the small core flux and do not significantly add to the measured losses.

Load losses are in turn broadly classified as I^2R losses due to Joule heating produced by current flow in the coils and as stray losses due to the stray flux as it encounters metal objects such as tank walls, clamps or bracing structures, and the coils themselves. Because the coil conductors are often stranded and transposed, the I^2R losses are usually determined by the d.c. resistance of the windings. The stray losses depend on the conductivity, permeability, and shape of the metal object encountered. These losses are primarily due to induced eddy currents in these objects. Even though the object may be made of ferromagnetic material, such as the tank walls and clamps, their dimensions are such that hysteresis losses tend to be small relative to eddy current losses.

Although losses are usually a small fraction of the transformed power (< 0.5% in large power transformers), they can produce localized heating which can compromise the operation of the transformer. Thus it is important to understand how these losses arise and to calculate them as accurately as possible so that, if necessary steps can be taken at the design stage to reduce them to a level which can be managed by the cooling system. Other incentives, such as the cost which the customer attaches to the losses, can make it worthwhile to find ways of lowering the losses.

Modern methods of analysis, such as finite element or boundary element methods, have facilitated the calculation of stray flux losses in complex geometries. These methods are not yet routine in design because they require a fair amount of geometric input for each new geometry. They can, however, provide useful insights in cases where analytic methods are not available or are very crude. Occasionally a parametric study using such methods can extend their usefulness beyond a specialized geometry. We will explore such methods, in particular the finite element method, when appropriate. However, we are largely concerned here with analytic methods which can provide useful formulas covering wide parameter variations.

10.2 NO-LOAD OR CORE LOSSES

Cores in power transformers are generally made of stacks of electrical steel laminations. These are usually in the range of 0.23-0.46 mm (9-18 mils) in thickness and up to about 1 meter (40 inches) wide or as wide as can be accommodated by the rolling mill. Modern electrical steels have a silicon content of about 3% which gives them a rather high resistivity, ~ 50×10^{-8} Ω-m. Although higher silicon content can produce even higher resistivity, the brittleness increases with silicon content and this makes it difficult to roll them in the mill as well as to handle them after. Special alloying, rolling, and annealing cycles produce the highly oriented Goss texture (cube on edge) with superior magnetic properties such as high permeability along the rolling direction. Thus it necessary to consider their orientation in relation to the flux direction when designing a core.

Although the thinness of the laminations and their high resistivity are desirable characteristics in reducing (classical) eddy current losses, the high degree of orientation (>95%) produces large magnetic domains parallel to the rolling direction as sketched in Fig. 10.1. The lines between domains with magnetizations pointing up and down are called domain walls. These are narrow transition regions where the magnetization vector rotates through 180°. During an a.c. cycle, the up domains increase in size at the expense of the down domains during one part of the cycle and the opposite occurs during another part of the cycle. This requires the domain walls to move in the direction shown in the figure for increasing up magnetization. As the domain walls move, they generate eddy current losses. These losses were calculated by Ref. [Pry58] for the idealized situation shown in the figure. They found that these losses were significantly higher than the losses obtained from a classical eddy current calculation which assumes a homogeneous mixture of many small domains. These non-classical losses depend on the size of the domains in the zero magnetization state where there are equal sized up and down domains. This is because the maximum distance the walls move and hence their velocity depends on the zero magnetization domain size. The larger this size and hence the greater the domain wall's velocity, the greater the loss.

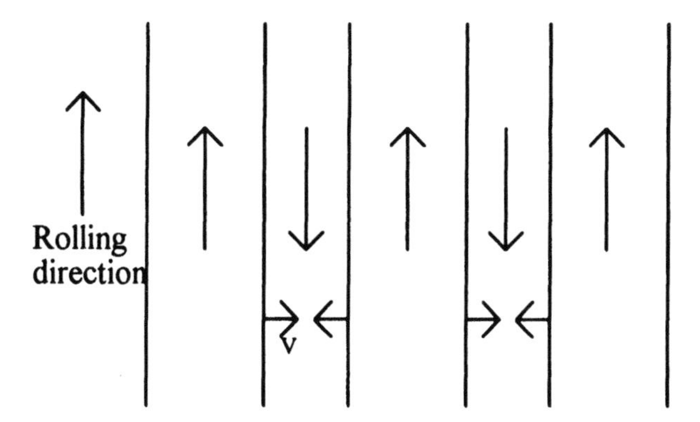

Figure 10.1 Idealized magnetic domain pattern in highly oriented electrical steel. The up and down arrows show the magnetization direction. The side pointing small arrows show the direction of domain wall motion for increasing up magnetization. v is their velocity.

In order to decrease the non-classical eddy current losses, it is therefore necessary to reduce the domain size. This is accomplished in practice by laser or mechanical scribing. A laser or mechanical stylus is rastered across the domains (perpendicular to their magnetization direction) at a certain spacing. This introduces localized stress at the surface since the scribe lines are not very deep. The domain size is dependent on the stress distribution in the laminations. Localized stresses help to refine the domains. Thus, after scribing, the laminations are not annealed since this would relieve the stress. Fig. 10.2 shows the domain pattern in an oriented electrical steel sample before and after laser scribing. The domain patterns are made visible by means of specialized optical techniques. One can clearly see the reduction in domain size as a result of laser scribing in this figure. The losses were reduced by ~ 12% as a result of laser scribing in this example.

Figure 10.2 Effect of laser scribing on the domain wall spacing of oriented electrical steel. Left side - before scribing, right side - after scribing. Courtesy of Armco Inc. With permission.

Another type of loss in electrical steels is hysteresis loss. This results from the domain walls encountering obstructions during their motion. At an obstruction, which can be a crystal imperfection, an occlusion or impurity, or even a localized stress concentration, the domain wall is pinned temporarily. However, because of the magnetizing force driving its motion, it eventually breaks away from the pinning site. This process occurs very suddenly and the resulting high wall velocity generates localized eddy currents. These localized eddy current losses are thought to be the essence of what are called hysteresis losses. Thus all losses in electrical steel are eddy current in nature. These hysteresis losses occur even at very low, essentially d.c., cycle rates. This is because, although the domain walls move very slowly until they encounter an obstacle, the breakaway process is still very sudden. Thus in loss separation studies, the hysteresis losses can be measured independently by going to low cycle rates, whereas the total loss, including hysteresis, is measured at high cycle rates. In high quality electrical steel, the hysteresis and eddy current losses contribute about equally to the total loss.

The manufacturer or supplier of electrical steel generally provides the user with loss curves which show the total loss per kilogram or pound as a function of induction at the frequency of interest, usually 50 or 60 Hz. One of these curves is shown in Fig. 10.3a. This curve is generally measured under ideal conditions, i.e. low stress on the laminations, and uniform, unidirectional, and sinusoidal flux in the laminations, so that it represents the absolute minimum loss per kg or lb to be expected in service. Another useful curve which the manufacturer can provide is a curve of the exciting power per unit weight versus induction at the frequency of interest. A sample curve is shown in Fig. 10.3b. Again this is an idealized curve, but it can be useful in estimating the power and current needed to energize the transformer.

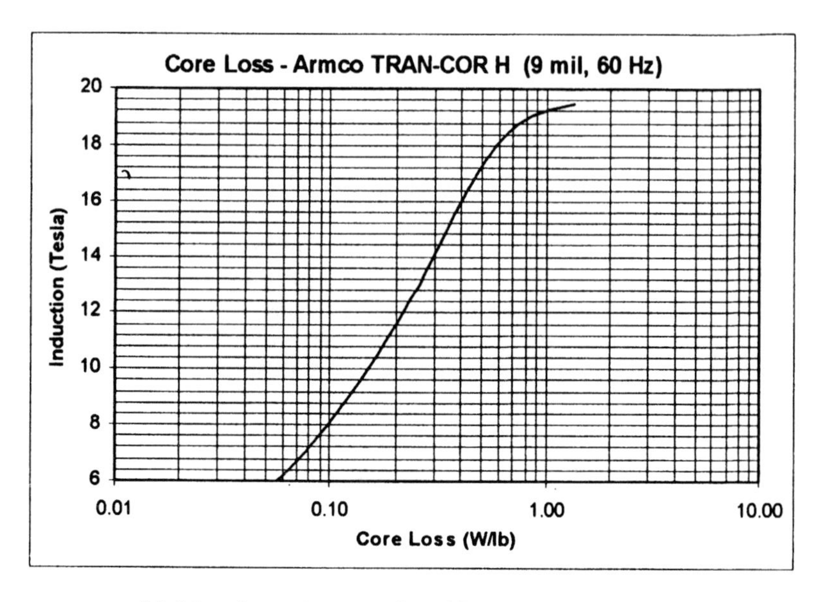

(a) Manufacturer's curve of specific core loss vs induction

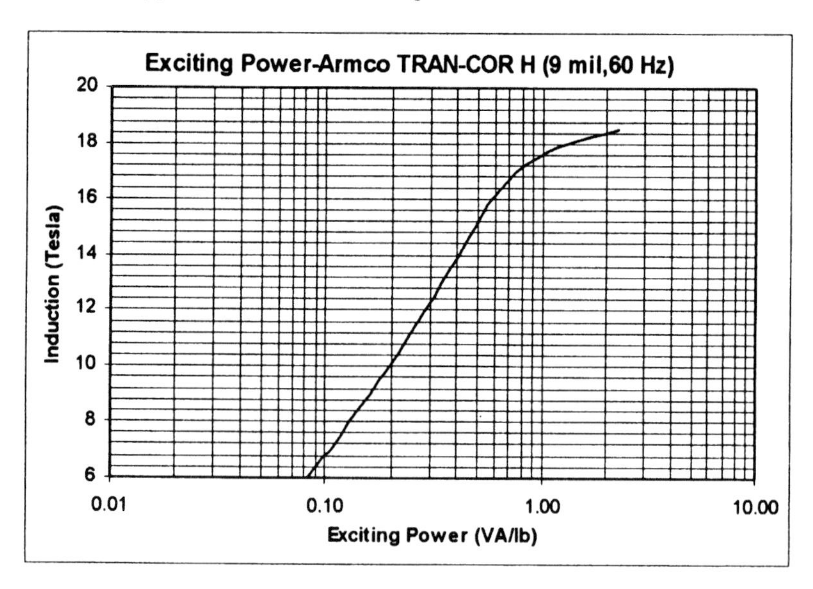

(b) Manufacturer's curve of specific exciting power vs induction

Figure 10.3 Graphs of core loss and exciting power based on a polynomial fit to data provided by Armco Inc. With permission. TRAN-COR H® is a registered trademark of Armco Inc.

10.2.1 Building Factor

As mentioned previously, the specific core losses at the operating induction provided by the manufacturer are minimum expected losses so that multiplying by the core weight produces a total core loss lower than what is measured in practice. The discrepancy is a result of the fact that stacked cores require joints where the induction must not only change direction but must bridge a gap between different laminations and also because stresses are produced in the steel due to cutting and stacking operations. There are other causes of this discrepancy such as burrs produced by cutting, but all of these can be lumped into a building factor which is simply a number which multiplies the ideal core loss to produce the measured core loss. These building factors are generally in the range of 1.2 - 1.4 and are roughly constant for a given core building practice.

Many attempts have been made to understand these extra losses, particularly in the region of the core joints, and have led to improved joint designs such as a step-lapped joint where the joint is made gradually in a step like manner. In a 3 phase core, studies have shown that near the joints where the flux changes direction by 90°, the induction vector rotates and higher losses are generated. Thus another approach to calculating the building factor is to apply a multiplier to the ideal losses for the amount of steel in the joint region only. This joint multiplier would be higher that the average multiplier and could be as high as ~ 1.7. The advantage of this approach is that cores having different fractions of their overall weight in the joint regions should receive more accurate average multipliers which reflect this difference.

10.2.2 Interlaminar Losses

The core laminations are coated with a glass-like insulating material. This is usually very thin, on the order of a few microns, to keep the space factor reasonably high (>96%). Like any other material, the coating is not a perfect insulator. Thus eddy currents, driven by the bulk flux in the core, can flow across the stacked laminations which comprise the core, i.e. normal to their surfaces. This is sketched in Fig. 10.4 for a rectangular cross-section core. Of course the eddy current paths are completed within the laminations where the resistance is much lower. The coating must be a good enough insulator to keep these losses low relative to the normal intralaminar losses. The insulative value of the coating is determined not only by the intrinsic resistivity of the coating material, which must be high, but also by its thickness. Although the thickness is generally not perfectly uniform, it should not vary so much that bald spots are produced. In high quality electrical steels, two types of coatings are generally applied, a first glass-like coating and a second coat of special composition designed to apply a favorable stress to the steel. The two coatings make the occurrence of bald spots unlikely.

The insulating value of the coating is determined by measuring the resistance across a stack of laminations or ideally a single lamination as shown in Fig. 10.4c. In terms of the parameters shown, the effective resistance across a lamination of area A is

$$R_{eff} = R_{coating} + R_{steel} = \frac{\rho_c t_c}{A} + \frac{\rho_s t_s}{A} = (\rho_c f_c + \rho_s f_s)\frac{t}{A} = \rho_{eff}\frac{t}{A} \quad (10.1)$$

where ρ_c is the coating resistivity and t_c its 2 sided thickness, ρ_s is the steel resistivity and t_s its thickness, $t = t_c + t_s$ is the combined thickness, f_c and f_s are fractional thicknesses of the coating and steel. It is often convenient to express the result in terms of a surface resistivity, $\sigma_{surf} = \rho_{eff} t$, where t is the thickness of one lamination. Its units are Ω-m^2 in the SI system.

We now estimate the interlaminar losses with the help of the geometry shown in Fig. 10.4b. Assume a rectangular stack of core steel of width w and height h and a uniform sinusoidal flux in the stack with peak induction B_o and angular frequency ω. We use Faraday's law

$$\oint \mathbf{E} \cdot d\mathbf{l} = -\frac{d\Phi}{dt} = -\omega(4xy)B_o \quad (10.2)$$

applied to the rectangle of area 4xy shown dotted in the figure. By symmetry and Lenz's law, the electric field points as shown on the two vertical sides of the rectangle. The electric field is nearly zero along the horizontal sides since these occur within the metallic laminations. Thus, from (10.2), we have

$$E(4y) = -\omega(4xy)B_o$$

so that the magnitude of the electric field E is given by

$$E = -\omega B_o x \quad (10.3)$$

pointing down on the right and up on the left sides of the rectangle. In these directions, we also have

$$E = \rho_{eff} J \quad (10.4)$$

where J is the current density and ρ_{eff} the effective resistivity perpendicular to the stack of laminations derived previously.

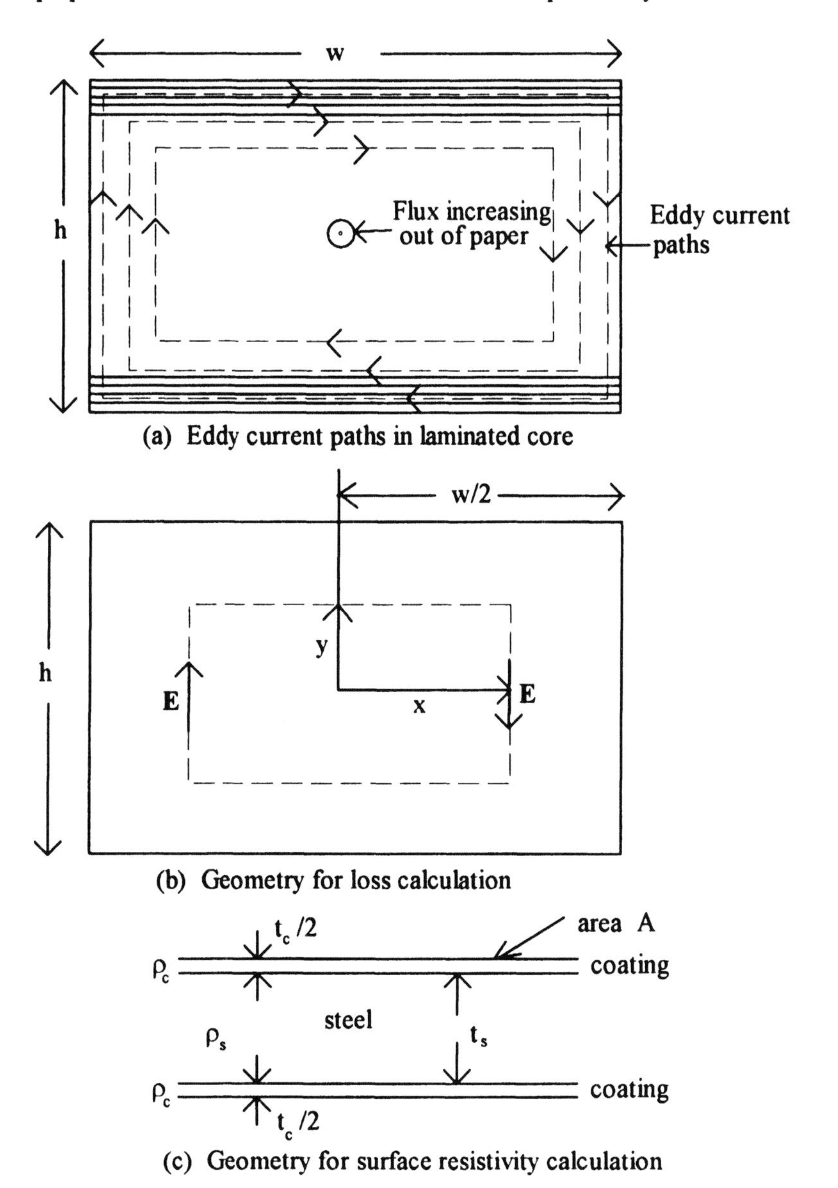

Figure 10.4 Interlaminar eddy currents produced by the bulk core flux

The interlaminar losses are given by

$$\text{Loss}_{\text{int}} = \frac{1}{2}\int \rho_{\text{eff}} J^2 dV \tag{10.5}$$

where the volume integral is over the lamination stack, assumed uniform in the direction into the paper (which we take to have unit length). The factor of 1/2 comes from time averaging J which is assumed to be expressed in terms of its peak value. Thus from (10.3) - (10.5), we get

$$\text{Loss}_{\text{int}} = \frac{\omega^2 B_o^2 h}{\rho_{\text{eff}}}\int_0^{w/2} x^2 dx = \frac{\omega^2 B_o^2 h w^3}{24\rho_{\text{eff}}} = \left(\frac{\pi^2}{6}\right)\frac{f^2 B_o^2 h w^3}{\rho_{\text{eff}}} \tag{10.6}$$

where we have used $\omega = 2\pi f$. The specific loss (loss per unit volume) is given by

$$P_{\text{int}} = \frac{\text{Loss}_{\text{int}}}{hw} = \left(\frac{\pi^2}{6}\right)\frac{f^2 B_o^2 w^2}{\rho_{\text{eff}}} \tag{10.7}$$

since we assumed unit length in the other dimension. To find the loss per unit weight or mass, divide by the density of the core steel in the appropriate units. Equation (10.7) is in the SI system where w is in meters, B_o in Tesla, ρ_{eff} in Ω-m, f in Hz, and P_{int} in Watts/m^3.

The interlaminar loss should be compared with the normal loss at the same peak induction. For typical values of the parameters, it is generally much smaller than the normal loss and can be ignored. As a numerical example, let $f = 60$ Hz, $w = 0.75$ m, $B_o = 1.7$ T, $\rho_{\text{eff}} = 20$ Ω-m. We get $P_{\text{int}} = 481$ W/m^3. The density of electrical steel is 7650 kg/m^3 (16870 lb/m^3) so that $P_{\text{int}} =$ 0.063 W/kg (0.029 W/lb). This is a fairly small loss compared with the normal losses at 1.7 T of ~ 1.3 W/kg = 0.60 W/lb. However, a high enough interlaminar resistance must be maintained to achieve these low losses.

10.3 LOAD LOSSES

10.3.1 I^2R Losses

I^2R losses in the coil conductors are generally the dominant source of load losses. They are normally computed using the d.c. value of resistivity. However, in the case of wires with large cross-sectional areas carrying a.c. current this normally requires that they be made of stranded and transposed conductors. To get a feeling for how a.c. current affects resistance, consider the resistance of an infinitely long cylinder of radius a , permeability μ , and d.c. conductivity $\sigma = 1/\rho$ where ρ is the d.c. resistivity. Let it carry current at an angular frequency $\omega = 2\pi f$. Then the ratio of a.c. to d.c. resistance is given by [Smy68]

$$\frac{R_{ac}}{R_{dc}} = \frac{x}{2}\left[\frac{\mathrm{ber}\,x\;\mathrm{bei}'x - \mathrm{ber}'x\;\mathrm{bei}\,x}{(\mathrm{ber}'x)^2 + (\mathrm{bei}'x)^2}\right] \quad , \quad x = a\sqrt{\omega\mu\sigma} \tag{10.8}$$

The ber and bei functions along with their derivatives ber' and bei' are given in Ref.[Dwi61]. This resistance ratio, which can also be regarded as the ratio of an effective a.c. to d.c. resistivity, is plotted in Fig. 10.5.

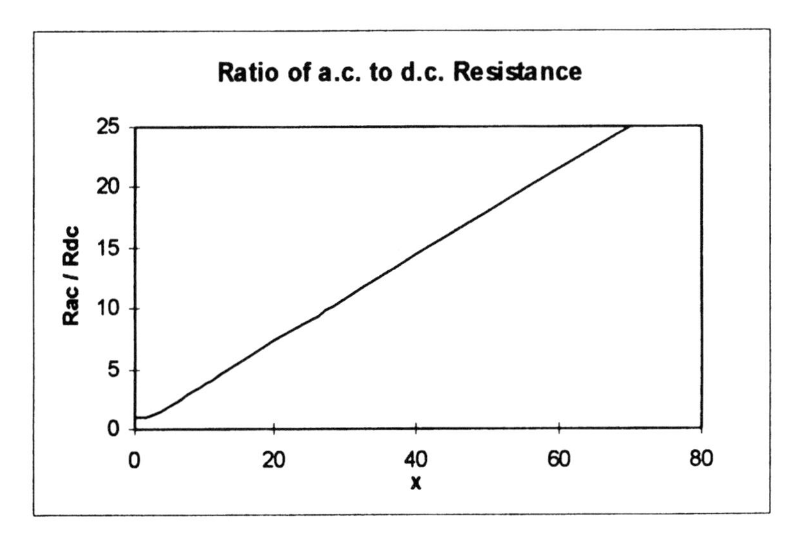

Figure 10.5 Plot of the a.c. to d.c. resistance ratio for an infinitely long cylinder.

As a numerical example, consider a copper cylinder with parameters: $a = 2.52$ cm = 0.6 in , $\rho = 2 \times 10^{-8}$ Ω-m , $\mu = 4\pi \times 10^{-7}$ H/m , f = 60 Hz resulting in $x = 1.95$. Then we find $R_{ac} / R_{dc} \approx 1.15$, i.e. a 15% effect. Making the conductor out of strands, insulated from each other, is not sufficient to eliminate this effect. In addition, the strands must be transposed so that each strand occupies a given region of the cross-sectional area as often as any other strand. This is accomplished in modern transposed cables which use typically 5 - 39 strands.

With this remedy or the use of small wire sizes, I^2R losses can be calculated using the d.c. resistance formula,

$$I^2R \text{ Loss} = \left(\frac{\rho\ell}{A}\right)I^2 \tag{10.9}$$

where ρ is the resistivity at the temperature of interest, ℓ the length of the conductor, A its cross-sectional area = the sum of the areas of all wires in parallel, and I the total current flowing into cross-sectional area A. The temperature dependent resistivity obeys the formula

$$\rho(T) = \rho_o\left[1 + \alpha(T - T_o)\right] \tag{10.10}$$

over a wide range of temperatures, where ρ_o is the resistivity at $T = T_o$ and α is the temperature coefficient of resistivity. For soft copper at T_o = 20 °C, $\rho_o = 1.72 \times 10^{-8}$ Ω-m and $\alpha = 0.0039$. For aluminum at T_o = 20 °C , $\rho_o = 2.83 \times 10^{-8}$ Ω-m and $\alpha = 0.0039$. These numbers apply to relatively pure materials. Alloying can change them considerably. The temperature dependence indicated in (10.10) is significant. For example both copper and aluminum resistivities and hence I^2R losses increase about 30% in going from 20 to 100 °C.

10.3.2 Stray Losses

These are losses caused by stray or leakage flux. Fig. 10.6 shows the leakage flux pattern produced by the coil currents in the bottom half of a single phase or leg of a transformer, assuming cylindrical symmetry about the center line. This was generated with a 2D finite element program. The main components, core, coils, tank, and clamp are shown. Shunts on the tank wall and clamp were given the material properties of transformer oil so they are not active. Fig. 10.7 shows the same plot but with the tank and clamp shunts or shields activated. These are made of the same laminated electrical steel as the core. The shunts or shields divert the flux from getting into the tank or clamp walls so that the stray losses in Fig. 10.7 are much less than those in Fig. 10.6. The stray flux pattern depends on the details of the winding sizes and spacings, the tank size, the clamp position, etc. The losses generated by this flux depend on whether shunts or shields are present as well as geometric and material parameters.

In addition to the coils' stray flux, there is also flux produced by the leads. This flux can generate losses, particularly if the leads are close to the tank wall or clamps. We should also mention losses in the tank wall depending on how the leads are taken out of the tank.

As Figs. 10.6 and 10.7 indicate, there is also stray flux within the coils themselves. This flux is less sensitive to the details of the tank and clamp position or whether shunts or shields are present. Therefore, other methods besides finite elements, such as Rabin's method which uses a simplified geometry, can be used to accurately calculate this flux in the coils. The coil flux generates eddy currents in the wires or individual strands of cable conductors. The losses depend on the strand size as well as its orientation relative to the induction vector and the induction vector's magnitude. The localized losses are therefore different at different positions in the coil.

There are other types of stray loss which occur either in case of an unusual design or when a manufacturing error occurs. In the latter category, extra losses are generated when a cross-over or transposition is missed or misplaced in a coil made of two or more wires or cables in parallel.

We will examine these various types of stray loss here, deriving analytic formulas or procedures for evaluating them where possible or relying on finite element studies or other numerical methods if necessary.

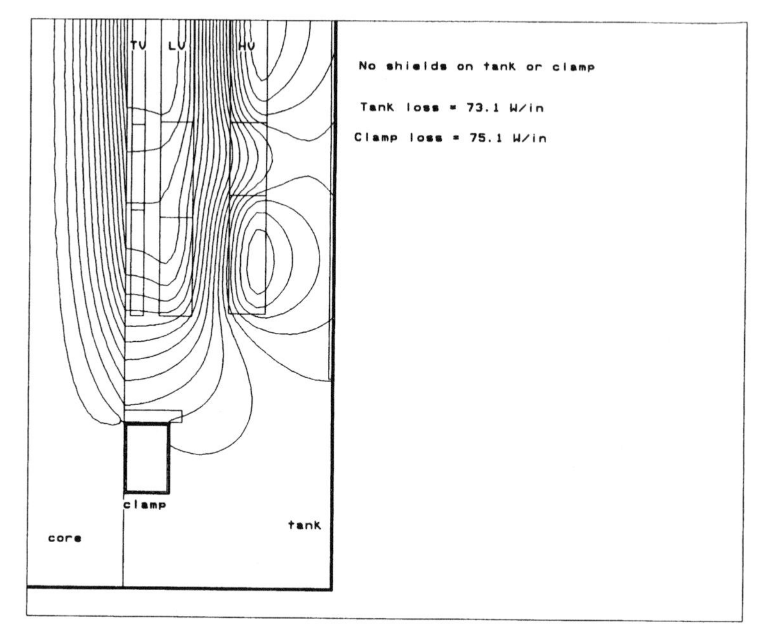

Figure 10.6 Stray flux in the lower half of a core leg with no shunts or shields

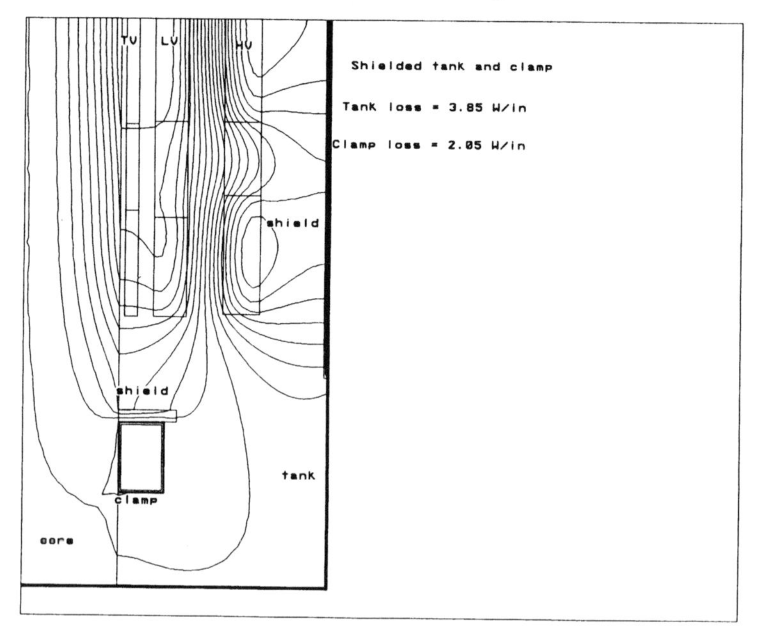

Figure 10.7 Stray flux in the lower half of a core leg with shunts or shields present

10.3.2.1 **Eddy Current Losses in the Coils**

In order to study the effect of stray flux on losses in the coils, we examine an individual wire or strand which could be part of a transposed cable. This is assumed to have a rectangular cross-section. The magnetic field at the site of this strand segment will point in a certain direction relative to the strand's orientation. This vector can be decomposed into components parallel to each side of the rectangular cross-section as shown in Fig. 10.8a. (In a transformer, there is little or no magnetic fields directed along the length of the wire.) We analyze the losses associated with each component of the magnetic field separately and add the results. This is accurate to the extent that the eddy currents associated with the different field components do not overlap. We will see in the following that the eddy currents tend to concentrate along the sides of the rectangular strand to which the field component is parallel. Thus the eddy current patterns associated with the two field components do not overlap significantly so that our method of analysis is reasonably accurate.

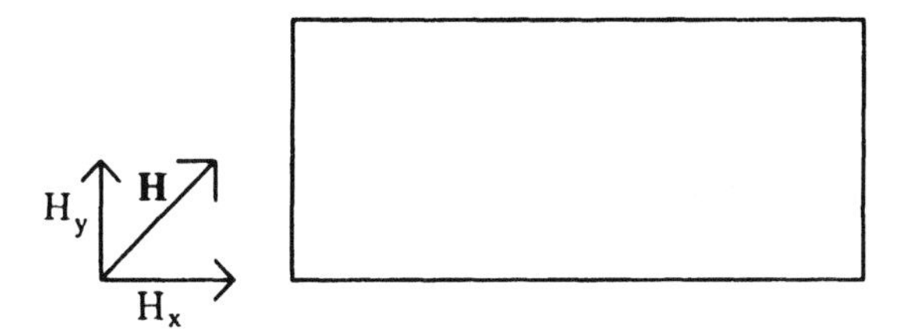

(a) Magnetic field at the location of a conducting strand.

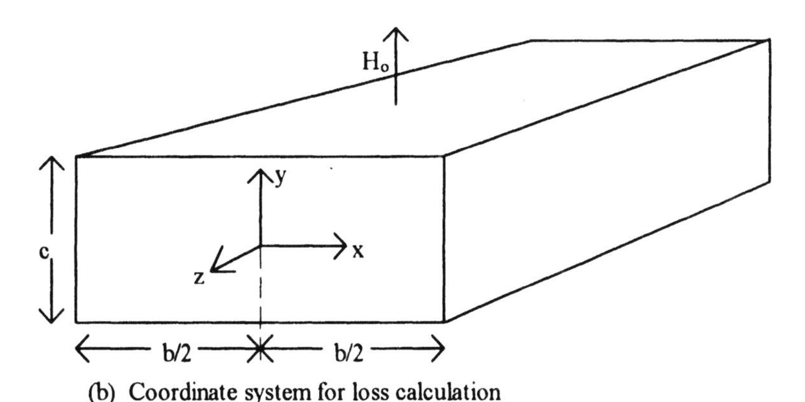

(b) Coordinate system for loss calculation

Figure 10.8 Geometry for calculating losses in a conducting strand due to an external magnetic field

Consider the losses associated with the y-component of an external magnetic field as shown in Fig. 10.8b, where the coordinate system and geometric parameters are indicated. We assume an idealized geometry where the strand is infinitely long in the z-direction. This implies that none of the electromagnetic fields have a z dependence. We further assume that the magnetic field, both external and internal, has only a y-component. Applying Maxwell's equations in this coordinate system and with these assumptions, we obtain in the SI system,

$$\nabla \times \mathbf{E} = -\frac{\partial \mathbf{B}}{\partial t} \quad \Rightarrow \quad \frac{\partial E_z}{\partial x} = \mu \frac{\partial H_y}{\partial t}$$

$$\nabla \times \mathbf{H} = \mathbf{J} \quad \Rightarrow \quad \frac{\partial H_y}{\partial x} = J_z \tag{10.11}$$

$$\nabla \cdot \mathbf{B} = 0 \quad \Rightarrow \quad \frac{\partial H_y}{\partial y} = 0$$

where μ is the permeability, $\mathbf{E}$ the electric field, and $\mathbf{H}$ the magnetic field. We have ignored the displacement current term which is only important at extremely high frequencies. In the metallic conductor, we have Ohm's law in the form

$$\mathbf{J} = \sigma \mathbf{E} \quad \Rightarrow \quad J_z = \sigma E_z \tag{10.12}$$

where σ is the electrical conductivity. There is only a z-component to $\mathbf{J}$ and $\mathbf{E}$. Combining (10.11) and (10.12), we obtain

$$\frac{\partial^2 H_y}{\partial x^2} = \mu\sigma \frac{\partial H_y}{\partial t} \tag{10.13}$$

where H_y is a function of x and t.

Let H_y have a sinusoidal time dependence of the form

$$H_y(x,t) = H_y(x)e^{j\omega t} \tag{10.14}$$

Then (10.13) becomes

$$\frac{\partial^2 H_y}{\partial x^2} = j\omega\mu\sigma H_y = k^2 H_y \quad , \quad k^2 = j\omega\mu\sigma \tag{10.15}$$

H_y is only a function of x in (10.15). Solving (10.15) with the boundary condition that $H_y = H_o$ at $x = \pm b/2$ where b is the strand width normal to the field direction and H_o is the peak amplitude of the external field, we get

$$H_y(x) = H_o \frac{\cosh(kx)}{\cosh(kb/2)} \tag{10.16}$$

and from (10.11)

$$J_z = -kH_o \frac{\sinh(kx)}{\cosh(kb/2)} \tag{10.17}$$

The eddy current loss per unit length in the z direction is

$$\frac{\text{Loss}_{ec}}{\text{unit length}} = \frac{c}{2\sigma}\int_{-b/2}^{b/2} |J|^2 dx = \frac{cH_o^2|k|^2}{\sigma|\cosh(kb/2)|^2}\int_0^{b/2} |\sinh(kx)|^2 dx \tag{10.18}$$

where c is the strand dimension along the field direction. The factor of 1/2 comes from taking a time average and using peak values of the field. The integration through only half the thickness is possible because of the symmetry of the integrand. Note that, from (10.15),

$$k = (1+j)\sqrt{\frac{\omega\mu\sigma}{2}} = (1+j)q \quad , \quad q = \sqrt{\frac{\omega\mu\sigma}{2}} \tag{10.19}$$

Using this expression for k, we have the identities.

$$|\sinh(kx)|^2 = \frac{1}{2}[\cosh(2qx) - \cos(2qx)]$$

$$|\cosh(kx)|^2 = \frac{1}{2}[\cosh(2qx) + \cos(2qx)] \tag{10.20}$$

Substituting into (10.18), integrating, and dividing by the cross-sectional area, we get the specific eddy current loss (loss/unit volume) as

$$P_{ec} = \frac{H_o^2 q}{\sigma b}\left[\frac{\sinh(qb) - \sin(qb)}{\cosh(qb) + \cos(qb)}\right] \tag{10.21}$$

This is in Watts/m^3 in the SI system. Simply divide by the density in this system to get the loss per unit mass or weight.

This last equation applies over a broad frequency range, up to where radiation effects start becoming important. At the low frequency end, which applies to transformers at power frequencies (small qb), this reduces to

$$P_{ec} \xrightarrow[\text{small qb}]{} \frac{H_o^2 q^4 b^2}{6\sigma} = \left(\frac{\pi^2}{6}\right) f^2 \mu^2 b^2 H_o^2 \sigma = \left(\frac{\pi^2}{6}\right)\frac{f^2 b^2 B_o^2}{\rho} \tag{10.22}$$

where we have used $\sigma = 1/\rho$, where ρ is the resistivity, $\omega = 2\pi f$, where f is the frequency, and $B_o = \mu H_o$.

As a numerical example, let $B_o = 0.05$T which is a typical leakage induction value in the coil region, $\rho = 2 \times 10^{-8}$ Ω-m, $\sigma = 5 \times 10^7$ (Ω-m)$^{-1}$, $\omega = 2\pi f = 2\pi(60)$ rads/sec , $b = 6.35 \times 10^{-3}$ m (0.25 in) , $\mu = 4\pi \times 10^{-7}$ H/m. Then $q = 108.8$ m^{-1} and $qb = 0.691$. This is small enough that the small qb limit should apply. Thus we get from (10.22), $P_{ec} = 2.985 \times 10^4$ W/m^3. The exact formula (10.21) yields $P_{ec} = 2.955 \times 10^4$ W/m^3. Using the density of copper $d_{Cu} = 8933$ kg/m^3, we obtain $P_{ec} = 3.35$ W/kg = 1.52 W/lb. The I^2R loss on a per volume basis associated with an rms current of 3×10^6 A/m^2 = 1935 A/in^2 (a typical value) in a material (copper) of the above resistivity, is 1.8×10^5 W/m^3 so that the eddy current loss amounts to about 17% of the I^2R losses in this case.

The losses given by (10.21) or (10.22) must be combined with the losses given by a similar formula with H_o or B_o referring to the peak value of the x-component of the field and with b and c interchanged. This will give the total eddy current loss density at the location of the strand. A method of calculating the magnetic field or induction at various locations in the coils is needed. From the axisymmetric field calculation (flux map) given in Figs. 10.6 and 10.7, we obtain values of the radial and axial components of the field. These replace the x , y components in the loss formulas given above. These loss densities will differ in different parts of the winding. To obtain the total eddy current loss, an average loss density can be obtained for the winding and this multiplied by the total weight or volume of the winding. However, in determining local winding temperatures and especially the hot spot temperature, a knowledge of how these losses are distributed is necessary.

From (10.17), (10.19), and (10.20), we can obtain the eddy current loss density as a function of position in the strand,

$$\frac{|J_z|^2}{2\sigma} = \frac{H_o^2 q^2}{\sigma}\left[\frac{\cosh(2qx) - \cos(2qx)}{\cosh(qb) + \cos(qb)}\right] \tag{10.23}$$

This vanishes at the center of the strand (x = 0) and is a maximum at the surface (x = ± b/2). The parameter q measures how fast this drops off from the surface. The fall off is more rapid the larger the value of q. The reciprocal of q is called the skin depth δ and is given by

$$\delta = \frac{1}{q} = \sqrt{\frac{2}{\omega\mu\sigma}} \tag{10.24}$$

For copper at ~ 60 °C, $\sigma = 5 \times 10^7\ (\Omega\text{-m})^{-1}$, and f = 60 Hz, we get δ = 0.92 cm = 0.36 in. For aluminum at ~60 °C, $\sigma = 3 \times 10^7\ (\Omega\text{-m})^{-1}$, and f = 60 Hz, we get δ = 1.19 cm = 0.47 in. Thus the skin depth is smaller for copper than aluminum which means that the eddy currents concentrate more towards the surface of copper than aluminum.

The high frequency limit of (10.21) (large qb) is

$$P_{ec} \xrightarrow[\text{large qb}]{} \frac{H_o^2 q}{\sigma b} \tag{10.25}$$

This increases as the square root of the frequency.

10.3.2.2 Tieplate Losses

The tieplate (also called flitch plate) is located just outside the core in the space between the core and innermost winding. It is a structural plate which connects the upper and lower clamps. Tension in this plate provides the clamping force necessary to hold the transformer together should a short circuit occur. It is usually made of magnetic steel or stainless steel and could be subdivided into several side by side vertical plates to help reduce the eddy current losses. Fig. 10.9 shows a schematic diagram of one of the tieplates associated with one leg. There is another on the opposite side of the core leg. These generally have a rectangular cross-section.

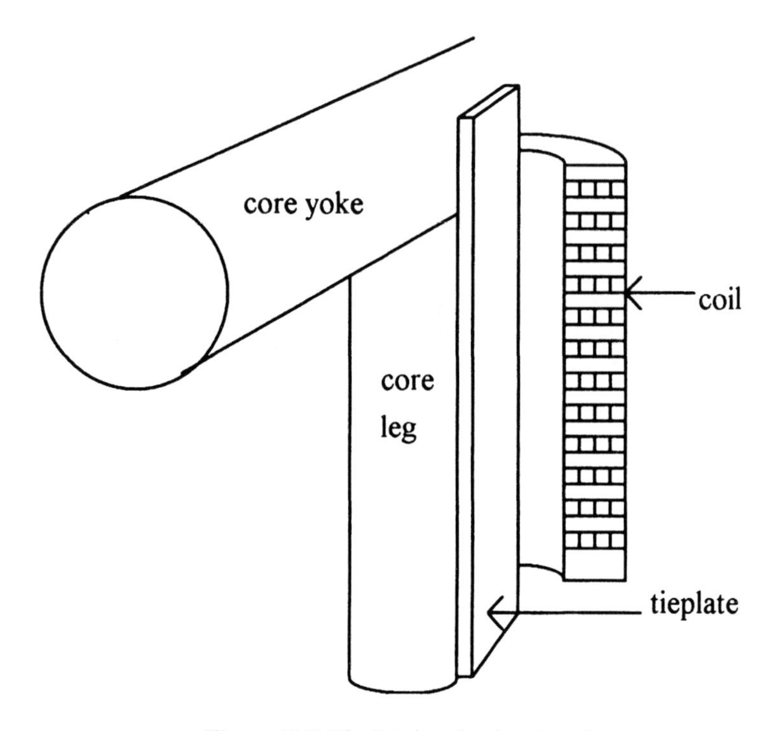

Figure 10.9 Tieplate location in a transformer

Since the flux plots in Figs. 10.6 and 10.7 are for a 2 dimensional, axisymmetric geometry, it was not possible to include the tieplate. (This would have made it a solid cylinder around the core.) However, the flux pattern shown in the figures should not be greatly altered by their presence since they occupy a fairly small fraction of the core's circumference. As the flux pattern in the figures show, the flux is primarily radial at the location of the tieplate. However, with an actual tieplate present, there will be some axial flux carried by the tieplate. This will depend on the permeability of the tieplate relative to that of the core. We can estimate the axial tieplate flux by reference to Fig. 10.10 where we show two side by side solids of permeabilities μ_1 , μ_2 and cross sectional areas A_1 , A_2 carrying flux. We assume that the coils producing this flux create a common magnetic field **H** at the location of the solids. (The field inside an ideal solenoid is a constant axial field.)

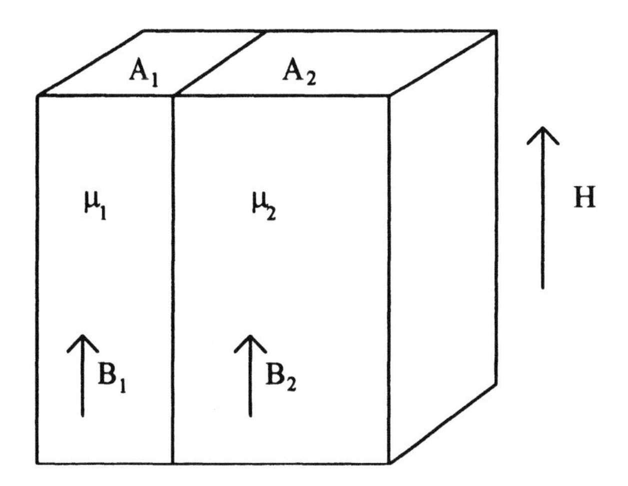

Figure 10.10 Dissimilar magnetic materials in a common magnetic field

The induction inside each solid is given by

$$B_1 = \mu_1 H \quad , \quad B_2 = \mu_2 H \tag{10.26}$$

so that

$$\frac{B_1}{B_2} = \frac{\mu_1}{\mu_2} \tag{10.27}$$

Using this last expression, we can estimate the induction in the tieplate. We can use the ratio of relative permeabilities in the above formula. These are a.c. permeabilities which we take to be ~ 5000 for the core and ~ 200 for a magnetic steel tieplate. With a core induction of 1.7 Tesla which is typical, we find that $B_{mag\ t.p.} = 0.04\ B_{core} = 0.068$ T. For a stainless steel tieplate of relative permeability = 1 , we obtain $B_{s.s.\ t.p.} = 0.0002\ B_{core} = 0.00034$ T. Thus the axial induction is not insignificant for a magnetic steel tieplate but ignorable for a stainless steel one.

Let us first look at the losses due to the radial induction since these are common to both magnetic and stainless steel tieplates. We have studied these losses using a 2D finite element analysis. Fig. 10.11 shows a flux plot for a magnetic steel tieplate, assuming a uniform 60 Hz sinusoidal flux density far from the plate. The plate is assumed to be infinitely long in the dimension perpendicular to the page. Only 1/2 the geometry is modeled, taking advantage of symmetry about the left hand axis. The plate rests directly on a material of high permeability, the core.

The loss density contours are shown in Fig. 10.12. This indicates that the eddy currents are concentrated near the surface due to the skin effect. Fig. 10.13 shows a similar flux plot for a stainless steel tieplate. Note that there is little eddy current screening so the flux penetrates the plate without much distortion. Fig. 10.14 shows the loss density contours for the stainless steel tieplate. There is much less surface concentration of the eddy currents. Although we only calculate eddy current losses with a finite element program, the hysteresis losses in a magnetic steel tieplate make up only a small fraction of the total losses for typical tieplate dimensions.

The finite element study was repeated for different tieplate widths (perpendicular to the flux direction) while keeping the thickness (along the flux direction) constant at 0.95 cm (3/8 in). The results are shown in Fig. 10.15, where the loss per unit length in the axial direction and per T^2 are plotted. To get the actual loss multiply the ordinate by the tieplate length and by the square of the radial induction in T^2. Fig. 10.16 shows the same information as plotted in Fig. 10.15 but on a log-log plot. This allows the extraction of the power dependence of the loss on the width of the tieplate. The material parameters assumed were, for magnetic steel: resistivity = 25×10^{-8} Ω-m and relative permeability = 200, for stainless steel: resistivity = 75×10^{-8} Ω-m and relative permeability = 1. We obtain for the losses

$$
\begin{aligned}
\text{Magnetic steel loss}(\mathrm{kW/m}) &= 2.65\times 10^{4}\,\mathrm{w}^{2.4}\mathrm{B}_{\mathrm{rms}}^{2} \quad &, \quad \mathrm{w-m}\,,\ \mathrm{B}_{\mathrm{rms}}-\mathrm{T}\\
(\mathrm{kW/in}) &= 0.10\,\mathrm{w}^{2.4}\mathrm{B}_{\mathrm{rms}}^{2} \quad &, \quad \mathrm{w-in}\,,\ \mathrm{B}_{\mathrm{rms}}-\mathrm{T}\\[1em]
\text{Stainless steel loss}(\mathrm{kW/m}) &= 1.44\times 10^{5}\,\mathrm{w}^{3}\mathrm{B}_{\mathrm{rms}}^{2} \quad &, \quad \mathrm{w-m}\,,\ \mathrm{B}_{\mathrm{rms}}-\mathrm{T}\\
(\mathrm{kW/in}) &= 0.06\,\mathrm{w}^{3}\mathrm{B}_{\mathrm{rms}}^{2} \quad &, \quad \mathrm{w-in}\,,\ \mathrm{B}_{\mathrm{rms}}-\mathrm{T}
\end{aligned}
\tag{10.28}
$$

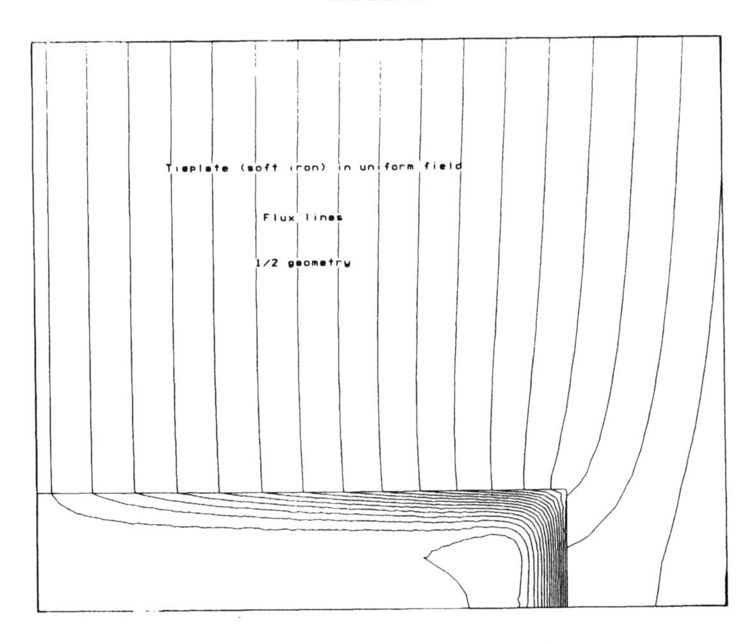

Figure 10.11 Flux lines for a magnetic steel tieplate in a uniform 60 Hz magnetic field normal to its surface. The tieplate rests on a high permeability core. Only 1/2 the geometry is modeled.

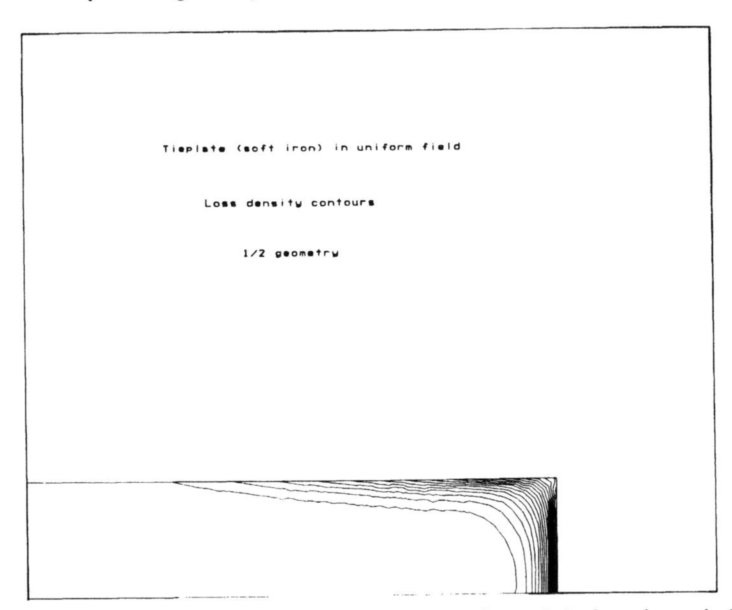

Figure 10.12 Loss density contours for the magnetic steel tieplate shown in Fig. 10.11.

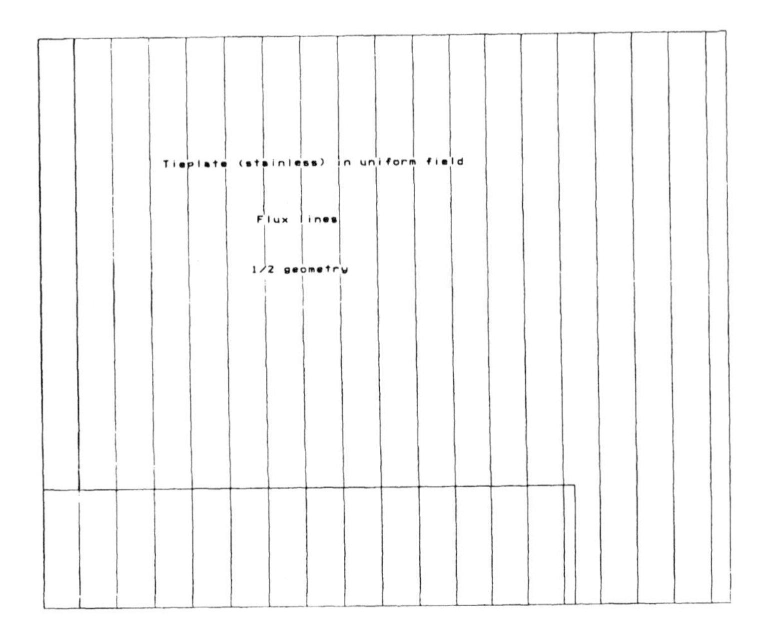

Figure 10.13 Flux lines for a stainless steel tieplate in a uniform 60 Hz field normal to its surface. The situation is otherwise the same as described in Fig. 10.11.

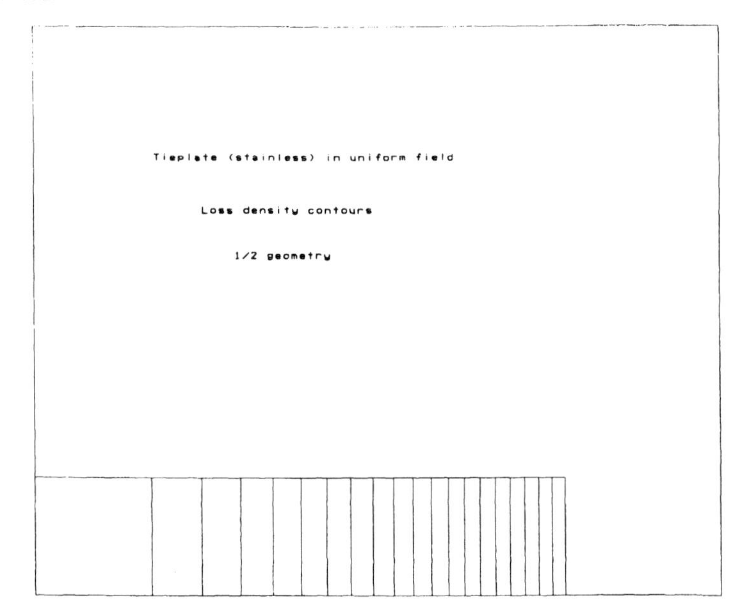

Figure 10.14 Loss density contours for the stainless steel tieplate shown in Fig. 10.13.

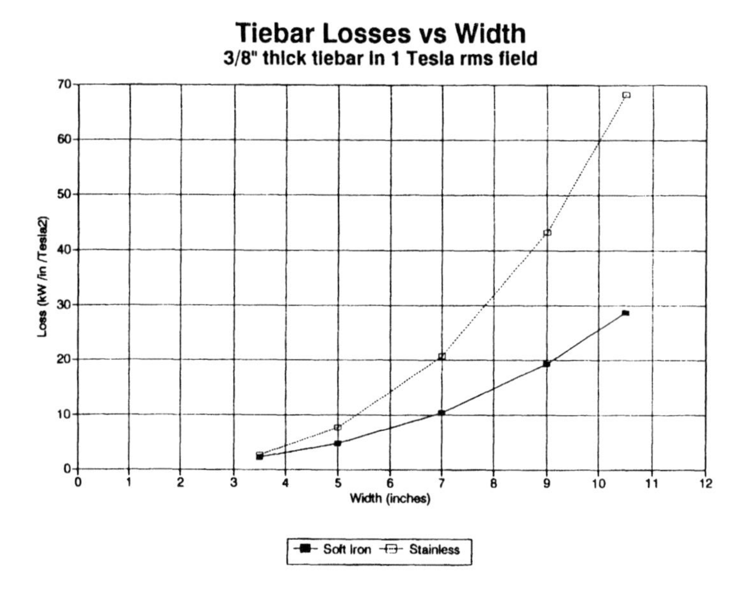

Figure 10.15 Losses for 0.95 cm (3/8 in) thick tieplates made of magnetic and stainless steel versus the tieplate width. The tieplates are in a uniform 60 Hz magnetic field directed normal to the tieplate surface.

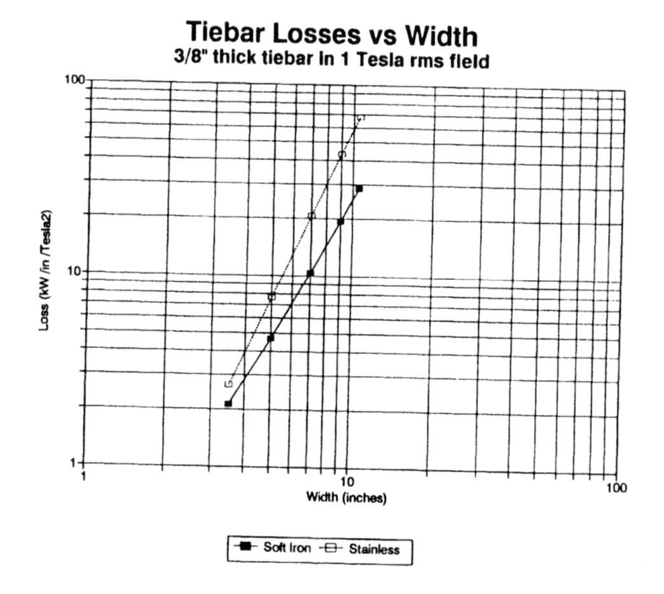

Figure 10.16 Same as Fig. 10.15 but on a log-log scale.

The stainless steel finite element study shows that the magnetic field and eddy currents inside the tieplate have nearly the same direction and geometry dependence as was assumed for the eddy current loss in conducting strands. Therefore if we multiply the loss density given in (10.21) or (10.22) by the plate's cross sectional area, we should get nearly the same result as given above for stainless steel. We note that w should be substituted for b in the previous formulas. Using the small qb formula (10.22), and putting in the stainless steel parameters, we obtain

$$\text{Loss}(\text{kW/m}) = 1.5 \times 10^5 \, w^3 B_{rms}^2$$

which is close to the result given above in SI units. We do not expect exact agreement since qb (= qw) for stainless at w = 0.127 m (5 in) is ~ 2.26 which is a little above the range of validity of the small qb formula.

In the case of magnetic steel, as Figs. 10.11 and 10.12 show, the conditions under which the strand eddy current losses were developed do not hold. The field is not strictly y-directed (radially directed) and it does not depend only on x (the width direction). We should note that, according to Fig. 10.15, the losses per unit length due to radial flux are higher in stainless steel tieplates than in magnetic steel plates for the same width and thickness. However, for magnetic steel tieplates, we need to add on the losses due to the axial flux they carry.

We can estimate the losses in the tieplate due to axial flux by resorting to an idealized geometry. We assume the tieplate is infinitely long and the flux is driven by a uniform axial magnetic field parallel to the tieplate's surface. We are also going to assume that its width is much greater than its thickness, in fact we assume an infinite width. Thus, as shown in Fig. 10.17, the only relevant dimension is the y-dimension through the sheet's thickness. The eddy currents will flow primarily in the x direction. We ignore their return paths in the y-direction which are small compared to the x-directed paths.

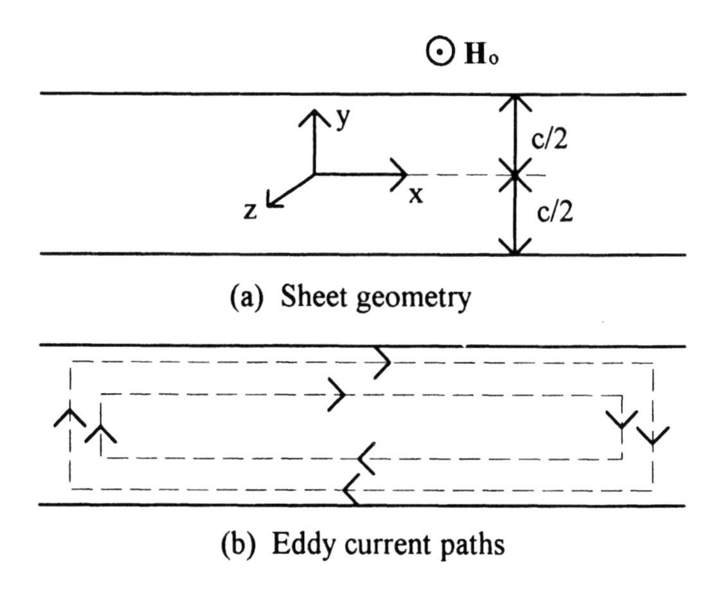

Figure 10.17 Geometry of an idealized plate or sheet driven by a uniform sinusoidal magnetic field paralled to its surface (perperdicular to page)

Applying Maxwell's equations and Ohm's law to this geometry, we obtain in SI units,

$$\nabla \times \mathbf{E} = -\frac{\partial \mathbf{B}}{\partial t} \quad \Rightarrow \quad \frac{\partial E_x}{\partial y} = \mu \frac{\partial H_z}{\partial t}$$

$$\nabla \times \mathbf{H} = \mathbf{J} \quad \Rightarrow \quad \frac{\partial H_z}{\partial y} = J_x \tag{10.29}$$

$$J = \sigma E \quad \Rightarrow \quad J_x = \sigma E_x$$

Combining these equations, we obtain

$$\frac{\partial^2 H_z}{\partial y^2} = \mu\sigma \frac{\partial H_z}{\partial t} \tag{10.30}$$

The last equation is identical in form to equation (10.13). Thus, using the previous results but altering the notation to fit the present geometry, we get

$$H_z(y) = H_o \frac{\cosh(ky)}{\cosh(kc/2)} \quad , \quad J_x(y) = -kH_o \frac{\sinh(ky)}{\cosh(kc/2)} \tag{10.31}$$

and

$$P_{tp,\,axial} = \frac{H_o^2 q}{\sigma c}\left[\frac{\sinh(qc) - \sin(qc)}{\cosh(qc) + \cos(qc)}\right] \, , \; k = (1+j)q \; , \; q = \sqrt{\frac{\omega\mu\sigma}{2}} \tag{10.32}$$

This is a loss per unit volume (W/m^3) in the tieplate due to axial flux. Since the eddy currents generating this loss are at right angles to the eddy currents associated with the radial flux normal to the surface, there is no interference between them and the two losses can be added.

For magnetic steel at 60 Hz with $\mu_r = 200$, $\mu = \mu_r 4\pi \times 10^{-7}$ H/m, $\sigma = 4 \times 10^6$ $(\Omega\text{-m})^{-1}$ and $c = 9.52 \times 10^{-3}$ m (0.375 in), we have $qc = 4.15$. This is large enough that (10.32) must be used without taking its small qc limit. Substituting the parameters just given for magnetic steel and using $B_o = 0.07$ T ($H_o = B_o/\mu$), we obtain $P_{tp,\,axial} = 925$ $W/m^3 = 0.12$ W/kg , where we have used 7800 kg/m^3 for the density of magnetic steel. The radial flux loss formula for magnetic steel given previously, equation (10.28), must be divided by the cross sectional area for comparison with the above loss. We find, using $B_{rms} = 0.1T$, $w = 0.127$ m (5 in) and the previous value for c, $P_{tp,\,radial} = 1552$ $W/m^3 = 0.20$ W/kg. Thus it appears that for a magnetic steel tieplate, the loss associated with the axial flux is comparable with the loss associated with the radial flux. In any event, these two losses must be added to get the total loss. Since stainless steel tieplates have virtually no axial flux loss, the net effect is that magnetic steel tieplates could have a higher loss than comparably sized stainless steel tieplates.

Since the axial flux loss is, to first order, independent of tieplate width, the width dependence of the total magnetic steel tieplate losses should not be as pronounced as that for stainless steel ones. Thus subdividing the tieplates either with axial slots or into separate plates will not be as effective for magnetic steel as compared with stainless steel tieplates in reducing the losses. However, one benefit of magnetic steel tieplates, not shared by stainless steel ones, is that they keep the radial flux from passing through them into the core steel. (Compare Figs. 10.11 and 10.13.) They do, however, concentrate the radial flux reaching the core to the regions near the outer edges of the tieplate where oil cooling should be more effective.

10.3.2.3 Tieplate and Core Losses Due to Unbalanced Currents

There is a rather specialized type of loss which can occur in transformers which have a large unbalanced net current flow. The net current is the algebraic sum of the currents flowing in all the windings. This, in contrast with the net Ampere-turns which are always nearly exactly balanced, can be unbalanced. To visualize the magnetic effect of a winding's current, consider a cylindrical (solenoidal) winding carrying a current which flows in at the bottom and out at the top. Outside the winding, the net upward current appears equally distributed around the cylinder. The magnetic field outside the cylinder associated with this current is the same as that produced by the current flowing along the centerline of the cylinder. In a core-form transformer, some of the windings carry current up and some down. The algebraic sum of all these currents can be considered as being carried by one cylinder of radius equal to a weighted average of the contributing windings, weighted by their current magnitudes. Outside this radius, the field is that of a straight wire along the centerline carrying the algebraic sum of the currents.

The field around a long straight wire carrying a current $\mathbf{I}$ is directed along concentric circles about the wire and has magnitude in SI units,

$$H_\varphi = \frac{\mathbf{I}}{2\pi r} \tag{10.33}$$

where r is the radial distance from the centerline and φ, the azimuthal angle, indicates that the field is azimuthally directed. We are using bold faced type here to indicate that the current is a phasor quantity, since we are considering a 3 phase transformer on a single 3 phase core. This field cuts through the transformer core windows as shown in Fig. 10.18. The alternating flux passing through the core window induces a voltage around the core structure which surrounds the window. The tieplates, including their connection to the upper and lower clamps, make a similar circuit around the core windows so that voltage is induced in them as well.

We can calculate the flux due to the coil in Fig. 10.18a through the left hand window, labeled 1 in the figure, Φ_{a1}, using (10.33)

$$\Phi_{a1} = \int B_\varphi dA = \frac{\mu_o h \mathbf{I}_a}{2\pi}\int_{r_w}^{d}\frac{dr}{r} = \frac{\mu_o h \mathbf{I}_a}{2\pi}\ln\left(\frac{d}{r_w}\right) \tag{10.34}$$

where μ_o is the permeability of oil (air) = $4\pi \times 10^{-7}$ H/m, h the effective winding height, d the leg center to center distance, and r_w the effective winding radius. h can be obtained as a weighted average of the contributing windings as was done for r_w. We have integrated all the way to the centerline of the center leg. This is an approximation as the flux lines will no doubt deviate from the ideal radial dependence given in (10.33) near the center leg. The flux through window 2 due to the phase a current, Φ_{a2}, is similarly

$$\Phi_{a2} = \int B_\varphi dA = \frac{\mu_o h I_a}{2\pi}\int_d^{2d}\frac{dr}{r} = \frac{\mu_o h I_a}{2\pi}\ln 2 \tag{10.35}$$

Using the same procedure, we find the flux through the two windows due to phases b and c shown in Fig. 10.18 b,c

$$\Phi_{b1} = -\frac{\mu_o h I_b}{2\pi}\ln\left(\frac{d}{r_w}\right) \quad , \quad \Phi_{b2} = \frac{\mu_o h I_b}{2\pi}\ln\left(\frac{d}{r_w}\right)$$

$$\Phi_{c1} = -\frac{\mu_o h I_c}{2\pi}\ln 2 \quad , \quad \Phi_{c2} = -\frac{\mu_o h I_c}{2\pi}\ln\left(\frac{d}{r_w}\right) \tag{10.36}$$

Thus the net fluxes through windows 1 and 2 are

$$\Phi_1 = \Phi_{a1} + \Phi_{b1} + \Phi_{c1} = \frac{\mu_o h}{2\pi}\ln\left(\frac{d}{r_w}\right)\left[I_a - I_b - \frac{\ln 2}{\ln(d/r_w)}I_c\right]$$

$$\Phi_2 = \Phi_{a2} + \Phi_{b2} + \Phi_{c2} = \frac{\mu_o h}{2\pi}\ln\left(\frac{d}{r_w}\right)\left[\frac{\ln 2}{\ln(d/r_w)}I_a + I_b - I_c\right] \tag{10.37}$$

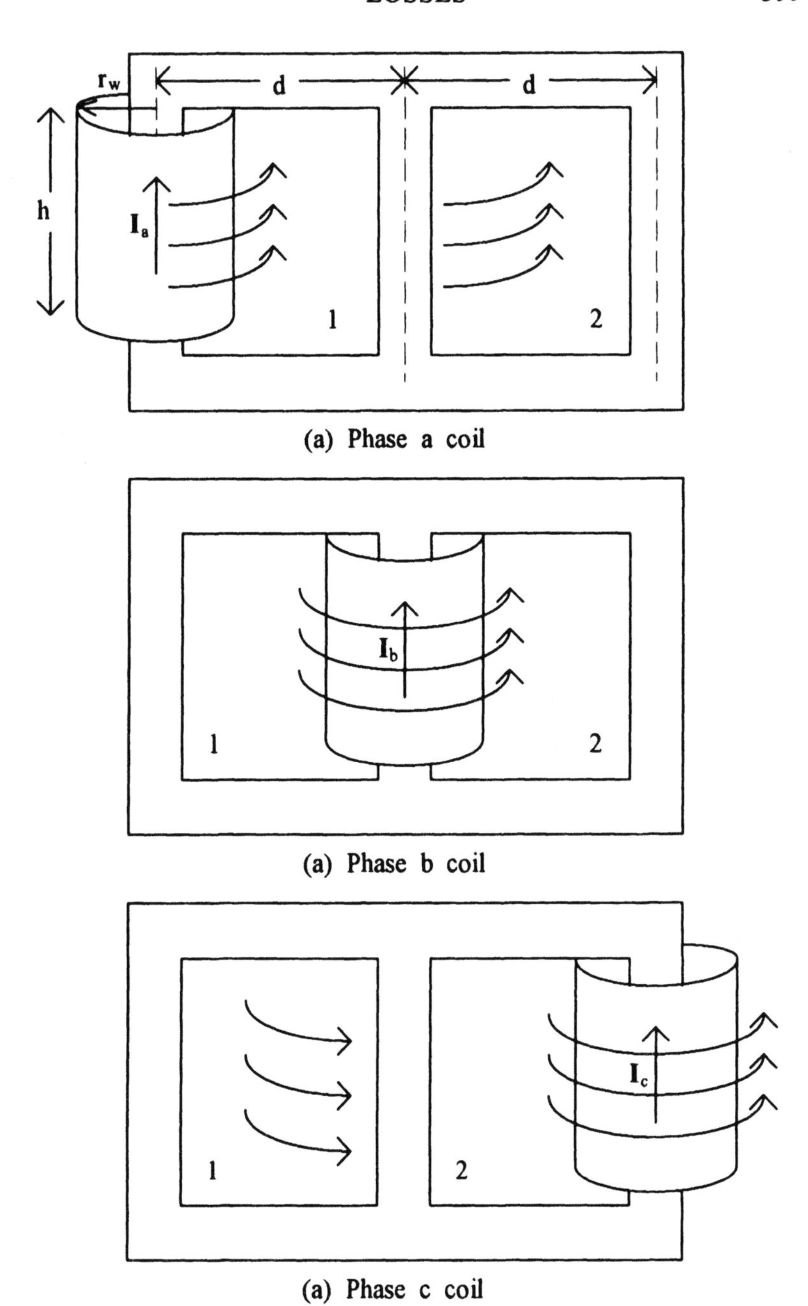

Figure 10.18 Field around a cylindrical winding located on a transformer leg and carrying a net upward current

Considering $\mathbf{I}_a$, $\mathbf{I}_b$, $\mathbf{I}_c$ to be a positive sequence set of currents and performing the phasor sums above, we get

$$\Phi_1 = \frac{\mu_o h \mathbf{I}_a}{2\pi} \ln\left(\frac{d}{r_w}\right)\sqrt{3+g^2}\, e^{j\theta_1} \quad , \quad \theta_1 = \tan^{-1}\left[\frac{\sqrt{3}(1-g)}{3+g}\right]$$

$$\Phi_2 = \frac{\mu_o h \mathbf{I}_a}{2\pi} \ln\left(\frac{d}{r_w}\right)\sqrt{3+g^2}\, e^{j\theta_2} \quad , \quad \theta_2 = -\tan^{-1}\left(\frac{\sqrt{3}}{g}\right) \quad , \quad g = \frac{\ln 2}{\ln(d/r_w)} \tag{10.38}$$

From Faraday's law, the voltages induced by the two fluxes are

$$\mathbf{V}_1 = -\frac{\mu_o h \omega \mathbf{I}_a}{2\pi} \ln\left(\frac{d}{r_w}\right)\sqrt{3+g^2}\, e^{j\theta_1} \quad , \quad \mathbf{V}_2 = \mathbf{V}_1\, e^{j(\theta_2-\theta_1)} \tag{10.39}$$

The direction of these voltages or emf's is given by Lenz's law, i.e. they try to oppose the driving flux. We can assume that $\mathbf{V}_1$ is the reference phasor and thus drop the minus sign and phase factor from its expression. Then $\mathbf{V}_2$ is given by (10.39) with a phase of $\theta_2 - \theta_1$ relative to the reference phasor.

The induced voltages will attempt to drive currents through the tieplates and core in loops surrounding the two windows. This will be opposed by the resistances of the tieplates and core and by the self and mutual inductances of the metallic window frames whether formed of core sections or of tieplate and clamp sections. We can treat these as lumped parameters, organized into the circuits of Fig. 10.19. There are 3 circuits involved since there are two tieplate circuits on either side of the core plus the core circuit. These are essentially isolated from each other except for coupling through the mutual inductances. Although the tieplate circuits share the top and bottom clamps in common, they are sufficiently symmetric that they can be regarded as separate circuits. We assume that magnetic coupling exists only between window 1 loops or window 2 loops but not between a window 1 loop and a window 2 loop. To make the circuit equations more symmetric, we have positioned V_2 in the circuit in such a way that it is necessary to take the negative of the expression in (10.39).

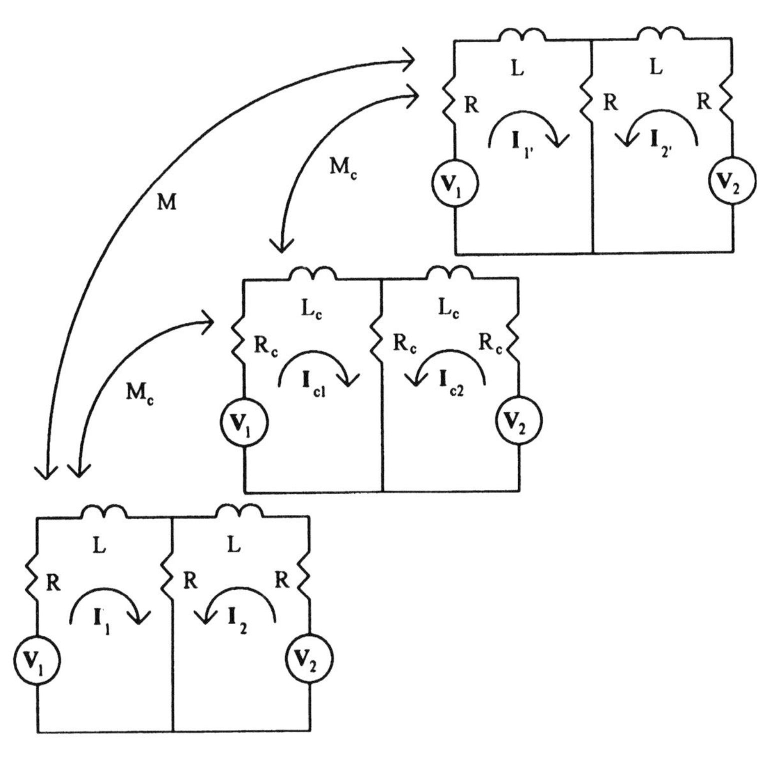

Figure 10.19 Equivalent circuits for tieplates and core driven by voltages induced through unbalanced currents

Thus we have for the voltage sources in the circuits of Fig 10.19, assuming $\mathbf{V}_1$ is the reference phasor,

$$\mathbf{V}_1 = \mu_o hf \mathbf{I}_a \ln\left(\frac{d}{r_w}\right)\sqrt{3+g^2} \quad , \quad \mathbf{V}_2 = -\mathbf{V}_1 e^{j\theta} = \mathbf{V}_1 e^{j(\theta+\pi)}$$

$$\theta = -\tan^{-1}\left(\frac{\sqrt{3}}{g}\right) - \tan^{-1}\left[\frac{\sqrt{3}(1-g)}{3+g}\right] \quad , \quad g = \frac{\ln 2}{\ln(d/r_w)} \tag{10.40}$$

where f is the frequency in Hz.

Using the notation of Fig. 10.19 and assuming sinusoidal conditions, we can write the circuit equations

$$\mathbf{V}_1 = (2\mathrm{R} + \mathrm{j}\omega\mathrm{L})\mathbf{I}_1 \quad + \quad \mathrm{R}\mathbf{I}_2 \quad + \quad \mathrm{j}\omega\mathrm{M}\mathbf{I}_{1'} \quad + \quad \mathrm{j}\omega\mathrm{M}_\mathrm{c}\mathbf{I}_{\mathrm{c}1}$$

$$\mathbf{V}_2 = \mathrm{R}\mathbf{I}_1 \quad + \quad (2\mathrm{R} + \mathrm{j}\omega\mathrm{L})\mathbf{I}_2 \quad + \quad \mathrm{j}\omega\mathrm{M}\mathbf{I}_{2'} \quad + \quad \mathrm{j}\omega\mathrm{M}_\mathrm{c}\mathbf{I}_{\mathrm{c}2}$$

$$\mathbf{V}_1 = \mathrm{j}\omega\mathrm{M}\mathbf{I}_1 \quad + \quad (2\mathrm{R} + \mathrm{j}\omega\mathrm{L})\mathbf{I}_{1'} \quad + \quad \mathrm{R}\mathbf{I}_{2'} \quad + \quad \mathrm{j}\omega\mathrm{M}_\mathrm{c}\mathbf{I}_{\mathrm{c}1} \tag{10.41}$$

$$\mathbf{V}_2 = \mathrm{j}\omega\mathrm{M}\mathbf{I}_2 \quad + \quad \mathrm{R}\mathbf{I}_{1'} \quad + \quad (2\mathrm{R} + \mathrm{j}\omega\mathrm{L})\mathbf{I}_{2'} \quad + \quad \mathrm{j}\omega\mathrm{M}_\mathrm{c}\mathbf{I}_{\mathrm{c}2}$$

$$\mathbf{V}_1 = \mathrm{j}\omega\mathrm{M}_\mathrm{c}\mathbf{I}_1 \quad + \quad \mathrm{j}\omega\mathrm{M}_\mathrm{c}\mathbf{I}_{1'} \quad + \quad (2\mathrm{R}_\mathrm{c} + \mathrm{j}\omega\mathrm{L}_\mathrm{c})\mathbf{I}_{\mathrm{c}1} \quad + \quad \mathrm{R}_\mathrm{c}\mathbf{I}_{\mathrm{c}2}$$

$$\mathbf{V}_2 = \mathrm{j}\omega\mathrm{M}_\mathrm{c}\mathbf{I}_2 \quad + \quad \mathrm{j}\omega\mathrm{M}_\mathrm{c}\mathbf{I}_{2'} \quad + \quad \mathrm{R}_\mathrm{c}\mathbf{I}_{\mathrm{c}1} \quad + \quad (2\mathrm{R}_\mathrm{c} + \mathrm{j}\omega\mathrm{L}_\mathrm{c})\mathbf{I}_{\mathrm{c}2}$$

These can be organized into a matrix equation and solved for the currents, using complex arithmetic. However, it is possible to solve them using real arithmetic by separating the vectors and matrix into real and imaginary parts. Thus, given a matrix equation

$$\mathbf{V} = M\,\mathbf{I} \tag{10.42}$$

where $\mathbf{V}$ and $\mathbf{I}$ are complex column vectors and M a complex matrix, write

$$\mathbf{V} = \mathbf{V}_{\mathrm{Re}} + \mathrm{j}\mathbf{V}_{\mathrm{Im}} \quad , \quad \mathbf{I} = \mathbf{I}_{\mathrm{Re}} + \mathrm{j}\mathbf{I}_{\mathrm{Im}} \quad , \quad M = M_{\mathrm{Re}} + \mathrm{j}M_{\mathrm{Im}} \tag{10.43}$$

Here bold faced type is doing double duty in indicating both vector and phasor quantities. Substituting this separation into real and imaginary parts into (10.42), we get

$$\begin{aligned}(\mathbf{V}_{\mathrm{Re}} + \mathrm{j}\mathbf{V}_{\mathrm{Im}}) &= (M_{\mathrm{Re}} + \mathrm{j}M_{\mathrm{Im}})(\mathbf{I}_{\mathrm{Re}} + \mathrm{j}\mathbf{I}_{\mathrm{Im}}) \\ &= (M_{\mathrm{Re}}\mathbf{I}_{\mathrm{Re}} - M_{\mathrm{Im}}\mathbf{I}_{\mathrm{Im}}) + \mathrm{j}(M_{\mathrm{Re}}\mathbf{I}_{\mathrm{Im}} + M_{\mathrm{Im}}\mathbf{I}_{\mathrm{Re}})\end{aligned} \tag{10.44}$$

This reduces to two separate equations,

$$\mathbf{V}_{\mathrm{Re}} = M_{\mathrm{Re}}\mathbf{I}_{\mathrm{Re}} - M_{\mathrm{Im}}\mathbf{I}_{\mathrm{Im}} \quad , \quad \mathbf{V}_{\mathrm{Im}} = M_{\mathrm{Re}}\mathbf{I}_{\mathrm{Im}} + M_{\mathrm{Im}}\mathbf{I}_{\mathrm{Re}} \tag{10.45}$$

which can be organized into a larger matrix equation

$$\begin{pmatrix} \mathbf{V}_{\mathrm{Re}} \\ \mathbf{V}_{\mathrm{Im}} \end{pmatrix} = \begin{pmatrix} M_{\mathrm{Re}} & -M_{\mathrm{Im}} \\ M_{\mathrm{Im}} & M_{\mathrm{Re}} \end{pmatrix} \begin{pmatrix} \mathbf{I}_{\mathrm{Re}} \\ \mathbf{I}_{\mathrm{Im}} \end{pmatrix} \tag{10.46}$$

Here the separate entries are real vectors and matrices. Thus we have doubled the dimension of the original matrix equation (10.41) from 6 to 12 but this is still small, considering the power of modern computers.

The remaining issues concern how to evaluate the resistances and the self and mutual inductances in (10.41). The resistance of a tieplate, R , can simply be taken as its d.c. resistance since it has a relatively small thickness. We can ignore the resistance of the clamps since these should have a much larger cross-sectional area than the tieplates. The core, with its fairly large radius and high permeability will have an enhanced a.c. resistance relative to its d.c. value. It can be estimated from Fig. 10.5, using an effective a.c. permeability and conductivity.

Since we do not expect extreme accuracy in this calculation, in view of the approximations already made, we can use approximate formulas for the self and mutual inductances. For example, Ref. [Gro73] gives a formula for the inductance of a rectangle of sides a and b made of wire with a circular cross-section of radius r and relative permeability μ_r which is, in SI units,

$$\begin{aligned} L = 4\times 10^{-7}\Big[& a\ln\left(\frac{2a}{r}\right) + b\ln\left(\frac{2b}{r}\right) + 2\sqrt{a^2+b^2} - a\sinh^{-1}\left(\frac{a}{b}\right) \\ & - b\sinh^{-1}\left(\frac{b}{a}\right) - 2(a+b) + \frac{\mu_r}{4}(a+b)\Big] \end{aligned} \tag{10.47}$$

L is in Henrys and lengths are in meters in the above formula. This formula can be applied directly to calculate the inductance of the core window. By defining an effective radius, it can be applied to the tieplate loop as well. In calculating the core inductance, remember that the core current generates magnetic field lines in the shape of concentric circles about the core centerline so that the appropriate relative permeability is roughly the effective permeability perpendicular to the laminations. For a stacking factor of 0.96 and infinitely permeable laminations, the effective perpendicular permeability is $\mu_r = 25$. Similarly, the effective tieplate permeability is close to 1.0 for both magnetic and stainless steel tieplates.

The mutual inductance terms are not quite so important so that an even cruder approximation may be used. Ref. [Gro73] gives an expression for the mutual inductance between two equal coaxial squares of thin wire which are close together. Letting s be the length of the side of the squares and d their separation, the mutual inductance, in SI units, is

$$M = 8 \times 10^{-7} s \left[\ln\left(\frac{s}{d}\right) - 0.7740 + \frac{d}{s} - 0.0429\left(\frac{d}{s}\right)^2 - 0.109\left(\frac{d}{s}\right)^4 \right] \quad (10.48)$$

In this equation, M is in Henrys and lengths in meters. With a little imagination, this can be applied to the present problem. Once (10.41) or its equivalent (10.46) is solved for the currents, the losses, which are I^2R type losses, can be calculated. As a numerical example, we found the core and tieplate losses due to an unbalanced current of 20,000 Amps rms at 60 Hz in a transformer with the following geometric parameters,

Winding height (h)	0.813 m (32 in)
Winding radius (r_w)	0.508 m (20 in)
Leg center-center distance (d)	1.727 m (68 in)
Core radius	0.483 m (19 in)
Tieplate and core height	5.08 m (200 in)
Tieplate width	0.229 m (9 in)
Tieplate thickness	9.525×10^{-3} m (0.375 in)

The tieplates were actually subdivided into three plates in the width direction but this does not affect the calculation. For magnetic steel tieplates, the calculated core loss was 11 Watts and the total tieplate loss was 1182 Watts. For stainless steel tieplates, the calculated core loss was 28 Watts and the total tieplate loss was 2293 Watts. For the magnetic steel case, the current in the core legs was about 200 Amps and in the tieplates about 600 Amps. For the stainless steel case, the current in the core legs was about 300 Amps and in the tieplates about 500 Amps. The stainless steel losses are higher mainly because of the higher resistivity of the material coupled with the fact that the impedances, which limit the currents, are mainly inductive and hence nearly the same for the two cases.

10.3.2.4 Tank and Clamp Losses

Tank and clamp losses are very difficult to calculate accurately. Here we are referring to the tank and clamp losses produced by the leakage flux from the coils, examples of which are shown in Figs. 10.6 and 10.7. The eddy current losses can be obtained from the finite element calculation, however the axisymmetric geometry is somewhat simplistic if we really want to model a 3 phase transformer in a rectangular tank. Modern 3D finite element and boundary element methods are being developed to solve eddy current problems and should be widely available in the near future. In fact, some of these codes are already on the market. They are not, however, in routine use largely because of the complexity of 3D versus 2D modeling as well as computation time and memory limitations. These problems should disappear with time and fully 3D calculations of losses should become routine.

In the meantime, much can be done with a 2D approach. We refer to a study in Ref. [Pav93] where several projection planes in the 3D geometry were chosen for analysis with a 2D finite element program. The losses calculated with this approach agreed very favorably with test results. As that study and our own show, 2D models allow one to quickly asses the impact on losses of design changes such as the addition of tank and/or clamp shunts made of laminated electrical steel or the effect on losses of aluminum or copper shields at various locations. In fact the losses in Fig. 10.7 with tank and clamp shunts present were dramatically reduced compared with those in Fig. 10.6 where no shunts are present. While the real losses may not show as quantitative a reduction, the qualitative effect is real. Another study which was done very quickly, using the parametric capability of the 2D modeling, was to asses the impact of extending the clamp shunts beyond the top surface of the clamp. We found that some extension was useful in reducing the losses caused by stray flux hitting the vertical side of the clamp.

An example of where a 3D approach is crucial in understanding the effect of design options on losses concerns the laminated steel shunts on the clamps. The side clamps extend along all three phases of a transformer as shown in the top view of Fig. 10.20. Should the clamp shunts be made of laminations stacked flat on top of the clamps or should the laminations be on edge, i.e. stacked perpendicularly to the top surface of the clamp? In addition, should the shunts extend uninterrupted along the full length of the clamps or can they be subdivided into sections which cover a region opposite each phase but with gaps in between? The 2D model of Fig. 10.7 cannot really answer these questions. In fact, in the 2D model the flux in the clamp shunts is forced to return to the core eventually, whereas in the 3D model one can imagine the clamp shunt flux from the 3 phases canceling itself out within the shunts, assuming the shunts are continuous along the sides. Because of the laminated nature of the shunt material, the magnetic permeability and electrical conductivity are both anisotropic. This will affect both the flux and eddy current patterns in the shunts in a way that only a 3D model which allows for these anisotropies can capture.

A simple and reasonably accurate method of obtaining losses from a 2D axisymmetric model is to take the total losses in the tank or clamp and divide by the circumference of the circle with radius equal to the radial distance to the center of the core. This produces a loss per unit length which can be multiplied by the tank perimeter or total clamp length to get the total loss. If only half the geometry is modeled as in Figs. 10.6 and 10.7, then a factor of 2 is needed to get the total loss. With tank shunts present, some correction will be needed if shunts are applied in packets with spaces in between. In addition, the tank radius will need to be an effective radius, considering the actual tank shape. The losses calculated in this way ignore hysteresis which should be only a small component of the total loss for typical tank wall or clamp dimensions.

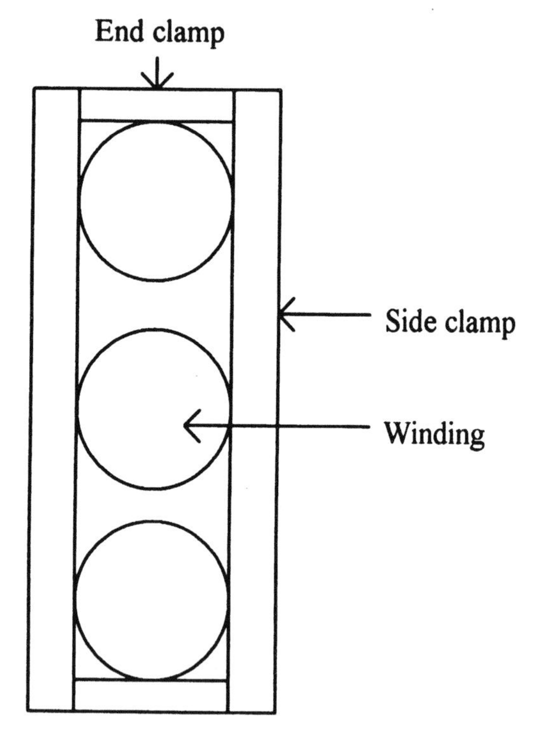

Figure 10.20 Geometry of clamp arrangement from a top view

10.3.2.5 **Tank Losses Due to Nearby Busbars**

When busbars carrying relatively high current pass close to the tank wall, their magnetic field induces eddy currents in the wall, creating losses. The busbars are usually parallel to the tank wall over a certain length. Because of the magnetic field direction, laminated magnetic shunts positioned near the busbar are not as effective at reducing these losses as are metallic shields made of aluminum or copper. Since busbars are usually present from all three phases, the question arises as to how the grouping of 2 or 3 busbars from different phases would affect the losses. Intuitively, we expect a reduction in the losses due to some cancellation in the magnetic field from different phases.

These loss issues can be studied by means of a 2D finite element program if we assume the busbars are infinitely long in the direction along their length. Thus we will calculate losses per unit length which can be multiplied by the total busbar length to get a reasonable approximation to the total loss. We can make this a parametric study by varying the distance of the busbars from the tank wall. Other parameters such as busbar dimensions could be varied, but this would greatly complicate the study. We have selected rather standard sized busbars and separation distances when a grouping of busbars from different phases is studied. Of course, a particular geometry can always be studied if desired. Because eddy current losses are proportional to the square of the current when linear magnetic materials are involved, it is only necessary to calculate the losses at one current.

Fig. 10.21 shows the geometry studied with the variable distance from the tank wall, d , indicated. We also show an aluminum shield which can be given the material properties of oil or air when the losses without shield are desired. The two busbar geometry applies to currents 120° apart and the three busbar geometry to a balanced 3 phase set of currents. The current magnitudes were set at 1 kA. The busbars were solid copper having cross sectional dimensions of 1.27 cm × 7.62 cm (0.5 in × 3 in) and separated from each other by 8.255 cm (3.25 in) in the 2 or 3 grouping cases. The aluminum shield was 1.27 cm (0.5 in) thick and 22.9 cm (9 cm) wide for the single busbar case and 30.5 cm (12 in) wide for the 2 and 3 busbar groupings. The tank wall was 0.95 cm (0.375 in) thick with material properties, $\mu_r = 200$, $\sigma = 4 \times 10^6 \ (\Omega\text{-m})^{-1}$. The study was done at 60 Hz.

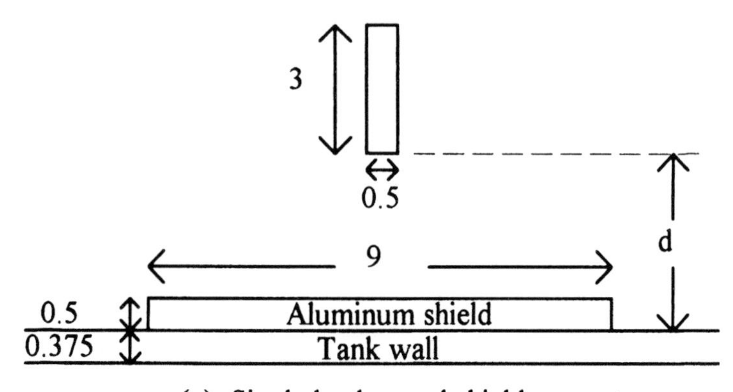

(a) Single busbar and shield geometry

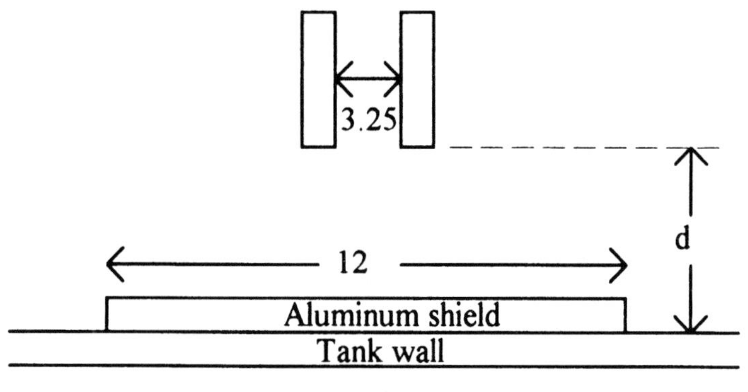

(a) 2 busbar and shield geometry

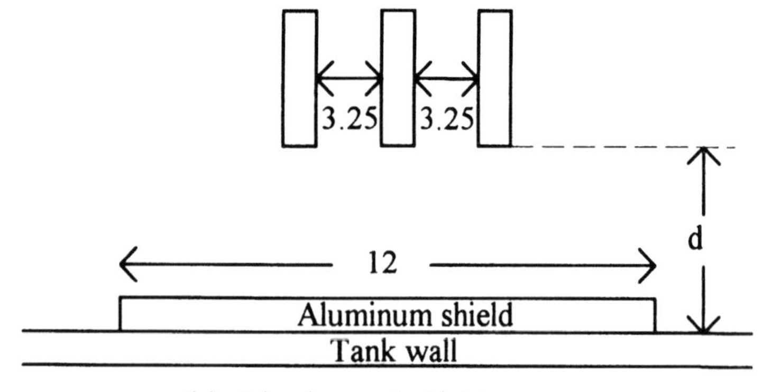

(a) 3 busbar and shield geometry

Figure 10.21 Geometry for loss study of busbars near the tank wall. The dimensions are in inches.

Since the busbars were modeled as solid copper and have a fairly large cross-sectional area, there were extra losses induced in them beyond the usual I^2R losses based on d.c. resistance. We found that these extra eddy current losses amounted to about 20% of the I^2R losses, regardless of whether one or groups of 2 or 3 busbars were studied. In the following figures, we plot the normalized losses which should be multiplied by the total length of the busbars or busbar group and by the current in kA squared to get the total loss, not including the loss in the busbars themselves. Thus the loss plotted is the tank loss when no shields are present and the tank plus shield loss when a shield is present.

Fig. 10.22 shows the normalized loss versus distance, with and without shield for the single busbar case. The figure shows that the losses without shield drop off with distance, d , while the losses with shield are relatively constant with distance. In the shielded case, the relative amounts of loss in the shield and tank wall change with distance, while their sum, which is plotted, remains almost constant. We see that shielding reduces the losses by about a factor of 5 at close distances and a factor of 2 - 3 at further distances compared with the unshielded losses.

Fig. 10.23 shows the normalized losses versus distance for the two busbar case. This figure has nearly the same features as the single busbar case. In fact, even the magnitude of the losses is nearly the same. However, since this figure applies to 2 busbars, this says that the losses can be cut in half by pairing 2 phases compared with leaving them separate. We also see in this case that shielding is very effective in reducing the losses.

Fig. 10.24 shows the normalized loss versus distance for a group of 3 busbars. We see that the magnitude of the loss is considerably reduced relative to the single or double busbar case. In fact, the losses shown should be divided by 3 to compare with the single busbar loss. In this case, shielding does not provide much improvement. This is because the 3 phase currents sum to zero at any instant of time, producing little net magnetic field at distances large relative to the conductor spacings. Fig 10.25 shows a flux plot of the 3 busbar case, where the flux cancellation can be directly observed.

It thus appears that loss reduction from busbars near bare tank walls can be achieved by pairing 2 or 3 phases together, the latter being preferable. Shielding is very effective in reducing the losses associated with 1 or 2 busbars from different phases but not for 3 busbars where the losses are small anyway.

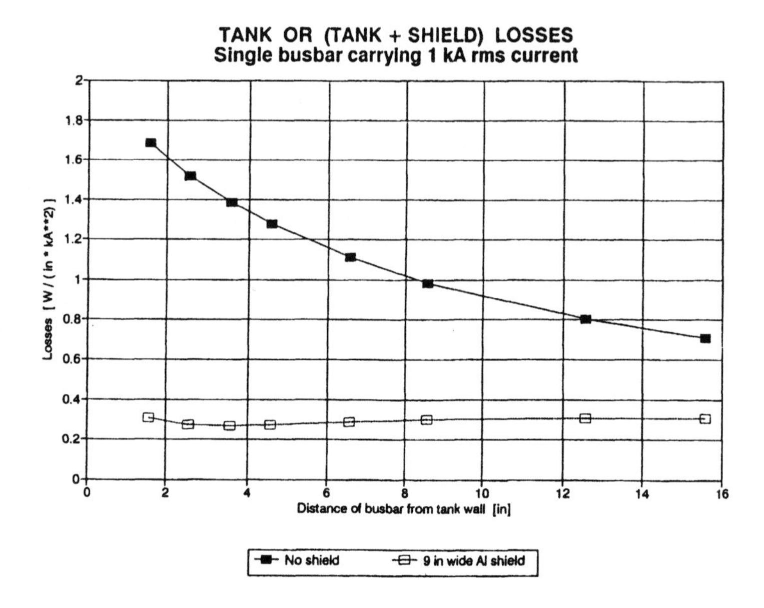

Figure 10.22 Stray losses due to a single busbar running parallel to the tank wall. The geometry is shown in Fig. 10.21a.

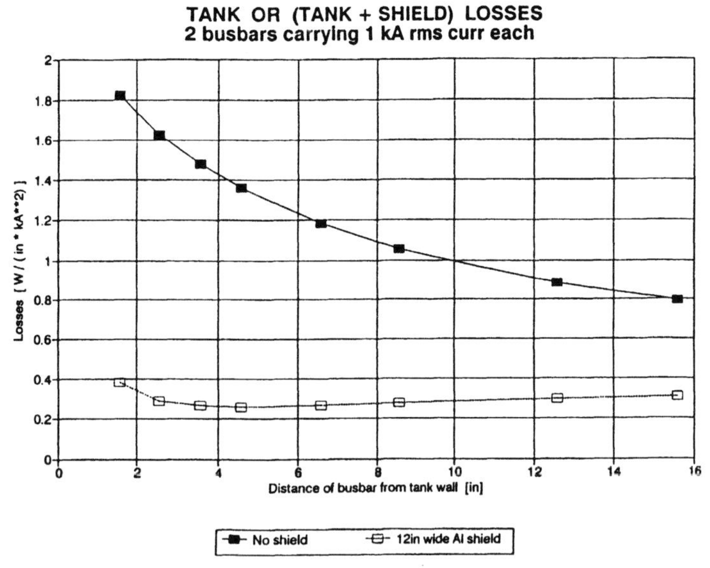

Figure 10.23 Stray losses due to two side by side busbars from different phases and running parallel to the tank wall. The geometry is shown in Fig. 10.21b.

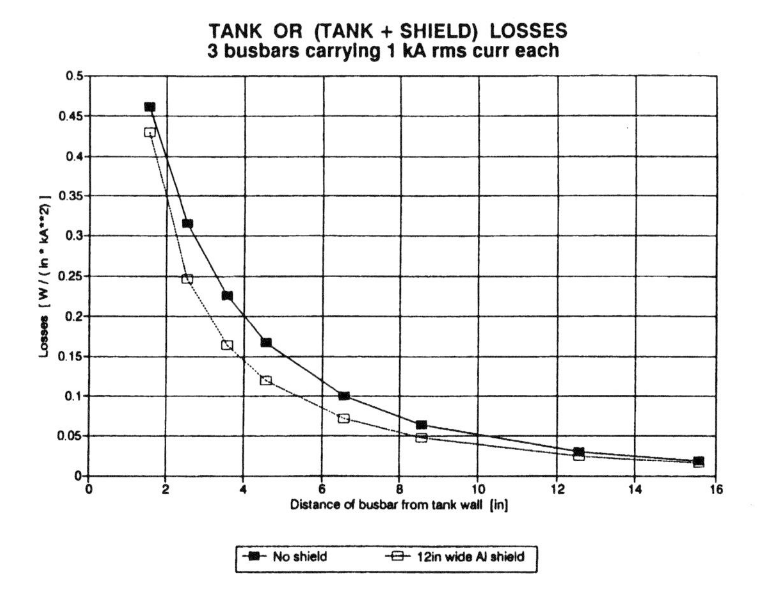

Figure 10.24 Stray losses due to a group of 3 busbars from different phases of a three phase system and running parallel to the tank wall. The geometry is shown in Fig. 10.21c.

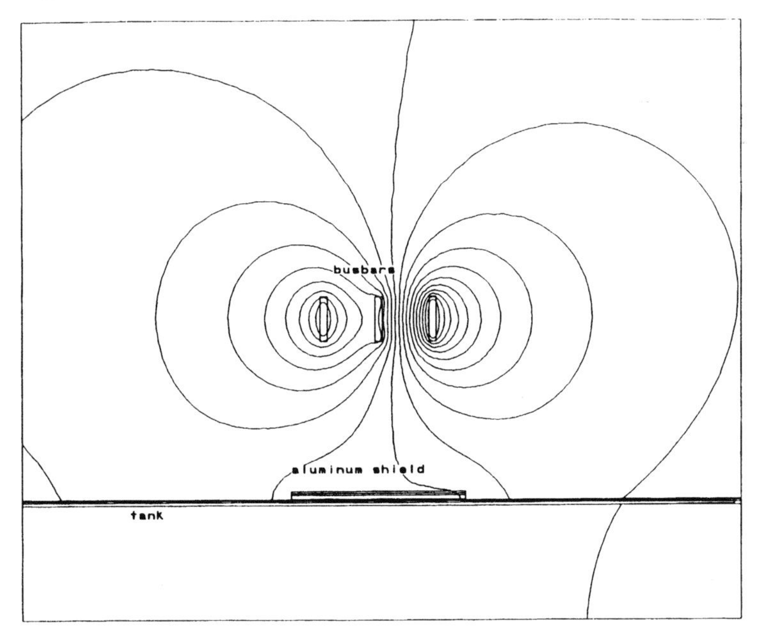

Figure 10.25 Flux plot from a balanced 3 phase set of busbars near tank wall with shield.

10.3.2.6 **Tank Losses Associated with the Bushings**

Current enters and leaves a transformer tank via the bushings. The bushings are designed to handle the voltage stresses associated with the voltage on the leads without breakdown as well as to dissipate heat due to the losses in the conductor which passes through the bushings. The conductor or lead, which must pass through the tank wall, creates a magnetic field which can generate eddy currents and accompanying losses in the tank wall near the lead. These losses must be calculated and appropriate steps taken to reduce them if necessary.

We can obtain a reasonable estimate of the tank losses due to a lead penetrating the tank wall by resorting to an idealized geometry as shown in Fig. 10.26. Thus we assume an infinitely long circular cross-section lead passing perpendicularly through the center of a circular hole in the tank wall. We can assume that the tank wall itself is a circle of large radius centered on the hole. Since the losses are expected to concentrate near the hole, the actual radial extent of the tank won't matter much. Also since most of the magnetic field which generates eddy currents in the tank comes from the portion of the lead near the tank, the infinite extent of the lead does not greatly affect the calculation.

The geometry in Fig. 10.26 is axisymmetric. Thus we need to work with Maxwell's equations in a cylindrical coordinate system. We assume the tank wall has permeability μ and conductivity σ. Combining Maxwell's equations with Ohm's law inside the tank wall and assuming that **H** varies harmonically in time as

$$\mathbf{H}(\mathbf{r},t) = \mathbf{H}(\mathbf{r})e^{j\omega t} \tag{10.49}$$

we get

$$\nabla^2\mathbf{H} = j\omega\mu\sigma\mathbf{H} \tag{10.50}$$

where **H** in (10.50) is only a function of position.

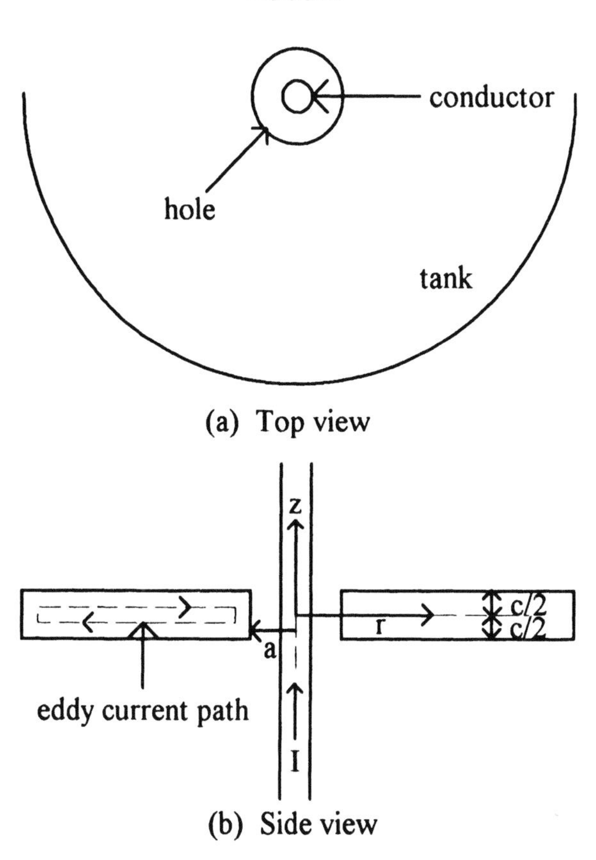

Figure 10.26 Idealized geometry and parameters used to calculate losses due to a lead passing through the tank

Because the problem is axisymmetric, **H** does not depend on φ, the azimuthal angle. We also assume that H has only a φ component. This is all that is needed to produce the expected eddy current pattern indicated in Fig. 10.26b, where the eddy currents approach and leave the hole radially and the paths are completed along short sections in the z-direction. The approximate solution we develop here will neglect these z-directed eddy currents which should not contribute much to the total loss. Thus we have

$$\mathbf{H}(r) = H_\varphi(r,z)\mathbf{a}_\varphi \tag{10.51}$$

where $\mathbf{a}_\varphi$ is the unit vector in the azimuthal direction. Expressing (10.50) in cylindrical coordinates and using (10.51), we get

$$\frac{1}{r}\frac{\partial}{\partial r}\left(r\frac{\partial H_\varphi}{\partial r}\right)-\frac{H_\varphi}{r^2}+\frac{\partial^2 H_\varphi}{\partial z^2}=j\omega\mu\sigma H_\varphi \tag{10.52}$$

We need to solve this equation subject to the boundary conditions

$$\begin{aligned} H_\varphi &= \frac{I}{2\pi a} \quad \text{at} \quad r=a \\ H_\varphi &= 0 \quad \text{at} \quad r=\infty \\ H_\varphi &= \frac{I}{2\pi r} \quad \text{at} \quad z=\pm\frac{c}{2} \end{aligned} \tag{10.53}$$

where a is the radius of the hole in the tank wall and c is the tank wall thickness. Once a solution is found, the current density $\mathbf{J}$ is given by

$$\mathbf{J}=\nabla\times\mathbf{H}=-\frac{\partial H_\varphi}{\partial z}\mathbf{r}+\frac{1}{r}\frac{\partial}{\partial r}\left(rH_\varphi\right)\mathbf{k} \tag{10.54}$$

where $\mathbf{r}$ and $\mathbf{k}$ are unit vectors in the r and z directions.

To solve (10.52), we use a separation of variables technique and write

$$H_\varphi(r,z)=R(r)Z(z) \tag{10.55}$$

Substituting into (10.52) and dividing by RZ, we obtain

$$\frac{1}{R}\left[\frac{1}{r}\frac{\partial}{\partial r}\left(r\frac{\partial R}{\partial r}\right)-\frac{R}{r^2}\right]+\frac{1}{Z}\frac{\partial^2 Z}{\partial z^2}=j\omega\mu\sigma \tag{10.56}$$

Thus we have two terms which are separately a function of r and z and whose sum is a constant. Hence, each term can be separately equated to a constant so long as their sum is $j\omega\mu\sigma$. We choose

$$\frac{1}{Z}\frac{\partial^2 Z}{\partial z^2}=j\omega\mu\sigma \quad , \quad \frac{1}{R}\left[\frac{1}{r}\frac{\partial}{\partial r}\left(r\frac{\partial R}{\partial r}\right)-\frac{R}{r^2}\right]=0 \tag{10.57}$$

Letting $k^2=j\omega\mu\sigma$ so that

$$k = (1+j)q \quad , \quad q = \sqrt{\frac{\omega\mu\sigma}{2}} \tag{10.58}$$

we can solve the first equation in (10.57) up to an overall multiplicative constant, Z_o ,

$$Z(z) = Z_o \cosh(kz) = Z_o\left[\cosh(qz)\cos(qz) + j\sinh(qz)\sin(qz)\right] \tag{10.59}$$

This equation takes into account the symmetry about the $z = 0$ plane. The second equation in (10.57) is solved up to an overall multiplicative constant, R_o , by

$$R(r) = R_o \frac{1}{r} \tag{10.60}$$

Thus the complete solution to (10.56), using the boundary conditions (10.53) is

$$H_\varphi = \frac{I}{2\pi r}\left[\frac{\cosh(qz)\cos(qz) + j\sinh(qz)\sin(qz)}{\cosh(q\,c/2)\cos(q\,c/2) + j\sinh(q\,c/2)\sin(q\,c/2)}\right] \tag{10.61}$$

This does not exactly satisfy the first boundary condition in (10.53) except in the limit of small qc/2.

Solving for the eddy currents, equation (10.54) , we see that the z-directed currents are zero when we substitute the solution (10.61). We expect that the losses contributed by these short paths will be small. Solving for the r-directed eddy currents, we obtain

$$J_r = -\frac{Iq}{2\pi r}\left\{\frac{\left[\sinh(qz)\cos(qz) - \cosh(qz)\sin(qz)\right] + j\left[\cosh(qz)\sin(qz) + \sinh(qz)\cos(qz)\right]}{\cosh(q\,c/2)\cos(q\,c/2) + j\sinh(q\,c/2)\sin(q\,c/2)}\right\} \tag{10.62}$$

and the loss density as a function of position, assuming I is an rms current, is

$$\frac{|J_r|^2}{\sigma} = \frac{2}{\sigma}\left(\frac{Iq}{2\pi r}\right)^2\left[\frac{\cosh(2qz) - \cos(2qz)}{\cosh(qc) + \cos(qc)}\right] \tag{10.63}$$

This drops off with radius as $1/r^2$ and so is highest near the opening. The total loss is given by

$$\text{Loss}_{\text{bush}} = 2\pi \int_a^b \int_{-c/2}^{c/2} \frac{|J_r|^2}{\sigma} r\,dr\,dz \tag{10.64}$$

The upper limit of the r integration, b, is chosen to be a large enough radius that the tank area of interest is covered. The result will not be too sensitive to the exact value chosen. Performing the integrations in (10.64), we obtain

$$\text{Loss}_{\text{bush}} = \frac{I^2 q}{\pi\sigma} \ln\left(\frac{b}{a}\right)\left[\frac{\sinh(qc) - \sin(qc)}{\cosh(qc) + \cos(qc)}\right] \tag{10.65}$$

For a magnetic steel tank wall, using $\mu_r = 200$, $\sigma = 4 \times 10^6\ (\Omega\text{-m})^{-1}$, $f = 60$ Hz, $c = 9.52 \times 10^{-3}$ m (0.375 in), we have $qc = 4.15$ as was found previously. Thus the small qc approximation to (10.65) cannot be used. However for a stainless steel tank wall of the same thickness and frequency, using $\mu_r = 1$, $\mu = 4\pi \times 10^{-7}$ H/m, $\sigma = 1.33 \times 10^6\ (\Omega\text{-m})^{-1}$, we have $qc = 0.17$ so the low qc limit of (10.65) may be used. This is

$$\text{Loss}_{\text{bush}} \xrightarrow[\text{small qc}]{} \left(\frac{\pi}{6}\right)\frac{I^2 f^2 \mu^2 c^3}{\rho} \ln\left(\frac{b}{a}\right) \tag{10.66}$$

where $\omega = 2\pi f$ and $\sigma = 1/\rho$ have been substituted. Applying these last two equations to the case where $I = 1000$ Amps rms (for normalization purposes), the hole radius $a = 0.165$ m (6.5 in), the outer radius $b = 0.91$ m (36 in), and using the above parameters for the two types of steel, we get $\text{Loss}_{\text{bush}}$(mag steel) = 61.9 Watts/$(\text{kA}_{\text{rms}})^2$ and $\text{Loss}_{\text{bush}}$(stainless) = 5.87×10^{-3} Watts/$(\text{kA}_{\text{rms}})^2$. These losses are associated with each bushing. Applying this to a situation where the lead is carrying 10 kA_{rms}, the loss in a magnetic steel tank wall would be 6190 Watts per bushing and in a stainless steel tank wall 0.587 Watts per bushing. These need to be multiplied by the number of bushings carrying the given current to get the total tank loss associated with the bushings. Thus when heavy currents are carried by the leads, it might be worth while to insert a stainless steel section of tank around the bushings, especially since these losses are concentrated in the part of the tank wall near the opening.

Another method of reducing these losses is to use a stainless steel insert only around part of the opening as shown in Fig. 10.27a. This will reduce the effective permeability as seen by the magnetic field which travels in concentric circles about the center of the opening. We can estimate this effect by considering the total flux which encircles the opening shown in Fig. 10.27b. By continuity, this flux crosses every imaginary radial cut we make centered on the opening. To a first approximation, therefore, we can assume that the induction B is the same everywhere along a circle as drawn in the figure. Letting the length of the circular path in the magnetic steel section be d_1 and the length in the stainless steel section be d_2, and using $B = \mu_1 H_1 = \mu_2 H_2$ along the path, we have from Ampere's law

$$\oint \mathbf{H} \cdot d\mathbf{l} = I \quad \Rightarrow \quad H_1 d_1 + H_2 d_2 = I$$
$$\Rightarrow \quad B\left(\frac{d_1}{\mu_1} + \frac{d_2}{\mu_2}\right) = I \quad \Rightarrow \quad B = \frac{I}{(d_1/\mu_1 + d_2/\mu_2)} \tag{10.67}$$

If this induction existed in a uniform material of permeability, μ_{eff}, we would have

$$B = \frac{\mu_{eff} I}{d} \quad , \quad d = d_1 + d_2 \tag{10.68}$$

Thus from (10.67) and (10.68), we obtain

$$\mu_{eff} = \frac{1}{(f_1/\mu_1 + f_2/\mu_2)} \quad , \quad f_1 = \frac{d_1}{d} \quad , \quad f_2 = \frac{d_2}{d} \tag{10.69}$$

The same analysis can be used to find the effective conductivity,

$$\sigma_{eff} = \frac{1}{(f_1/\sigma_1 + f_2/\sigma_2)} \tag{10.70}$$

For a strip of uniform width as shown in Fig. 10.27, f_1 and f_2 vary with radius so that an averaging process should be used for μ_{eff} and σ_{eff}. In view of the approximate nature of this calculation, we will simply assume that f_1 and f_2 are evaluated at some average radius, weighted towards smaller radii. Thus, if we take d_2 = 0.127 m (5 in) and r_{ave} = 0.254 m (10 in), we get f_1 = 0.92 , f_2 = 0.08 , and using the material parameters given previously, we obtain μ_{eff} = 11.8 (relative permeability) and $\sigma_{eff} = 3.45 \times 10^6 \ (\Omega\text{-m})^{-1}$. Using these parameters, we get qc = 0.935 so that, from (10.65) we obtain $Loss_{bush} = 2.05 \ Watts/(kA_{rms})^2$ for a single bushing. This is a considerable reduction from the loss without a stainless steel insert. At a current of 10 kA_{rms}, this is a loss of 205 Watts per bushing.

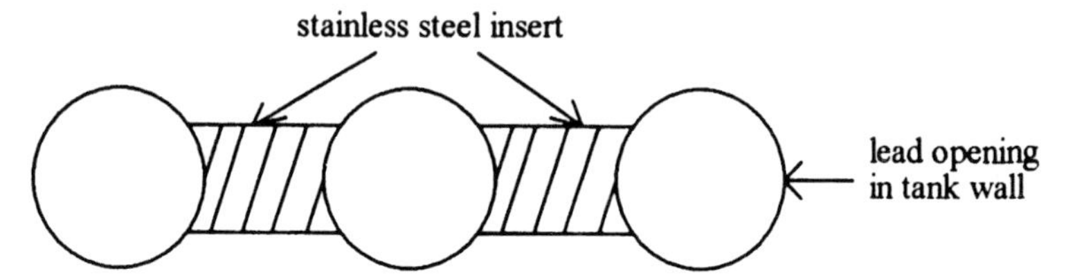

(a) Openings for leads in a three phase tank showing stainless steel inserts

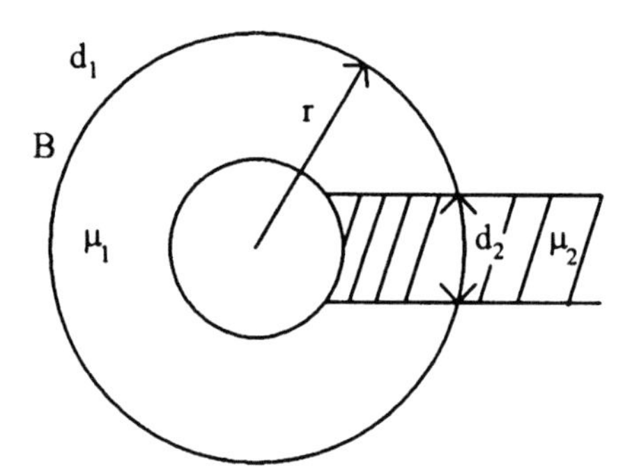

(b) Geometry for calculating the effective permeability

Figure 10.27 Openings for three phase leads in the tank wall with stainless steel strips inserted to reduce the permeability

10.3.3 Winding Losses Due to Missing or Unbalanced Cross-Overs

When two or more wires or cables in parallel are used to construct a winding, it is necessary to interchange their positions at suitable points along the winding in order to cancel out induced voltages produced by the stray flux. Otherwise, any net induced voltage can drive currents around the loop established when the parallel turns are joined at either end of the winding and these circulating currents will produce extra losses. The interchange points are called cross-overs and their number will depend on the number of parallel strands and on the symmetry of the stray flux pattern. An extreme example of cross-overs is provided by transposed cable where cross-overs occur among the individual strands comprising the cable at relatively short equally spaced intervals along the cable. Thus a single cable can be wound into a coil without any need for additional cross-overs within the cable.

In order to estimate the losses associated with missing or unbalanced transpositions, we make some simplifying assumptions. We assume that the windings are uniformly wound and that the magnetic field pattern is as shown in Fig. 10.28 for an inner and outer winding. Thus we are ignoring end effects. Although this assumption is reasonably accurate for coils long compared with their radii, it loses some accuracy for short coils or coil sections. However, it should be clear in the following development where improvements can be made for more accuracy.

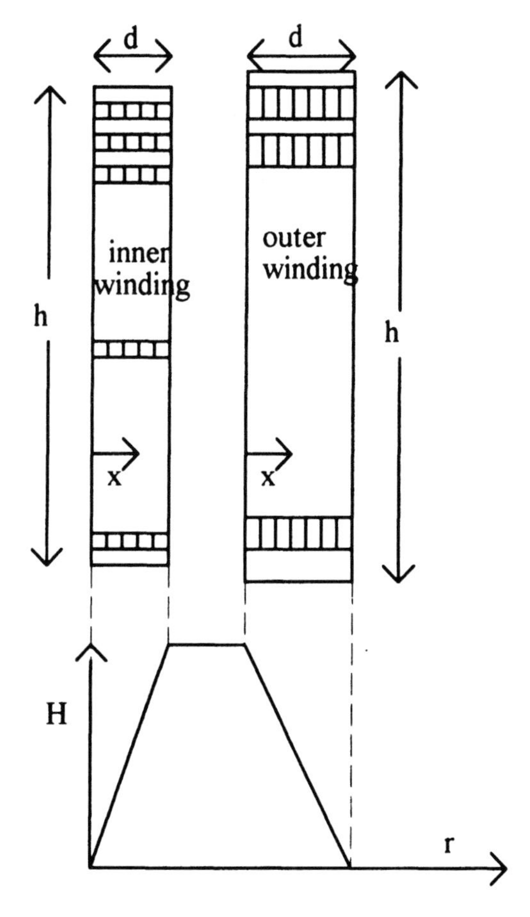

Figure 10.28 Idealized magnetic field pattern for an inner-outer winding pair. The parameters h and d can differ for the two windings.

Since the permeability in the coil region is close to the permeability of vacuum or air, μ_o , we have for the B field of an inner coil , B_{in} , or outer coil , B_{out} , in SI units,

$$B_{in} = \mu_o H_{in} = \frac{\mu_o NI}{h}\left(\frac{x}{d}\right) \quad , \quad B_{out} = \mu_o H_{out} = \frac{\mu_o NI}{h}\left(1-\frac{x}{d}\right) \tag{10.71}$$

where h is the coil height, d its radial build, N the total number of electrical turns in the winding, I the current, and x the distance through the radial build, starting at the inner radius. These parameters can differ for the inner and outer coils.

We assume that there are n radial turns in parallel. Any axial turns in parallel do not see a different flux pattern from their radial counterpart to first order. It is therefore unnecessary to consider them in calculating voltage imbalances. The voltage induced in the ith parallel radial turn by the leakage flux is different from that seen by the jth for $i \neq j$. Therefore it is necessary to interchange parallel radial turns so that each one occupies the position of all the others as the winding proceeds along the coil axis. If an interchange is missing or misplaced, different voltages will be induced in the parallel turns and the voltage differences will drive circulating currents around the loops formed by the top and bottom connections.

Let us calculate the voltage induced in a single turn in the ith position. This will equal the time derivative of the flux linkage, λ_i . Referring to Fig. 10.29, the flux linkage for the ith turn is

$$\lambda_i = 2\pi \int_0^{(i-1)t} B(r_o + x)dx + 2\pi \int_{(i-1)t}^{it} B(r_o + x)\left(i - \frac{x}{t}\right)dx \tag{10.72}$$

Here B_{in} or B_{out} should be substituted for B to obtain $\lambda_{i,in}$ or $\lambda_{i,out}$ for inner and outer coils. t is the turn radial thickness and $d = nt$. r_o is the inner radius of the coil. In the first term above, the integration is up to the inner radius of the turn so all the flux is linked by the whole turn. In the second term, the flux linkage changes from 1 at the start of the integral to 0 at the end and this is taken care of by the factor (i-x/t). For B_{out} we are ignoring the flux for radii $< r_o$. This is the same for all the turns and will not contribute to voltage differences between parallel radial turns.

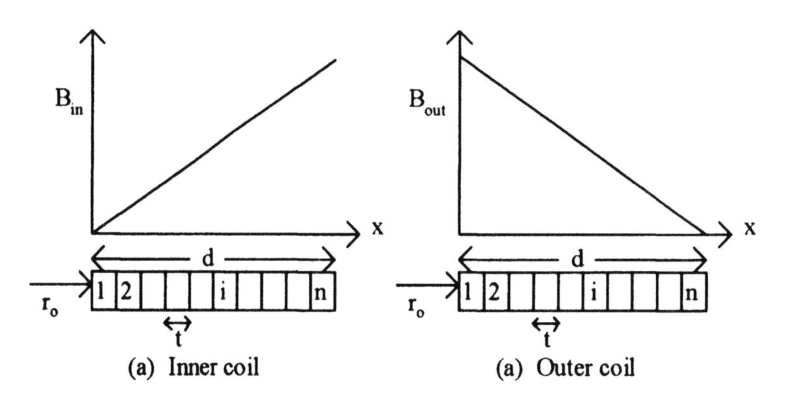

Figure 10.29 Radial parallel turn numbering scheme and geometric parameters for determining induced voltages

Substituting (10.71) into (10.72) and integrating, we obtain

$$\lambda_{i,in} = \frac{\pi\mu_o N I r_o t^2}{3hd}\left\{1 + 3i(i-1) + \frac{t}{2r_o}\left[2i\left(2i^2 - 3i + 2\right) - 1\right]\right\}$$

$$\lambda_{i,out} = \frac{\pi\mu_o N I r_o t}{h}\left\{2i - 1 - \frac{t}{3d}\left(1 - \frac{d}{r_o}\right)\left[3i(i-1) + 1\right] - \frac{t^2}{6dr_o}\left[2i\left(2i^2 - 3i + 2\right) - 1\right]\right\}$$

(10.73)

These are the flux linkages per turn. If there are N_u uncompensated electrical turns due to missed or unbalanced cross-overs, then (10.73) should be multiplied by N_u to get the total flux linkage corresponding to parallel radial turn i. Assuming sinusoidal currents of angular frequency $\omega = 2\pi f$, we obtain for the induced voltages,

$$V_{i,in} = \frac{\pi\mu_o \omega N_u N I r_o t^2}{3hd}\left\{1 + 3i(i-1) + \frac{t}{2r_o}\left[2i\left(2i^2 - 3i + 2\right) - 1\right]\right\}$$

$$V_{i,out} = \frac{\pi\mu_o \omega N_u N I r_o t}{h}\left\{2i - 1 - \frac{t}{3d}\left(1 - \frac{d}{r_o}\right)\left[3i(i-1) + 1\right] - \frac{t^2}{6dr_o}\left[2i\left(2i^2 - 3i + 2\right) - 1\right]\right\}$$

(10.74)

The number of unbalanced electrical turns needs to be estimated for each particular case. For instance, if there are n parallel radial turns, there should be a minimum of n-1 equally spaced cross-overs, assuming uniform flux along the winding. The number of electrical turns between cross-overs in this case is N/n. The simplest case to consider is that where a cross-over is misplaced by a fraction f of N/n. Then the two adjacent sections contain (1+f)N/n and (1-f)N/n turns. If a cross-over is completely missed, then $f = 1$. We can even imagine situations where $f > 1$.

Perhaps the best way to handle the variety of situations which may occur is to calculate the induced voltages and resulting induced currents in each section of coil between cross-overs, including the sections between the coil ends and the nearest cross-over. In this way the whole coil is covered. If the sections are balanced, the induced currents will automatically sum to zero. Otherwise, there will be net unbalanced currents in the radial turns which will result in losses. We must identify the positions of the radial turns in the starting configuration which may, without loss of generality, be numbered consecutively starting at the inner radius and follow these turns in the configurations which result after the cross-overs. We then sum the currents for each turn in the starting configuration as it advances through the coil, recognizing the fact that its radial position is different in the different configurations. In carrying this out, we will assume that the cross-overs are accomplished by shifting the turns to the right by one turn position. Other cross-over schemes could be accommodated with little additional effort. We also refer to the turns in the different sections as N_u even though these no longer refer to the net unbalanced turns. We will also refer to the section involved as the unbalanced section.

Now we must consider the circuit model to which these induced voltages should be applied. This is sketched in Fig. 10.30. The parallel radial turns are joined at the winding ends. Since we are only interested in circulating currents, the main winding voltage drop across the coil and its associated load current are eliminated. Each parallel radial turn, i , has a resistance, R_i , a self inductance, L_i , and is mutually coupled to all other turns as indicated by the mutual inductances, M_{ik}. Before evaluating these parameters, let's look at the circuit equations. Assuming sinusoidal conditions, we have

$$\mathbf{V}_i = (R_i + j\omega L_i)\mathbf{I}_i + j\omega \sum_{k \neq i}^{n} M_{ik}\mathbf{I}_k \quad , \quad i = 1,\cdots,n \quad ; \quad \sum_{k=1}^{n} \mathbf{I}_k = 0 \qquad (10.75)$$

We are using bold faced type here to denote phasors. Singling out the i = 1 equation, we can rewrite the last equation in (10.75)

$$\mathbf{I}_1 = -\sum_{k=2}^{n} \mathbf{I}_k \tag{10.76}$$

Subtracting the $\mathbf{V}_1$ equation from the voltage equations in (10.75) and using (10.76), we obtain

$$\mathbf{V}_i - \mathbf{V}_1 = \left[R_1 + R_i + j\omega(L_1 + L_i - 2M_{1i})\right]\mathbf{I}_i + \sum_{k \neq 1,i}^{n}\left[R_1 + j\omega(L_1 + M_{ik} - M_{1i} - M_{1k})\right]\mathbf{I}_k \tag{10.77}$$

where $i = 2,\cdots,n$. We have used $M_{ik} = M_{ki}$ which is valid for linear systems.

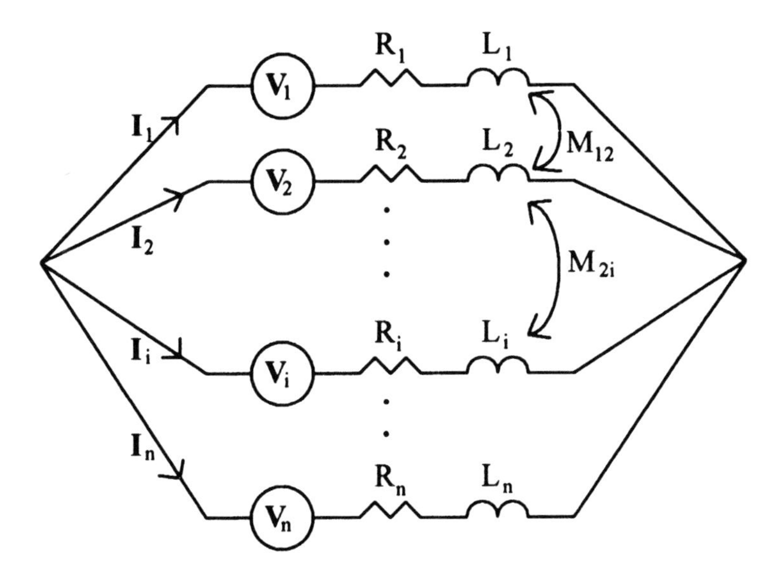

Figure 10.30 Circuit model of n parallel radial turns in a winding

The expression in (10.77), $L_1 + L_i - 2M_{1i}$, is the two winding leakage inductance between turns 1 and i. Because any two turns are tightly coupled over a fully balanced winding where their positions are shifting and interchanging, this would average to zero. Thus to a good approximation it need only be evaluated over the unbalanced section. Similarly the term, $L_1 + M_{ik} - M_{1i} - M_{1k}$, would average to zero over a balanced winding. This is because the M terms would average to a common value, M, and the residual, $L_1 - M$, is the single winding leakage impedance between turn 1 and any other turn and it will average to zero because of the close coupling of the turns. Therefore, this expression need only be evaluated over the unbalanced section. However, the resistances in (10.77) must be evaluated over the full length of the winding since they do not depend on the turn coupling. This will have nearly the same value for all the turns.

In order to evaluate the self and mutual inductances in the circuit model, we will assume that the unbalanced section is long enough axially that the same approximations as were used for the induced voltage calculation are also valid here. Thus the magnetic field for a coil of N_u turns located at the ith radial position is shown in Fig. 10.31. Analytically, the B-field is

$$B = \mu_o H = \begin{cases} \dfrac{\mu_o N_u I}{h_u} & \text{for } x \le 0 \text{ or } r \le r_o + (i-1)t \\ \dfrac{\mu_o N_u I}{h_u}\left(1 - \dfrac{x}{t}\right) & \text{for } 0 < x \le t \\ 0 & \text{for } x > t \end{cases} \tag{10.78}$$

where h_u is the axial height of the unbalanced section ($= h\, N_u / N$).

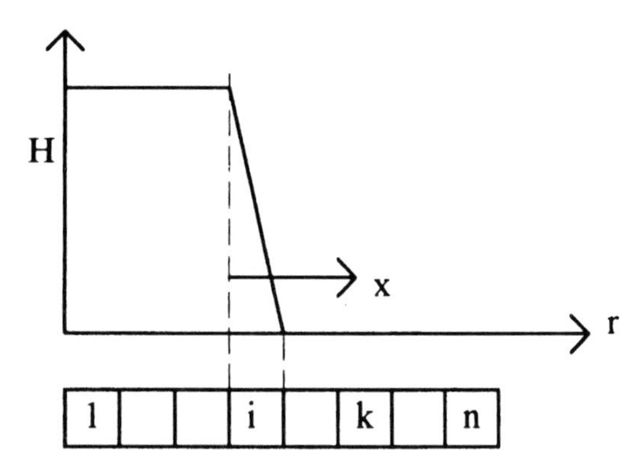

Figure 10.31 Magnetic field for a coil consisting of N_u turns at radial position i

The flux linkage to itself, λ_{ii}, is

$$\lambda_{ii} = \pi\left[r_o + (i-1)t\right]^2 \frac{\mu_o N_u^2 I}{h_u} + \frac{2\pi\mu_o N_u^2 I}{h_u}\int_0^t \left[r_o + (i-1)t + x\right]\left(1 - \frac{x}{t}\right)^2 dx$$

(10.79)

The turn linkage factor N_u (1-x/t) has been included in the second term. Performing the integral, we obtain

$$\lambda_{ii} = \frac{\pi\mu_o N_u^2 I}{h_u}\left\{[r_o + (i-1)t]^2 + \frac{2}{3}t[r_o + (i-1)t] + \frac{t^2}{6}\right\} \tag{10.80}$$

From this, the self inductance, L_i, is

$$L_i = \frac{\lambda_{ii}}{I} = \frac{\pi\mu_o N_u^2}{h_u}\left\{[r_o + (i-1)t]^2 + \frac{2}{3}t[r_o + (i-1)t] + \frac{t^2}{6}\right\} \tag{10.81}$$

so that

$$L_1 = \frac{\pi\mu_o N_u^2}{h_u}\left(r_o^2 + \frac{2}{3}r_o t + \frac{t^2}{6}\right) \tag{10.82}$$

The mutual inductance between turns i and k can be found, again using Fig. 10.31 and (10.78). Let turn k be outside of turn i. Then the flux linked to turn k from flux produced by turn i, λ_{ik}, is

$$\begin{aligned}\lambda_{ik} &= \pi[r_o + (i-1)t]^2 \frac{\mu_o N_u^2 I}{h_u} + \frac{2\pi\mu_o N_u^2 I}{h_u}\int_0^t [r_o + (i-1)t + x]\left(1 - \frac{x}{t}\right)dx \\ &= \frac{\pi\mu_o N_u^2 I}{h_u}\left\{[r_o + (i-1)t]^2 + t[r_o + (i-1)t] + \frac{t^2}{3}\right\}\end{aligned}$$

(10.83)

so that the mutual inductance is

$$M_{ik} = \frac{\lambda_{ik}}{I} = \frac{\pi\mu_o N_u^2}{h_u}\left\{[r_o + (i-1)t]^2 + t[r_o + (i-1)t] + \frac{t^2}{3}\right\} \tag{10.84}$$

This same expression would have been obtained if k were the coil creating the magnetic field. Thus in (10.84), the terms involving i on the right hand side refer to the inner winding of the pair. This insures that $M_{ik} = M_{ki}$. From (10.84), we obtain

$$M_{1k} = M_{1i} = \frac{\pi\mu_o N_u^2}{h_u}\left(r_o^2 + r_o t + \frac{t^2}{3}\right) \tag{10.85}$$

since 1 is always the inner winding in the unbalanced section.

Using these formulas, we find for the two winding leakage impedance term in (10.77),

$$L_1 + L_i - 2M_{1i} = \frac{2\pi\mu_o N_u^2}{h_u}\left(r_o(i-1)t + (i-1)^2\frac{t^2}{2} + (i-1)\frac{t^2}{3} - \frac{r_o t}{3} - \frac{t^2}{6}\right) \tag{10.86}$$

This formula holds for $i > 1$. This same expression could have been obtained by energy methods. The second expression in (10.77) can also be obtained from the above formulas,

$$L_1 + M_{ik} - M_{1i} - M_{1k} = \frac{2\pi\mu_o N_u^2}{h_u}\left(r_o(i-1)t + (i-1)^2\frac{t^2}{2} + (i-1)\frac{t^2}{2} - \frac{r_o t}{6} - \frac{t^2}{12}\right) \tag{10.87}$$

This last formula holds for $i < k$ and $i, k > 1$. If $i > k$, substitute k for i on the right hand side.

The resistances, R_i and R_1 in (10.77) are roughly the same for all turns. In the standard formula $R = \rho\ell/A$, ℓ is the length of a parallel radial turn over the whole winding and A is its cross sectional area, including the area of any corresponding axially displaced turns in parallel with it.

Let us cast (10.77) in matrix form by defining a matrix X with elements

$$X_{ii} = \omega\left(L_1 + L_{i+1} - 2M_{1,i+1}\right), \quad i = 1,\cdots,n-1$$

$$X_{ik} = \omega\left(L_1 + M_{i+1,k+1} - M_{1,i+1} - M_{1,k+1}\right), \quad i,k = 1,\cdots,n-1, \quad i \neq k \tag{10.88}$$

i.e. we are renumbering the equations so that they start from 1 rather than 2. Similarly, define a resistance matrix R,

$$R_{ii} = R_1 + R_{i+1} \quad , \quad i = 1,\cdots,n-1$$
$$R_{ik} = R_1 \quad , \quad i,k = 1,\cdots,n-1 \quad , \quad i \neq k \tag{10.89}$$

Let us also define a vector of voltages, $\mathbf{E}$, and currents, $\mathbf{K}$, by

$$\mathbf{E}_i = \mathbf{V}_{i+1} - \mathbf{V}_1 \quad , \quad \mathbf{K}_i = \mathbf{I}_{i+1} \quad , \quad i = 1,\cdots,n-1 \tag{10.90}$$

Here bold faced type is doing dual service in distinguishing vectors and phasors. Then we can write (10.77) in matrix form

$$\mathbf{E} = (R + jX)\mathbf{K} \tag{10.91}$$

Separating $\mathbf{E}$ and $\mathbf{K}$ into real and imaginary parts,

$$\mathbf{E} = \mathbf{E}_{Re} + j\mathbf{E}_{Im} \quad , \quad \mathbf{K} = \mathbf{K}_{Re} + j\mathbf{K}_{Im} \tag{10.92}$$

we can write (10.91) as a real matrix equation

$$\begin{pmatrix} \mathbf{E}_{Re} \\ \mathbf{E}_{Im} \end{pmatrix} = \begin{pmatrix} R & -X \\ X & R \end{pmatrix} \begin{pmatrix} \mathbf{K}_{Re} \\ \mathbf{K}_{Im} \end{pmatrix} \tag{10.93}$$

where the elements in this equation are vectors or matrices. From (10.74), we see that the voltages depend on the main coil current which can be taken to be real so we can set $\mathbf{E}_{Im} = 0$.

These equations are solved for the complex currents $\mathbf{K}_i = \mathbf{I}_{i+1}$ and then $\mathbf{I}_1$ is given by
(10.76). These currents apply to the standard ordering, $1,\cdots,n$. These same equations, with different N_u and h_u are solved for the other configurations making up the winding and the currents corresponding to the position of turn i in each configuration are added for $i = 1,\cdots,n$. Thus we obtain

$$\mathbf{I}_{net,i} = \sum_{\text{all configurations}} \mathbf{I}_i \quad , \quad i = 1,\cdots,n \tag{10.94}$$

Here bold faced type indicates phasor quantities. Once these currents are obtained, the losses are found from

$$\text{Loss}_{\text{cross-over}} = \sum_{i=1}^{n} R_i \left| \mathbf{I}_{\text{net},i} \right|^2 \tag{10.95}$$

As a numerical example, we solved an inner coil problem with a cross-over shifted by 4 electrical turns and with the following parameters:

$r_o = 0.567$ m (22.3 in)
$h = 2.60$ m (102.5 in)
$d = 8.25$ cm (3.25 in)
$t = 2.06$ cm (0.813 in)
$N = 80$
$n = 4$
$N_u = 4$
$N/n = 20$
A = Resistance area = 3.71 cm^2 (0.575 in^2)
$I = 6449$ Amps
$\rho = 2 \times 10^{-8}$ Ω-m

We obtained for the losses due to circulating currents 12.3 kW. This should be compared with the total I^2R losses of 171 kW. Thus the circulating current loss amounts to about 7% of the I^2R loss in this case.

The above calculation could be improved by using more accurate expressions for the inductances and mutual inductances, reflecting the fact that the unbalanced sections are not necessarily axially very long.

11. THERMAL MODEL OF A CORE FORM POWER TRANSFORMER AND RELATED THERMAL CALCULATIONS

Summary A model of a core form power transformer is presented which utilizes a detailed network of oil flow paths through the coils and radiators. Along each path segment, oil velocities, temperatures, and temperature rises are computed. The oil flows may be either thermally or pump driven. Throughout the disk coils, the flow is assumed directed by means of oil flow washers. Temperatures are computed for each disk along the disk coils. Coils with non-directed oil flow are also treated but in less detail. The model includes temperature dependent oil viscosity, resistivity, and oil density as well as both temperature and velocity dependent heat transfer and friction coefficients. The resulting non-linear system of equations is solved iteratively. The radiator oil flow is also treated by means of a similar network model. Radiator cooling can be by natural convection or fans. Tank cooling by convection and radiation is also included. Iterations are performed in a back and forth manner between the coils and radiators until steady state is reached where the electric power losses equal the losses to the surroundings. Some assumptions regarding the temperature distribution of the tank oil and tank oil flows are made in order to tie the coil and radiator flows together. Although detailed output information is available such as path temperatures and velocities, average coil, coil hot spot, average oil, and top oil temperatures are also computed and compared with test data. In the transient version, time constants and times at overload until a particular hot spot or top oil temperature is reached can be obtained. Other thermal issues such as the loss of life determination, cable, tieplate, and tank wall temperatures are also addressed.

11.1 INTRODUCTION

A thermal model of an oil cooled power transformer is presented here, along with details of the computer implementation and experimental verification. Any such model, particularly of such a complex system, is necessarily approximate. Thus, the model assumes that the oil flows in definite paths and ignores local circulation or eddy patterns which may arise. Since we assume that the oil flow through the disk coils is guided by means of oil flow washers, this assumption should be fairly accurate for these types of coils. Although recent studies have shown that irregular eddy flow patterns may exist in non-directed oil flow cooling in vertical ducts [Pie92], these types of coils occur to a very limited extent in our designs. Such patterns may also occur in the bulk tank oil. We assume these are small compared with the major or average convective cooling flow in the tank. We further assume that the convective flow in the tank results in a linear temperature profile from the bottom of the radiators to the top of the coils in the tank oil external to the coils. The model likewise ignores localized heating which may occur, for example, due to high current carrying leads near the tank wall. It accounts for these types of stray losses in only an average way. However, a localized distribution of eddy current losses in the coils is allowed for, along with the normal I^2R losses.

The radiators we model consist of a collection of radiator plates spaced equally along inlet and outlet pipes. The plates contain several vertical ducts in parallel. The ducts have oblong shaped, nearly rectangular, cross-sections. Fans, vertically mounted, may or may not be present (or turned on). In addition to radiator cooling, cooling also occurs from the tank walls by both natural convection to the surrounding air and by radiation. Although the oil may be pumped through the radiators, most of our designs are without pumps so that the oil flow in the radiators and coils is laminar. This determines the expressions used for the heat transfer and friction coefficients along the oil flow paths.

Our model of a disk winding is similar to that of Oliver [Oli80]. However, since we consider the whole transformer, we need to reconcile the oil flows and heat generated by the individual coils plus stray losses with the radiator and tank cooling in an overall iteration scheme in order to arrive at a steady state condition. We also consider transient heating.

Previous thermal models of whole transformers have focused on developing analytic formulas with adjustable parameters to predict overall temperature rises of the oil and coils [IEE81, Blu51, Eas65, Tay58, Aub92, Pie92a]. While these produce acceptable results on average or for a standard design, they are less reliable when confronting a new or untried design. The approach taken here is to develop a model to describe the basic physical processes occurring in the unit so that reliance on parameter fitting is minimized. Such an approach can accommodate future improvements in terms of a more detailed description of the basic processes or the addition of new features as a result of a design change.

11.2 THERMAL MODEL OF A DISK COIL WITH DIRECTED OIL FLOW

This section is essentially a reprint of our published paper [Del99]. from the Proceedings of the 1999 IEEE Transmission and Distribution Conference, New Orleans, LA, 11-16 April, 1999, pp. 914-919.

The disk coil is assumed to be subdivided into directed oil flow cooling paths as shown in Fig. 11.1. We number the disks, nodes, and paths, using the scheme shown. The geometry is really cylindrical and the inner radius R_{in} is indicated. Only one section (region between two oil flow washers) and part of a second is shown, but there can be as many sections as desired. Each section can contain different numbers of disks and the number of turns per disk, insulation thickness, etc. can vary from section to section. The duct sizes can vary within a section as well as from section to section.

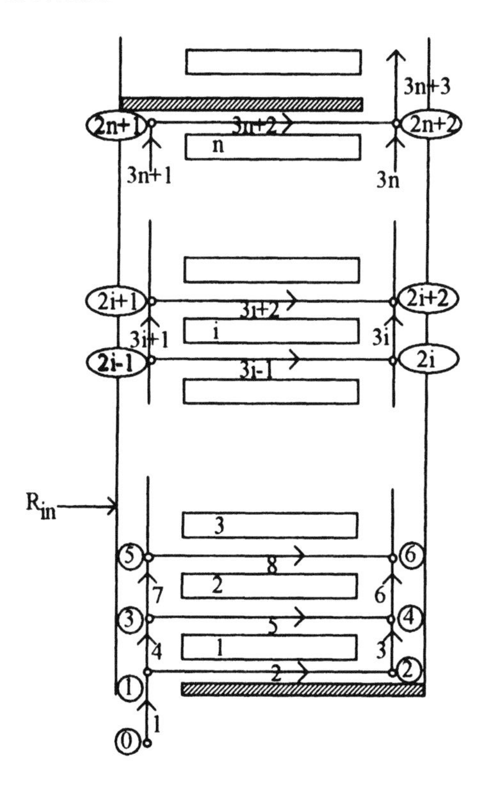

Figure 11.1 Disk, node, and path numbering scheme for a disk coil with directed oil flow

Fig. 11.2 shows the cross-sectional areas, A, and hydraulic diameters, D, of the various paths. Normally the vertical duct geometric parameters, labeled 1 and 2, are the same throughout the coil but the horizontal duct values can differ along the coil.

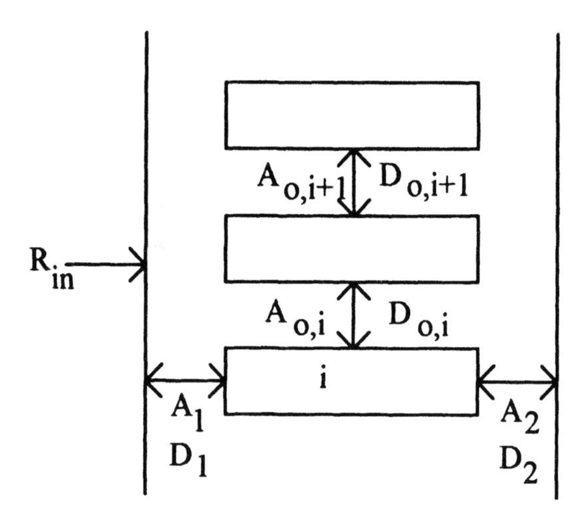

Figure 11.2 Cross-sectional areas and hydraulic diameters of different oil paths. The vertical ducts are assumed to be uniform along the coil.

In Fig. 11.3 we have indicated the unknowns which must be solved for at each node and path. These include the nodal temperatures, T, the nodal pressures, P, the path oil velocities, v, the path oil temperature rises, ΔT, and the disk temperatures, T_c. These are labeled with their corresponding node, path or disk number. Note that we are not allowing for a temperature profile along a single disk but are assuming that each disk is at a uniform temperature. This is an approximation which could be refined if more detail is required.

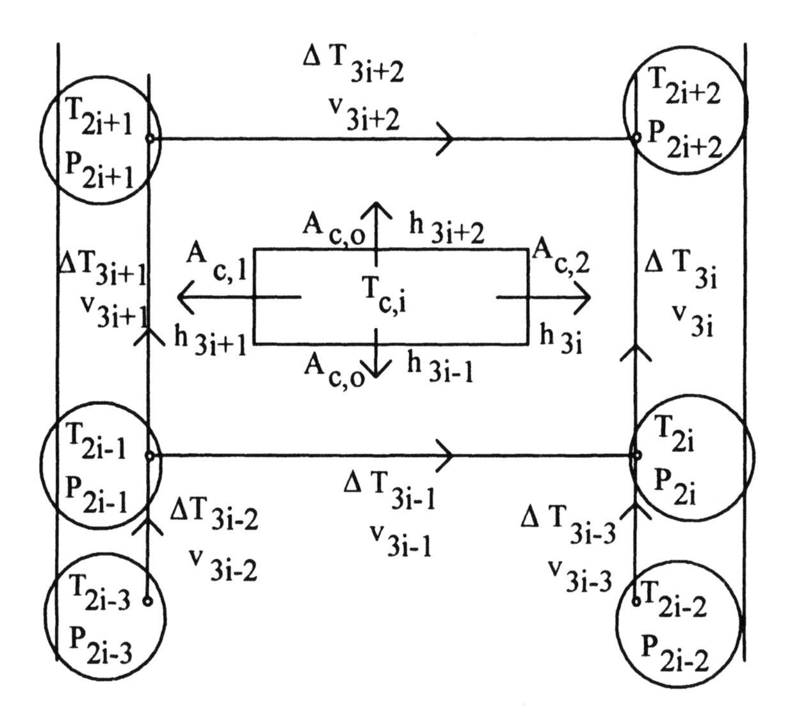

Figure 11.3 Numbering scheme for disk temperatures, node temperatures, node pressures, path velocities, and path temperature changes. Also indicated are the heat transfer surface areas, A_c , and the heat transfer coefficients, h, for these surfaces.

11.2.1 Oil Pressures and Velocities

Along a given path, the oil velocities are uniform since the cross-sectional area is assumed to remain constant. Ignoring gravitational effects, the pressure drop along a given path is only required to overcome friction Treating a generic path and labeling the pressures at the beginning and end of the path P_1 and P_2 respectively, we can write, using standard notation [Dai73],

$$P_1 - P_2 = \frac{1}{2}\rho f \frac{L}{D} v^2 \tag{11.1}$$

where ρ is the fluid density, f the friction coefficient, L the path length, D the hydraulic diameter, and v the fluid velocity. We note that the hydraulic diameter is given by

$$D = 4 \times \text{cross-sectional area / wetted perimeter}$$

MKS units are used throughout this report. For laminar flow in circular ducts, $f = 64 / Re_D$,
where Re_D is the Reynolds number, given by

$$Re_D = \frac{\rho v D}{\mu} \tag{11.2}$$

where μ is the fluid viscosity. For laminar flow in non-circular ducts, the number 64 in the expression for f changes. In particular, for rectangular ducts with sides a and b with $a < b$, we can write

$$f = \frac{K(a/b)}{Re_D} \tag{11.3}$$

where $K(a/b)$ is given approximately by

$$K(a/b) = 56.91 + 40.31(e^{-3.5\,a/b} - 0.0302) \tag{11.4}$$

This expression is based on a fit to a table given in reference [Ols80].
Substituting (11.2) and (11.3) into (11.1), we obtain

$$P_1 - P_2 = \frac{1}{2}\frac{\mu K L}{D^2} v \tag{11.5}$$

where $K = K(a/b)$ is implied. This equation is linear in the pressures and velocities. However, the oil viscosity is temperature dependent and this will necessitate an iterative solution. Based on a table of transformer oil viscosities verses temperature given in [Kre80], we achieved a good fit to the table with the expression

$$\mu = \frac{6900}{(T+50)^3} \tag{11.6}$$

with T in °C and μ in Ns/m^2.

For non-laminar flow ($Re_D > 2000$), the expression for f is more complicated and (11.5) would no longer be linear in v. Note that we are ignoring the extra friction arising from flow branching and direction changing at the nodes. There is considerable uncertainty in the literature as to what these additional frictional effects are for laminar flow. Instead we have chosen to allow the friction coefficient in (11.3) to be multiplied by a correction factor which is the same for all the branches. This correction factor is close to 1 in practice for disk coils.

We get an equation of type (11.5) for all the paths in Fig. 11.1. Note that we need only solve for the unknowns in one section at a time since the values at the topmost node and path can be taken as input to the next section. There are 3n+2 paths in the section, where n is the number of disks. However, the number of pressure unknowns is 2n+2, i.e. the number of nodes. We do not include the node 0, where the pressure must be input. This pressure will be given by the requirement that the overall pressure drop through the coil is determined by buoyancy effects as will be discussed below. The number of velocity unknowns is 3n+2 or the number of paths. Hence 2n+2 additional equations are required to solve for the pressures and oil velocities. We will treat the temperatures as fixed during the solution process. A separate set of equations will be derived for the temperatures and the two systems will ultimately be coupled through back and forth iteration.

The additional equations needed to solve for the pressures and velocity unknowns come from conservation of mass at the nodes. The mass of fluid of velocity v flowing through a duct of cross-sectional area A per unit time, dM/dt, is given by

$$\frac{dM}{dt} = \rho A v \tag{11.7}$$

Since the fluid is nearly incompressible, we can consider conservation of volume instead, where $Q = Av$ is the volume flow per unit time. Referring to Fig. 11.3 and using the areas indicated in Fig. 11.2, we obtain for a typical inner (smallest radii) node, 2i-1 ,

$$A_1 v_{3i-2} = A_1 v_{3i+1} + A_{o,i} v_{3i-1} \tag{11.8}$$

and for a typical outer node, 2i ,

$$A_{o,i} v_{3i-1} + A_2 v_{3i-3} = A_2 v_{3i} \tag{11.9}$$

However, the nodes at the beginning and end require special treatment (see Fig. 11.1). We want to calculate the velocity on path 3n+3 which is input to the next coil. We assume that the input velocity at the bottom of the section, v_1 on path 1, is known by some other means (overall energy balance). This still leaves 3n+2 unknown velocities to solve for but v_{3n+3} will be re-labeled v_1 and v_1 will be called v_0, a known input velocity. Thus at node 1 we have

$$A_1 v_o = A_1 v_4 + A_{o,1} v_2 \tag{11.10}$$

At node 2,

$$A_{o,1} v_2 = A_2 v_3 \tag{11.11}$$

At node 2n+1

$$A_1 v_{3n+1} = A_{o,n+1} v_{3n+2} \tag{11.12}$$

and at node 2n+2

$$A_{o,n+1} v_{3n+2} + A_2 v_{3n} = A_2 v_1 \tag{11.13}$$

We have 5n+4 equations to solve for the pressure and velocity unknowns, where n is the number of disks in the section. For n = 30, which is about as large a number of disks as would be used in one section, this yields 154 equations. Since the equations are quite sparse, a linear equation solver which uses sparsity techniques was used. Note that the cross-sectional areas must be calculated taking into account the area lost due to key spacers or vertical spacer sticks. These also influence the hydraulic diameters.

The overall pressure drop through the coil is produced by the difference in buoyancy between the hot oil inside the coil's cooling ducts and the cooler tank oil outside the coil. Thus

$$\Delta P_{coil} = (\rho_{ave,out} - \rho_{ave,in}) g H \tag{11.14}$$

where $\rho_{ave,in}$ is the average oil density inside the coil and $\rho_{ave,out}$ the average density immediately outside the coil, g is the acceleration of gravity, and H the coil height. Letting β be the volume coefficient of thermal expansion, we have

$$\frac{1}{\rho}\frac{d\rho}{dT} = -\beta \tag{11.15}$$

and thus, since $\beta = 6.8\times10^{-4}$ /°K for transformer oil, we have to a good approximation over a reasonably large temperature range, $\Delta\rho = -\beta\rho\Delta T$ so that (11.14) can be written

$$\Delta P_{coil} = \beta\rho g H(T_{ave,in} - T_{ave,out}) \tag{11.16}$$

Because we are only considering pressures which produce oil flows, the pressure at the top of the coil when steady state is achieved should be zero. This means that $\Delta P_{coil} = P_0$, the pressure at node 0 in Fig. 11.1.

The oil velocity into the coil along path 1 is determined by the overall energy balance. The energy per unit time acquired by the oil must equal the energy per unit time lost by the coil. In steady state, the latter is just the total resistive loss of the coil. Thus

$$\rho c v_o A_1 \Delta T_{oil} = \sum_{i=1}^{\#disks} I^2 R_i \tag{11.17}$$

where c is the specific heat of the oil, ΔT_{oil} is the increase in temperature of the oil after passing through the coil, and R_i is the temperature dependent resistance of disk i , including eddy current effects. We use

$$R_i = \gamma_o(1+\alpha\Delta T_i)(1+ecf_i)\ell_i/A_{turn} \tag{11.18}$$

where γ_0 is the resistivity at some standard temperature, T_{std}, ΔT_i is the temperature rise of the disk above the standard temperature, α the temperature coefficient of resistivity, ecf_i the fraction of the normal losses due to eddy currents for disk i , ℓ_i the length of the cable or wire in disk i , and A_{turn} the cross-sectional current-carrying area of the cable or wire. Note that ecf_i can vary from disk to disk to account for the effects of non-uniform stray flux along the coil. For a given temperature distribution for the disks and for known input and output oil temperatures, (11.17) can be solved for v_0. These temperatures will be given by a subsequent analysis and a back and forth iteration is required to achieve consistency. Using these results, the path 1 equation can be written

$$P_o - P_1 = \frac{1}{2}\frac{\mu K L}{D^2} v_o \tag{11.19}$$

which contains only P_1 as an unknown.

After solving these equations for the first section, we proceed to the next section, taking $P_{0,second} = P_{2n+2}$ and $v_{0,second} = v_1 = v_0\ (A_1\ /\ A_2)$. We proceed similarly from section to section. It is necessary to interchange A_1 and A_2 since the oil enters alternate vertical ducts in adjacent sections. This strategy keeps the equations looking the same from section to section. Some sort of relaxation technique must be used to keep the iteration process from becoming unstable. Thus P_0 and v_0 must vary gradually from iteration to iteration.

11.2.2 Oil Nodal Temperatures and Path Temperature Rises

For a given path, the temperature rise of the oil, ΔT, is determined from an energy balance. Thus the energy rise of the oil per unit time is

$$c\Delta T \frac{dM}{dt} = c\rho Av\Delta T \tag{11.20}$$

where a constant oil specific heat has been assumed. The energy lost per unit time through a surface of a conductor is given by

$$hA_c(T_c - T_b) \tag{11.21}$$

where h is the heat transfer coefficient, A_c the surface area, T_c the conductor temperature, and T_b the average (bulk) oil temperature in the adjacent duct. h should include the effects of conductor insulation as well as convective heat transfer and can be expressed

$$h = h_{conv} \Big/ \left(1 + \frac{h_{conv}\tau_{insul}}{k_{insul}}\right) \tag{11.22}$$

where h_{conv} is the convection heat transfer coefficient, τ_{insul} is the insulation thickness, and k_{insul} is the thermal conductivity of the insulation. While k_{insul} is nearly constant over the temperature range of interest, h_{conv} varies with temperature and oil velocity. For laminar flow in ducts, we use [Kre80]

$$h_{conv} = 1.86 \frac{k}{D}\left(Re_D\, Pr \frac{D}{L}\right)^{0.33}\left(\frac{\mu}{\mu_s}\right)^{0.14} \tag{11.23}$$

where k is the thermal conductivity of the oil, D the hydraulic diameter, L the duct length, Re_D the Reynolds number (11.2), Pr the Prandtl number ($Pr = \mu c / k$), μ the viscosity of the bulk oil, and μ_s the oil viscosity at the conductor surface. Equation (11.23) applies when Re_D Pr D/L > 10. A correction must be applied for smaller values. The major temperature variation comes from the viscosity (11.6) and the velocity dependence comes from the Reynold's number. The other parameters are nearly constant over the temperatures of interest. For transformer oil, we use [Kre80] $\rho = 867$ kg/m^3, $c = 1880$ J / kg°C, $k = 0.11$ W/m°C.

Equating energy gains and losses per unit time (equations (11.20) and (11.21)), we have for a horizontal duct between 2 disks (path 3i-1 in Fig. 11.3)

$$c\rho A_{o,i} v_{3i-1} \Delta T_{3i-1} = h_{3i-1} A_{c,o} \left[(T_{c,i} - T_{b,3i-1}) + (T_{c,i-1} - T_{b,3i-1}) \right] \quad (11.24)$$

where losses occur through 2 conductor surfaces and the bulk duct oil temperature, $T_{b,3i-1}$, is approximately given by

$$T_{b,3i-1} = T_{2i-1} + \Delta T_{3i-1}/2 \quad (11.25)$$

with T_{2i-1} the nodal oil temperature at the duct's entrance. For horizontal ducts at the beginning and end of a section, only one conductor surface contributes to the right side of (11.24). For the vertical ducts, we have

$$\begin{aligned} c\rho A_1 v_{3i+1} \Delta T_{3i+1} &= h_{3i+1} A_{c,1} \left(T_{c,i} - T_{b,3i+1} \right) \\ c\rho A_2 v_{3i} \Delta T_{3i} &= h_{3i} A_{c,2} \left(T_{c,i} - T_{b,3i} \right) \end{aligned} \quad (11.26)$$

for the left and right ducts 3i+1 and 3i respectively. We obtain 3n+2 equations in this manner. However, along the entrance and exit ducts (ducts 1 and 3n+3) we have $\Delta T = 0$ since no heat is going into the oil here. Hence this equation and unknown can be eliminated.

We obtain further equations from energy balance at the nodes. Thus the thermal energy which the oil carries into a node must equal that leaving. The thermal energy per unit time carried by a mass of oil moving with velocity v normal to a surface of area A is given by

$$\rho c A v (T - T_{ref}) \tag{11.27}$$

assuming c remains constant from some reference temperature T_{ref} up to the temperature of interest T. Balancing these energy flows at node 2i-1 in Fig. 11.3, we obtain

$$\rho c A_1 v_{3i-2}(T_{2i-3} + \Delta T_{3i-2} - T_{ref})$$
$$= \rho c A_1 v_{3i+1}(T_{2i-1} - T_{ref}) + \rho c A_{o,i} v_{3i-1}(T_{2i-1} - T_{ref})$$

Canceling ρc and using (11.8) we get

$$T_{2i-3} + \Delta T_{3i-2} = T_{2i-1} \tag{11.28}$$

Similarly at node 2i we obtain,

$$A_{o,i} v_{3i-1}(T_{2i-1} + \Delta T_{3i-1}) + A_2 v_{3i-3}(T_{2i-2} + \Delta T_{3i-3}) = A_2 v_{3i} T_{2i} \tag{11.29}$$

These equations are slightly modified for a few nodes at the beginning and end of a section, as was done for the mass conservation equations.

The temperature at node 0 or 1 will be taken as an input. It is the bottom tank oil temperature and will be obtained from an overall balance of heat generation and loss. The calculated temperature at the topmost node of the section, 2n+2 , will be taken as input to the next section so that the calculation can proceed section by section up the coil as was the case for the pressure-velocity equations. The temperature dependence of the heat transfer coefficients makes these equations non-linear. In addition, this coefficient depends on the oil velocity which requires iteration between these equations and the pressure-velocity equations. Another set of unknowns which appear in these equations are the disk temperatures T_c. These have not been dealt with yet. They are probably best solved for separately.

11.2.3 Disk Temperatures

The disk temperatures can be found by equating the I^2R losses to the heat passing through the disk surface in the steady state. Thus, for the generic disk shown in Fig. 11.3, we have, using previously defined parameters,

$$I^2R_i = h_{3i-1}A_{c,o}\left(T_{c,i} - T_{b,3i-1}\right) + h_{3i}A_{c,2}\left(T_{c,i} - T_{b,3i}\right) + h_{3i+1}A_{c,1}\left(T_{c,i} - T_{b,3i+1}\right) + h_{3i+2}A_{c,o}\left(T_{c,i} - T_{b,3i+2}\right) \tag{11.30}$$

Solving for $T_{c,i}$, we find

$$T_{c,i} = \frac{I^2R_i + h_{3i-1}A_{c,o}T_{b,3i-1} + h_{3i}A_{c,2}T_{b,3i} + h_{3i+1}A_{c,1}T_{b,3i+1} + h_{3i+2}A_{c,o}T_{b,3i+2}}{h_{3i-1}A_{c,o} + h_{3i}A_{c,2} + h_{3i+1}A_{c,1} + h_{3i+2}A_{c,o}} \tag{11.31}$$

We can use (11.25) to express the T_b in terms of previously calculated quantities and (11.18) for the R_i. Thus the disk temperatures can be obtained directly once the other unknowns are solved for. The equations must, however, all be iterated together until convergence to the steady-state is achieved. This coil steady-state depends on the average oil temperature of the tank oil outside the coil and the bottom tank oil which enters the coil These temperatures are obtained from an overall energy balance involving heat generation from all the coils and heat lost through the radiators and tank surfaces. This necessitates an overall iteration strategy involving all the coils, the radiators, and the tank. Before discussing this, we need to model coils having vertical cooling ducts with non-directed flow and the radiator and tank cooling.

11.3 THERMAL MODEL FOR COILS WITHOUT DIRECTED OIL FLOW

Our treatment of non-directed oil flow coils is fairly simplistic. These are used mainly as tap coils in our designs and thus carry a small fraction of the total power. As shown in Fig. 11.4, there are inner and outer vertical oil flow channels with cross sectional areas A_1, A_2 and hydraulic diameters D_1, D_2. The oil velocities in the two channels can differ. They are labeled v_1 and v_2 in the figure. The oil temperature is assumed to vary linearly from T_o at the coil bottom to $T_o + \Delta T_1$ and $T_o + \Delta T_2$ at the top of the inner and outer channel respectively. The conductor temperature is also assumed to vary linearly with an average value of T_c.

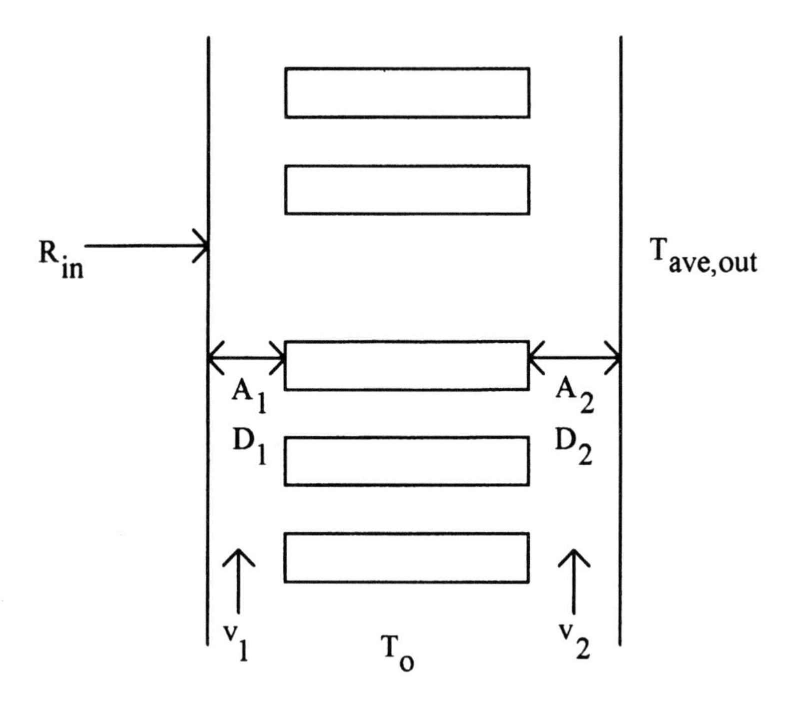

Figure 11.4 Thermal model of a non-directed oil flow coil

The thermal pressure drops in the two channels are given by (11.16) which becomes, in terms of the variables defined above

$$\Delta P_1 = \beta\rho g H\left(T_o + \Delta T_1/2 - T_{ave,out}\right)$$
$$\Delta P_2 = \beta\rho g H\left(T_o + \Delta T_2/2 - T_{ave,out}\right) \tag{11.32}$$

These pressure drops need only overcome the fluid friction in the two channels which, for laminar flow, is given by (11.5). In terms of the present parameters, this becomes

$$\Delta P_1 = \frac{1}{2}\frac{\mu_1 K_1 H}{D_1^2} v_1 \quad , \quad \Delta P_2 = \frac{1}{2}\frac{\mu_2 K_2 H}{D_2^2} v_2 \tag{11.33}$$

Equating (11.32) and (11.33), we obtain

$$\frac{\mu_1 K_1}{\beta \rho g D_1^2} v_1 - \Delta T_1 = 2\left(T_o - T_{ave,out}\right)$$

$$\frac{\mu_2 K_2}{\beta \rho g D_2^2} v_2 - \Delta T_2 = 2\left(T_o - T_{ave,out}\right) \tag{11.34}$$

The overall energy balance equation, analogous to (11.17) is

$$\rho c v_1 A_1 \Delta T_1 + \rho c v_2 A_2 \Delta T_2 = I^2 R \tag{11.35}$$

where R is the total resistance of the coil and is a function of T_C by a formula analogous to (11.18). Further thermal equations come from the energy balance between the surface heat loss by the coil to the separate oil flow paths. Thus, we have approximately

$$h_1 A_{c,1}\left[T_c - \left(T_o + \Delta T_1/2\right)\right] = \rho c A_1 v_1 \Delta T_1$$

$$h_2 A_{c,2}\left[T_c - \left(T_o + \Delta T_2/2\right)\right] = \rho c A_2 v_2 \Delta T_2 \tag{11.36}$$

where h_1 , h_2 are the surface heat transfer coefficients as given by (11.22) and $A_{c,1}$, $A_{c,2}$ are the conductor surface areas across which heat flows into the two oil channels. We take these areas to be half the total surface area of the conductor. This assumes the oil meanders into the horizontal spaces between the conductors on its way up the coil. We use (11.23) for h_{conv} with a smaller effective value for L than the coil height.

Equations (11.34), (11.35), and (11.36) are 5 equations in the 5 unknowns v_1 , v_2 , ΔT_1 , ΔT_2 , T_C . The quantities T_0 and $T_{ave,out}$ will be determined from an overall energy balance for the transformer and are considered as known here. These equations are non-linear because the μ's and h's depend on temperature and velocity and also because products of unknowns such as $v\Delta T$ occur. We use a Newton-Raphson iteration scheme to solve these equations. If desired, the pressure drops can be obtained from (11.33).

11.4 RADIATOR THERMAL MODEL

Our radiators are fairly typical in that they consist of a series of vertical plates containing narrow oil channels. The plates are uniformly spaced and attached to inlet and outlet pipes at the top and bottom. These pipes are attached to the transformer tank and must be below the top oil level for non-pumped flow. Fans may be present. They blow air horizontally through one or more radiators stacked side by side. The oil flow paths for a radiator are similar to those of a disk winding section turned on its side as shown in Fig. 11.5. In fact, the same node and path numbering scheme is used. In some of the paths, however, the positive flow direction is reversed. The analysis is very similar the that given previously for disk coils with directed oil flow and will only be sketched here.

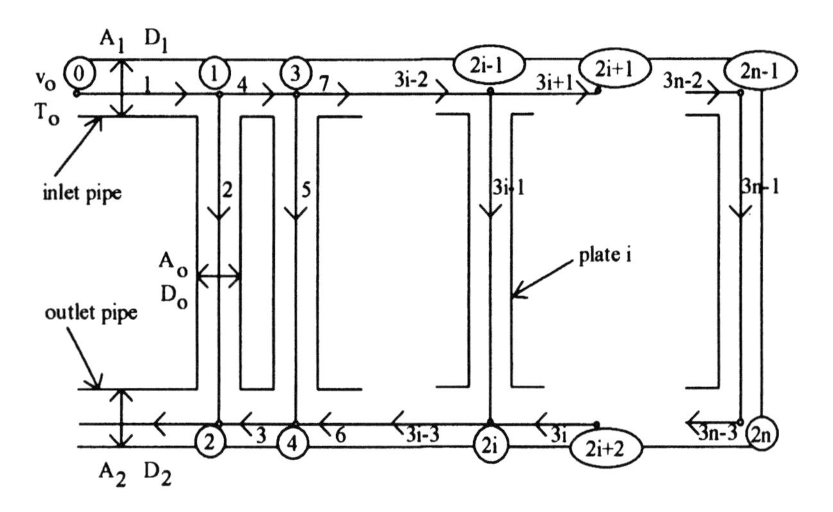

Figure 11.5 Node and path numbering scheme for a radiator containing n plates

The pressure differences along a path are balanced by the frictional resistance in steady state so that, for laminar flow, equation (11.5) holds along each path, where the velocity and pressure unknowns will be labeled by the appropriate path and node numbers. For n plates along a radiator, there are 2n nodes and 3n-1 oil paths as shown in Fig. 11.5. Thus we need 2n more equations. These are given as before by mass continuity at the nodes, similar to equations (11.7) - (11.13). Some differences will occur because the positive flow direction is changed for some of the paths. Also, similar to what was done for the coils, we determine the overall pressure difference across the radiator from buoyancy considerations so that an equation like (11.16) holds. However, $T_{ave,out}$ will differ from that used for the coils since the tank oil adjacent to the radiators will be hotter than that adjacent to the coils because the average radiator vertical position is above that of the coils. An overall energy balance is needed to obtain the input oil velocity so that an equation similar to (11.17) holds with the right hand side replaced by the total heat lost by the radiators.

In the radiator cooling process, the oil temperature drops as it passes downward through a radiator plate, giving up its heat to the air through the radiator surface. Fig. 11.6a shows a simplified drawing of a plate with some of the parameters labeled.

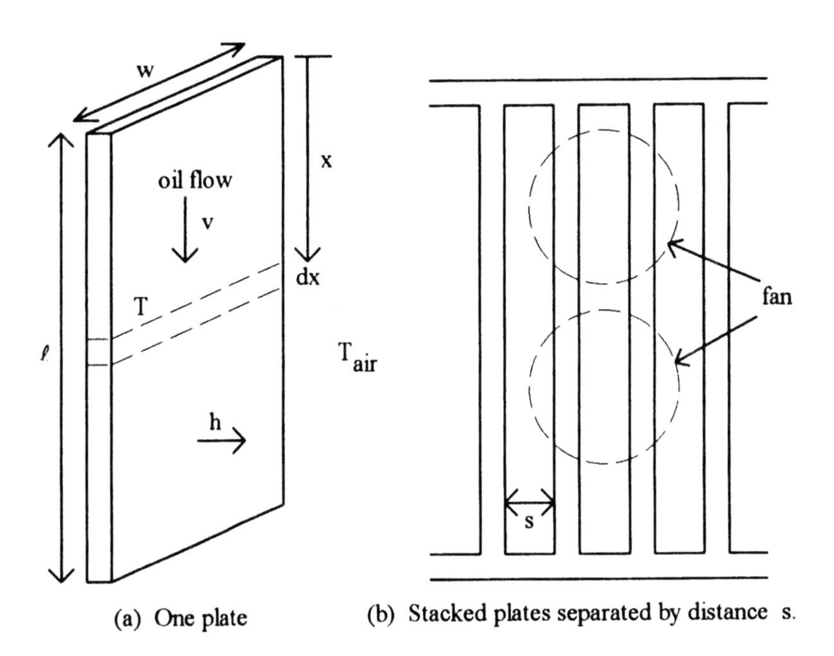

Figure 11.6 Parameters used in radiator cooling calculation

We consider a thin horizontal strip of radiator surface of area 2w dx $= 2w\,\ell\,(dx / \ell) = 2A_S\,(dx / \ell)$, where A_S is the area of one side of the plate and the factor of 2 accounts for both sides. ℓ is the plate height and w its width. The heat lost through this surface is $2hA_S(T - T_{air})(dx / \ell)$, where h is the heat transfer coefficient, T is the oil temperature at position x, and T_{air} is the ambient air temperature. The heat lost by the oil in flowing past the distance dx is $-\rho c v A_R dT$, where v is the oil velocity and A_R the cross-sectional area through which the oil flows through the plate. Equating these expressions and rearranging, we get

$$\frac{2hA_s}{\rho cvA_R\ell}dx = -\frac{dT}{T - T_{air}} \tag{11.37}$$

Assuming that h is constant here (it will be evaluated at the average temperature), we can integrate to obtain

$$\Delta T = \left(T_{top} - T_{air}\right)\left[1 - \exp\left(-\frac{2hA_s}{\rho cvA_R}\right)\right] \tag{11.38}$$

where ΔT is the temperature drop across the plate and T_{top} the top oil temperature.

We get an equation like this for each plate so that for plate i , v should be labeled v_{3i-1} , T_{top} as T_{2i-1} , h as h_i , and ΔT as ΔT_i . This gives n equations. The new unknowns are the n ΔT's and 2n nodal temperatures. However, the top nodal temperatures are all equal to the oil input temperature T_o since we are neglecting any cooling along the input and output pipes. At the bottom nodes, using a treatment similar to that used to get (11.29), we obtain for node 2i

$$A_R v_{3i-1}\left(T_o - \Delta T_i\right) + A_2 v_{3i} T_{2i+2} = A_2 v_{3i-3} T_{2i} \tag{11.39}$$

A slight modification of this is needed for node 2n. Thus we obtain sufficient equations to solve for all the unknowns. Iteration is required between the pressure-velocity equations and the T - ΔT equations since they are interdependent. Iteration is also required because of the non-linearities.

The expression for the surface heat transfer coefficient, h , depends on whether or not fans are used. For natural convection (no fans), we use an expression which applies to a row of vertical plates separated by a distance s [Roh85]

$$h = \frac{k_{air}}{s}\left[\left(\frac{24}{Ra}\right)^{1.9} + \left(\frac{1}{0.62Ra^{1/4}}\right)^{1.9}\right]^{-\frac{1}{1.9}} \tag{11.40}$$

Ra is the Rayleigh number, given by the product of the Grashof and Prandtl numbers

$$Ra = Gr\,Pr = \frac{g\beta_{air} c_{air} \Delta T \rho_{air}^2}{k_{air}\mu_{air}} \frac{s^4}{\ell} \tag{11.41}$$

where the compressibility β, specific heat c, density ρ, thermal conductivity k, and viscosity μ all apply to air at the temperature $(T_S + T_{air})/2$ where T_S is the average surface temperature of the plate. Also $\Delta T = T_S - T_{air}$. The quantities c, ρ, k, and μ for air vary with temperature [Kre80] and this must be taken into account in the above formulas since T_S can differ from plate to plate. However, we find in practice that a similar expression given later for tank cooling, (11.44) with ℓ replacing L, works better for our radiators so it is used.

When fans are blowing, we use the heat transfer coefficient for turbulent flow of a fluid through a long narrow channel of width s

$$h = 0.023 \frac{k_{air}}{D} Re_D^{0.8} Pr^{0.33} \tag{11.42}$$

where the hydraulic diameter $D = 2s$. The Reynolds number is given by (11.2) with ρ and μ for air, $D = 2s$, and v an average velocity determined by the characteristics of the fans and the number of radiators stacked together. The Prandtl number for air is nearly constant throughout the temperature range of interest and is $Pr = 0.71$.

A simple way of parametrizing the radiator air velocity with fans present which works well in practice is

$$v_n = \left(\frac{1}{n}\right)^p v_1 \tag{11.43}$$

where n is the number of stacked radiators cooled by a given fan bank, v_n is the air velocity flowing past the radiator surfaces, p is an exponent to be determined by test results, and v_1 is the fan velocity produced when only one radiator is present. We find in our designs that $p = 0.58$ works well. In addition v_1 was taken to be the nominal fan velocity as specified by the manufacturer. (This can be determined from the flow capacity and the fan's area.) We assumed that the radiator surface cooled by fans was proportional to the fraction of the fan's area covering the radiator side (Fig. 11.6b). The remaining surface was assumed cooled by convective cooling.

11.5 TANK COOLING

Cooling from the tank occurs by means of natural convection and radiation. Of course the radiators also cool to some extent by radiation, however this is small compared with convective cooling. This is because the full radiator surface does not participate in radiative cooling. Because of the nearness of the plates to each other, much of a plate's radiant energy is re-absorbed by a neighboring plate. The effective cooling area for radiation is really an outer surface envelope and as such is best lumped with the tank cooling. Thus the effective tank area for radiative cooling is an outer envelope, including the radiators. (A string pulled tautly around the tank plus radiators would lie on the cooling surface.) However, for convective cooling, the normal tank surface area is involved.

For natural convection in air from the tank walls of height L, we use for the heat transfer coefficient [Roh85]

$$h_{conv,tank} = \frac{k_{air}}{L}\left[\left(\frac{2.8}{\ln\left(1+\frac{5.44}{Ra_L^{1/4}}\right)}\right)^6 + 1.18\times 10^{-6} Ra_L^2\right]^{1/6} \tag{11.44}$$

where Ra is the Rayleigh number which, in this context is given by

$$Ra_L = \frac{g\beta_{air} c_{air} \Delta T \rho_{air}^2 L^3}{h_{air}\mu_{air}} \tag{11.45}$$

where the temperature dependencies are evaluated at $(T_s + T_{air})/2$ and $\Delta T = T_s - T_{air}$ as before. The Rayleigh number in this formula should be restricted to the range $1 < Ra_L < 10^{12}$. This is satisfied for the temperatures and tank dimensions of interest. T_s is the average tank wall temperature and T_{air} the ambient air temperature. Thus the heat lost from the tank walls per unit time due to convection, $W_{conv,tank}$, is given by

$$W_{conv,tank} = h_{conv,tank} A_{conv,tank} \left(T_s - T_{air}\right) \tag{11.46}$$

where $A_{conv,tank}$ is the area of the tank's lateral walls.

The top surface contributes to the heat loss according to a formula similar to (11.46) but with a heat transfer coefficient given by [Kre80]

$$h_{conv,top} = 0.15 \frac{k_{air}}{B} Ra_B^{0.333} \tag{11.47}$$

and with the tank top area in place of the side area. In (11.47), B is the tank width which must also be used in the Rayleigh number (11.45) instead of L. In this formula the Rayleigh number should be restricted to the range $8 \times 10^6 < Ra_B < 10^{11}$ which is also satisfied for typical tank widths in power transformers. On the bottom, we assume the ground acts as an insulator.

Radiant heat loss per unit time from the tank, $W_{rad,tank}$, is given by the Stephan-Boltzmann law

$$W_{rad,tank} = \sigma E A_{rad,tank} \left(T_{K,s}^4 - T_{K,air}^4\right) \tag{11.48}$$

where $\sigma = 5.67 \times 10^{-8}\ W / m^2\ °K^4$ is the Stephan-Boltzmann constant, E is the surface emissivity ($E \approx 0.95$ for gray paint), $A_{rad,tank}$ is the effective tank area for radiation, $T_{K,s}$ is the average tank surface temperature in °K, and $T_{K,air}$ is the ambient air temperature in °K. (11.48) can be written to resemble (11.46) with a convection coefficient given by

$$h_{rad,tank} = \sigma E \left(T_{K,s} + T_{K,air}\right)\left(T_{K,s}^2 + T_{K,air}^2\right) \tag{11.49}$$

This is temperature dependent but so is $h_{conv,tank}$.

The total power loss by the tank is given by the sum of (11.46), together with the corresponding expression for the top heat loss, and (11.48). This is then added to the radiator loss to get the total power loss from the transformer. The radiator loss is the sum from all the radiators. Although we modeled only one radiator, assuming they are all identical, the total radiator loss is just the number of radiators times the loss from one. Otherwise we must take into account differences among the radiators. In equilibrium, the total power loss must match the total power dissipation of the coils + core + the stray losses in the tank walls, brackets, leads, etc.

11.6 OIL MIXING IN THE TANK

Perhaps the most complex part of the oil flow in transformers occurs in the tank. The lack of constraining channels or baffles means that the oil is free to take irregular paths such as along localized circulations or eddies. Nevertheless, there is undoubtedly some overall order in the temperature distribution and flow pattern. As with any attempt to model a complicated system, we make some idealized assumptions here in an effort to describe the average behavior of the tank oil.

As shown in Fig. 11.7, we assume that the cold oil at the bottom of the tank has a uniform temperature, T_{bot}, between the bottom radiator discharge pipe to the tank bottom and that the top oil is at a uniform temperature, T_{top}, from the top of the coils to the top oil level. We further assume that the temperature variation between T_{bot} and T_{top} is linear. This allows us to calculate the average oil temperature along a column of oil adjacent to and of equal height as the coils and likewise for the radiators, determining $T_{ave,out,coils}$ and $T_{ave,out,rads}$. These two temperatures will differ because the average radiator vertical position is above that of the coils. These average temperatures are used in determining the thermal pressure drop across the coils and radiators as discussed previously.

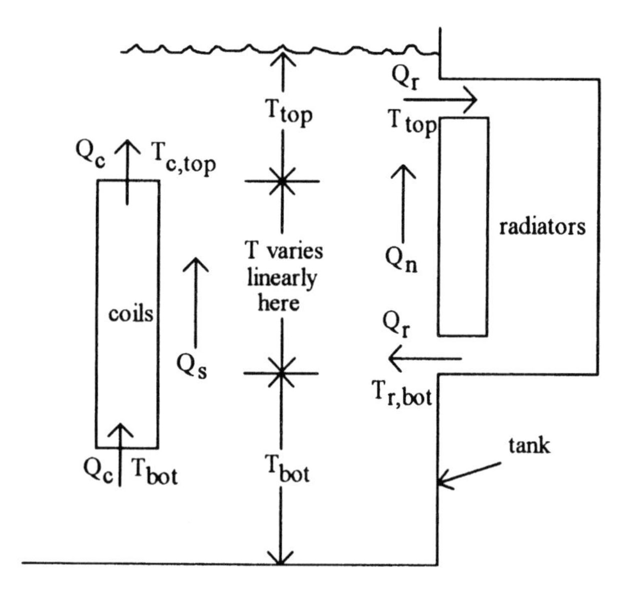

Figure 11.7 Assumed oil temperature distribution inside tank. The oil flows, Q, as well as the flow weighted temperatures are also indicated.

In Fig. 11.7, we have labeled the volumetric oil flows Q with subscripts c for coils, r for radiators, s for stray loss oil flows, and n for net. Thus

$$Q_n = Q_r - Q_c - Q_s \tag{11.50}$$

Q_s accounts for the oil flow necessary to cool the core, brackets, tank walls, etc. Since there is no constraint, mechanical or otherwise, to force the radiator and coil + stray loss flows to be identical, Q_n can differ from zero. The flow Q_c refers to the sum of the flows from all the coils and Q_r the sum from all the radiators.

We have also indicated temperatures associated with some of the flows. The oil flowing into the bottom of the coils is at temperature T_{bot} and the oil flowing into the top of the radiators is at temperature T_{top}. The temperature of the oil flowing out of the top of the coils, $T_{c,top}$, is a flow weighted temperature of the oil from all of the coils. Thus

$$Q_c T_{c,top} = \sum_{i=1}^{\#\,coils} Q_{c,i} T_{c,top,i} \tag{11.51}$$

where i labels the individual coil flows and top temperatures of the oil emerging from the coils. Since we are assuming that all radiators are identical, $T_{r,bot}$ is the bottom temperature of the oil exiting a radiator. If the radiators were not all alike, we would use a flow weighted average for this temperature also.

We assume, for simplicity, that the volumetric flow Q_s, associated with the stray power loss W_s, results in a temperature change of $\Delta T = T_{top} - T_{bot}$ for the oil participating in this flow. Thus Q_s is given by

$$Q_s = \frac{W_s}{\rho c \left(T_{top} - T_{bot}\right)} \tag{11.52}$$

This is inherently an upward flow and, like the coil flow, is fed by the radiators. We do not attribute any downward oil flow to the tank cooling, but assume that this merely affects the average oil temperature in the tank.

We assume that any net flow, Q_n, if positive, results in the transport of cold radiator oil at temperature $T_{r,bot}$ to the top of the tank and, if negative, results in the transport of hot coil oil at temperature

$T_{c,top}$ to the bottom of the tank. These assumptions imply that for $Q_n >$ 0,

$$T_{top} = \frac{T_{c,top}Q_c + T_{r,bot}Q_n}{(Q_r - Q_s)}$$

$$T_{bot} = T_{r,bot} \tag{11.53}$$

while for $Q_n < 0$,

$$T_{top} = T_{c,top}$$

$$T_{bot} = \frac{T_{r,bot}Q_r - T_{c,top}Q_n}{(Q_c + Q_s)} \tag{11.54}$$

Another way of handling the net flow Q_n is to assume that it raises the level of the bottom oil layer at temperature T_{bot} if positive and that it allows the top oil layer at temperature T_{top} to expand downward if negative. This will continue until $Q_n = 0$ at equilibrium.

After each calculation of the coil and radiator flows and temperatures, the quantities T_{top}, T_{bot}, $T_{ave,out,coils}$, $T_{ave,out,rads}$, and Q_s can be determined. Also the losses in the coils and from the radiators can be obtained. From these, updated pressure drops across the coils and radiators can be calculated as well as updated values of the oil velocities flowing into the coils and radiators. The iterations continue until the temperatures reach their steady state values and the losses generated equal the losses dissipated to the atmosphere to within some acceptable tolerance. This requires several levels of iteration. The coil and radiator iterations assume that the tank oil temperature distribution is known and these in turn influence the tank oil temperatures. A relaxation technique is required to keep the iteration process from becoming unstable. This simply means that the starting parameter values for the new iteration are some weighted average of the previous and newly calculated values.

11.7 TIME DEPENDENCE

The basic assumption we make in dealing with time dependent conditions is that at each instant of time the oil flow is in equilibrium with the heat (power) transferred by the conductors to the oil and with the power loss from the radiators at that instant. The velocities and oil temperatures will change with time but in such a way that equilibrium is maintained at each time step. This is referred to as a quasi-static approximation. The conductor heating, or cooling, on the other hand, is transient. We assume, however, that the temperature of a disk, which is part of a directed oil flow coil, is uniform throughout the disk so that only the time dependence of the average disk temperature is treated. A similar assumption is made for non-directed oil flow coils.

The heat generated in a conductor per unit time is I^2R, where R is its resistance (including eddy current effects) at the instantaneous temperature T_C and I its current. Here conductor refers to a single disk for a directed oil flow coil and to the entire coil for a non-directed oil flow coil. The transient thermal equation for this conductor is

$$\rho_{cond} c_{cond} V \frac{dT_c}{dt} = I^2 R - \sum_i h_i A_{c,i} \left(T_c - T_{b,i} \right) \tag{11.55}$$

The left hand side is the heat stored in the conductor of volume V per unit time. This equals the heat generated inside the conductor per unit time minus the heat lost through its surface per unit time. The sum is over all surfaces of the conductor. The other symbols have their usual meaning with the subscript cond indicating that they refer to the conductor properties. The usual temperature and velocity dependencies occur in some of the parameters in (11.55). We solve these equations by a Runge-Kutta technique. These equations replace the steady state disk coil equations for the T_C (11.31) and, in slightly modified form, one of the non-directed oil flow coil equations.

The previous equation accounts for heat (energy) storage in the current carrying conductors as time progresses. Thermal energy is also stored in the rest of the transformer. We assume, for simplicity, that the remainder of the transformer is at a temperature given by the average oil temperature. This includes the core, the tank, the radiators, the brackets or braces, coils not carrying current, the insulation, the main tank oil, etc. Thus, the conservation of energy (power) requires that

$$\left(\sum_i c_i m_i\right)\frac{dT_{ave,oil}}{dt} = \rho c Q_c\left(T_{c,top} - T_{bot}\right) + W_s - W_{rads+tank} \tag{11.56}$$

where c_i is the specific heat and m_i the mass of the parts of the transformer, apart from the current carrying coils, which store heat. Equation (11.56) states that the heat absorbed by the various parts of a transformer per unit time, except the current carrying coils, at a particular instant equals the power flowing out of the coils in the form of heated oil plus the stray power losses minus the power dissipated by the radiators and tank to the atmosphere. Here the stray losses include core losses, tank losses, and any losses occurring outside the coils. Equation (11.56) is solved using a trapezoidal time stepping method.

Thus transient cooling can be handled by making relatively minor modifications to the steady state treatment, using the assumptions given above. So far the computer program we have developed treats the case where the MVA of the transformer is suddenly changed from one level to another. A steady state calculation is performed at the first MVA level. Then the transient calculation begins with currents and stray losses appropriate to the second MVA level. Also the fans may be switched on or off for the transient calculation. As expected, the solution approaches the steady state values appropriate to the second MVA level after a sufficiently long time. With a little extra programming, one could input any desired transformer transient loading schedule and calculate the transient behavior.

11.8 PUMPED FLOW

The type of pumped flow considered here is one in which the radiator oil is pumped and some type of baffle arrangement is used to channel some or all of the pumped oil through the coils. Thus the oil velocity into the entrance radiator pipes due to the pumps, v_0, is given by

$$v_o = \frac{Q_{pump}}{A_1 N_{rad}} \tag{11.57}$$

where Q_{pump} is the volume of oil per unit time which the pump can handle under the given conditions, A_1, is the area of the radiator entrance pipe, and N_{rad} is the number of radiators. In addition to the pressure drop across the radiators due to the pump, the thermal pressure drop is also present and this adds a contribution, although small, to the velocity in (11.57).

We assume that a fraction f of the pumped oil through the radiators bypasses the coils. This could be due to inefficient baffling or it could be by design. We also assume that part of the oil flow from the radiators is used to cool the stray losses. This flow, Q_S, is given by

$$Q_s = \frac{W_s}{W_t}(1-f)Q_{pump} \tag{11.58}$$

where W_S is the stray power loss, W_t, the total loss, and f the fraction of the flow bypassing the coils. In addition to the thermal pressure drop across the coils, a pressure drop due to the pump is present. At equilibrium, this is determined by the requirement that

$$Q_c = (1-f)Q_{pump} - Q_s \tag{11.59}$$

Thus after each iteration, the flow from all the coils is determined and the pressure drop across the coils is adjusted until (11.59) is satisfied at steady-state.

For pumped flow, it is necessary to check for non-laminar flow conditions and to adjust the heat transfer and friction coefficients accordingly.

11.9 COMPARISON WITH TEST RESULTS

Computer codes, based on the analysis given in this report, were written to perform steady state, transient, and pumped oil flow calculations. Although we did not measure detailed temperature profiles along a coil, the codes calculate temperatures of the all the coil disks and of the oil in all the ducts as well as duct oil velocities. Figures 11.8, 11.9, 11.10 show these profiles for one such coil having directed oil flow washers. The oil ducts referred to in the figures are the horizontal ducts and the conductor temperature is the average disk temperature. There is thinning in this coil, i.e. increased duct size at two locations, near disk numbers 35 and 105 , and this can be seen in the profiles. The kinks in the profiles are an indication of the location of the oil flow washers.

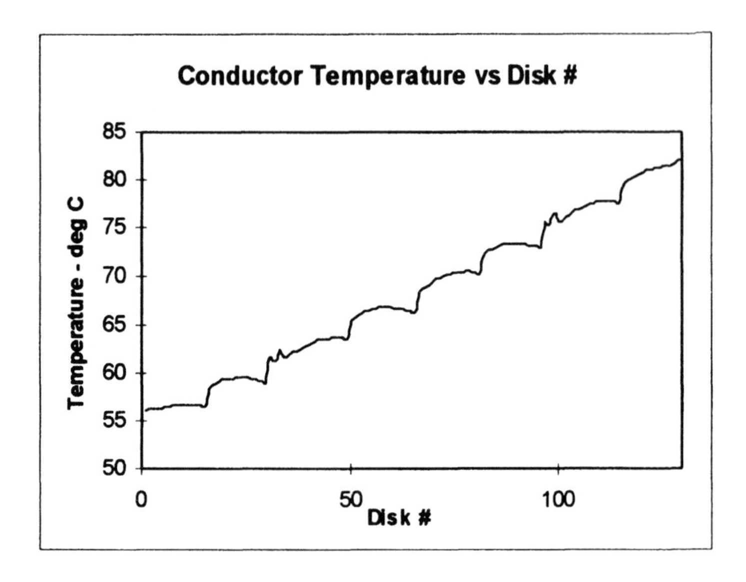

Figure 11.8 Calculated temperatures of the disks along a disk coil with directed oil flow washers.

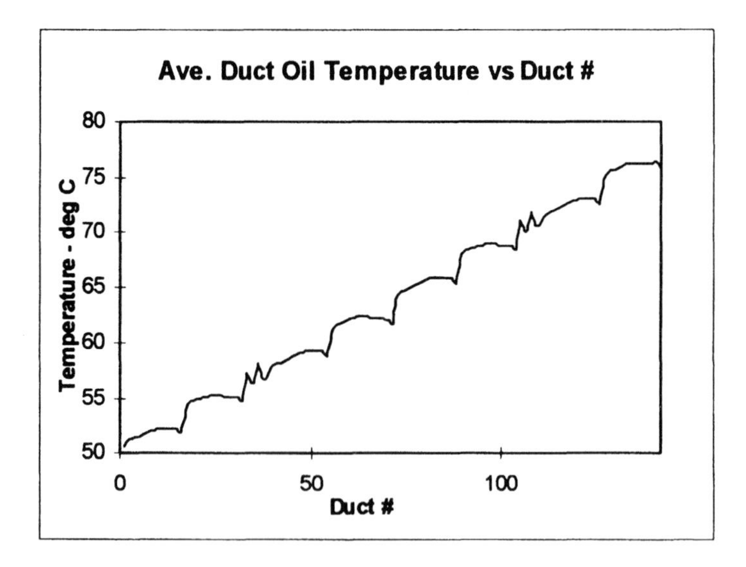

Figure 11.9 Calculated average oil temperatures in the horizontal ducts along a disk coil with directed oil flow washers. The temperature is assumed to vary linearly in the ducts.

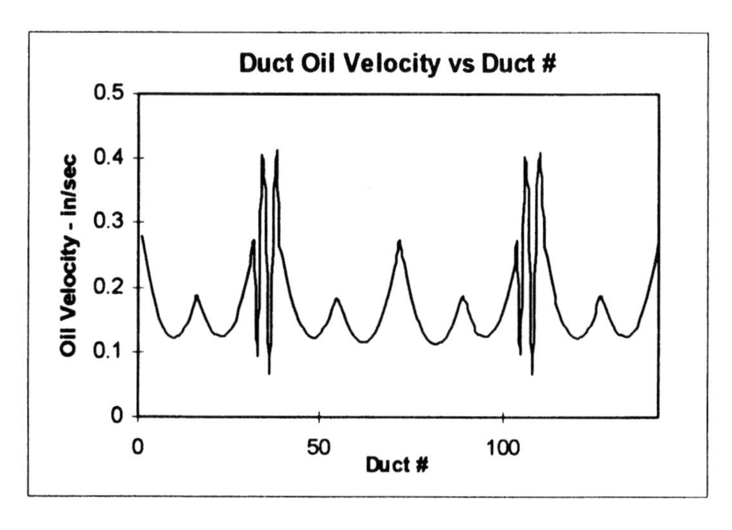

Figure 11.10 Calculated velocities of the oil in the horizontal ducts along a disk coil with directed oil flow washers.

As mentioned previously, some leeway was allowed for in the friction factors for the coils and radiators by including an overall multiplying factor. This accounts, in an average way, for non ideal conditions in real devices and must be determined experimentally. By comparing test data with calculations, we determined that the best agreement is achieved with a multiplying factor of 1.0 for the coils and 2.0 for the radiators. Thus the coil friction is close to the theoretical value whereas that for the radiators is twice as high.

Temperature data normally recorded in our standard heat runs are: (1) mean oil temperature rise, (2) top oil temperature rise, (3) temperature drop across the radiators, and (4) average temperature rise of the windings. These are usually measured under both OA and FA conditions. The rises are with respect to the ambient air temperature. A statistical analysis of such data, taken on a number of transformers with MVAs ranging from 12 to 320, was performed to determine how well the calculations and test results agreed. This is shown in Table 11.1 where the mean and standard deviations refer to the differences between the calculated and measured quantity. Thus a mean of 0 and a standard deviation of 0 would indicate a perfect fit. Non zero results for these statistical measures reflect both the limitations of the model and some uncertainty in the measured quantities. Table 11.1 shows generally good agreement between the calculated and measured results. Essentially only one parameter was adjusted in order to improve the agreement, namely the overall friction factor multiplier for the radiators.

Table 11.1 Comparison of Calculated and Measured Quantities

	OA		FA	
	Mean	Std. Dev.	Mean	Std. Dev.
Mean oil rise	0.86	2.58	0.19	5.11
Top oil rise	1.56	2.36	2.11	4.82
Drop across radiators	-2.23	4.51	-0.83	3.44
Average winding rise	-0.56	3.19	-0.14	2.38

Figures 11.11 and 11.12 show representative output from the transient program in graphical form. In this example, the transformer is operating at steady state at time 0 at the first MVA value (93 MVA) and suddenly the loading corresponding to the second MVA (117 MVA) is applied. The time evolution of the oil temperature is shown in Fig. 11.11 while that of one of the coils is shown in Fig. 11.12. The program calculates time constants and m and n values from this information. Fig. 11.13 shows a direct comparison of the calculated and measured hot spot temperature of a coil in which a fiber optics temperature probe was imbedded. Although absolute temperatures are off by a few degrees, the shapes of the curves are very similar, indicating that the time constants are nearly the same.

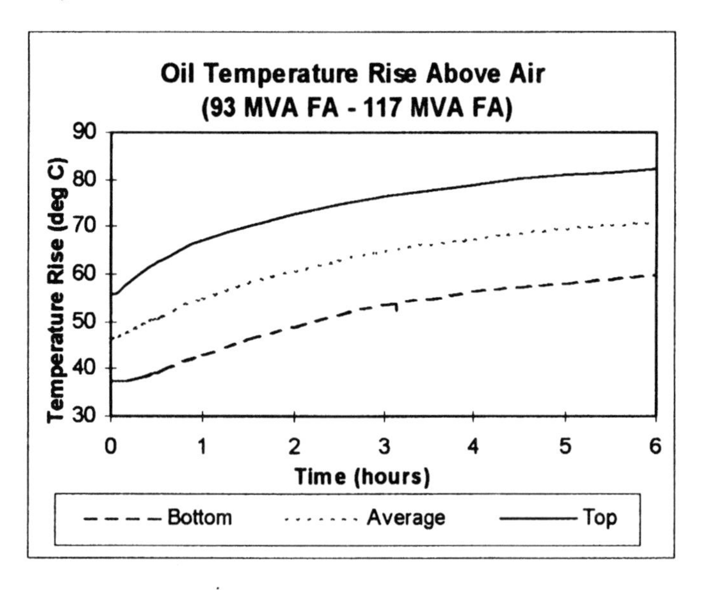

Figure 11.11 Time evolution of the bottom, average, and top tank oil temperatures starting from a steady state loading of 93 MVA at time 0, when a sudden application of a 117 MVA loading is applied.

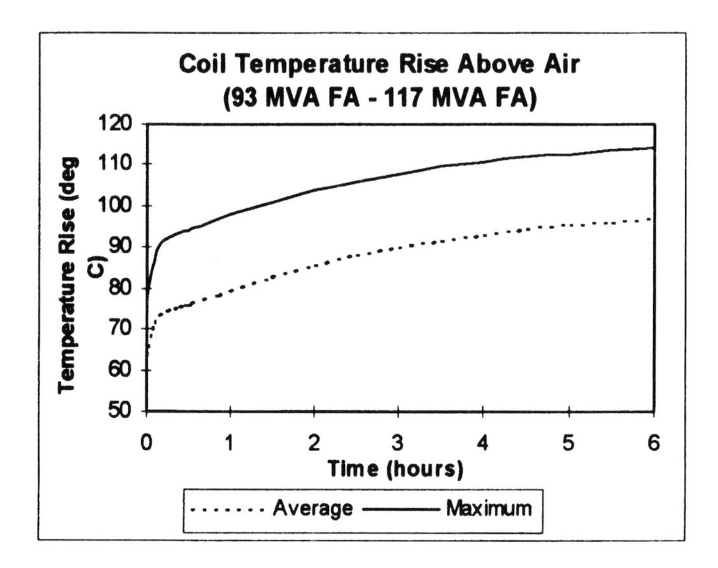

Figure 11.12 Time evolution of the average and maximum coil temperature starting from a steady state loading of 93 MVA at time 0, when a sudden application of a 117 MVA loading is applied.

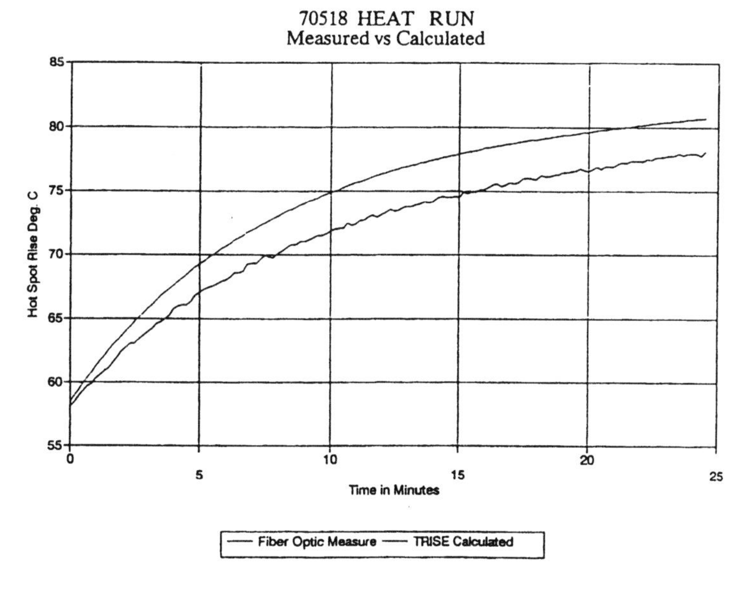

Figure 11.13 Comparison of calculated and measured temperature versus time of the coil's hot spot. The coil is operating at steady state at time 0, when a sudden additional loading is applied. The hot spot is measured by means of a fiber optics probe.

11.10 DETERMINING M AND N EXPONENTS

In order to extimate the temperature rises which occur under overload conditions, approximate empirical methods have been developed which make use of the m and n exponents. Although a detailed thermal model such as the one presented above should make these approximations unnecessary, these exponents are widely used to obtain quick estimates of the overload temperatures for transformers in service. Although they are frequently measured in heat run tests, they can also be obtained from either the steady-state or transient calculations of the type presented above.

The n exponent is used to estimate the top oil temperature rise above the ambient air temperature, ΔT_{to}, under overload conditions from a knowledge of this quantity under rated conditions. For this estimation, the rated transformer total losses must be known as well as the losses at the overload condition. The latter losses can be estimated from a knowledge of how the losses change with increased loading and will be discussed below. The formula used to obtain the overload temperature rise above ambient is

$$\frac{\Delta T_{to,2}}{\Delta T_{to,1}} = \left(\frac{\text{Total Loss}_2}{\text{Total Loss}_1} \right)^n \tag{11.60}$$

Thus if 2 refers to the overload condition and 1 to the rated condition, we assume that the 1 subscripted quantities are known either by measurement or calculation.

In order to estimate the losses under overload conditions, we must separate the losses into I^2R resistive losses in the coils and stray losses, which include the core, tank, clamp, and additional eddy current losses which occur in the coils due to the stray flux. The I^2R loss can be calculated fairly accurately. By subtracting it from the total measured loss, a good estimate can be obtained for the stray loss. Alternatively, core loss, eddy current loss, tank loss, and clamp loss can be obtained from design formulas. All of these losses depend on on the square of the load current in either winding. (The stray losses depend on the square of the stray flux which is proportional to the square of the current, assuming the materials are operating in their linear range.) Since the load current is proportional to the MVA of the unit, these losses are proportional to the MVA squared. The I^2R losses are proportional to the resistivity of the winding conductor which is temperature dependent while the stray losses are inversely proportional the the resistivity of the material in which they occur. Thus the total loss at the higher loading can be obtained from the rated loss by

$$\text{Total Loss}_2 = \left(\frac{\text{MVA}_2}{\text{MVA}_1}\right)^2 \left[I^2\text{R Loss}_1 \left(\frac{\rho_2}{\rho_1}\right)_{\text{coil}} + \text{Stray Loss}_1 \left(\frac{\rho_1}{\rho_2}\right)_{\text{stray}} \right] \tag{11.61}$$

The resistivity ratios are temperature dependent so that some iteration may be required with (11.60) to arrive at a self consistent solution. The stray resistivity ratio is meant to be a weighted average over the materials involved in the stray loss. However since the temperature dependence of the coil, tank, and clamp materials is nearly the same, this ratio will be nearly the same as that of the coils.

We should note that the n coefficient is generally measured or calculated under conditions where the cooling is the same for the rated and overload conditions. Thus if fans are turned on for the rated loading, they are also assumed to be on for the overload condition. Similarly if pumps are turned on at the 1 rating, they must also be on for the 2 rating for the above formulas to work. Although n can be determined from measurements or calculations performed at two ratings, it is often desireable to determine it from 3 different loadings and do a best fit to (11.60) based on these. The 3 ratings generally chosen are usually 70%, 100%, and 125%. The 70%, 100% or 100%, 125% combinations are chosen when only 2 ratings are used. Measurements and calculations show that n based on the 70%, 100% combination can be fairly different from n based on the 100%, 125% combination. This suggests that n is not strictly a constant for a given transformer and it should be determined for loadings in the range where it is expected to be used. A typical value of n for large power transformers is 0.9.

The m exponent is defined for each winding and relates the winding gradient to the winding current. The winding gradient at a particular steady state loading is the temperature difference between the mean winding temperature and the mean oil temperature in the tank, i.e.

$$\text{Gradient} = \text{Mean winding temperature} - \text{Mean Tank oil temperature} \tag{11.62}$$

Letting 1 and 2 designate rated and overload conditions, the m coefficient for a particular winding relates the gradients to the winding currents by

$$\left(\frac{\text{Gradient}_2}{\text{Gradient}_1}\right)=\left(\frac{\text{Current}_2}{\text{Current}_1}\right)^{2m} \tag{11.63}$$

Since the currents are proportional the the MVA's, the MVA ratio could be substituted in the above formula for the current ratio. The m exponents can be determined from 2 or 3 MVA loadings and, like the n exponent, depend to some extent on the MVA range covered. They also can vary considerably for different windings. A typical value for large power transformers is about 0.8.

The winding gradient is often used to estimate the winding's maximum temperature. A common procedure is to add some multiple of the gradient to the top oil temperature to arrive at the maximum temperature for that winding, for example

$$\text{Max winding temp} = \text{Top oil temp} + 1.1 \times \text{Winding Gradient} \tag{11.64}$$

The factor of 1.1 can differ among manufacturers or transformer types and should be determined experimentally. Formulas such as (11.64) become unnecessary when detailed temperature calculations are available or maximum temperatures are measured directly by fiber optics or possibly other types of temperature probes, inserted into the winding at its most probable location. Detailed temperature calculations such as described above can provide some guidance as to where these probes should be inserted.

11.11 LOSS OF LIFE CALCULATION

Although transformers can fail for a number of reasons such as from the application of excessive electrical or mechanical stress, even a highly protected unit will eventually fail due to aging of its insulation. While the presence of moisture and oxygen affect the rate of insulation aging, these are usually limited to acceptable levels in modern transformers so that the main factor affecting insulation aging is temperature. The insulation's maximum temperature, called the hot spot temperature, is thus the critical temperature governing aging. This is usually the highest of the maximum temperatures of the different windings, although it could also be located on the leads which connect the windings to the bushings or to each other.

Numerous experimental studies have shown that the rate of insulation aging as measured by various parameters such as tensile strength or degree of polymerization (DP) follows an Arrhenius relationship,

$$K = Ae^{-B/T} \tag{11.65}$$

Here K is a reaction rate constant (fractional change in quantity per unit time), A and B are parameters, and T is the absolute temperature. For standard cellulose pased paper insulation, $B = 15000$ °K. This value for B is an average from different studies using different properties to determine aging [McN91]. If X is the property used to measure aging, e.g. tensile strength of DP, then we have

$$X(t) = X(t=0)e^{-Kt} \tag{11.66}$$

The time required for the property X to drop to some fraction f of its initial value if held at a constant temperature T, t_T , is thus determined by

$$\frac{X(t=t_T)}{X(t=0)} = f = e^{-Kt_f} \tag{11.67}$$

Taking logarithms, we obtain

$$t_T = -\frac{\ln f}{K} = -\frac{\ln f}{A}e^{B/T} \tag{11.68}$$

where (11.65) has been used.

There are several standard or normal insulation lifetimes which have found some acceptance. For these, the insulation hot spot is assumed to age at a constant temperature of 110 °C. The insulation is also assumed to be well dried and oxygen free. Under these conditions, the time required for the insulation to retain 50% of its initial tensile strength is 7.42 years (65000 hours), to retain 25% of its initial tensile strength is 15.41 years (135000 hours), and to retain a DP level of 200 is 17.12 years (150000 hours). Any of these criteria could be taken as a measure of a normal lifetime, depending on the experience or the degree of conservatism of the user. However, if the transformer is operated at a lower temperature continuously, the actual lifetime can be considerably longer than this. For example, if the hot spot is kept at 95 °C continuously, the transformer insulation will take 36.6 years to be left with 50% of its initial tensile strength. Conversely, if operated at higher temperatures, the actual lifetime will be shortened.

Depending on the loading, the hot spot temperature will vary over the course of a single day so that this must be taken into account in determining the lifetime. A revealing way of doing this is to compare the time required to produce a fractional loss of some material property such as tensile strength at a given temperature with the time required to produce the same loss of the property at the reference temperature of 110 °C. Thus the time required to produce the fractional loss f in the material property at temperature T is given by (11.68) and the time to produce the same fractional loss at the reference temperature, designated T_o, is given by the same formula with T_o replacing T. This latter time will be designated t_{To}. Hence we have for the given fractional loss,

$$\frac{t_{To}}{t_T} = e^{(B/T_o - B/T)} = AAF(T) \tag{11.69}$$

We have defined the aging acceleration factor, AAF, in (11.69). Thus it requires AAF(T) times as much time at the reference temperature than at temperature T to produce the same fractional loss of life as determined by property X. Since the T's are in °K, we have

$$AAF(T) = e^{\left[15000/383 - 15000/\left(T(°C)+273\right)\right]} \tag{11.70}$$

AAF is > or < 1 depending on whether T(°C) is > or < 110 °C. This is plotted in Fig. 11.14.

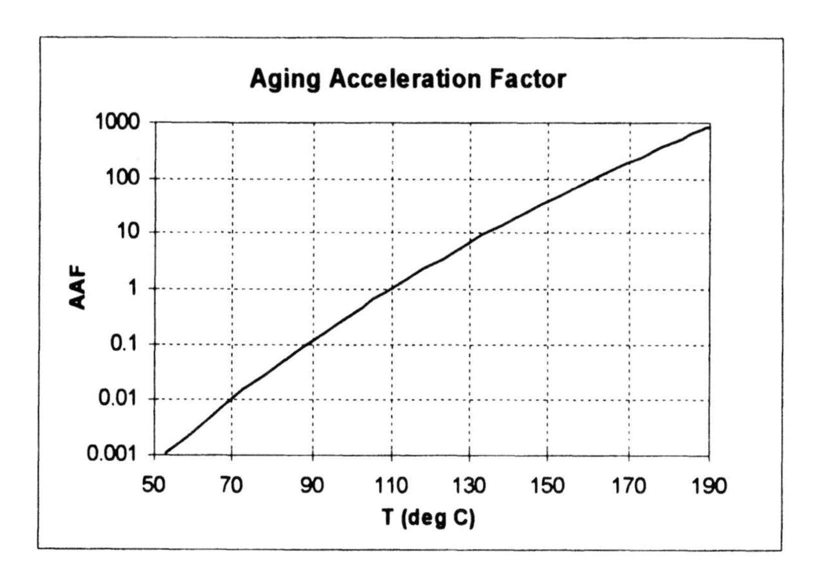

Figure 11.14 Aging Acceleration Factor (AAF) relative to 110 °C vs temperature.

A method of using the daily variations of hot spot temperature to compute aging is to subdivide the day into (not necessarily equal) time intervals over which the hot spot temperature is reasonably constant. Let AAF_i be the aging acceleration factor for time interval Δt_i measured in hours. Then for a full day

$$AAF_{\text{day ave}} = \frac{1}{24}\sum_{i=1}^{N} AAF_i \Delta t_i \tag{11.71}$$

where N is the total number of time intervals for that day.

The actual loss of life will depend on the definition of lifetime. If the 50% retained tensile strength criterion is uned, then in 1 day at 110 °C insulation hot spot temperature, the transformer loses the fraction $24/65000 = 3.69 \times 10^{-4}$ of its life. The fraction lost with variable hot spot temperature is therefore $AAF_{\text{day ave}} \times 3.69 \times 10^{-4}$. For the 25% retained tensile strength criterion, the loss of life for 1 day is $AAF_{\text{day ave}} \times 1.78 \times 10^{-4}$.

A cautionary note must be sounded before too literal a use is made of this procedure. Gas bubbles can start to form in the oil next to the insulation when the insulation temperature reaches about 140 °C. These bubbles can lead to dielectric breakdown which could end the transformer's life, rendering the calculations inapplicable.

There is bound to be some uncertainty involved in determining the hot spot temperature unless fiber optics or other probes are used and properly positioned. Short of this, it should be estimated as best as possible, based on top oil temperature, winding gradients, etc. Also a winding temperature indicator, if properly calibrated, can be used.

Since the AAF is independent of the choice made for the normal transformer lifetime, the fractional loss of life can be easily recalculated if a different choice of normal lifetime is made based on additional experience or knowledge.

11.12 CABLE AND LEAD TEMPERATURE CALCULATION

Although the conductors in the coils are usually cooled sufficiently to meet the required hot spot limits, it is also necessary to insure that the lead and cable temperatures remain below these limits as well. Even though these are generally not in critical regions of electric stress to cause a breakdown if gassing occurs, the gassing itself can trigger alarms which could put the transformer out of service. It is thus necessary for the design engineer to insure that the leads and cables are sized to meet the temperature requirements.

In general, no more insulation (paper) should be used on the leads and cables than is necessary for voltage standoff. Given this minimum paper thickness, the current carrying area of the lead or cable should then be chosen so that the temperature limits are met when rated current is flowing. A method for calculating the lead or cable temperature rise is given here. It allows for the possibility that the lead is brazed to a cable and that some heat conduction can occur to the attached cable, acting as a heat sink. It also considers the case of a lead inside of duct or tube or a lead in the bulk transformer oil.

We treat a cylindrical geometry. In the event the lead is not cylindrical, an effective diameter should be calculated so that it may be approximated by a cylinder. The cylinder is assumed to be long so that end effects may be neglected. The geometric parameters are shown in Fig. 11.15.

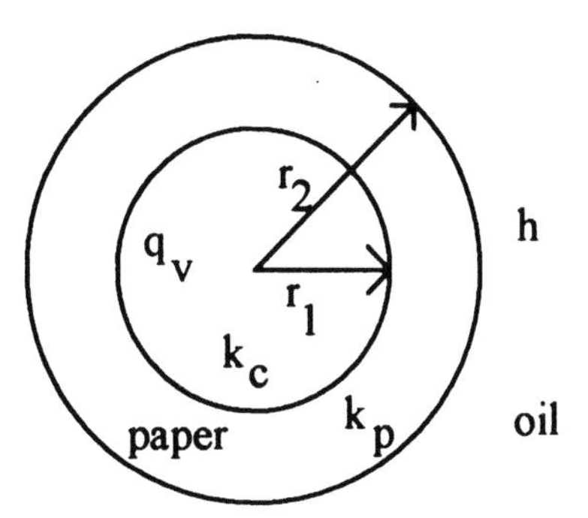

Figure 11.15 Cylindrical geometry of a paper wrapped conductor surrounded by oil

The convective heat transfer coefficient of the oil is denoted by h . The thermal conductivities of the conductor and paper are denoted k_c and k_p respectively and the power generated per unit volume inside the conductor is denoted by q_v . This is given by the Joule loss density , ρJ^2 , where ρ is the resistivity and J the rms current density. In addition, to account for extra losses due to stray flux, we should multiply this by $(1 + f)$ where f is the stray flux loss contribution expressed as a fraction of the Joule losses.

There are two cases to consider for the convective heat transfer coefficient of the oil. They are the case of a horizontal cylinder in free tank oil and the case of a horizontal cylinder inside of a channel. (The heat transfer coefficient for a vertical cylinder is higher so the resulting temperature rises will be lower than for a horizontal cylinder.) These are given by [Kre80]:

Natural convection from a horizontal cylinder of diameter D in bulk fluid:

$$Nu_D = \left[0.60 + 0.387 \left\{ \frac{Gr_D \, Pr}{\left[1 + (0.56/Pr)^{9/16} \right]^{16/9}} \right\}^{1/6} \right]^2 \qquad (11.72)$$

Mixed natural convection and laminar flow in horizontal ducts:

$$\mathrm{Nu_D} = 1.75\left[\mathrm{Gz} + 0.012\left(\mathrm{Gz}\,\mathrm{Gr_D^{1/3}}\right)^{4/3}\right]^{1/3}\left(\frac{\mu_b}{\mu_s}\right)^{0.14} \tag{11.73}$$

where the symbols have the following meaning:

Nusselt number $\mathrm{Nu_D} = \dfrac{hD}{k}$

Reynold's number $\mathrm{Re_D} = \dfrac{\rho_m v D}{\mu}$

Prandtl number $\mathrm{Pr} = \dfrac{\mu c}{k}$

Grashof number $\mathrm{Gr_D} = \dfrac{g\beta\rho_m^2(T_s - T_b)D^3}{\mu^2}$

Graetz number $\mathrm{Gz} = \mathrm{Re_D}\,\mathrm{Pr}\left(\dfrac{D}{L}\right)$

v = fluid velocity
D = hydraulic diameter for flow inside ducts = 4 × flow area / wetted perimeter
= outer cylinder diameter for external flow
L = duct or cylinder length
g = acceleration of gravity = 9.8 m/sec^2
h = convective heat transfer coefficient
k = thermal conductivity of transformer oil = 0.11 W/m°C
c = specific heat of transformer oil = 1880 J/kg °C
μ = oil viscosity = 6900.0 / (T + 50)^3 Ns/m^2 , T = temp in °C
ρ_m = oil mass density = 867 exp(-0.00068 (T - 40)) kg/m^3 , T = temp in °C
β = volume expansion coefficient of oil = 0.00068/°C
Subscripts s and b label the oil at the outside surface of the paper and the bulk oil

In steady state, the thermal equations governing this situation are

Inside conductor,

$$\frac{1}{r}\frac{d}{dr}\left(r\frac{dT_c}{dr}\right)+\frac{q_v}{k_c}=0 \tag{11.74}$$

Inside paper,

$$\frac{1}{r}\frac{d}{dr}\left(r\frac{dT_p}{dr}\right)=0 \tag{11.75}$$

The boundary conditions are:

At $r = 0$, $$\frac{dT_c}{dr}=0$$

At $r = r_1$, $$T_c = T_p \quad \text{and} \quad k_c\frac{dT_c}{dr}=k_p\frac{dT_p}{dr}$$

At $r = r_2$, $$-k_p\frac{dT_p}{dr}=h(T_s-T_b)$$

where c labels the temperature inside the conductor and p inside the paper. The solution is given by

$$T_c=\frac{q_v}{4k_c}\left(r_1^2-r^2\right)+\frac{q_v r_1^2}{2k_p}\left[\ln\left(\frac{r_2}{r_1}\right)+\frac{k_p}{hr_2}\right]+T_b \tag{11.76}$$

$$T_p=\frac{q_v r_1^2}{2k_p}\left[\ln\left(\frac{r_2}{r}\right)+\frac{k_p}{hr_2}\right]+T_b \tag{11.77}$$

The highest temperature occurs at $r = r_1$ and is given, in terms of temperature rise above the bulk oil, by

$$T_c(r_1)-T_b=\frac{q_v r_1}{2}\left[\frac{r_1}{k_p}\ln\left(\frac{r_2}{r_1}\right)+\frac{r_1}{hr_2}\right] \tag{11.78}$$

The quantity $q_v r_1/2$ is a surface heat flux, q_s, since

$$q_s = \frac{q_v \pi r_1^2 L}{2\pi r_1 L} = \frac{q_v r_1}{2} \tag{11.79}$$

where L is the cylinder length which cancels. Thus (11.78) has the form

$$q_s = h_{eff}\left(T_c(r_1) - T_b\right) \tag{11.80}$$

where

$$h_{eff} = \frac{h}{\frac{r_1}{r_2} + \frac{r_1 h}{k_p} \ln\left(\frac{r_2}{r_1}\right)} \tag{11.81}$$

is an effective heat transfer coefficient which takes into account conduction through the paper. In the case of a thin paper layer of thickness τ where $\tau << r_1$ and $r_2 = r_1 + \tau$, (11.81) reduces to

$$h_{eff} = \frac{h}{1 + \frac{\tau h}{k_p}} \tag{11.82}$$

This last expression also applies to a planar geometry. Note that the surface heat flux in (11.81) is based on the surface area of a cylinder of radius r_1, i.e. that of the metallic part of the cable. We also need the temperature rise of the surface oil in order to compute the Grashof number. This is given by

$$T_p(r_2) - T_b = \frac{q_v r_1}{2}\left(\frac{r_1}{h r_2}\right) \tag{11.83}$$

so that, comparing (11.78) and (11.83), we can write

$$\frac{T_p(r_2) - T_b}{T_c(r_1) - T_b} = \frac{1}{1 + \frac{r_2 h}{k_p} \ln\left(\frac{r_2}{r_1}\right)} \tag{11.84}$$

so that the temperature rise of the surface paper can be found once the maximum conductor temperature rise is calculated.

In case another cable is brazed to the conductor which may act as a heat sink, we must consider heat conduction along the conductor to the other cable. The situation is depicted in Fig. 11.16.

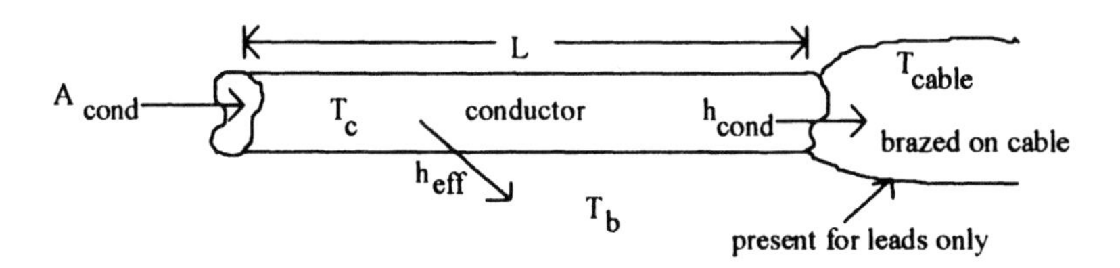

Figure 11.16 Geometry and thermal parameters for a cable or lead with another brazed to it

Letting W = the total power dissipated inside the conductor of length L, we have

$$W = h_{eff} A_{side}(T_c - T_b) + \underbrace{h_{cond} A_{cond}(T_c - T_{cable})}_{\text{present for leads only}}$$

$$= h_{eff} A_{side}(T_c - T_b) + h_{cond} A_{cond}(T_c - \underbrace{T_b + T_b}_{\text{add and subtract } T_b} - T_{cable}) \quad (11.85)$$

$$= h_{eff} A_{side}(T_c - T_b) + h_{cond} A_{cond}(T_c - T_b) - h_{cond} A_{cond}(T_{cable} - T_b)$$

where

T_c = conductor surface temperature
T_b = bulk fluid temperature
T_{cable} = brazed on cable temperature
p = perimeter of the conductor's cooling surface
A_{side} = pL = conductor's side area through which heat flows to the cooling fluid
A_{cond} = conductor's current carrying area
h_{cond} = k_{cu}/L = heat transfer coefficient for heat to flow from the conductor to the brazed on cable. For copper k_{cu} = 400 W/m°C.

In order to use the above expression to obtain the lead temperature rise, the temperature rise of the cable must be calculated first. Since the cable is normally very long, its temperature rise can be obtained by neglecting its brazed connection.

Although we have indicated the numerical value of most of the material parameters in the preceding section, the missing ones will be given here. They are:

Thermal conductivity of paper $k_p = 0.16$ W/m°C
Resistivity of copper $\rho = 1.72 \times 10^{-8} (1 + 0.004 (T - 20))$, T in °C
v = oil velocity in duct = 0.1 in/sec = 0.00254 m/sec (This is a conservative value. If a better value can by obtained, e.g. from an oil flow calculated in the cooling program, then it should be used.)

In addition, it should be noted that many of these material parameters are temperature dependent so that it is necessary specify a bulk oil temperature and to iterate the calculations. The appropriate temperature to use for evaluating the oil parameters is $(T_s + T_b) / 2$, i.e. the average of the surface and bulk oil temperatures. For the conductor, its temperature is nearly uniform so its maximum temperature can be used. The Grashof number depends on the temperature difference $T_s - T_b$ so it must be recalculated during the iteration process along with the other material parameters.

These formulas have been applied to some standard cable sizes to obtain the maximum continuous currents permissible for a hot spot temperature rise of about 25 °C. We assumed an oil temperature rise of 55 °C above an ambient air temperature of 30 °C resulting in a bulk oil temperature of 85 °C. Thus a rise of 25 °C would result in a hot spot temperature of 110 °C. The results are shown in Table 11.2. Note that the conductor area does not fill 100% of the geometrical area of the cable since the conductor is stranded.

Table 11.2 Maximum Continuous rms Currents for Standard Cable Sizes producing a Temperature Rise Above the Surrounding Oil of 25 °C

Cable Size	Cond Area	Cond Diam	Max Continuous Current (rms Amps)		
AWG/ MCM	sq. inches	inches	0.094 in paper	0.25 in paper	0.50 in paper
6	0.02062	0.186	125	100	85
4	0.03278	0.235	170	135	114
2	0.05213	0.296	230	180	152
1 / 0	0.08289	0.374	320	245	205
2 / 0	0.1045	0.420	375	285	235
3 / 0	0.1318	0.533	465	345	285
4 / 0	0.1662	0.599	545	405	330
300	0.2356	0.714	700	515	415
350	0.2749	0.773	780	570	460
400	0.3142	0.825	860	625	505
500	0.3927	0.923	1010	730	585
600	0.4713	1.022	1155	830	660

11.13 TANK WALL TEMPERATURE CALCULATION

Heating in the tank wall above the adjacent oil temperature is due to eddy currents induced by the stray leakage flux from the main coils and flux from any nearby leads or busses. Generally a chief cause for concern is the presence of high current carrying leads near the tank wall. We have indicated how these may be calculated in Chapter 10. Here we are going to assume that these losses are known and that they are uniformly distributed throughout the tank wall. We will also assume that the tank wall surface dimensions involved are large compared with the tank wall thickness so that only 1 spatial dimension through the wall thickness is important. Fig. 11.17 shows the geometric and other relevent parameters.

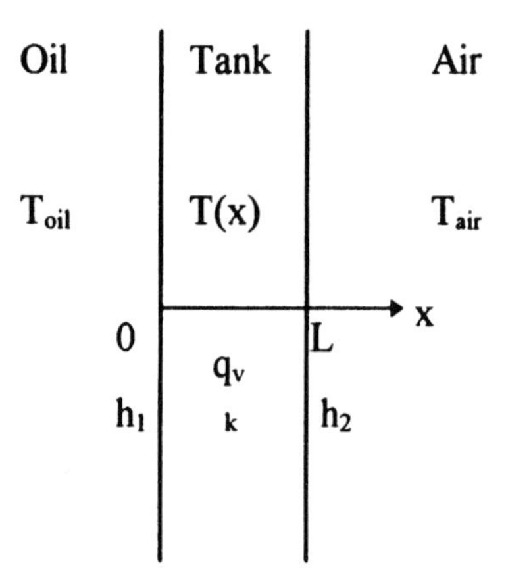

Figure 11.17 Tank wall geometry and thermal parameters

Here L is the wall thickness, q_v the losses generated in the tank wall per unit volume, k the thermal conductivity of the tank wall material, h_1 the heat transfor coefficient from the tank wall to the oil, h_2 the heat transfer coefficient from the tank to the air, and T(x) the temperature distribution in the tank wall.

The steady state thermal equation for this situation is given by

$$\frac{d^2T}{dx} + \frac{q_v}{k} = 0 \tag{11.86}$$

with the boundary conditions

$$k\left.\frac{dT}{dx}\right|_{x=0} = h_1\left(T(0) - T_{oil}\right) \quad , \quad k\left.\frac{dT}{dx}\right|_{x=L} = -h_2\left(T(L) - T_{air}\right) \tag{11.87}$$

The different signs in the two boundary conditions are necessary to properly account for the direction of heat flow. The solution can be expressed as

$$T(x) = T_{oil} - \frac{q_v L^2}{2k}\left(\frac{x}{L}\right)^2 + \left[\frac{\frac{q_v L^2}{2k}\left(1 + \frac{2k}{h_2 L}\right) - \left(T_{oil} - T_{air}\right)}{1 + \frac{k}{h_1 L}\left(1 + \frac{h_1}{h_2}\right)}\right]\left(\frac{x}{L} + \frac{k}{h_1 L}\right) \tag{11.88}$$

This expression can be simplified by noticing that thermal transfer within the tank wall is much larger that that at the surface. This means that k is much larger than h_1L or h_2L so that, ignoring terms ≤ 1 relative to k/h_1L or k/h_2L, (11.88) becomes

$$T(x) = T_{tank} = \frac{q_v L}{(h_1 + h_2)} + \frac{h_1 T_{oil} + h_2 T_{air}}{(h_1 + h_2)} \tag{11.89}$$

This is independent of x so that the tank wall temperature is essentially uniform throughout and we have written T_{tank} for it. The last term on the right is the weighted average of the oil and air temperatures, weighted by their respective heat transfer coefficients. The first term on the right accounts for the heat generated in the tank wall. We can use the heat transfer coefficients given in Section 11.5 on Tank Cooling. Another expression which could be used for heat transfer due to natural convection from one side of a vertical plate of height H immersed in a fluid (air or oil) is given by [Kre80]

$$h = 0.021 \frac{k}{H} (Gr_H \, Pr)^{2/5} \tag{11.90}$$

where the Grashof and Prandtl numbers have been defined in Section 11.12. This expression holds for $10^9 < Gr_H Pr < 10^{13}$. To this, one would have to add the radiation term to the air heat transfer coefficient.

As a numeric example, let $h_1 = h_{oil} = 70$ W/m^2°C , $h_2 = h_{air} + h_{rad} =$ 13 W/m^2°C , $k = k_{steel} = 40$ W/m°C , $L = 9.525 \times 10^{-3}$ m (0.375 in). From these we obtain $k/h_1L = 60$ and $k/h_2L = 323$. Both of these are much larger than 1 so that the approximate formula (11.89) may be used. Letting $q_v = 2 \times 10^5$ W/m^3 , $T_{oil} = 85$ °C , $T_{air} = 30$ °C , we obtain $T_{tank} =$ 99.3 °C.

11.14 TIEPLATE TEMPERATURE CALCULATION

Tieplate losses were calculated in Chapter 10. These losses must be obtained by some method, such as that described there, in order to calculate the tieplate's surface temperature. In order to make this temperature calculation tractable analytically, we make some idealized assumptions which appear to be reasonable. We assume the plate is infinite in two dimensions and that the losses fall off exponentially from the surface facing the coils in the other dimension. The exponential drop-off is determined by the skin depth and has the form $e^{-x/\tau}$ where x is the distance from the free surface and τ is the skin depth. The surface facing the coils is assumed to be cooled convectively by the oil with a heat transfer coefficient h. The boundary condition on the other surface facing the core will depend on the construction details which may differ among manufacturers. We will assume that there is a cooling gap here but that the heat transfer coefficient may be different from that at the other surface. To allow for an insulated core facing surface, this heat transfer coefficient may simply be set to zero. This geometry and the cooling assumptions are identical to that assumed for the tank wall temperature calculation. However here we are allowing for an exponential drop off of the loss density which is more realistic in the case of soft steel tieplates. Thus we are using the geometric and thermal parameters of Fig. 11.17 except that air now refers to oil cooling on the surface at a distance $x = L$, where L is the tieplate thickness. Tieplates have a finite width and we are ignoring the enhanced loss density at the corners which is evident in the finite element study for soft steel of Chapter 10. However there is also extra cooling at the corner from the side surface which could mitigate the effect of this extra loss density. One could address this problem by means of a thermal finite element program. However, the analytic solution presented below should apply to temperatures away from the corner and these may possibly be higher or comparable to the corner temperature.

This is a 1-dimensional steady-state heat transfer problem. The differential equation to solve is the same as (11.86) in the previous section. However, here we have $q_V = q_{VO}e^{-x/\tau}$, where q_{VO} is the loss density at the surface at $x = 0$ and τ is the skin depth, rather than the constant assumption for q_v used for tank cooling. Thus the side facing the coils is at $x = 0$. The boundary conditions are the same as (11.87) with the air subscript replaced by oil. The solution is

$$T(x) - T_{oil} = \frac{q_{vo}\tau}{h_1}\left\{1 + \frac{h_1\tau}{k}\left(1 - e^{-x/\tau}\right) - \left(1 + \frac{h_1 x}{k}\right)\left[\frac{e^{-L/\tau}\left(1 - \frac{h_2\tau}{k}\right) + \frac{h_2\tau}{k} + \frac{h_2}{h_1}}{1 + \frac{h_2 L}{k} + \frac{h_2}{h_1}}\right]\right\} \tag{11.91}$$

We have expressed the solution is this form so that the case of an insulated back side ($h_2 = 0$) could be easily obtained. From (11.91) we can obtain the surface temperature rise at the $x = 0$ surface,

$$T(x=0)-T_{oil}=\frac{q_{vo}\tau}{h_1}\left\{1-\left[\frac{e^{-L/\tau}\left(1-\frac{h_2\tau}{k}\right)+\frac{h_2\tau}{k}+\frac{h_2}{h_1}}{1+\frac{h_2L}{k}+\frac{h_2}{h_1}}\right]\right\} \tag{11.92}$$

As a numerical example, let us consider the surface temperature rise above the surrounding oil for soft iron and stainless steel tieplates using the following parameters:

$h_1 = 70\ W/m^2{}^\circ C$
$h_2 = 20\ W/m^2{}^\circ C$
$L = 9.525\times10^{-3}$ m (3/8 in)
$k = 40\ W/m^\circ C$ (soft iron)
$k = 15\ W/m^\circ C$ (stainless steel)
$\tau = 2.297\times10^{-3}$ m (0.0904 in) (soft iron)
$\tau = 5.627\times10^{-2}$ m (2.215 in) (stainless steel)

We would like to compare these two materials when the total losses in them are the same. This means that the losses per unit surface area should be the same since we are assuming the geometry is unlimited in the plane of the surface. Letting q_A be the loss per unit area, we can obtain q_{vo} in terms of this by means of

$$q_{vo}=\frac{q_A}{\tau\left(1-e^{-L/\tau}\right)} \tag{11.93}$$

Using (11.92) and (11.93) and the numerical data given above, we obtain the surface temperature rise above the ambient oil for the two types of tieplates,

$T(x = 0) - T_{oil} = 0.0111\ q_A$ (soft iron)
$T(x = 0) - T_{oil} = 0.0111\ q_A$ (stainless steel)

with q_A in W/m^2 and temperatures in °C. Thus we see that for the same total losses, soft or magnetic iron tieplates have the same surface temperature rise as stainless steel tieplates. Even though the surface loss density is higher in soft iron tieplates than in stainless steel tieplates for the same total loss, the higher thermal conductivity of the soft iron rapidly equalizes the temperature within the material so that extreme temperature differences cannot develop. If the back surface were insulated ($h_2 = 0$), all the losses would have to leave the front surface in the steady state. In this case also the surface temperatures of the soft iron and stainless steel tieplates would be identical for the same total loss. As h_2 increases up to h_1, the front surface temperatures gradually differ for the two cases but not significantly.

11.15 CORE STEEL TEMPERATURE CALCULATION

Because core steel is made up of thin stacked insulated laminations, it has an anisotropic thermal conductivity. The thermal conductivity in the plane of the laminations is much higher than the thermal conductivity perpendicular to this plane or in the stacking direction. The steady state heat conduction equation for this situation is, in rectangular coordinates,

$$k_x \frac{\partial^2 T}{\partial x^2} + k_y \frac{\partial^2 T}{\partial y^2} + q_v = 0 \tag{11.94}$$

where the thermal conductivities k can differ in the x and y directions are are labelled accordingly. The loss per unit volume, q_v, is assumed to be a constant here. Again there are finite element programs which can solve this equation for complex geometries such as a stepped core. We will derive a simple but approximate analytical solution here for a rectangular geometry as shown in Fig. 11.18.

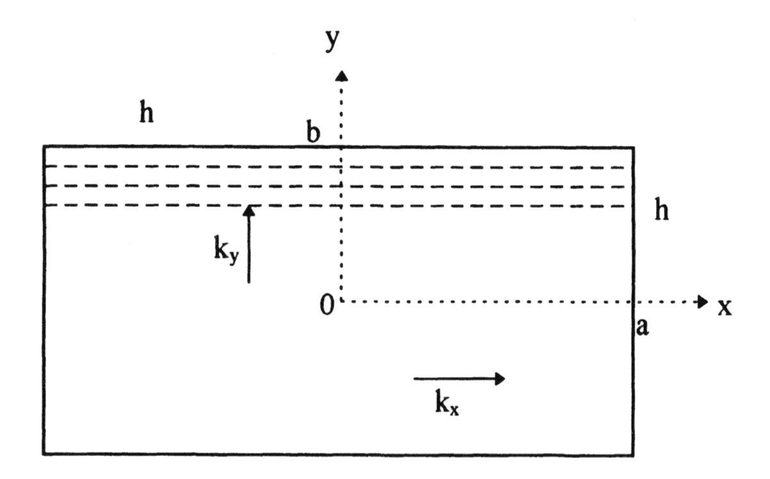

Figure 11.18 Rectangular geometry for anisotropic thermal calculation

This could apply to one step of a core, say the central step, with cooling ducts on the long sides or to all or part of a core between two cooling ducts by suitably calculating effective dimensions for an approximating rectangle from the steps involved.

We look for a solution of (11.94) of the form

$$T = A + Bx + Cx^2 + Dy + Ey^2 \tag{11.95}$$

satisfying the boundary conditions at the center

$$\frac{\partial T}{\partial x} = \frac{\partial T}{\partial y} = 0 \qquad \text{at} \quad x = y = 0 \tag{11.96}$$

This is required by symmetry. At the outer surfaces, we only approximately satisfy the boundary conditions for convective cooling

$$\begin{aligned} -k_x \frac{\partial T}{\partial x} &= h(T - T_{oil}) \qquad \text{at} \quad x = a\,,\, y = 0 \\ -k_y \frac{\partial T}{\partial y} &= h(T - T_{oil}) \qquad \text{at} \quad x = 0\,,\, y = b \end{aligned} \tag{11.97}$$

This means that we are satisfying the convective boundary condition exactly only at the surface points on the two axes. However, this is where we will evaluate the surface temperature. The solution is

$$T(x,y) - T_{oil} = \frac{\dfrac{q_v}{h}}{\left[\dfrac{1}{a\left(1+\dfrac{ha}{2k_x}\right)} + \dfrac{1}{b\left(1+\dfrac{hb}{2k_y}\right)}\right]} \left[1 - \left(\frac{\dfrac{ha}{2k_x}}{1+\dfrac{ha}{2k_x}}\right)\left(\frac{x}{a}\right)^2 - \left(\frac{\dfrac{hb}{2k_y}}{1+\dfrac{hb}{2k_y}}\right)\left(\frac{y}{b}\right)^2\right] \tag{11.98}$$

Note that a and b are 1/2 the rectangle dimensions. We can estimate the thermal conductivity in the direction perpendicular to the lamination plane (the y-direction here) from a knowledge of the coating thickness or stacking factor and the thermal properties of the coating. For stacked silicon steel we use $k_x = 30$ W/m°C and $k_y = 4$ W/m°C. Letting $h = 70$ W/m^2°C for convective oil cooling and assuming an effective stack or rectangle size of 0.762 m (30 in) × 0.254 m (10 in) and using $q_v = 10^4$ W/m^3 (1.3 W/kg, 0.6 W/lb), we find for the maximum internal temperature rise above the surrounding oil $T(0,0) - T_{oil} = 25.8$ °C. The surface temperature rise in the x or long dimension is $T(a,0) - T_{oil} = 17.8$ °C and the surface temperature rise at the surface at $y = b$ in the short direction perperdicular to the stack is $T(0,b) - T_{oil} = 12.2$ °C. Note that the surface in the direction of lower thermal conductivity is actually cooler than the surface towards which heat is more easily conducted. This is because, the central temperature being the same, there is a smaller thermal gradient in the high conductivity direction relative to the lower one.

12. LOAD TAP CHANGERS

Summary Load tap changers are a major component of many transformers. They are used to change the voltage of the primary or secondary side of a transformer to compensate for voltage variations in the power system mainly caused by variable loading. We discuss generically the various types of load tap changers and some of the important ways they are connected to the primary or secondary windings as well as the advantages and disadvantages of these connection schemes. We also address some of the maintanence issues which are of interest to the user.

12.1 INTRODUCTION

The flicker of the house lights while having dinner is a sign of a load tap changer (LTC) tap change. When most people get home from work, they start using electricity for cooking and lighting. This increases the electrical load on the distribution network, which causes the voltage to sag below its nominal value. The latter signals the LTC to change taps to adjust the voltage back to its nominal value with some tolerance.

There are many applications of transformers in modern electric power systems: from generator step-up to system interconnection to distribution to arc-furnace to HVDC converters to mention just a few. The role of a transformer is to convert the electrical energy from one voltage level to another. As power systems become larger and more complex, power transformers play a major role in how efficient and stable the system is. For each transformer installed in a network, there is an ideal (optimal) voltage ratio for an optimal operation of the system. Unfortunately, this optimal voltage ratio varies depending on the operating conditions of the total network. Early in the history of electric power systems, it became evident that for power systems to operate satisfactorily, transformer voltage ratios needed to be adjustable without interrupting the flow of energy. This is the role of a load tap changer.

12.2 GENERAL DESCRIPTION OF LTC

A load tap changer is a device that connects different taps of tapped windings of transformers without interrupting the load. It must be capable of switching from one tap position to another without at any time interrupting the flow of the current to the load and without at any time creating a short-circuit between any two taps of the transformer winding. Tap changing transformers are used to control the voltage or the phase angle or both in a regulated circuit.

A LTC is made of 4 elements:

1) A selector switch which allows the selection of the active tap
2) A change-over switch, referred as a reversing switch when it reverses the polarity of the tapped winding, used to double the number of positions available
3) A transition mechanism, including an arcing or diverter switch, which effects the transition from one tap to the other
4) A driving mechanism which includes a motor and gear box and controls to drive the system

There are 2 generic types of tap changers:

<u>In-tank</u>

The cover mounted, in-tank tap changer, known as the Jansen type, sits in the main transformer oil together with the core-and-coil assembly. Its selector and change-over switches are at the bottom of the tap changer in the main oil. The arcing (diverter) switch is located in a separate compartment at the top, usually within a sealed cylinder made of fiberglass or other similar material. All arcing is confined to this compartment. They exist in single phase or three phase neutral end Wye (Y) connected versions. For three phase fully insulated applications, three single phase tap changers must be used. This type of tap changer is used for higher voltages or current levels.

<u>Separate compartment</u>

The side mounted, separate compartment types have their own box and are assembled separately from the transformer. They are bolted to the side of the tank and connected to the transformer tapped windings through a connecting board. Their selector, change-over, and arcing switches are located in an oil compartment completely isolated from the main transformer oil. Some of them have two compartments, one for the arcing switch and one for the selector and change-over switches. Others have everything in one compartment only. They are available in three-phase assembly, either Wye connected for application at the neutral end of a three-phase transformer, or fully insulated for applications at the line end.

12.3 TYPES OF REGULATION

The main types of regulation are illustrated in Fig. 12.1 and are discussed below.

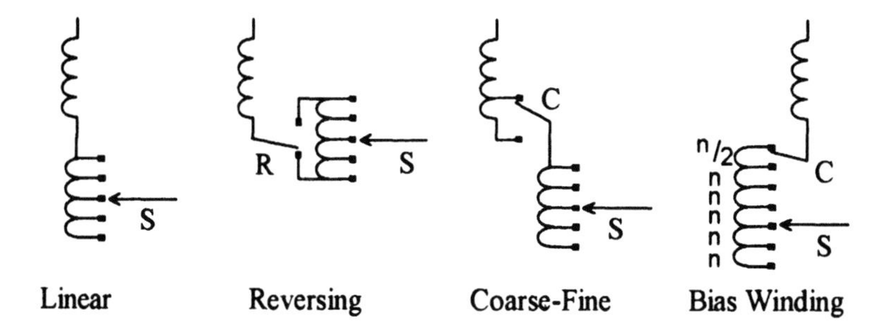

Figure 12.1 LTC main regulation types. S = Selector switch, C = Change-over switch, R = Reversing switch

Linear

In linear switching, tapped turns are added in series with the main winding and their voltage adds to the voltage of the main winding. No change-over switch is needed for this type. The tapped winding is totally by-passed in the minimum voltage position. The rated position can be any one of the tap positions.

Plus - Minus (Reversing)

In a reversing type of regulation, the whole tapped winding can be connected in additive or reversed polarity with respect to the main winding. The tapped turns can add or subtract their voltage with respect to the main winding. The tapped winding is totally by-passed in the neutral (mid-range) voltage position. The rated position is normally the mid one. The total number of positions available is twice the number of sections in the tapped winding plus one.

Coarse - Fine

The coarse-fine regulation can be defined as a two-stage linear regulation where the first or coarse stage contains a large number of turns which can be totally by-passed by the change-over selector. These turns are shown as a single loop in the top coil of the figure. Fine regulation is achieved with the selector switch. Normally, the coarse section contains as many turns as the tapped winding plus one section. In that way, the total number of positions available is twice the number of sections in the tapped winding plus one.

Bias winding

The bias winding type of regulation is similar to the coarse-fine except that the number of turns in the bias winding is half the turns of one section of the tapped winding. It is used to provide half steps between the main tap steps. Thus the total number of positions available is twice the number of sections in the tapped winding plus one. The bias winding technique can be combined with the reversing scheme to provide twice the number of positions (four times the number of sections in the tapped winding plus one) at the expense of adding one more switch and increasing the complexity of the switching and driving mechanism.

Although not indicated in the figure, the tap selector switch is often a circular type of switch to allow a smooth transition between the tap voltages when the change-over or reversing switch operates.

12.4 PRINCIPLE OF OPERATION

A load tap changer must be capable of switching from one tap position to another without interrupting the flow of current to the load at any time. It must therefore follow a "make before break" switching sequence. On the other hand, it cannot at any time create a short-circuit between any two taps of the transformer winding. This means that during this make before break interval, there must be something to prevent the shorting of the turns. Two ways are primarily used to accomplish this.

12.4.1 Resistive Switching

Fig. 12.2 shows an example of a six step switching sequence. This particular type of tap changer has two selector switches and an arcing switch with four contacts.

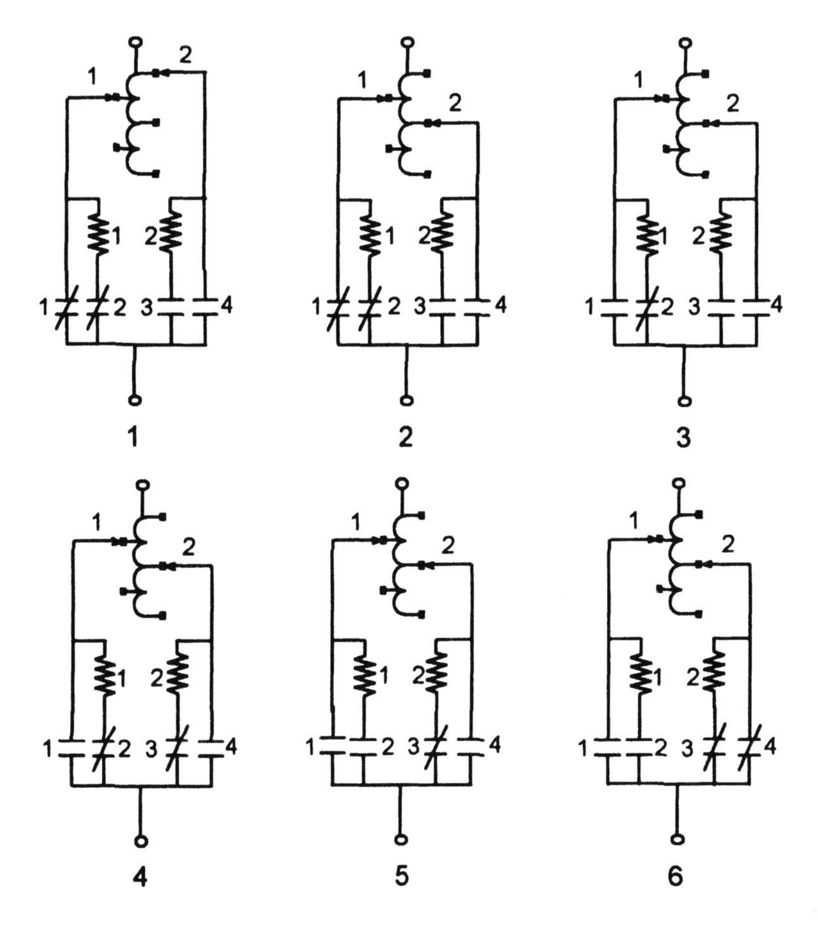

Figure 12.2 Sequence of operations involved in tap changers using resistive switching. The selector switches are the top most switches. The arcing switches are closed when a diagonal (shorting) line is present and open otherwise.

1) Step 1 of the figure above shows the steady state of the TC just before the switching operation. The load current is flowing through the contact #1 and the selector #1. Let us call that position tap #2.
2) At step 2, the non-conducting selector has moved 2 steps down, from position 1 to position 3. The current is still flowing in contact #1 and the selector #1.
3) At step 3, the contact #1 opens and the current flows through resistor #1 and selector #1.
4) At step 4, the contact #3 closes and the current splits between the 2 resistors and the 2 selectors. There is also a circulating

current flowing into the loop limited by the 2 resistors and the reactance of the loop.

5) At step 5, the contact #2 opens, breaking the current in the resistor and selector #1. It is at that moment that arcing occurs. The arc remains on until the current crosses the zero line. At worst, it can last one half cycle which, at 60 Hz, is 8 msec. The average arcing time is around 5 to 6 msec.
6) At step #6, the contact #4 closes and the load current by-passes the resistor #2 and goes directly to the selector #2. The TC has reached the steady state for tap position #3.

This type of transition requires that the resistors be capable of withstanding the full load current plus the circulating current during the transition (from step 3 to step 5). In order to reduce the energy absorption requirement for the resistors, the time of the complete transition has to be minimized. These tap changers have normally very fast transitions.

12.4.2 Reactive Switching with Preventative Autotransformer

Another widely used way of handling the transition from one tap to another is to use reactors instead of resistors. These reactors do not have to dissipate as much energy as the resistors. They mainly use reactive energy, which does not produce any heat. Therefore, they can be designed to withstand the full load plus the circulating current for long periods of time, even continuously. Two reactors per phase are needed. They are normally wound on a common gapped core, making them mutually coupled. When they are connected in series, they act as an auto-transformer. When they are connected in parallel, they act as a single reactor. Transformer designers take advantage of this feature. They use the reactors not only to prevent the load current from being interrupted or to prevent sections of a tapped winding from being shorted, but also to act as a transformer and provide intermediate voltage steps in between two consecutive tap sections of the main transformer. The total number of positions available with this scheme is twice the number of sections in the tapped winding plus one. Here is how it works.

The method of operation is diagrammed in Fig. 12.3. The tap changer has two selectors, two reactors (actually one reactor with 2 windings), two by-pass switches (#1 and #2) and an arcing switch (#3).

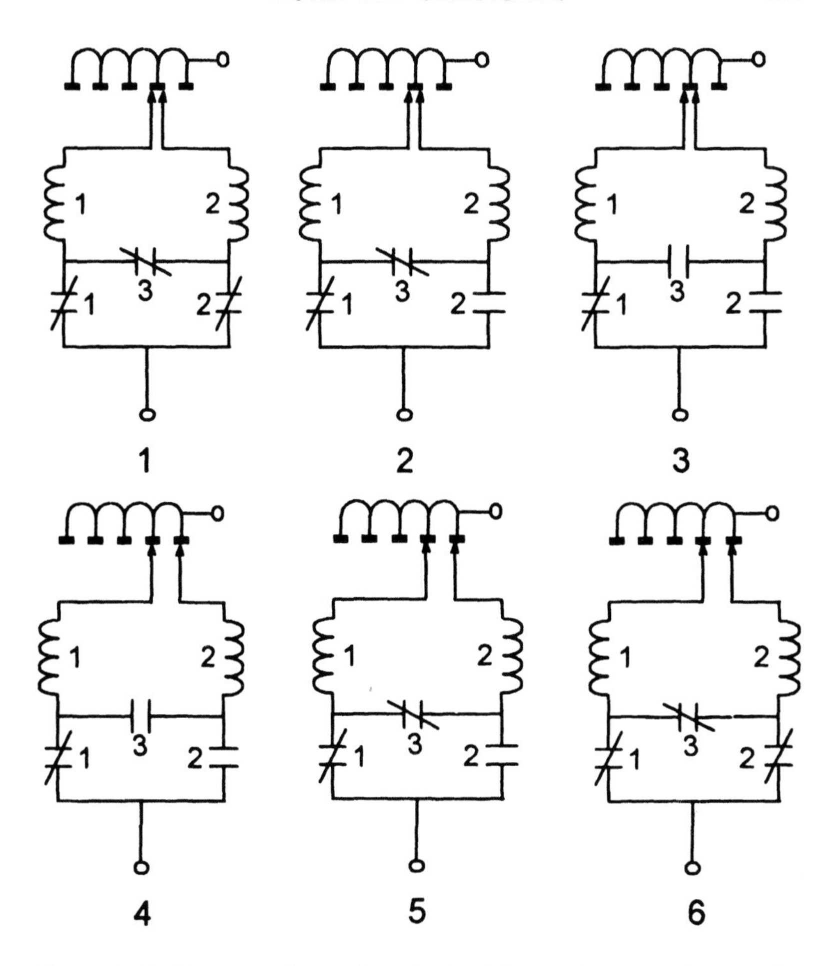

Figure 12.3 Sequence of operations involved in tap changers using reactive switching

1) In step 1 of the figure above, the tap-changer is in the steady state mode on tap position #2. The load current flows in the 2 selectors, the 2 reactors and the by-pass switches. Although the arcing switch is closed, no current flows through it.
2) In step 2, the by-pass switch opens. The current flows through the switch #1, and then splits up between the 2 reactors and the 2 selectors.
3) In step 3, the arcing switch opens. All current flows through the reactor 1 and the selector 1. Arcing occurs at that step because the current in the reactor is interrupted.

4) In step 4, the right selector moves one step to the right, while it carries no current.
5) In step 5, the arcing switch closes, causing current to flow again in the reactor #2 and in selector #2. Note that the two selectors are on different taps. This causes a circulating current to flow into the loop.
6) In step 6, the by-pass switch #2 closes and a steady state condition is reached, which is in this case tap position #1. Because the 2 ends of the reactors are on 2 different taps, they are in series between the 2 taps. The voltage at their mid point is half way between the 2 taps. This state is called the "bridging position", where the reactors bridge between 2 taps.

When combined with a reversing switch, this scheme provides four times as many voltage positions as tapped sections in the winding plus one. The price to pay is that the reactor has to be designed to withstand continuously the full load current plus the circulating current; the tapped winding has also to withstand the full load plus the circulating current; and the reactor introduces losses and draws more magnetizing current from the source, specially on bridging positions. It might also add audible noise to the transformer.

12.5 CONNECTION SCHEMES

12.5.1 Full Transformers

Normally, the primary winding of transformers is fed at a constant voltage and the role of the tap changer is to add or subtract turns in order to vary the voltage ratio of the transformer to maintain its output at a constant voltage despite fluctuations in the load current. In principle, this can be accomplished by changing the number of active turns either in the primary or secondary winding. There are subtle differences in the two ways of connecting the tap changer which transformer and system designers have to be aware of.

Fixed volts/turn

The most natural way to add a tapped winding to a transformer is to connect it in series with the regulated side. As shown in Fig. 12.4, the primary winding of the transformer (H1 - H2) is fed at a constant voltage and it has a fixed number of turns. So the volts/turn of the transformer is constant. The voltage across X1-X2 varies with the number of turns. If all the taps have equal number of turns, then the voltage increase is equal for each step. If the X winding is Wye connected, then the tapped winding and the tap changer can be placed at the low potential neutral end and do not require a high insulation level. The price to pay is that the tap changer has to be capable of carrying the current of the X winding. For high current windings, the cost of such a tap changer could become prohibitive. Another disadvantage is that low voltage windings often have few turns. The design of the tapped winding might become impractical if not impossible considering that fractional turns cannot be used.

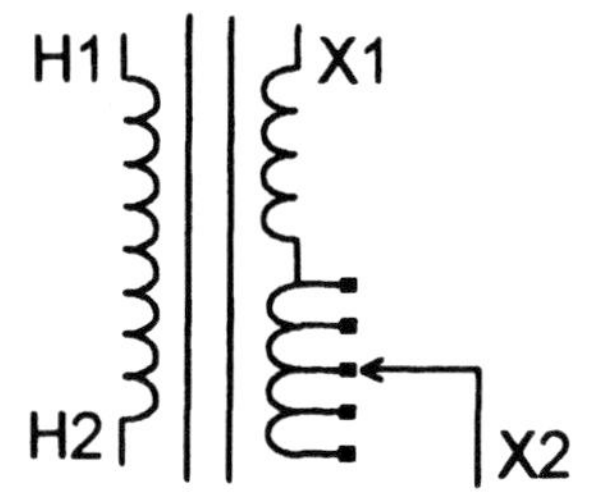

Figure 12.4 Fixed volts/turn tap changing scheme

Variable volts/turn

One way to solve the problem of high current low voltage windings is to put the tapped winding in the high voltage side. The low voltage side can still be regulated in this way. This is particularly applicable if the HV is Wye connected since the tap changer could be placed at the neutral end and would not require a high insulation level while not carrying a high current. This scheme is illustrated in Fig. 12.5

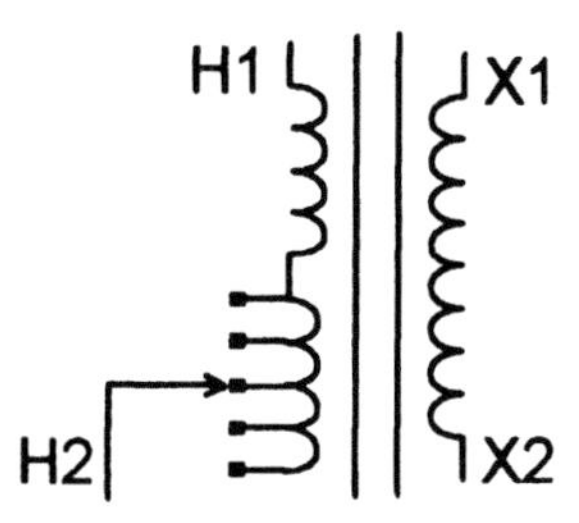

Figure 12.5 Variable volts/turn tap changing scheme

Although the solution might look very attractive, it has its disadvantages:

- If we consider that the voltage across H1-H2 is constant, varying the number of turns in that winding implies that the volts/turn and thus the flux in the core varies. It means that the core has to be designed for the minimum turn position. The core would be bigger than its fixed volt/turn counterpart and would really be used efficiently only at that minimum turn position. At any other position, it operates at lower flux densities.
- If the flux varies in the core, the no-load losses, the exciting current, the impedance, and the sound level of the transformer will vary also. If a transformer is designed to meet certain guaranteed losses and sound level at rated position which is generally the mid position, it is likely that at the minimum turn position, it will exceed these guaranteed values significantly.
- In a variable flux transformer, the voltage variation per step is not constant even if the number of turns per step is constant.
- If there is a third winding in the transformer used to feed a different circuit, its voltage will vary as the tap changer moves, which might not be desirable.

12.5.2 Autotransformers

Autotransformers with tap changers present special challenges for the transformer designer. There are three ways of connecting a tap changer in an auto-transformer without using an auxiliary transformer. These are illustrated in the following figures.

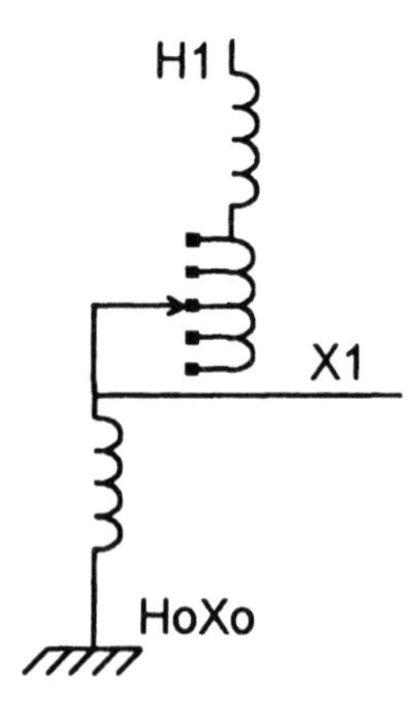

Figure 12.6 Taps in the series winding of an autotransformer

The connection shown in Fig. 12.6 is used when the high voltage of the transformer is to be varied while keeping the low voltage constant. In this case, we are assuming that the LV side is supplied by a fixed voltage source. The voltage at the HV terminal varies linearly with the number of turns added or subtracted to the series winding. The flux in the core is constant. In this connection, the tap changer and the tapped winding must be isolated for the voltage level of the LV line terminal plus the voltage across the tapped section. They are directly exposed to any voltage surge coming through the LV line so extra precautions have to be taken at the design stage.

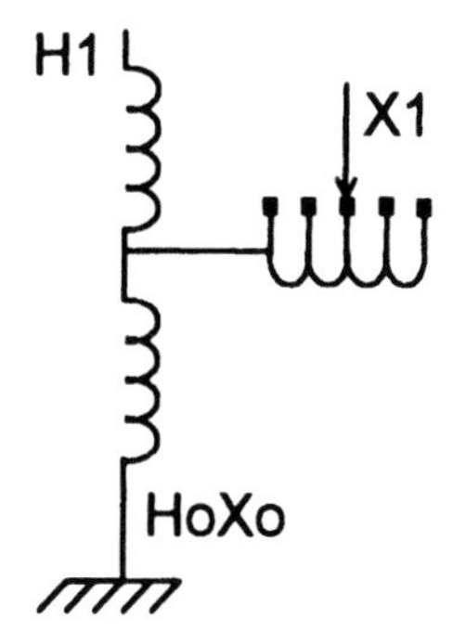

Figure 12.7 Taps in the LV line from the auto connection

Fig. 12.7 shows the connection used when the high voltage side of the transformer is to be kept constant while the low voltage varies. In this case, the HV side is fed from a constant voltage source. The voltage at the LV terminal varies linearly with the number of turns added or subtracted in series with the LV line. As in the previous case, the flux in the core is constant, the tap changer and the tapped winding must be isolated for the voltage level of the LV line terminal plus the voltage across the tapped section. They are directly exposed to any voltage surge coming through the LV line so extra precautions have to be taken at the design stage. Finally, the tap changer must be designed to carry the full load current of the LV terminal.

In the previous 2 cases, if a three-phase transformer is designed, the tap changer must have full insulation between the phases. If the voltage on the LV is above 138 kV, then three single-phase, Jansen type tap changers must be used.

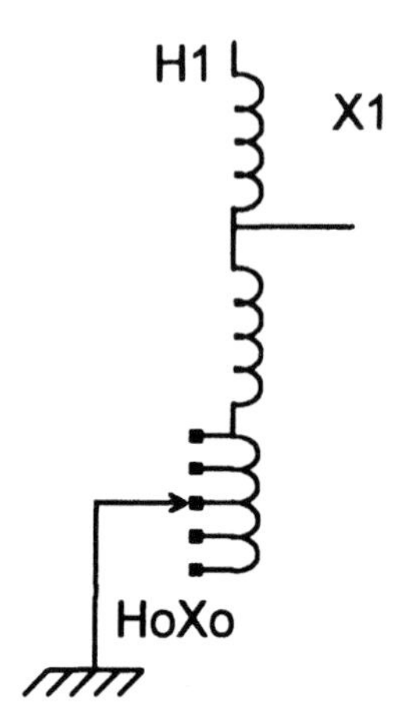

Figure 12.8 Taps in the common winding of an autotransformer, at the neutral end

The connection shown in Fig. 12.8 can be used when the LV is to be varied while the HV is to be kept constant. The advantages of this connection are:

- The level of insulation for the tapped winding and the tap changer is low
- A Wye connected three-phase tap changer can be used for a three-phase transformer
- The current in the common winding is lower than the current in the LV line terminal; so a smaller tap changer can be selected.

As usual, there is a price to pay to obtain these benefits. If we assume that the voltage between H1 and HoXo is constant, varying the number of turns implies varying the volts/turn, and therefore the flux in the core. It is a variable flux design. As explained earlier, in a variable flux transformer, the losses, the exciting current, the impedance, and the sound level vary with the tap position, and a bigger core must be selected for proper operation at the minimum turn position. If there is a tertiary winding, its voltage will vary as the tap changer changes position.

In an auto-transformer connection, the turns in the common branch are part of both the HV and LV circuits. This means that if we add turns in the common branch, we add turns to both the HV and LV circuits. Because of this, the number of turns required to achieve a specified regulating range is higher than for the other 2 cases. Here is an example.

Let us suppose that we want 10 % voltage regulation. We need a ratio that varies by 10 % when we add all the turns of the tapped winding. Suppose that we have 100 turns in the series winding and 100 turns in the common winding. The LV circuit has 100 turns and the HV circuit has 200 turns with no taps in circuit. It gives a ratio of 2:1. Let us consider all three cases:

1) For the tapped turns in the series winding, if we want a 10% change in the HV, we need to add or remove 20 turns to the HV circuit. The size or rating of the tapped winding would be 10% of the HV x the full load current of the HV = 10% of the total MVA of the transformer.
2) For the case of taps in the LV line from the auto point, we need to add or remove 10 turns to the LV circuit to achieve 10 % change in the LV. The size of the winding would be 10% of the LV voltage x the full current of the LV = 10% of the total MVA of the transformer. The ratio of LV to HV is 0.5 +/- .05 or 0.45 - 0.55
3) In the case of the taps in the neutral end, assuming that the HV is constant, in order to increase the LV by 10 %, we need approximately 23 turns in order to get a ratio of 123 / 223 = 0.552. If we subtract the 23 turns (we add them in reverse polarity), we obtain a ratio of 77 / 177 = 0.435. The variation of the voltage is not symmetrical. The size of the tapped winding is 23% of the common winding or 11.5% of the total MVA of the transformer. It is bigger than the other 2 cases and the volts per step in the tap changer is also higher. The designer of the transformer has to weigh all these factors when selecting the connection of the tap winding.

12.5.3 Use of Auxiliary Transformer

When a transformer with a relatively high current LV winding is specified and regulation is desired on the LV side, it may be economical to use an auxiliary transformer to achieve regulation. This gives the designer a greater flexibility in the design of the tapped winding and the savings on the main transformer often outweighs the cost of the auxiliary transformer. There are 2 popular way of using auxiliary transformers.

The Series Voltage Regulator

In this connection, the tapped winding is located on the auxiliary transformer which is inserted in series with the low voltage line terminal, adding or subtracting its voltage with the LV winding of the main transformer. The main transformer has only 2 windings and can be optimized independently from the series voltage regulator (SVR). This is shown in Fig. 12.9.

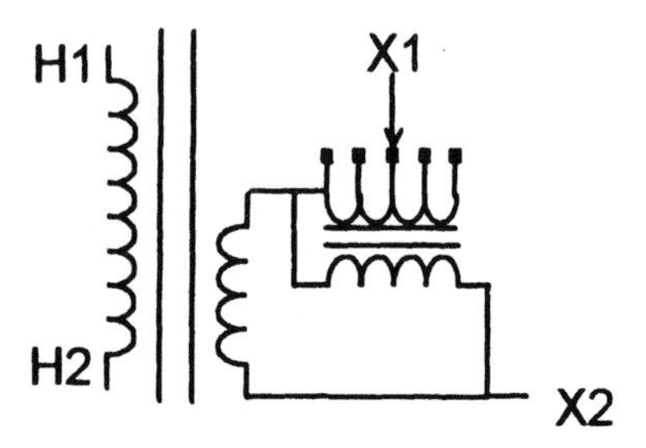

Figure 12.9 Tap changing scheme using a series voltage regulator

The Series (booster) Transformer

In this connection, shown in Fig. 12.10, the tapped winding is a separate winding on the main transformer. The advantages of this connection are that there is much more flexibility in the design of the tapped winding. The number of turns per step can be selected so that the tap changer carries a smaller current. The main saving in this case is on the tap changer itself and in the greater flexibility in the optimization of the main transformer.

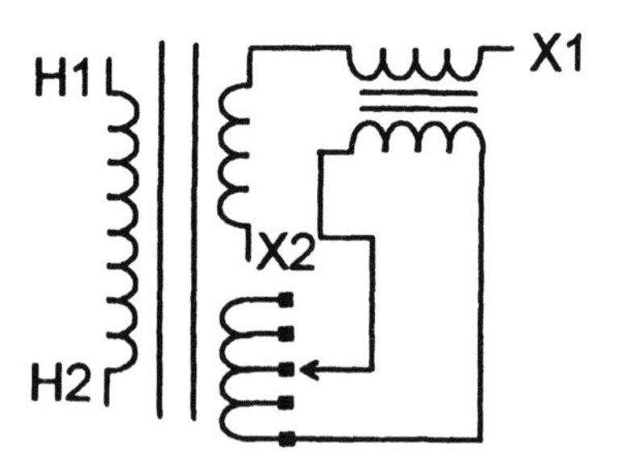

Figure 12.10 Tap changing scheme using a series (booster) transformer

Auxiliary transformers can also be used with auto-transformers pretty much in the same way. There are also plenty of possible connections for using auxiliary transformers. The choice is only limited by the imagination of the designer.

12.5.4 Phase shifting transformers

A phase shifting transformer is ideally a transformer with a 1:1 voltage ratio and which has the ability to change the phase angle of its output voltage relative to its input. This feature is used by electric power companies to help transfer power more efficiently within the electrical grid. In these transformers, the voltage from the tapped windings is added in quadrature with the input line terminals, producing an effective phase shift in the output. There are quite a few possible connections to achieve this and some of these are discussed in Chapter 13. Let us just mention here that in a phase shifting transformer, the tap changer is a key element and the size and rating of the tapped windings are much higher than for conventional transformers.

12.6 GENERAL MAINTENANCE

The reason for the maintenance on a tap changer is to insure its reliability. Any device with moving parts, some of which are used to interrupt currents and voltages requires maintenance. It would be useless to repeat here what can be found in the manufacturers instruction books. We will only mention general principles. The factors affecting the reliability of tap changers include:

- Oil quality
- Contact pressure
- Contact resistance and temperature
- Timing of movements
- Load currents
- Number of operations

The maintenance of the controls and the drive mechanism are normally not a problem. The gears and other moving parts might wear and require replacement, but in general, they are designed to last very long and do not need frequent inspection. The main areas of concern are the switches and the connecting board if present. The contacts have to be inspected and verified periodically. The frequency of these inspections depends mainly on the number of operations, the current flowing through the contacts and the cleanliness of the oil.

The inspection and replacement of contacts require that the transformer be taken out of service. Since it is always costly to take a transformer out of service, it is desirable to extend the period between inspections as much as possible. Besides recommendations that can be found in most instruction books for tap changers, here are some of the tactics used by utilities to reduce their maintenance costs and the failure rate of their tap changers.

The process of changing tap position includes arcing, which generates gases and carbon particles. The movement of contacts also promotes erosion of the metal and generates small metal particles. In the long run, these particles can affect the dielectric properties of the oil. Moreover, the carbon particles tend to aggregate on contacts and increase the contact resistance. They also accumulate on connecting boards and may lead to tracking on the surface of the board. For these reasons, the reliability of a tap changer and its overall performance can be improved by a simple filtering device which would keep the oil clean. This is specially important if the arcing switch is located in the same compartment as the selector and change-over switches. Some utilities have introduced oil filters for load tap changers and they have found that with these filters, they can extend their time between inspections. However in some tap changers, the arcing occurs in a separate vacuum chamber so that oil comtamination is reduced or eleminated.

In a tap changer, contacts that operate frequently remain clean. The ones that do not move tend to oxidize or collect carbon or both. With time, their contact resistance increases, which produces more heat and the situation deteriorates. This is typical of change-over selectors. Some of those operate only twice a year. Many cases of damage to these switches have been reported. Several utilities now force their tap changers to ride over the mid position at least once a month and they claim a much better performance of their tap changers since they have introduced the procedure.

Instead of taking all transformers out of service for regular and frequent inspections of tap changers, some utilities have introduced monitoring procedures for tap changers without de-energizing their transformers. One of those procedures is dissolved gas analysis (DGA). They regularly take oil samples from tap changer compartments and monitor the gasses. After a while, they know the pattern of all their tap changers. Whenever one changes pattern, they go and inspect it. They report that this procedure has helped them reduce the frequency of their internal inspections while alerting them to dangerous problems before they lead to a catastrophic failure.

Another trick is to have inspection windows in the tap changer box. When used in conjunction with an oil filter, they can look at their contacts without opening the box and take the transformer out of service if required.

Another possible procedure is to monitor the temperature on the tap changer compartment regularly with infrared measuring devices. It is very inexpensive and easy to do. They claim that if anything goes wrong in a tap changer, it typically increases the temperature of the contacts to a point that you can detect it by checking the temperature on the outside of the wall of the box. Their rule is that if the temperature on the surface of the tap changer compartment is higher than the tank of the transformer by more than 7 to 10 degrees, something is going wrong inside the transformer. They would then inspect the tap changer. Quite a few problems have been detected this way.

13. PHASE SHIFTING TRANSFORMERS

Summary Equivalent circuit models are derived for three commonly used types of phase shifting transformer. The winding configurations are chosen to have a positive phase shift and no change in voltage magnitude at no-load. The equivalent impedance as seen from the input or output is expressed in terms of two winding impedances. This is done for positive, negative, and zero sequence circuits. Negative sequence circuits have a negative phase shift and zero sequence circuits have no phase shift. The model is then used to study regulation effects and to calculate fault currents for the major fault types.

13.1 INTRODUCTION

Phase shifting transformers are used in power systems to help control power flow and line losses. They shift the input voltage and current phases by an angle which can be adjusted by means of a tap changer. They operate by adding a voltage at $\pm 90°$ to the input voltage, i.e. in quadrature. For 3 phase transformers, the quadrature voltage to be added to a given phase voltage can be derived from the other phases. The many ways of doing this give rise to a large number of configurations for these transformers. We will deal with only a few common types here. Phase shifting capability can be combined with voltage magnitude control in the same transformer. This results in a more complex unit, involving two sets of tap changers, which we no not discuss here.

As a simplified example of their utility, consider the circuit in Fig. 13.1 where a power source at voltage $\mathbf{V}$ is feeding a load along two parallel paths. (We use bold faced symbols for phasor quantities.) With the voltage source $\mathbf{E} = 0$, the currents will divide according to the line impedances Z_1 and Z_2,

$$\frac{\mathbf{I}_1}{\mathbf{I}_2} = \frac{Z_2}{Z_1} \tag{13.1}$$

With **E** inserted, we find,

$$\mathbf{I}_1 = \mathbf{V}\frac{Z_2}{K} - \mathbf{E}\frac{Z_L}{K} \quad , \quad \mathbf{I}_2 = \mathbf{V}\frac{Z_1}{K} + \mathbf{E}\frac{(Z_L + Z_1)}{K}$$

$$K = Z_1 Z_2 + Z_1 Z_L + Z_2 Z_L \tag{13.2}$$

Since, normally $|Z_L| >> |Z_1|$, the effect of **E** is to add a current to $\mathbf{I}_2$ and subtract a current from $\mathbf{I}_1$, i.e. introduce a circulating current of

$$\mathbf{I}_{circ} = \mathbf{E}\frac{Z_L}{K} \tag{13.3}$$

The magnitude as well as the phase of **E** controls the amount of circulating current.

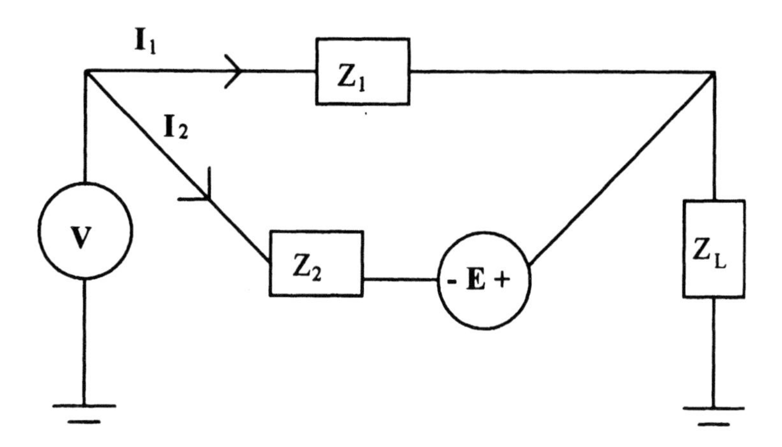

Figure 13.1 Addition of voltage E to control circulating current

In the approximation that Z_L is much larger than Z_1 or Z_2 and assuming $Z_1 = Z_2 = Z$, we have

$$\mathbf{I}_1 \approx \frac{\mathbf{V}}{2Z_L} - \frac{\mathbf{E}}{2Z} \quad , \quad \mathbf{I}_2 \approx \frac{\mathbf{V}}{2Z_L} + \frac{\mathbf{E}}{2Z} \tag{13.4}$$

If Z_L is resistive, i.e. a unity power factor load, and Z is primarily inductive or capacitive, then

$$\mathbf{I}_{\text{circ}} = \pm j \frac{\mathbf{E}}{2|Z|} \tag{13.5}$$

Thus if $\mathbf{E}$ is $\pm 90°$ out of phase with $\mathbf{V}$, then a circulating current in phase with the primary current results. This can be useful in controlling the flow of real power along each of the parallel lines. A phase shifting transformer could add the required $\mathbf{E}$.

As another example, consider feeding a common load from two voltage sources which could be out of phase with each other as shown in Fig. 13.2 Solving for the currents, we get

$$\mathbf{I}_1 = \mathbf{V}_1 \frac{(Z_2 + Z_L)}{K} - \mathbf{V}_2 \frac{Z_L}{K} \quad , \quad \mathbf{I}_2 = \mathbf{V}_2 \frac{(Z_1 + Z_L)}{K} - \mathbf{V}_1 \frac{Z_L}{K} \tag{13.6}$$

The current into the load, $\mathbf{I}_L$, is

$$\mathbf{I}_L = \mathbf{I}_1 + \mathbf{I}_2 = \mathbf{V}_1 \frac{Z_2}{K} + \mathbf{V}_2 \frac{Z_1}{K} \tag{13.7}$$

and the voltage across the load, $\mathbf{V}_L$, is

$$\mathbf{V}_L = \mathbf{V}_1 - \mathbf{I}_1 Z_1 = \frac{Z_L}{K}(\mathbf{V}_1 Z_2 + \mathbf{V}_2 Z_1) \tag{13.8}$$

Thus the complex power delivered to the load is

$$\mathbf{V}_L \mathbf{I}_L^* = \frac{Z_L}{|K|^2} |\mathbf{V}_1 Z_2 + \mathbf{V}_2 Z_1|^2 \tag{13.9}$$

where * denotes complex conjugation.

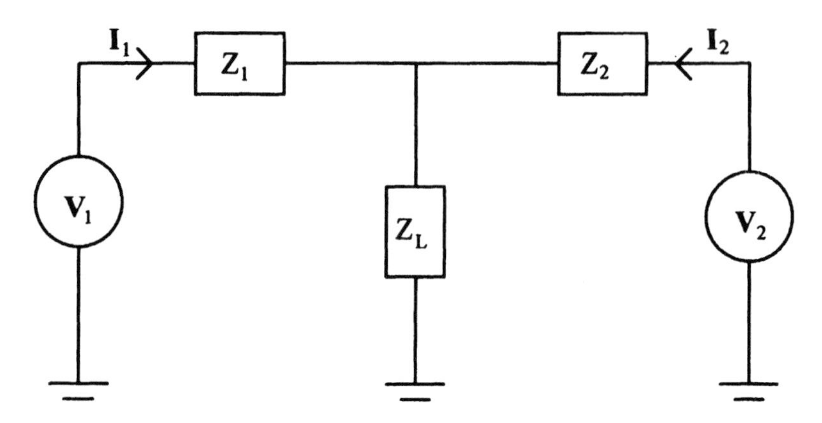

Figure 13.2 Two possibly out of phase sources feeding a common load

Consider the case where $Z_1 = Z_2 = Z$, $\mathbf{V}_1 = V$, $\mathbf{V}_2 = Ve^{j\theta}$. Then (13.9) becomes

$$V_L I_L^* = \frac{Z_L |Z|^2 V^2}{|K|^2} 2(1+\cos\theta) \tag{13.10}$$

Thus it can be seen that the maximum power is transferred when $\theta = 0$. In this case a phase shifting transformer could be used to adjust the phase of $\mathbf{V}_2$ so that it equals that of $\mathbf{V}_1$.

In modern power systems which are becoming more and more interconnected, the need for these devices is growing. Other methods for introducing a quadrature voltage are being developed, for instance by means of power electronic circuits in conjunction with ac-dc converters. These can act much faster than on-load tap changers. However, at present they are more costly than phase shifting transformers and are used primarily when response time is important. (Electronic on-load tap changers are also being developed for fast response time applications.)

In this report, we develop a circuit model description for three common types of phase shifting transformer. This is useful in order to understand the regulation behavior of such devices, i.e. how much the output voltage magnitude and phase change when the unit is loaded as compared with the no load voltage output. In the process, we also find how the phase angle depends on tap position, a relationship which can be non-linear. In addition to the positive sequence circuit model which describes normal operation, we also determine the negative and zero sequence circuit models for use in short circuit fault current analysis.

The latter analysis is carried out for the standard types of fault for two of the phase shifting transformers.

Very little has been published on this subject in the open literature beyond general interconnection diagrams and how they are used in specific power grids. However, several references which emphasize basic principles are Refs.[Hob39, Cle39]. Other useful references are [Kra98, Wes64].

13.2 BASIC PRINCIPLES

We neglect exciting current and model the individual phases in terms of their leakage impedances. For a two winding phase, we use the circuit model shown in Fig. 13.3. Z_{12} is the two winding leakage impedance referred to side 1. Most of the development described here is carried out in terms of impedances in Ohms. Because of differences in per unit bases for the input and winding quantities, per unit quantities are not as convenient in the analysis. At the appropriate place, we indicate where per unit quantities might prove useful. The currents are assumed to flow into their respective windings. With N_1 the number of turns on side 1 and N_2 the number of turns on side 2, the ideal transformer voltages satisfy

$$\frac{\mathbf{E}_1}{\mathbf{E}_2} = \frac{N_1}{N_2} \tag{13.11}$$

and the currents satisfy

$$\frac{\mathbf{I}_1}{\mathbf{I}_2} = -\frac{N_2}{N_1} \tag{13.12}$$

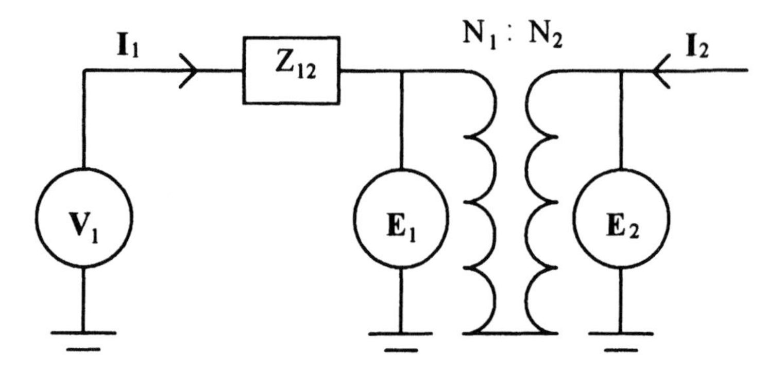

Figure 13.3 Model of a two winding transformer phase

For 3 windings per phase, we use the model shown in Fig. 13.4. In this case, the ideal transformer voltages satisfy

$$\frac{\mathbf{E}_1}{N_1} = \frac{\mathbf{E}_2}{N_2} = \frac{\mathbf{E}_3}{N_3} \tag{13.13}$$

and the currents satisfy

$$N_1\mathbf{I}_1 + N_2\mathbf{I}_2 + N_3\mathbf{I}_3 = 0 \tag{13.14}$$

The single winding impedances are given in terms of the 2 winding impedances by

$$Z_1 = \frac{1}{2}\left[Z_{12} + Z_{13} - \left(\frac{N_1}{N_2}\right)^2 Z_{23}\right]$$

$$Z_2 = \frac{1}{2}\left(\frac{N_2}{N_2}\right)^2\left[Z_{12} + \left(\frac{N_1}{N_2}\right)^2 Z_{23} - Z_{13}\right] \tag{13.15}$$

$$Z_3 = \frac{1}{2}\left(\frac{N_3}{N_1}\right)^2\left[Z_{13} + \left(\frac{N_1}{N_2}\right)^2 Z_{23} - Z_{12}\right]$$

Here the two winding impedances are referred to the winding corresponding to the first subscript and the second subscript refers to the winding which would be shorted when measuring the impedance. To refer impedances to the opposite winding, use

$$Z_{ij} = \left(\frac{N_i}{N_j}\right)^2 Z_{ji} \tag{13.16}$$

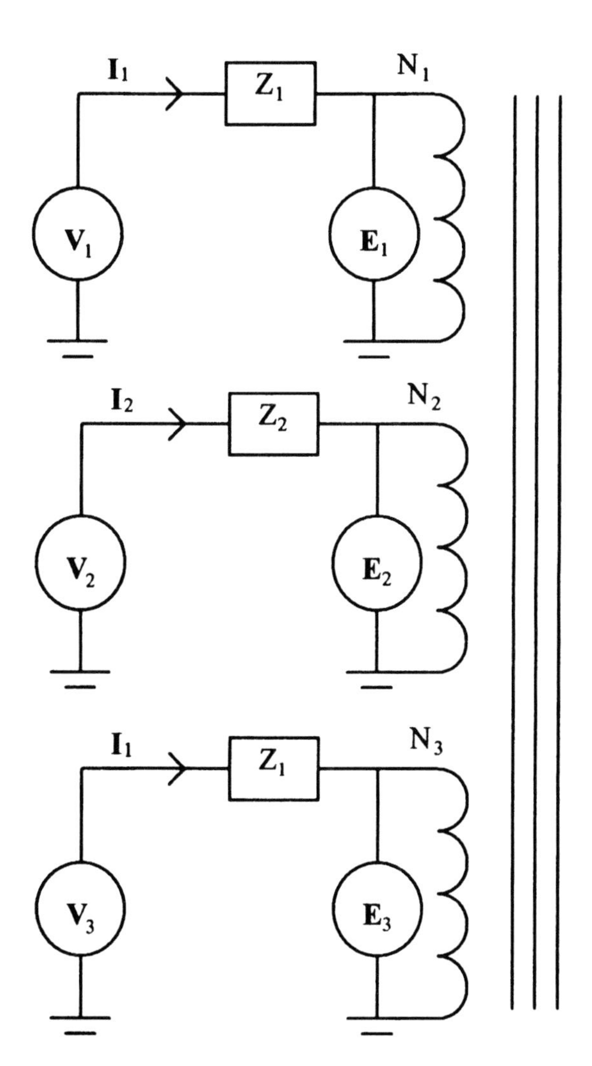

Figure 13.4 Model of a 3 winding transformer phase. The vertical lines connecting the diagrams represent the common core.

For a 3 phase system, the positive sequence quantities correspond to the ordering of the unit phasors shown in Fig. 13.5. Letting

$$\alpha = e^{j120^\circ} = -\frac{1}{2} + j\frac{\sqrt{3}}{2} \tag{13.17}$$

the ordering is 1, α^2, α. Negative sequence ordering is 1, α, α^2. This is obtained by interchanging 2 phases. Note that

$$\alpha^2 = e^{j240^\circ} = e^{-j120^\circ} = \alpha^* = -\frac{1}{2} - j\frac{\sqrt{3}}{2} \tag{13.18}$$

Zero sequence quantities are all in phase with each other.

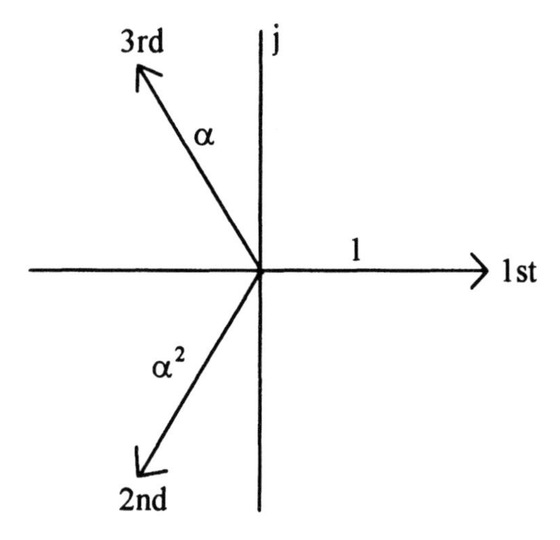

Figure 13.5 Positive sequence unit phasors

Our interconnections will be chosen to produce a positive phase shift at a positive tap setting. By interchanging two phases, a negative phase shift can be produced at a positive tap setting. Interchanging two phases is equivalent to imputing a negative sequence set of voltages. This implies that negative sequence circuits have the opposite phase shift to positive sequence circuits. Zero sequence circuits have zero phase shift. By positive phase shift we mean that output voltages and currents lead input voltages and currents, i.e. output quantities are rotated counter clockwise on a phasor diagram relative to input quantities.

13.3 SQUASHED DELTA PHASE SHIFTING TRANSFORMER

One of the simplest phase shifters to analyze is the squashed delta configuration shown in Fig. 13.6. In the figure S labels the source or input quantities and L the load or output quantities. The input and output set of voltages and corresponding currents form a balanced positive sequence set. The input and output voltage phasor diagram is shown in Fig 13.6b. These are voltages to ground. The currents form a similar set but are not shown. Similarly, the internal voltages and corresponding currents form a positive sequence set as shown in Fig. 13.6c for the voltages. Note that Figs 13.6b and 13.6c could be rotated relative to each other if shown on a common phasor diagram. Information on their relative orientation is not contained in the figure although the phase order is consistent between the figures. This applies as well to the corresponding current phasors. a and a' , etc. refer to windings on the same leg, with the prime labeling the tapped winding. The 2 winding impedances will be referred to the unprimed coil, i.e. $Z_{aa'}$, etc. Since these impedances are all the same by symmetry, we need only this one symbol. This transformer can be designed with a single 3 phase core and is referred to generically as a single core design. It is also best adapted to phase angle shifts in one direction.

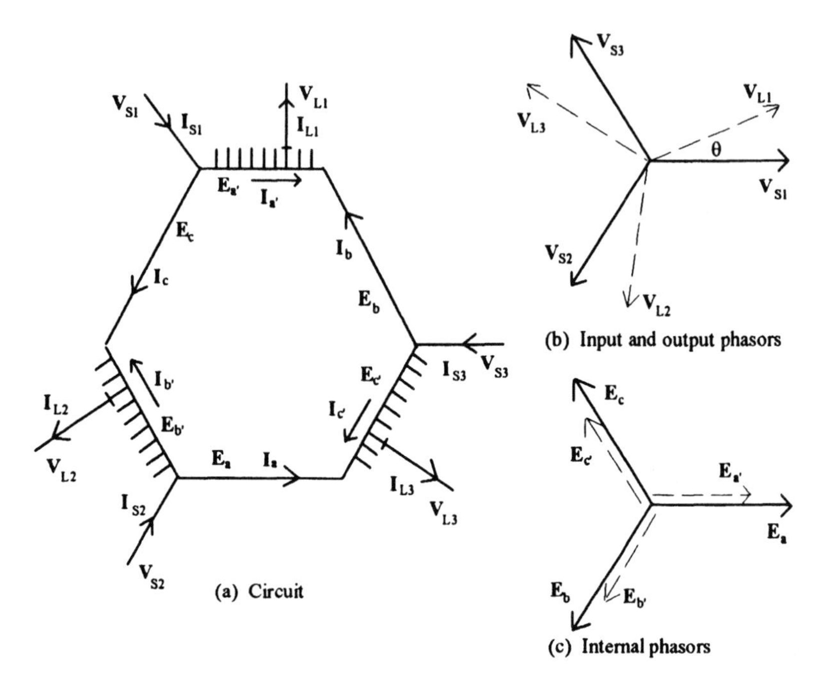

Figure 13.6 Squashed delta configuration with phase quantities labelled. The ideal transformer voltages, E , increase in the opposite direction of the assumed current flow. θ is a positive phase shift angle.

Using the two winding phase model described in the last section, adapted to the present labeling scheme, and concentrating on one input-output pair, we can write

$$\mathbf{V}_{S1} - \mathbf{V}_{L1} = \mathbf{E}_{a'} \quad , \quad \mathbf{V}_{S1} - \mathbf{V}_{L2} = \mathbf{I}_c Z_{aa'} + \mathbf{E}_c$$
$$\mathbf{I}_{S1} = \mathbf{I}_c + \mathbf{I}_{a'} \quad , \quad \mathbf{I}_{L1} = \mathbf{I}_{a'} + \mathbf{I}_b \tag{13.19}$$

We also have the transformer relations,

$$\mathbf{E}_{a'} = \frac{N_{a'}}{N_a}\mathbf{E}_a = \frac{1}{n}\mathbf{E}_a = \frac{e^{-j120°}}{n}\mathbf{E}_c$$
$$\mathbf{I}_{a'} = -\frac{N_a}{N_{a'}} = -n\mathbf{I}_a = -ne^{-j120°}\mathbf{I}_c \tag{13.20}$$

where we have defined $n = N_a/N_{a'}$, the turns ratio. We also have the following relationships,

$$\mathbf{V}_{S1} - \mathbf{V}_{S2} = (1-\alpha^2)\mathbf{V}_{S1} = \sqrt{3}e^{j30^\circ}\mathbf{V}_{S1} \tag{13.21}$$

$$\mathbf{V}_{S2} - \mathbf{V}_{L2} = \mathbf{E}_{b'} = e^{-j120^\circ}\mathbf{E}_{a'} = \frac{e^{j120^\circ}}{n}\mathbf{E}_c$$

It is worthwhile going into the details of the solution of these equations since we will need some of the intermediate results later. We assume that $\mathbf{V}_{S1}$ and $\mathbf{I}_{S1}$ are given. From (13.19) and (13.20), we obtain

$$\mathbf{I}_c = \frac{\mathbf{I}_{S1}}{1-ne^{-j120^\circ}} \quad , \quad \mathbf{I}_{a'} = -\frac{ne^{-j120^\circ}}{1-ne^{-j120^\circ}}\mathbf{I}_{S1} = \frac{n}{n-e^{j120^\circ}}\mathbf{I}_{S1} \tag{13.22}$$

We also have

$$\mathbf{I}_{L1} = \mathbf{I}_{a'} + \mathbf{I}_b = \mathbf{I}_{a'} + e^{j120^\circ}\mathbf{I}_c = \left(\frac{n-e^{-j120^\circ}}{n-e^{j120^\circ}}\right)\mathbf{I}_{S1} = e^{j\theta}\mathbf{I}_{S1}$$

$$(13.23)$$

where $\theta = 2\tan^{-1}\left(\frac{\sqrt{3}}{2n+1}\right)$

Since n is determined by the tap position, this shows that θ is a non-linear function of the tap position, assuming the taps are evenly spaced.

Adding equations (13.21) and using (13.19),

$$\mathbf{V}_{S1} - \mathbf{V}_{L2} = \sqrt{3}e^{j30^\circ}\mathbf{V}_{S1} + \frac{e^{j120^\circ}}{n}\mathbf{E}_c = \mathbf{I}_c Z_{aa'} + \mathbf{E}_c$$

Solving for E_c, using (13.22),

$$\mathbf{E}_c = \frac{n\sqrt{3}e^{j30^\circ}}{n-e^{j120^\circ}}\mathbf{V}_{S1} + \frac{ne^{j120^\circ}}{\left(n-e^{j120^\circ}\right)^2}\mathbf{I}_{S1}Z_{aa'} \tag{13.24}$$

Combining with the first of equations (13.19) and (13.20), we obtain

$$\mathbf{V}_{L1} = \left[\mathbf{V}_{S1} - \mathbf{I}_{S1}\left(\frac{Z_{aa'}}{n^2+n+1}\right)\right]e^{j\theta} \tag{13.25}$$

where θ is given in (13.23). Although the current is shifted by θ in all cases, in general, the voltage is shifted by θ only under no load conditions.

The circuit model suggested by (13.25) is shown in Fig. 13.7. By symmetry, this applies to all three phases with appropriate labeling. The equivalent impedance shown in Fig. 13.7 is given by

$$Z_{eq} = \frac{Z_{aa'}}{n^2+n+1} \tag{13.26}$$

Fig. 13.7 is a positive sequence circuit model. The negative sequence model is obtained simply by changing θ to $-\theta$ with Z_{eq} unchanged. Note that Z_{eq} depends on the tap setting.

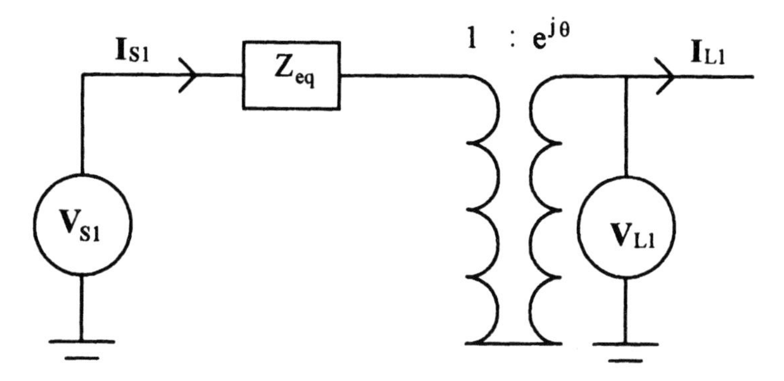

Figure 13.7 Circuit model of one phase of a phase shifting transformer. This is for positive sequence. For negative sequence, change θ to $-\theta$.

While the input power per phase is $P_{in} = \mathbf{V}_{S1}\mathbf{I}_{S1}^*$, the transformed or winding power per phase, P_{wdg}, is somewhat less. Ignoring impedance drops, we have from (13.22) and (13.24),

$$P_{wdg} = E_c I_c = j\left(\frac{n\sqrt{3}}{n^2+n+1}\right)P_{in} \tag{13.27}$$

where the factor of j shows that the transformed power is in quadrature with the input power. As a numerical example, let $\theta = 30°$. Then, from (13.23) and (13.27),

$$n = \frac{N_a}{N_{a'}} = 2.732 \quad , \quad |P_{wdg}| = 0.4227|P_{in}|$$

We can express (13.25) in per unit terms, however because of the difference between the input power and voltage base and that of the windings, we must be careful to specify the base used. For the input base, use the rated input power, $P_{in,b}$, the rated input voltage, $V_{in,b}$, and the rated input current which can be derived from the power and voltage base, $I_{in,b} = P_{in,b}/V_{in,b}$. Similarly the input impedance base can be derived from the power and voltage base, $Z_{in,b} = V_{in,b}/I_{in,b} = (V_{in,b})^2/P_{in,b}$. Note that, because the transformation ratio is 1:1 in terms of magnitude, the output base values are the same as the input base values. Thus (13.25) can be written

$$\frac{\mathbf{V}_{L1}}{V_{in,b}} = \left[\frac{\mathbf{V}_{S1}}{V_{in,b}} - \frac{\mathbf{I}_{S1}}{I_{in,b}}\frac{I_{in,b}}{V_{in,b}}Z_{eq}\right]e^{j\theta} = \left[\frac{\mathbf{V}_{S1}}{V_{in,b}} - \frac{\mathbf{I}_{S1}}{I_{in,b}}\frac{Z_{eq}}{Z_{in,b}}\right]e^{j\theta}$$

or (13.28)

$$\mathbf{v}_{L1} = \left[\mathbf{v}_{S1} - \mathbf{i}_{S1}z_{eq}\right]e^{j\theta}$$

where small letters are used to denote per unit quantities. Note that all the quantities in (13.28) are on a consistent base.

However, it is often more convenient to express winding quantities such as impedances on a winding basis. Assuming rated input power and voltage, the winding impedance base can be obtained from (13.27) and (13.24), ignoring impedance drops,

$$Z_{wdg,b} = \frac{|\mathbf{E}_c|^2}{|P_{wdg}|} = \sqrt{3}\,n\,Z_{in,b} \tag{13.29}$$

Thus, z_{eq} can be expressed as

$$z_{eq} = \frac{Z_{eq}}{Z_{in,b}} = \left(\frac{1}{n^2+n+1}\right)\left(\frac{Z_{wdg,g}}{Z_{in,b}}\right)\left(\frac{Z_{aa'}}{Z_{wdg,b}}\right) = \left(\frac{\sqrt{3}\,n}{n^2+n+1}\right)z_{aa'} \tag{13.30}$$

where, as the development in (13.30) indicates, z_{eq} is on an input base and $z_{aa'}$ is on a winding base. Note that the winding base depends on the turns ratio as indicated in (13.29). Thus this base is perhaps most useful when n refers to the maximum phase angle. At the other extreme, with $\theta = 0$, we have $n = \infty$, so that $Z_{eq} = z_{eq} = 0$. In this case, the input is directly connected to the output, bypassing the coils.

The zero sequence circuit model may be derived with reference to Fig. 13.6 but assuming all quantities are zero sequence. Thus Fig. 13.6b,c should be replaced by diagrams with all phasors in parallel. Using a zero subscript for zero sequence quantities, we can write

$$\begin{aligned} \mathbf{V}_{S1,0} - \mathbf{V}_{L1,0} &= \mathbf{E}_{a',0} \quad , \quad \mathbf{V}_{S1,0} - \mathbf{V}_{L2,0} = \mathbf{I}_{c,0}\mathbf{Z}_{aa',0} + \mathbf{E}_{c,0} \\ \mathbf{I}_{S1,0} &= \mathbf{I}_{c,0} + \mathbf{I}_{a',0} \quad , \quad \mathbf{I}_{L1,0} = \mathbf{I}_{a',0} + \mathbf{I}_{b,0} \end{aligned} \tag{13.31}$$

and

$$\begin{aligned} \mathbf{E}_{a',0} &= \frac{N_{a'}}{N_a}\mathbf{E}_{a,0} = \frac{1}{n}\mathbf{E}_{a,0} = \frac{1}{n}\mathbf{E}_{c,0} \\ \mathbf{I}_{a',0} &= -\frac{N_a}{N_{a'}} = -n\mathbf{I}_{a,0} = -n\mathbf{I}_{c,0} \end{aligned} \tag{13.32}$$

Since $\mathbf{I}_{b,0} = \mathbf{I}_{c,0}$, (13.31) shows that

$$\mathbf{I}_{L1,0} = \mathbf{I}_{S1,0} \tag{13.33}$$

so the current is not phase shifted. We also have

$$\mathbf{V}_{S1,0} - \mathbf{V}_{S2,0} = 0 \quad , \quad \mathbf{V}_{S2,0} - \mathbf{V}_{L2,0} = \mathbf{E}_{b',0} = \frac{1}{n}E_{b,0} = \frac{1}{n}\mathbf{E}_{c,0} \tag{13.34}$$

Solving the above zero sequence equations, we obtain

$$\mathbf{I}_{c,0} = \frac{\mathbf{I}_{S1,0}}{1-n} \quad , \quad \mathbf{E}_{c,0} = -\frac{n}{n-1}\mathbf{I}_{c,0}\mathbf{Z}_{aa',0} = \frac{n}{(n-1)^2}\mathbf{I}_{S1,0}\mathbf{Z}_{aa',0} \tag{13.35}$$

and

$$\mathbf{V}_{L1,0} = \mathbf{V}_{S1,0} - \mathbf{I}_{S1,0} \frac{Z_{aa',0}}{(n-1)^2} \tag{13.36}$$

The circuit model for this last equation is shown in Fig. 13.8, where we have defined

$$Z_{eq,0} = \frac{Z_{aa',0}}{(n-1)^2} \tag{13.37}$$

This has a different dependence on n from the positive sequence circuit. It becomes infinite when $n = 1$. This is reasonable, because from (13.32) $\mathbf{I}_{a',0} = -\mathbf{I}_{c,0}$ when $n = 1$ so that from (13.31) $\mathbf{I}_{S1,0} = 0$. Thus no zero sequence current can flow into the squashed delta transformer when $n = 1$. Internal current can however circulate around the delta. From (13.23) , $\theta = 60°$ when $n = 1$.

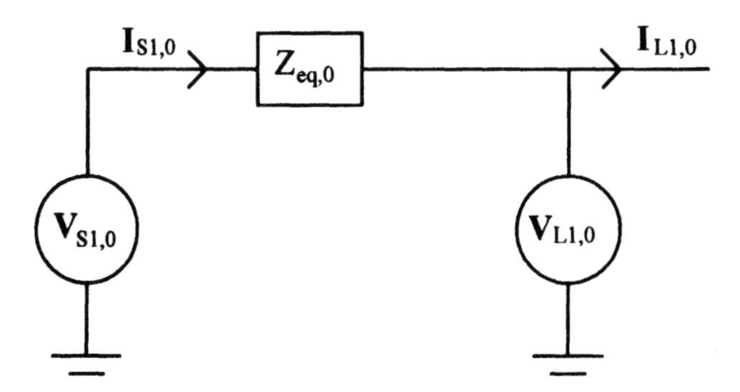

Figure 13.8 Zero sequence circuit model of one phase of a squashed delta phase shifting transformer

There is no phase angle shift in the voltage at no load. Equation (13.36) can be written in per unit terms referred to the same input base as used for positive sequence

$$\mathbf{v}_{L1,0} = \mathbf{v}_{S1,0} - \mathbf{i}_{S1,0} z_{eq,0} \tag{13.38}$$

Using the positive sequence winding base for the two winding zero sequence impedance, $Z_{aa',0}$ we can write

$$z_{eq,0} = \frac{\sqrt{3}\,n}{(n-1)^2} z_{aa',0} \qquad (13.39)$$

As before, in (13.39), the two z's are on different bases and this expression is primarily useful when the winding base is for the maximum phase shift angle.

We postpone a discussion of regulation effects and short circuit current calculations until other types of phase shifting transformers are treated. This is because the positive, negative, and zero sequence circuit diagrams for all the cases treated will have the same appearance, although Z_{eq} or z_{eq} and θ will differ among the various types.

13.4 STANDARD DELTA PHASE SHIFTING TRANSFORMER

As opposed to the squashed delta design, the standard delta design utilizes an unsquashed delta winding but is still a single core design. The connection diagram is given in Fig. 13.9. The tapped windings are on the same core as the corresponding parallel windings on the delta in the figure. The taps are symmetrically placed with respect to the point of contact at the delta vertex. This assures that there is no change in current or no-load voltage magnitude from input to output. It also means that $\mathbf{E}_{a'} = \mathbf{E}_{a''}$, etc. for the other phases. Each phase really consists of three windings, the two tap windings and the winding opposite and parallel in the figure, where primes and double primes are used to distinguish them. Thus a 3 winding per phase model is needed. Z_a , $Z_{a'}$, $Z_{a''}$ will be used to label the single winding impedances for phase a. Since all the phases have equivalent impedances by symmetry, these same designations will be used for the other phases as well. The same remarks apply to the phasor diagrams as for the squashed delta case. Although not shown, the current phasor diagrams have the same sequence order as their corresponding voltage phasor diagrams, however the two diagrams could be rotated relative to each other.

In Fig. 13.9a, $\mathbf{V}_1$, $\mathbf{V}_2$, $\mathbf{V}_3$ designate the phasor voltages to ground at the delta vertices. Thus we have, using $\mathbf{E}_{a'} = \mathbf{E}_{a''}$, etc. and $\mathbf{I}_{a'} = \mathbf{I}_{S1}$, $\mathbf{I}_{a''} = \mathbf{I}_{L1}$, etc.,

$$\mathbf{V}_{S1} - \mathbf{V}_1 = \mathbf{I}_{S1} Z_{a'} + \mathbf{E}_{a'} \quad , \quad \mathbf{V}_1 - \mathbf{V}_{L1} = \mathbf{I}_{L1} Z_{a''} + \mathbf{E}_{a'}$$

$$\mathbf{V}_2 - \mathbf{V}_3 = \left(\alpha^2 - \alpha\right)\mathbf{V}_1 = -j\sqrt{3}\,\mathbf{V}_1 = \mathbf{I}_a Z_a + \mathbf{E}_a \tag{13.40}$$

$$\mathbf{I}_{S1} - \mathbf{I}_{L1} = \mathbf{I}_c - \mathbf{I}_b = \left(\alpha^2 - \alpha\right)\mathbf{I}_a = j\sqrt{3}\,\mathbf{I}_a$$

In addition, we have the transformer relations

$$\frac{\mathbf{E}_a}{\mathbf{E}_{a'}} = \frac{N_a}{N_{a'}} = n$$

$$N_a \mathbf{I}_a + N_{a'} \mathbf{I}_{S1} + N_{a'} \mathbf{I}_{L1} = 0 \qquad \text{or} \qquad n\mathbf{I}_a + \mathbf{I}_{S1} + \mathbf{I}_{L1} = 0 \tag{13.41}$$

where we have defined the turns ratio n as the ratio between the turns in one of the delta windings to the turns in one of the tap windings. Both tap windings have the same number of turns. Solving for $\mathbf{I}_a$ in the last equation in (13.40) and inserting into the last equation in (13.41), we obtain

$$\mathbf{I}_{L1} = e^{j\theta}\mathbf{I}_{S1} \qquad \text{with} \qquad \theta = 2\tan^{-1}\left(\frac{\sqrt{3}}{n}\right) \tag{13.42}$$

We also have

$$\mathbf{I}_a = -\frac{2}{n\left(1 - j\frac{\sqrt{3}}{n}\right)}\mathbf{I}_{S1} \tag{13.43}$$

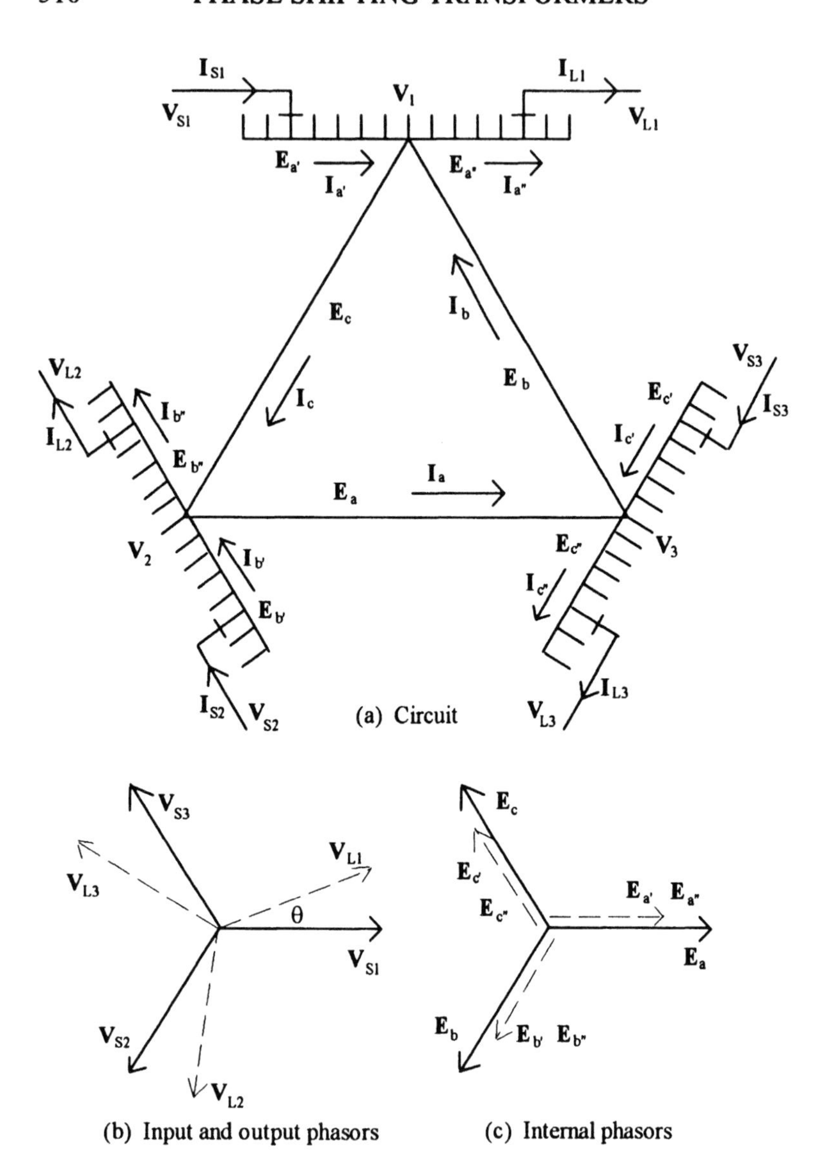

Figure 13.9 Standard delta configuration with phase quantities labelled. The ideal transformer voltages, E , increase in the opposite direction of the assumed current flow. The taps are symmetrically positioned.

From the above equations, we obtain

$$\mathbf{E}_a = -\frac{j\sqrt{3}}{\left(1-j\frac{\sqrt{3}}{n}\right)}\mathbf{V}_{S1} + \frac{j\sqrt{3}}{\left(1-j\frac{\sqrt{3}}{n}\right)}\left[Z_{a'} - \frac{j2\,Z_a}{\sqrt{3}\,n\left(1-j\frac{\sqrt{3}}{n}\right)}\right]\mathbf{I}_{S1} \tag{13.44}$$

and

$$\mathbf{V}_{L1} = \left[\mathbf{V}_{S1} - \mathbf{I}_{S1}\left(Z_{a'} + Z_{a''} + \frac{4Z_a}{n^2+3}\right)\right]e^{j\theta} \tag{13.45}$$

with θ as given in (13.42). This is the no-load phase angle shift. This conforms to the circuit model shown in Fig. 13.7 with

$$Z_{eq} = Z_{a'} + Z_{a''} + \frac{4Z_a}{n^2+3} \tag{13.46}$$

or in terms of 2 winding impedances, using (13.15),

$$Z_{eq} = Z_{a'a''} + \frac{2\left(Z_{aa'} + Z_{aa''} - n^2 Z_{a'a''}\right)}{n^2+3} \tag{13.47}$$

This is the positive sequence circuit. As before the negative sequence circuit is found by changing θ to $-\theta$ without change in Z_{eq}.

Let us again determine the winding power per phase, P_{wdg}, in terms of the input power per phase, $P_{in} = \mathbf{V}_{S1}\mathbf{I}_{S1}^*$. Again, ignoring impedance drops, we get

$$P_{wdg} = \mathbf{E}_a\mathbf{I}_a^* = j\left(\frac{2\sqrt{3}\,n}{n^2+3}\right)P_{in} \tag{13.48}$$

where j indicates that the transformed power is in quadrature with the input power. Using trigonometric identities, it can be shown that

$$P_{wdg} = j\sin\theta\,P_{in} \tag{13.49}$$

with θ as given in (13.42). (This relation does not hold for the squashed delta design.) Thus for $\theta = 30°$, $|P_{wdg}| = 0.5|P_{in}|$. In per unit terms, based on input quantities, (13.45) can be cast in the form of (13.28). If we wish to use a winding base for the 2 winding impedances given in (13.47) then we must determine the relation between their bases and the input base for impedances. The 2 winding impedances given above are referred to either the a or a' winding. Their bases are given by.

$$Z_{a\ wdg,b} = \frac{|E_a|^2}{|P_{wdg}|} = \frac{\sqrt{3}\,n}{2} Z_{in,b} \quad , \quad Z_{a'\ wdg,b} = \frac{|E_{a'}|^2}{|P_{wdg}|} = \frac{1}{n^2} Z_{a\ wdg,b} \qquad (13.50)$$

Thus, in per unit terms, Z_{eq} in (13.47) becomes

$$z_{eq} = \frac{\sqrt{3}\,n}{2}\left[\frac{z_{a'a''}}{n^2} + \frac{2\left(z_{aa'} + z_{aa''} - z_{a'a''}\right)}{n^2 + 3}\right] \qquad (13.51)$$

As before z_{eq} is on an input base and the 2 winding impedances on their winding base. Again this is primarily useful when n refers to the maximum phase angle. At zero phase shift, $n = \infty$, (13.47) indicates that $Z_{eq} = Z_{a'a''}$. However, at this tap position, the tap windings are effectively out of the circuit so that $Z_{a'a''} = 0$ and the input is directly connected to the output.

The zero sequence circuit model can be derived with reference to Fig. 13.9 but with all phasors taken to be zero sequence. Rewriting (13.40) and (13.41) with this in mind and appending a zero subscript, we get

$$\mathbf{V}_{S1,0} - \mathbf{V}_{1,0} = \mathbf{I}_{S1,0} Z_{a',0} + \mathbf{E}_{a',0} \quad , \quad \mathbf{V}_{1,0} - \mathbf{V}_{L1,0} = \mathbf{I}_{L1,0} Z_{a'',0} + \mathbf{E}_{a',0}$$

$$\mathbf{V}_{2,0} - \mathbf{V}_{3,0} = 0 = \mathbf{I}_{a,0} Z_{a,0} + \mathbf{E}_{a,0} \quad , \quad \mathbf{I}_{S1,0} - \mathbf{I}_{L1,0} = \mathbf{I}_{c,0} - \mathbf{I}_{b,0} = 0$$

$$\frac{\mathbf{E}_{a,0}}{\mathbf{E}_{a',0}} = \frac{N_a}{N_{a'}} = n \quad , \quad n\mathbf{I}_{a,0} + \mathbf{I}_{S1,0} + \mathbf{I}_{L1,0} = 0$$

$$(13.52)$$

Solving, we find

$$\mathbf{I}_{L1,0} = \mathbf{I}_{S1,0} \qquad (13.53)$$

so that the zero sequence current undergoes no phase shift. We also have

$$\mathbf{E}_{a,0} = -\mathbf{I}_{a,0}Z_{a,0} \quad , \quad \mathbf{I}_{a,0} = -\frac{2}{n}\mathbf{I}_{S1,0} \tag{13.54}$$

and

$$\mathbf{V}_{L1,0} = \mathbf{V}_{S1,0} - \mathbf{I}_{S1,0}\left(Z_{a',0} + Z_{a'',0} + \frac{4}{n^2}Z_{a,0}\right) \tag{13.55}$$

Thus the voltage undergoes no phase shift at no-load. The circuit model of Fig. 13.8 applies with

$$Z_{eq,0} = Z_{a',0} + Z_{a'',0} + \frac{4}{n^2}Z_{a,0} \tag{13.56}$$

or, in terms of 2 winding impedances,

$$Z_{eq,0} = Z_{a'a'',0} + \frac{2}{n^2}\left(Z_{aa',0} + Z_{aa'',0} - n^2 Z_{a'a'',0}\right) \tag{13.57}$$

On a per unit basis, using rated input quantities, (13.55) has the same appearance as (13.38). The equivalent per unit impedance on an input base with the 2 winding impedances on their positive sequence winding bases is given by

$$z_{eq,0} = \frac{\sqrt{3}}{2n}\left[2\left(z_{aa',0} + z_{aa'',0}\right) - z_{a'a'',0}\right] \tag{13.58}$$

Again this is primarily useful when n refers to the maximum phase angle shift. At zero phase shift, the output is directly connected to the input as was the case for positive sequence.

13.5 TWO CORE PHASE SHIFTING TRANSFORMER

For large power applications, phase shifters are often designed as two units, the series unit and the excitor unit, each having its own core and associated coils. Depending on size, the two units can be inside the same tank or be housed in separate tanks. This construction is largely dictated by tap changer limitations. A commonly used circuit diagram is shown in Fig. 13.10. The phasor diagrams refer to positive sequence quantities and although the phase ordering is consistent among the diagrams, their relative orientation is not specified. Currents have the same phase ordering as their associated voltages. A 3 winding model is needed for the series unit, while a 2 winding model applies to the excitor unit. We have used two different labeling schemes for the series and excitor units. Letter subscripts are used for series quantities and number subscripts for excitor quantities. Primes and double primes are used to distinguish different windings associated with the same phase. Note that the input-output coils in the figure are part of the series unit but are attached to the excitor unit at their midpoints. The input and output voltages are voltages to ground as before. We assume the input voltage and current phasors are given.

Following an analysis similar to that of the last section for the series unit, we can write, using $\mathbf{E}_{a'} = \mathbf{E}_{a''}$, etc.,

$$\mathbf{V}_{S1} - \mathbf{V}_1 = \mathbf{I}_{S1} Z_{a'} + \mathbf{E}_{a'} \quad , \quad \mathbf{V}_1 - \mathbf{V}_{L1} = \mathbf{I}_{L1} Z_{a''} + \mathbf{E}_{a'}$$

$$\mathbf{E}_{2'} - \mathbf{E}_{3'} = \left(\alpha^2 - \alpha\right)\mathbf{E}_{1'} = -j\sqrt{3}\,\mathbf{E}_{1'} = \mathbf{I}_a Z_a + \mathbf{E}_a \tag{13.59}$$

$$\mathbf{V}_1 = \mathbf{I}_1 Z_{11'} + \mathbf{E}_1 \quad , \quad \mathbf{I}_1 = \mathbf{I}_{S1} - \mathbf{I}_{L1} \quad , \quad -\mathbf{I}_{2'} = \mathbf{I}_a - \mathbf{I}_c$$

We also have the transformer relations,

$$\frac{\mathbf{E}_a}{\mathbf{E}_{a'}} = \frac{N_a}{N_{a'}} = n_s \quad , \quad \frac{\mathbf{E}_1}{\mathbf{E}_{1'}} = \frac{N_1}{N_{1'}} = n_e \quad , \quad \frac{\mathbf{I}_{1'}}{\mathbf{I}_1} = -\frac{N_1}{N_{1'}} = -n_e$$

$$N_a \mathbf{I}_a + N_{a'} \mathbf{I}_{S1} + N_{a'} \mathbf{I}_{L1} = 0 \quad \text{or} \quad n_s \mathbf{I}_a + \mathbf{I}_{S1} + \mathbf{I}_{L1} = 0 \tag{13.60}$$

where we have defined the turns ratio of the series unit , n_s , as the ratio of the turns in a coil of the delta to the turns in the first or second half of the input-output winding, i.e. from the input to the midpoint or from the output to the midpoint. We have also defined the excitor winding ratio, n_e , as the ratio of the turns in the winding connected to the midpoint of the input-output winding to the turns in the tapped winding. This latter ratio will depend on the tap position. From (13.59) , (13.60), and the phasor diagrams, we obtain

$$-\mathbf{I}_{2'} = -e^{-j120^\circ}\mathbf{I}_{1'} = n_e e^{-j120^\circ}\mathbf{I}_1 = \mathbf{I}_a - \mathbf{I}_c = (1-\alpha)\mathbf{I}_a = \sqrt{3}\,e^{-j30^\circ}\mathbf{I}_a$$

which implies

$$\mathbf{I}_1 = j\frac{\sqrt{3}}{n_e}\mathbf{I}_a \tag{13.61}$$

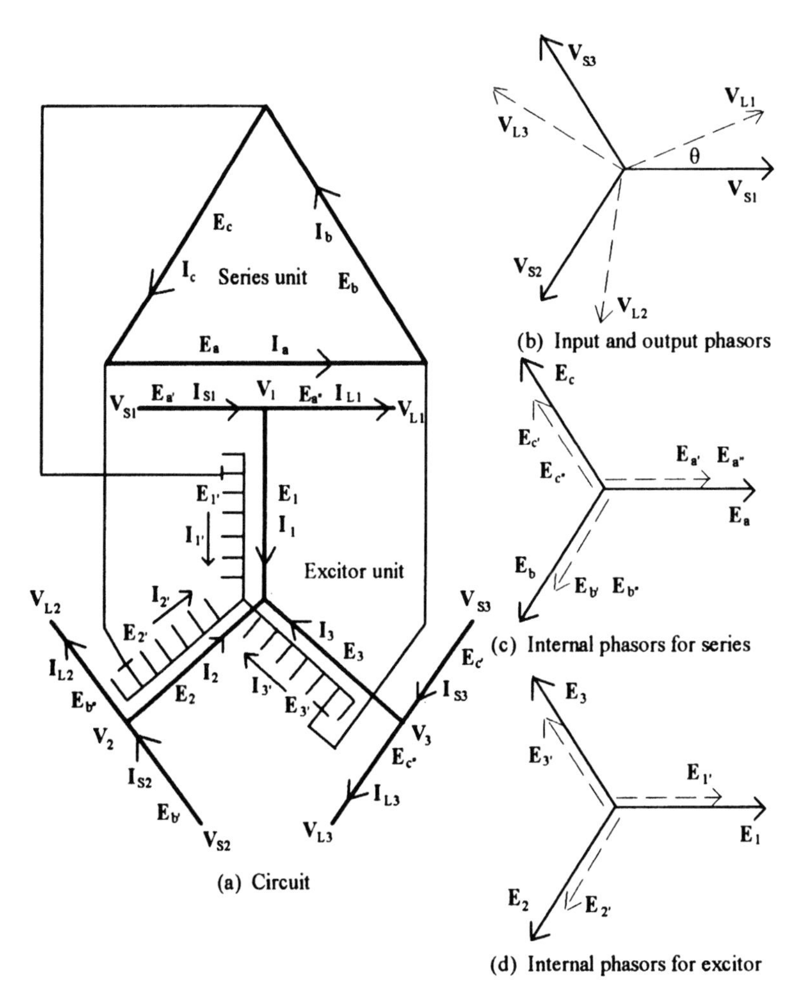

Figure 13.10 Circuit diagram of a 2 core phase shifting transformer. The ideal transformer voltages, the E's , increase in the opposite direction to the assumed current flow. The input-output coil is equally divided.

Substituting $\mathbf{I}_1$ from (13.59) and $\mathbf{I}_a$ from (13.60), we get

$$\mathbf{I}_{L1} = \left[\frac{1+j\left(\frac{\sqrt{3}}{n_e n_s}\right)}{1-j\left(\frac{\sqrt{3}}{n_e n_s}\right)}\right]\mathbf{I}_{S1} = e^{j\theta}\mathbf{I}_{S1} \quad \text{where} \quad \theta = 2\tan^{-1}\left(\frac{\sqrt{3}}{n_e n_s}\right) \tag{13.62}$$

Thus, in terms of the known input current, we can write, using (13.59) and (13.60),

$$\mathbf{I}_1 = -\frac{j2\left(\frac{\sqrt{3}}{n_e n_s}\right)}{\left[1-j\left(\frac{\sqrt{3}}{n_e n_s}\right)\right]}\mathbf{I}_{S1} \quad , \quad \mathbf{I}_a = -\frac{2}{n_s\left[1-j\left(\frac{\sqrt{3}}{n_e n_s}\right)\right]}\mathbf{I}_{S1} \tag{13.63}$$

From the initial set of equations, we also obtain

$$\mathbf{E}_1 = j\frac{n_e}{\sqrt{3}}\left(\mathbf{I}_a Z_a + \mathbf{E}_a\right) \tag{13.64}$$

Substituting (13.61) and (13.64) into the $\mathbf{V}_1$ equation in (13.59), we find

$$\mathbf{V}_1 = \left[Z_{11'} + \frac{n_e^2}{3}Z_a\right]\mathbf{I}_1 + j\frac{n_e}{\sqrt{3}}\mathbf{E}_a \tag{13.65}$$

Substituting into the first equation in (13.59) and using (13.60) and the first equation of (13.63),

$$\mathbf{E}_a = -\frac{j\left(\frac{\sqrt{3}}{n_e}\right)}{\left[1-j\left(\frac{\sqrt{3}}{n_e n_s}\right)\right]}\mathbf{V}_{S1} + \left[\frac{\frac{6n_s}{(n_e n_s)^2}}{\left[\left[1-j\left(\frac{\sqrt{3}}{n_e n_s}\right)\right]\right]^2}\left(Z_{11'} + \frac{n_e^2}{3}Z_a\right) + \frac{j\left(\frac{\sqrt{3}}{n_e}\right)}{\left[1-j\left(\frac{\sqrt{3}}{n_e n_s}\right)\right]}Z_{a'}\right]\mathbf{I}_{S1} \tag{13.66}$$

Adding the first 2 equations in (13.59) and using (13.66), we get

$$\mathbf{V}_{L1} = \left\{ \mathbf{V}_{S1} - \mathbf{I}_{S1} \left[Z_{a'} + Z_{a''} + \frac{12}{(n_e n_s)^2 + 3} \left(Z_{11'} + \frac{n_e^2}{3} Z_a \right) \right] \right\} e^{j\theta} \tag{13.67}$$

where θ is given in (13.62). This can be represented with the same circuit model as Fig. 13.7, with

$$Z_{eq} = Z_{a'} + Z_{a''} + \frac{12}{(n_e n_s)^2 + 3} \left(Z_{11'} + \frac{n_e^2}{3} Z_a \right) \tag{13.68}$$

or, in terms of 2 winding impedances,

$$Z_{eq} = Z_{a'a''} + \frac{12}{(n_e n_s)^2 + 3} \left[Z_{11'} + \frac{n_e^2}{6} \left(Z_{aa'} + Z_{aa''} - n_s^2 Z_{a'a''} \right) \right] \tag{13.69}$$

Using (13.63) and (13.64) and ignoring impedance drops, we can show that the winding power per phase is the same for the series and excitor units, i.e.,

$$P_{wdg} = \mathbf{E}_1 \mathbf{I}_1^* = \mathbf{E}_a \mathbf{I}_a^* = j \left[\frac{2\sqrt{3}\, n_e n_s}{(n_e n_s)^2 + 3} \right] P_{in} \tag{13.70}$$

In terms of the phase shift θ given in (13.62), (13.70) can be written

$$P_{wdg} = j \sin\theta \, P_{in} \tag{13.71}$$

as was the case for the standard delta phase shifter.

In per unit terms, based on rated input power and voltage, (13.67) can be cast in the form of (13.28). To express the two winding impedances on their winding bases, we need to find the impedance winding bases for both the series and excitor windings. Although the winding power is the same for these, their voltages are different. Using the above formulas, we find

$$Z_{a\,wdg,b} = \frac{|\mathbf{E}_a|^2}{|P_{wdg}|} = \frac{\sqrt{3}\,n_s}{2n_e} Z_{in,b} \quad , \quad Z_{a'\,wdg,b} = \frac{|\mathbf{E}_{a'}|^2}{|P_{wdg}|} = \frac{1}{n_s^2} Z_{a\,wdg,b}$$

$$Z_{1\,wdg,b} = \frac{|\mathbf{E}_1|^2}{|P_{wdg}|} = \frac{n_e n_s}{2\sqrt{3}} Z_{in,b} = \frac{n_e^2}{3} Z_{a\,wdg,b} \tag{13.72}$$

Using these relations, the per unit equivalent impedance can be written

$$z_{eq} = \frac{\sqrt{3}\,n_s}{2n_e}\left\{ z_{a'a''} + \frac{4n_e^2}{(n_e n_s)^2 + 3}\left[z_{11'} + \frac{1}{2}\left(z_{aa'} + z_{aa''} - z_{a'a''}\right)\right]\right\} \tag{13.73}$$

where as before z_{eq} is on an input base and the two winding impedances are on their winding bases. Again this is primarily useful when n_e refers to the maximum phase angle shift. Also the negative sequence circuit has the same equivalent impedance but a phase angle shift in the opposite direction to the positive sequence circuit.

The zero sequence circuit model is derived with reference to Fig. 13.10 by assuming all quantities are zero sequence. Thus we simply rewrite the basic equations, appending a zero subscript,

$$\mathbf{V}_{S1,0} - \mathbf{V}_{1,0} = \mathbf{I}_{S1,0} Z_{a',0} + \mathbf{E}_{a',0} \quad , \quad \mathbf{V}_{1,0} - \mathbf{V}_{L1,0} = \mathbf{I}_{L1,0} Z_{a'',0} + \mathbf{E}_{a',0}$$

$$\mathbf{E}_{2',0} - \mathbf{E}_{3',0} = 0 = \mathbf{I}_{a,0} Z_{a,0} + \mathbf{E}_{a,0} \tag{13.74}$$

$$\mathbf{V}_{1,0} = \mathbf{I}_{1,0} Z_{11',0} + \mathbf{E}_{1,0} \quad , \quad \mathbf{I}_{1,0} = \mathbf{I}_{S1,0} - \mathbf{I}_{L1,0} \quad , \quad -\mathbf{I}_{2',0} = \mathbf{I}_{a,0} - \mathbf{I}_{c,0} = 0$$

and

$$\frac{\mathbf{E}_{a,0}}{\mathbf{E}_{a',0}} = \frac{N_a}{N_{a'}} = n_s \quad , \quad \frac{\mathbf{E}_{1,0}}{\mathbf{E}_{1',0}} = \frac{N_1}{N_{1'}} = n_e \quad , \quad \frac{\mathbf{I}_{1',0}}{\mathbf{I}_{1,0}} = -\frac{N_1}{N_{1'}} = -n_e$$

$$N_a \mathbf{I}_{a,0} + N_{a'} \mathbf{I}_{S1,0} + N_{a'} \mathbf{I}_{L1,0} = 0 \quad \text{or} \quad n_s \mathbf{I}_{a,0} + \mathbf{I}_{S1,0} + \mathbf{I}_{L1,0} = 0 \tag{13.75}$$

Solving, we find,

$$\mathbf{I}_{1,0} = 0 \quad , \quad \mathbf{I}_{a,0} = -\frac{2}{n_s}\mathbf{I}_{S1,0} \quad , \quad \mathbf{I}_{L1,0} = \mathbf{I}_{S1,0}$$

$$\mathbf{E}_{a,0} = \frac{2}{n_s}\mathbf{I}_{S1,0}Z_{a,0} \quad , \quad \mathbf{V}_{1,0} = \mathbf{E}_{1,0} \tag{13.76}$$

Notice that, even if both Y windings of the excitor were grounded, no zero sequence current flows into the excitor because the secondary current from the tap winding would have to flow into the closed delta of the series unit and this is not possible for zero sequence currents.

From the above formulas, we obtain

$$\mathbf{V}_{L1,0} = \mathbf{V}_{S1,0} - \mathbf{I}_{S1,0}\left(Z_{a',0} + Z_{a'',0} + \frac{4}{n_s^2}Z_{a,0}\right) \tag{13.77}$$

Thus we see that the zero sequence current and no-load voltage have no phase angle shift and the circuit of Fig. 13.8 applies with

$$Z_{eq,0} = Z_{a',0} + Z_{a'',0} + \frac{4}{n_s^2}Z_{a,0} \tag{13.78}$$

or, in terms of 2 winding impedances,

$$Z_{eq,0} = \frac{2}{n_s^2}\left(Z_{aa',0} + Z_{aa'',0}\right) - Z_{a'a'',0} \tag{13.79}$$

Using the same basis as for positive sequence, the per unit version of (13.79) is

$$z_{eq,0} = \frac{\sqrt{3}}{2n_e n_s}\left[2\left(z_{aa',0} + z_{aa'',0}\right) - z_{a'a'',0}\right] \tag{13.80}$$

where, as before, $z_{eq,0}$ is on an input base and the two winding impedances are on their winding bases.

13.6 REGULATION EFFECTS

Because all the phase shifting transformers examined here have the same basic positive sequence (as well as negative and zero sequence) circuit, Fig. 13.7, with different expressions for Z_{eq} or z_{eq} and θ, the effect of a load on the output can be studied in common. The relevant circuit model is shown in Fig. 13.11.

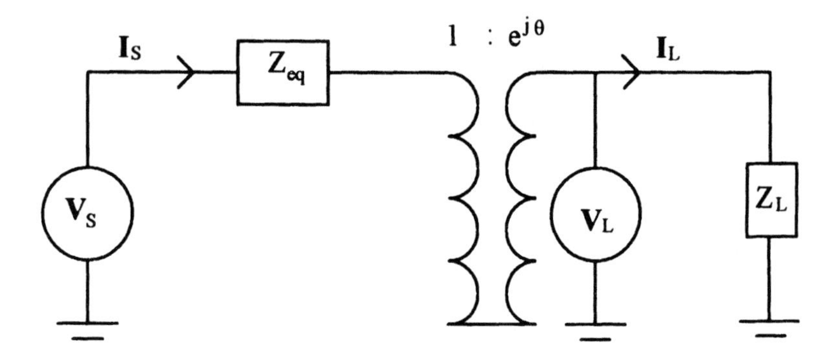

Figure 13.11 One phase of a phase shifting transformer under load

We are assuming a balanced positive sequence system. Since all phases are identical, we drop the phase subscript. We could just as well use per unit quantities for this development since a common base, employing rated input quantities, can be used for the various quantities in the figure. This is accomplished by simply changing form upper case to lower case letters. From the figure, we see that

$$\mathbf{V}_L = \left(\mathbf{V}_S - \mathbf{I}_S Z_{eq}\right) e^{j\theta} = \mathbf{I}_L Z_L = e^{j\theta} \mathbf{I}_S Z_L \tag{13.81}$$

where Z_L is the load impedance. Solving for $\mathbf{I}_S$, we find

$$\mathbf{I}_S = \frac{\mathbf{V}_S}{\left(Z_{eq} + Z_L\right)} \tag{13.82}$$

so that

$$\mathbf{V}_L = \mathbf{V}_S \left[1 - \frac{Z_{eq}}{(Z_{eq} + Z_L)} \right] e^{j\theta} = \mathbf{V}_S \left[\frac{1}{1 + \dfrac{Z_{eq}}{Z_L}} \right] e^{j\theta} \tag{13.83}$$

Thus any phase angle or magnitude shift from no load conditions is due to a non-zero Z_{eq} / Z_L . Since Z_{eq} is almost entirely inductive, a purely inductive or capacitive load will not affect θ but will result only in a magnitude change in the voltage. On the other hand, a resistive or complex load will lead to both magnitude and phase angle shifts.

Following convention, let the load current lag the load voltage by an angle θ_L , i.e. if θ_V is the voltage phasor angle, then

$$\mathbf{V}_L = V_L e^{j\theta_V} \quad , \quad \mathbf{I}_L = I_L e^{j(\theta_V - \theta_L)} \quad \text{so that} \quad Z_L = \frac{\mathbf{V}_L}{\mathbf{I}_L} = \frac{V_L}{I_L} e^{j\theta_L} \tag{13.84}$$

Note the distinction between bold faced phasor quantities and normal type used for magnitudes in the above equation. A positive value of θ_L is characteristic of a load with an inductive component. Let us also set

$$Z_{eq} = R + jX \tag{13.85}$$

Then

$$\frac{Z_{eq}}{Z_L} = \frac{I_L}{V_L} \left[(R\cos\theta_L + X\sin\theta_L) + j(X\cos\theta_L - R\sin\theta_L) \right] \tag{13.86}$$

Under no-load , NL , conditions ($Z_L = \infty$), (13.83) shows that

$$\mathbf{V}_{L,NL} = \mathbf{V}_S e^{j\theta}$$

Thus taking ratios and using (13.86), we see that

$$\frac{\mathbf{V}_L}{\mathbf{V}_{L,NL}} = \frac{1}{1 + \dfrac{Z_{eq}}{Z_L}} = \frac{V_L \exp\left\{ -j\tan^{-1}\left[\dfrac{I_L(X\cos\theta_L - R\sin\theta_L)}{V_L + I_L(R\cos\theta_L + X\sin\theta_L)} \right] \right\}}{\sqrt{\left[V_L + I_L(R\cos\theta_L + X\sin\theta_L) \right]^2 + \left[I_L(X\cos\theta_L - R\sin\theta_L) \right]^2}} \tag{13.87}$$

Thus the presence of a load will lower the magnitude and shift the phase generally in a negative direction since R is usually small compared with X and θ_L is generally positive and closer to zero that to 90°.

13.7 FAULT CURRENT ANALYSIS

We briefly review the sequence circuit method of fault current analysis here. This will also help establish notation. Although we do not work on a per unit basis, the same formulas apply with the substitution of lower case letters for the upper case ones if the input base is used. Our phase designations here are not necessarily related to those chosen for the transformers previously analyzed. Let $\mathbf{V}_{fa}$, $\mathbf{V}_{fb}$, $\mathbf{V}_{fc}$ be the voltages at the fault point for phases a , b , c and let the currents leaving the fault be $\mathbf{I}_{fa}$, $\mathbf{I}_{fb}$, $\mathbf{I}_{fc}$. These voltages and currents do not necessarily form a balanced set. The positive, negative, and zero sequence phasors corresponding to phase a will be denoted $\mathbf{V}_{fa,1}$, $\mathbf{V}_{fa,2}$, $\mathbf{V}_{fa,0}$ and $\mathbf{I}_{fa,1}$, $\mathbf{I}_{fa,2}$, $\mathbf{I}_{fa,0}$ respectively. Each one of these has a corresponding member for phases b and c so that $\mathbf{V}_{fa,1}$, $\mathbf{V}_{fb,1}$, $\mathbf{V}_{fc,1}$ form a balanced positive sequence set, $\mathbf{V}_{fa,2}$, $\mathbf{V}_{fb,2}$, $\mathbf{V}_{fc,2}$ form a balanced negative sequence set, and $\mathbf{V}_{fa,0}$, $\mathbf{V}_{fb,0}$, $\mathbf{V}_{fc,0}$ form a balanced zero sequence set. Therefore, once $\mathbf{V}_{fa,1}$ is known, $\mathbf{V}_{fb,1}$ and $\mathbf{V}_{fc,1}$ are automatically determined and similarly for $\mathbf{V}_{fa,2}$ and $\mathbf{V}_{fa,0}$. The same applies to currents. Thus $\mathbf{V}_{fa,1}$, $\mathbf{V}_{fa,2}$, $\mathbf{V}_{fa,0}$ contain all the information available in the original set $\mathbf{V}_{fa}$, $\mathbf{V}_{fb}$, $\mathbf{V}_{fc}$. It is therefore natural that they are related by some transformation. This is given by

$$\begin{pmatrix} \mathbf{V}_{fa} \\ \mathbf{V}_{fb} \\ \mathbf{V}_{fc} \end{pmatrix} = \begin{pmatrix} 1 & 1 & 1 \\ 1 & \alpha^2 & \alpha \\ 1 & \alpha & \alpha^2 \end{pmatrix} \begin{pmatrix} \mathbf{V}_{fa,0} \\ \mathbf{V}_{fa,1} \\ \mathbf{V}_{fa,2} \end{pmatrix} \tag{13.88}$$

and its inverse

$$\begin{pmatrix} \mathbf{V}_{fa,0} \\ \mathbf{V}_{fa,1} \\ \mathbf{V}_{fa,2} \end{pmatrix} = \begin{pmatrix} 1 & 1 & 1 \\ 1 & \alpha & \alpha^2 \\ 1 & \alpha^2 & \alpha \end{pmatrix} \begin{pmatrix} \mathbf{V}_{fa} \\ \mathbf{V}_{fb} \\ \mathbf{V}_{fc} \end{pmatrix} \tag{13.89}$$

where α is given in (13.17). The above equations also apply to currents with appropriate change in notation.

Since the sequence systems are balanced systems, it is only necessary to study one phase of each system. Quite generally, they can be modeled as a Thevenin equivalent system, containing an ideal voltage source in series with a Thevenin equivalent impedance as shown in Fig. 13.12. A voltage source is shown only in the positive sequence circuit since 3 phase sources are generally positive sequence.

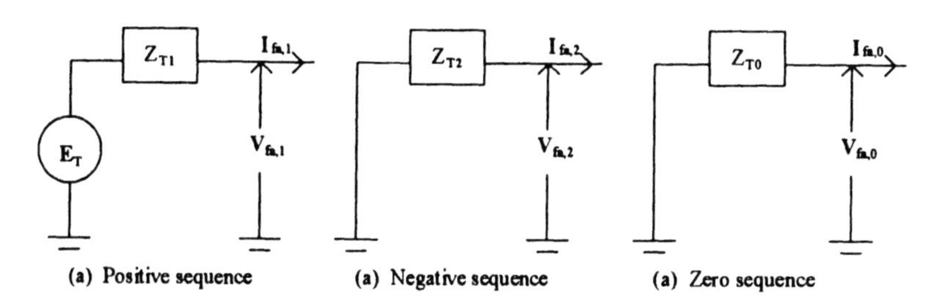

Figure 13.12 Thevenin equivalent swquence circuits

From Fig. 13.12, we have

$$\mathbf{V}_{fa,1} = \mathbf{E}_T - \mathbf{I}_{fa,1}Z_{T1} \quad , \quad \mathbf{V}_{fa,2} = -\mathbf{I}_{fa,2}Z_{T2} \quad , \quad \mathbf{V}_{fa,0} = -\mathbf{I}_{fa,0}Z_{T0} \qquad (13.90)$$

The faults of interest are:

(1) 3 phase line to ground fault
(2) Single phase line to ground fault
(3) Line to line fault
(4) Double line to ground fault

For fault type (1), we have $\mathbf{V}_{fa} = \mathbf{V}_{fb} = \mathbf{V}_{fc} = 0$ so that, using (13.88) - (13.90),

$$\mathbf{I}_{fa,1} = \frac{\mathbf{E}_T}{Z_{T1}} \quad , \quad \mathbf{I}_{fa,2} = \mathbf{I}_{fa,0} = 0 \qquad (13.91)$$

This simply means that the fault currents form a balanced positive sequence set as expected from the symmetry of the fault.

For fault type (2), we assume that the a phase line is shorted so that $\mathbf{V}_{fa} = 0$, $\mathbf{I}_{fb} = \mathbf{I}_{fc} = 0$. From (13.88) - (13.90),

$$\mathbf{I}_{fa,1} = \mathbf{I}_{fa,2} = \mathbf{I}_{fa,0} = \frac{\mathbf{E}_T}{Z_{T1} + Z_{T2} + Z_{T0}} \tag{13.92}$$

For fault type (3), assume the b and c lines are shorted together so that $\mathbf{V}_{fb} = \mathbf{V}_{fc}$, $\mathbf{I}_{fa} = 0$, $\mathbf{I}_{fc} = -\mathbf{I}_{fb}$. From (13.88) - (13.90), we obtain

$$\mathbf{I}_{fa,1} = -\mathbf{I}_{fa,2} = \frac{\mathbf{E}_T}{Z_{T1} + Z_{T2}} \quad , \quad \mathbf{I}_{fa,0} = 0 \tag{13.93}$$

For fault type (4), we assume the b and c lines are shorted to ground so that $\mathbf{V}_{fb} = \mathbf{V}_{fc} = 0$, $\mathbf{I}_{fa} = 0$. Then from (13.88) - (13.90),

$$\mathbf{I}_{fa,1} = \frac{\mathbf{E}_T(Z_{T2} + Z_{T0})}{Z_{T1}Z_{T2} + Z_{T1}Z_{T0} + Z_{T2}Z_{T0}}$$

$$\mathbf{I}_{fa,2} = -\frac{\mathbf{E}_T Z_{T0}}{Z_{T1}Z_{T2} + Z_{T1}Z_{T0} + Z_{T2}Z_{T0}} \tag{13.94}$$

$$\mathbf{I}_{fa,0} = -\frac{\mathbf{E}_T Z_{T2}}{Z_{T1}Z_{T2} + Z_{T1}Z_{T0} + Z_{T2}Z_{T0}}$$

Because voltages and currents are untransformed in magnitude and transformed by the same phase shift across the ideal transformer in Fig. 13.7, impedances can be brought to one side or the other of the circuit without change. We make use of this fact by shifting system impedances on the load side to the source side and consider faults at the source terminal. Faults on the load side can be found from the source side results by slight notational changes in the results. Thus the sequence circuit diagrams shown in Fig. 13.13 can be used. Z_{SS} and Z_{SL} are system impedances on the source and load sides respectively. Note that $\mathbf{E}_{out}$ is the load voltage multiplied by $e^{-j\theta}$ to bring it across the ideal transformer. Positive and negative sequence impedances are not distinguished with subscripts since they are the same to a good approximation.

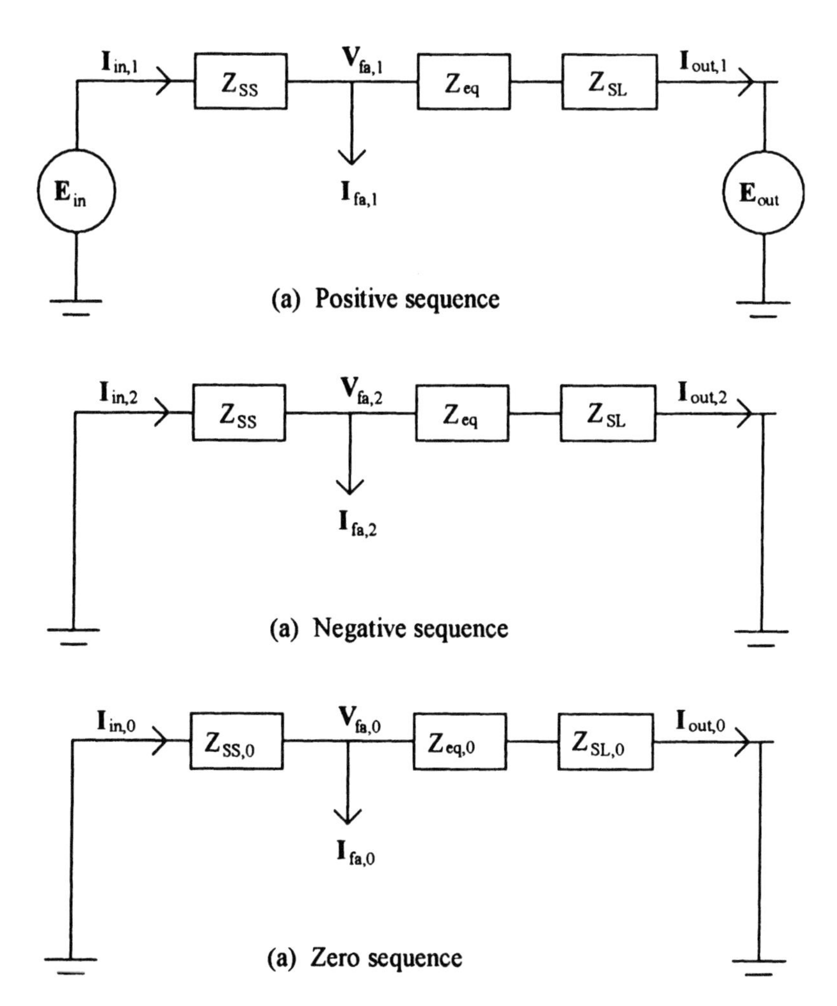

Figure 13.13 Sequence circuits showing system impedances and transformer equivalent impedances. The fault is on the source side terminal.

We assume for simplicity that no current flows in the pre-fault condition. Because fault currents are usually much larger than pre-fault currents, this is a good assumption. The Thevenin equivalent voltage and impedances referred to the fault point are,

$$\mathbf{E}_T = \mathbf{E}_{in} = \mathbf{E}_{out} \quad , \quad Z_{T1} = Z_{T2} = \frac{Z_{SS}\left(Z_{eq} + Z_{SL}\right)}{Z_{eq} + Z_{SS} + Z_{SL}} \quad , \quad Z_{T0} = \frac{Z_{SS,0}\left(Z_{eq,o} + Z_{SL,0}\right)}{Z_{eq,0} + Z_{SS,0} + Z_{SL,0}} \tag{13.95}$$

By substituting these into previous formulas, we obtain the fault sequence currents in terms of known quantities. The fault sequence currents into the transformer source side terminal are $\mathbf{I}_{out,1}$, $\mathbf{I}_{out,2}$, $\mathbf{I}_{out,0}$. From Fig. 13.13 , (13.90) and $Z_{T1} = Z_{T2}$, we obtain

$$\mathbf{I}_{out,1} = -\mathbf{I}_{fa,1}\frac{Z_{T1}}{\left(Z_{eq}+Z_{SL}\right)} \quad , \quad \mathbf{I}_{out,2} = -\mathbf{I}_{fa,2}\frac{Z_{T1}}{\left(Z_{eq}+Z_{SL}\right)} \quad , \quad \mathbf{I}_{out,0} = -\mathbf{I}_{fa,0}\frac{Z_{T0}}{\left(Z_{eq,0}+Z_{SL,0}\right)} \tag{13.96}$$

This last set of equations are standard equations for fault currents into a two winding transformer. However, in a phase shifting transformer, the winding currents are quite different from the terminal currents. In addition, positive and negative sequence currents experience opposite phase shifts. This makes the analysis of fault currents inside these transformers quite different from fault currents in a standard two winding transformer.

We have already considered currents in the windings of a phase shifting transformer when we derived their equivalent impedance for positive, negative, and zero sequence circuits. We simply make use of these results now, together with the fact that negative sequence currents inside a phase shifting transformer can be obtained from positive sequence currents by changing their phase shift to its negative. Note that I_{out} above equals I_{S1} in the previous formulas. Although we have singled out the a phase in the above analysis, the above equations could refer to any phase in our previous results for the different phase shifting transformers, provided we maintain the correct phase ordering. We now apply the above results to two of our previously studied phase shifting transformers. Because of the length and similarity of the formulas, we omit results for the two core design.

13.7.1 Squashed Delta Fault Currents

For the squashed delta, we find from (13.22) and (13.35), letting the c phase in that analysis correspond to the a phase here,

$$\mathbf{I}_{c,1} = -\frac{\mathbf{I}_{out,1}e^{j\left(120^\circ+\theta\right)}}{\sqrt{n^2+n+1}} \quad , \quad \mathbf{I}_{c,2} = -\frac{\mathbf{I}_{out,2}e^{-j\left(120^\circ+\theta\right)}}{\sqrt{n^2+n+1}} \quad , \quad \mathbf{I}_{c,0} = -\frac{\mathbf{I}_{out,0}}{n-1} \tag{13.97}$$

where θ is given in (13.23). The currents in the tapped winding are obtained from these by multiplying by the negative of the turns ratio. Using (13.96), these can be written

$$\mathbf{I}_{c,1} = \frac{\mathbf{I}_{fa,1} Z_{T1} e^{j(120^\circ+\theta)}}{\left(Z_{eq} + Z_{SL}\right)\sqrt{n^2+n+1}} \quad , \quad \mathbf{I}_{c,2} = \frac{\mathbf{I}_{fa,2} Z_{T1} e^{-j(120^\circ+\theta)}}{\left(Z_{eq} + Z_{SL}\right)\sqrt{n^2+n+1}}$$

$$\mathbf{I}_{c,0} = \frac{\mathbf{I}_{fa,0} Z_{T0}}{\left(Z_{eq,o} + Z_{SL,0}\right)(n-1)} \tag{13.98}$$

The actual phase currents can be found from (13.88) applied to currents and the appropriate interpretation of subscripts,

$$\mathbf{I}_c = \frac{Z_{T1}}{\left(Z_{eq} + Z_{SL}\right)\sqrt{n^2+n+1}}\left[\mathbf{I}_{fa,1} e^{j(120^\circ+\theta)} + \mathbf{I}_{fa,2} e^{-j(120^\circ+\theta)}\right] + \frac{Z_{T0}}{\left(Z_{eq,0} + Z_{SL,0}\right)(n-1)}\mathbf{I}_{fa,0}$$

$$\mathbf{I}_a = \frac{Z_{T1}}{\left(Z_{eq} + Z_{SL}\right)\sqrt{n^2+n+1}}\left[\alpha^2\mathbf{I}_{fa,1} e^{j(120^\circ+\theta)} + \alpha\mathbf{I}_{fa,2} e^{-j(120^\circ+\theta)}\right] + \frac{Z_{T0}}{\left(Z_{eq,0} + Z_{SL,0}\right)(n-1)}\mathbf{I}_{fa,0}$$

$$\mathbf{I}_b = \frac{Z_{T1}}{\left(Z_{eq} + Z_{SL}\right)\sqrt{n^2+n+1}}\left[\alpha\mathbf{I}_{fa,1} e^{j(120^\circ+\theta)} + \alpha^2\mathbf{I}_{fa,2} e^{-j(120^\circ+\theta)}\right] + \frac{Z_{T0}}{\left(Z_{eq,0} + Z_{SL,0}\right)(n-1)}\mathbf{I}_{fa,0} \tag{13.99}$$

For a 3 phase fault, these formulas become, using (13.91),

$$\mathbf{I}_c = \frac{\mathbf{E}_T e^{j(120^\circ+\theta)}}{\left(Z_{eq} + Z_{SL}\right)\sqrt{n^2+n+1}} \quad , \quad \mathbf{I}_a = \alpha^2\mathbf{I}_c \quad , \quad \mathbf{I}_b = \alpha\mathbf{I}_c \tag{13.100}$$

i.e. a balanced positive sequence set as expected. The overall phase in the above formula can be disregarded when analyzing forces and stresses.

For a single line to ground fault, (13.99) becomes, using (13.92)

$$\mathbf{I}_c = -\frac{\mathbf{E}_T}{(2Z_{T1}+Z_{T0})}\left\{\frac{Z_{T1}}{(Z_{eq}+Z_{SL})\sqrt{n^2+n+1}}\left(\cos\frac{\theta}{2}+\sqrt{3}\sin\frac{\theta}{2}\right)+\frac{Z_{T0}}{(Z_{eq,0}+Z_{SL,0})(n-1)}\right\}$$

$$\mathbf{I}_a = \frac{\mathbf{E}_T}{(2Z_{T1}+Z_{T0})}\left\{\frac{Z_{T1}}{(Z_{eq}+Z_{SL})\sqrt{n^2+n+1}}\left(2\cos\frac{\theta}{2}\right)-\frac{Z_{T0}}{(Z_{eq,0}+Z_{SL,0})(n-1)}\right\}$$

$$\mathbf{I}_b = -\frac{\mathbf{E}_T}{(2Z_{T1}+Z_{T0})}\left\{\frac{Z_{T1}}{(Z_{eq}+Z_{SL})\sqrt{n^2+n+1}}\left(\cos\frac{\theta}{2}-\sqrt{3}\sin\frac{\theta}{2}\right)+\frac{Z_{T0}}{(Z_{eq,0}+Z_{SL,0})(n-1)}\right\} \tag{13.101}$$

For a line to line fault, (13.99) becomes, using (13.93) and $Z_{T1} = Z_{T2}$,

$$\mathbf{I}_c = j\mathbf{E}_T\left\{\frac{1}{2(Z_{eq}+Z_{SL})\sqrt{n^2+n+1}}\left(\sqrt{3}\cos\frac{\theta}{2}-\sin\frac{\theta}{2}\right)\right\}$$

$$\mathbf{I}_a = j\mathbf{E}_T\left\{\frac{1}{2(Z_{eq}+Z_{SL})\sqrt{n^2+n+1}}\left(2\sin\frac{\theta}{2}\right)\right\} \tag{13.102}$$

$$\mathbf{I}_b = -j\mathbf{E}_T\left\{\frac{1}{2(Z_{eq}+Z_{SL})\sqrt{n^2+n+1}}\left(\sqrt{3}\cos\frac{\theta}{2}+\sin\frac{\theta}{2}\right)\right\}$$

The factor of j can be ignored in the analysis of winding forces and stresses.

For a double line to ground fault, (13.99) becomes, using (13.94) and $Z_{T1} = Z_{T2}$,

$$\mathbf{I}_c = \frac{\mathbf{E}_T}{(Z_{T1}+2Z_{T0})}\left\{\frac{1}{(Z_{eq}+Z_{SL})\sqrt{n^2+n+1}}\left[-\frac{Z_{T1}}{2}\left(\cos\frac{\theta}{2}+\sqrt{3}\sin\frac{\theta}{2}\right)\right.\right.$$

$$\left.\left.+j\left(\frac{Z_{T1}}{2}+Z_{T0}\right)\left(\sqrt{3}\cos\frac{\theta}{2}-\sin\frac{\theta}{2}\right)\right]+\frac{Z_{T0}}{(Z_{eq,0}+Z_{SL,0})(n-1)}\right\}$$

$$\mathbf{I}_a = \frac{\mathbf{E}_T}{(Z_{T1}+2Z_{T0})}\left\{\frac{1}{(Z_{eq}+Z_{SL})\sqrt{n^2+n+1}}\left[Z_{T1}\left(\cos\frac{\theta}{2}\right)\right.\right.$$

$$\left.\left.+j(Z_{T1}+2Z_{T0})\left(\sin\frac{\theta}{2}\right)\right]+\frac{Z_{T0}}{(Z_{eq,0}+Z_{SL,0})(n-1)}\right\}$$

$$\mathbf{I}_b = \frac{\mathbf{E}_T}{(Z_{T1}+2Z_{T0})}\left\{\frac{1}{(Z_{eq}+Z_{SL})\sqrt{n^2+n+1}}\left[-\frac{Z_{T1}}{2}\left(\cos\frac{\theta}{2}-\sqrt{3}\sin\frac{\theta}{2}\right)\right.\right.$$

$$\left.\left.-j\left(\frac{Z_{T1}}{2}+Z_{T0}\right)\left(\sqrt{3}\cos\frac{\theta}{2}+\sin\frac{\theta}{2}\right)\right]+\frac{Z_{T0}}{(Z_{eq,0}+Z_{SL,0})(n-1)}\right\} \quad (13.103)$$

In this case, the fault currents are complex.

13.7.2 Standard Delta Fault Currents

For the standard delta phase shifting transformer, we have from (13.43) and (13.54),

$$\mathbf{I}_{a,1} = -\frac{2\mathbf{I}_{out,1}}{\sqrt{n^2+3}}e^{j\frac{\theta}{2}} \quad , \quad \mathbf{I}_{a,2} = -\frac{2\mathbf{I}_{out,2}}{\sqrt{n^2+3}}e^{-j\frac{\theta}{2}} \quad , \quad \mathbf{I}_{a,0} = -\frac{2}{n}\mathbf{I}_{out,0} \quad (13.104)$$

with θ given in (13.42). Substituting from (13.96), these become

$$\mathbf{I}_{a,1} = \frac{2Z_{T1}}{\left(Z_{eq} + Z_{SL}\right)\sqrt{n^2+3}}\mathbf{I}_{fa,1}e^{j\frac{\theta}{2}}$$

$$\mathbf{I}_{a,2} = \frac{2Z_{T1}}{\left(Z_{eq} + Z_{SL}\right)\sqrt{n^2+3}}\mathbf{I}_{fa,2}e^{-j\frac{\theta}{2}} \tag{13.105}$$

$$\mathbf{I}_{a,0} = \frac{2Z_{T0}}{n\left(Z_{eq,0} + Z_{SL,0}\right)}\mathbf{I}_{fa,0}$$

Since each phase has 3 windings, we must also specify the currents in the other windings on the leg. One of these, the a' winding carries the input current, designated $\mathbf{I}_{out}$ here. Using (13.96), we can write

$$\mathbf{I}_{a',1} = -\frac{Z_{T1}}{\left(Z_{eq} + Z_{SL}\right)}\mathbf{I}_{fa,1}$$

$$\mathbf{I}_{a',2} = -\frac{Z_{T1}}{\left(Z_{eq} + Z_{SL}\right)}\mathbf{I}_{fa,2} \tag{13.106}$$

$$\mathbf{I}_{a',0} = -\frac{Z_{T0}}{\left(Z_{eq,0} + Z_{SL,0}\right)}\mathbf{I}_{fa,0}$$

The other winding, a", carries the output current. Again, using (13.96), this can be written

$$\mathbf{I}_{a'',1} = -\frac{Z_{T1}e^{j\theta}}{\left(Z_{eq} + Z_{SL}\right)}\mathbf{I}_{fa,1}$$

$$\mathbf{I}_{a'',2} = -\frac{Z_{T1}e^{-j\theta}}{\left(Z_{eq} + Z_{SL}\right)}\mathbf{I}_{fa,2} \tag{13.107}$$

$$\mathbf{I}_{a'',0} = -\frac{Z_{T0}}{\left(Z_{eq,0} + Z_{SL,0}\right)}\mathbf{I}_{fa,0}$$

The phase currents can be found from the transformation given by (13.88) applied to currents, with subscripts appropriate to the present context,

$$\mathbf{I}_a = \frac{2Z_{T1}}{\left(Z_{eq}+Z_{SL}\right)\sqrt{n^2+3}}\left[\mathbf{I}_{fa,1}e^{j\frac{\theta}{2}}+\mathbf{I}_{fa,2}e^{-j\frac{\theta}{2}}\right]+\frac{2Z_{T0}}{n\left(Z_{eq,0}+Z_{SL,0}\right)}\mathbf{I}_{fa,0}$$

$$\mathbf{I}_b = \frac{2Z_{T1}}{\left(Z_{eq}+Z_{SL}\right)\sqrt{n^2+3}}\left[\alpha^2\mathbf{I}_{fa,1}e^{j\frac{\theta}{2}}+\alpha\mathbf{I}_{fa,2}e^{-j\frac{\theta}{2}}\right]+\frac{2Z_{T0}}{n\left(Z_{eq,0}+Z_{SL,0}\right)}\mathbf{I}_{fa,0}$$

$$\mathbf{I}_c = \frac{2Z_{T1}}{\left(Z_{eq}+Z_{SL}\right)\sqrt{n^2+3}}\left[\alpha\mathbf{I}_{fa,1}e^{j\frac{\theta}{2}}+\alpha^2\mathbf{I}_{fa,2}e^{-j\frac{\theta}{2}}\right]+\frac{2Z_{T0}}{n\left(Z_{eq,0}+Z_{SL,0}\right)}\mathbf{I}_{fa,0} \tag{13.108}$$

and

$$\mathbf{I}_{a'} = -\frac{Z_{T1}}{\left(Z_{eq}+Z_{SL}\right)}\left(\mathbf{I}_{fa,1}+\mathbf{I}_{fa,2}\right)-\frac{Z_{T0}}{\left(Z_{eq,0}+Z_{SL,0}\right)}\mathbf{I}_{fa,0}$$

$$\mathbf{I}_{b'} = -\frac{Z_{T1}}{\left(Z_{eq}+Z_{SL}\right)}\left(\alpha^2\mathbf{I}_{fa,1}+\alpha\mathbf{I}_{fa,2}\right)-\frac{Z_{T0}}{\left(Z_{eq,0}+Z_{SL,0}\right)}\mathbf{I}_{fa,0} \tag{13.109}$$

$$\mathbf{I}_{c'} = -\frac{Z_{T1}}{\left(Z_{eq}+Z_{SL}\right)}\left(\alpha\mathbf{I}_{fa,1}+\alpha^2\mathbf{I}_{fa,2}\right)-\frac{Z_{T0}}{\left(Z_{eq,0}+Z_{SL,0}\right)}\mathbf{I}_{fa,0}$$

and

$$\mathbf{I}_{a''} = -\frac{Z_{T1}}{\left(Z_{eq}+Z_{SL}\right)}\left(\mathbf{I}_{fa,1}e^{j\theta}+\mathbf{I}_{fa,2}e^{-j\theta}\right)-\frac{Z_{T0}}{\left(Z_{eq,0}+Z_{SL,0}\right)}\mathbf{I}_{fa,0}$$

$$\mathbf{I}_{b''} = -\frac{Z_{T1}}{\left(Z_{eq}+Z_{SL}\right)}\left(\alpha^2\mathbf{I}_{fa,1}e^{j\theta}+\alpha\mathbf{I}_{fa,2}e^{-j\theta}\right)-\frac{Z_{T0}}{\left(Z_{eq,0}+Z_{SL,0}\right)}\mathbf{I}_{fa,0} \tag{13.110}$$

$$\mathbf{I}_{c''} = -\frac{Z_{T1}}{\left(Z_{eq}+Z_{SL}\right)}\left(\alpha\mathbf{I}_{fa,1}e^{j\theta}+\alpha^2\mathbf{I}_{fa,2}e^{-j\theta}\right)-\frac{Z_{T0}}{\left(Z_{eq,0}+Z_{SL,0}\right)}\mathbf{I}_{fa,0}$$

Substituting $\mathbf{I}_{fa,1}$, $\mathbf{I}_{fa,2}$, $\mathbf{I}_{fa,0}$ for the various fault types we obtain,
For a 3 phase fault:

$$\mathbf{I}_a = \mathbf{E}_T \frac{2e^{j\frac{\theta}{2}}}{(Z_{eq} + Z_{SL})\sqrt{n^2+3}} \quad , \quad \mathbf{I}_b = \alpha^2 \mathbf{I}_a \quad , \quad \mathbf{I}_c = \alpha \mathbf{I}_a$$

$$\mathbf{I}_{a'} = -\mathbf{E}_T \frac{1}{(Z_{eq} + Z_{SL})} \quad , \quad \mathbf{I}_{b'} = \alpha^2 \mathbf{I}_{a'} \quad , \quad \mathbf{I}_{c'} = \alpha \mathbf{I}_{a'} \qquad (13.111)$$

$$\mathbf{I}_{a''} = -\mathbf{E}_T \frac{e^{j\theta}}{(Z_{eq} + Z_{SL})} \quad , \quad \mathbf{I}_{b''} = \alpha^2 \mathbf{I}_{a''} \quad , \quad \mathbf{I}_{c''} = \alpha \mathbf{I}_{a''}$$

For a single line to ground fault:

$$\mathbf{I}_a = \frac{\mathbf{E}_T}{(2Z_{T1} + Z_{T0})} \left[\frac{2Z_{T1}}{(Z_{eq} + Z_{SL})\sqrt{n^2+3}} \left(2\cos\frac{\theta}{2} \right) + \frac{2Z_{T0}}{n(Z_{eq,0} + Z_{SL,0})} \right]$$

$$\mathbf{I}_b = \frac{\mathbf{E}_T}{(2Z_{T1} + Z_{T0})} \left[\frac{2Z_{T1}}{(Z_{eq} + Z_{SL})\sqrt{n^2+3}} \left(-\cos\frac{\theta}{2} + \sqrt{3}\sin\frac{\theta}{2} \right) + \frac{2Z_{T0}}{n(Z_{eq,0} + Z_{SL,0})} \right]$$

$$\mathbf{I}_c = \frac{\mathbf{E}_T}{(2Z_{T1} + Z_{T0})} \left[\frac{2Z_{T1}}{(Z_{eq} + Z_{SL})\sqrt{n^2+3}} \left(-\cos\frac{\theta}{2} - \sqrt{3}\sin\frac{\theta}{2} \right) + \frac{2Z_{T0}}{n(Z_{eq,0} + Z_{SL,0})} \right]$$

(13.112)

and

$$\mathbf{I}_{a'} = -\frac{\mathbf{E}_T}{(2Z_{T1} + Z_{T0})}\left[\frac{2Z_{T1}}{(Z_{eq} + Z_{SL})} + \frac{Z_{T0}}{(Z_{eq,0} + Z_{SL,0})}\right]$$

$$\mathbf{I}_{b'} = \frac{\mathbf{E}_T}{(2Z_{T1} + Z_{T0})}\left[\frac{Z_{T1}}{(Z_{eq} + Z_{SL})} - \frac{Z_{T0}}{(Z_{eq,0} + Z_{SL,0})}\right] \quad (13.113)$$

$$\mathbf{I}_{c'} = \frac{\mathbf{E}_T}{(2Z_{T1} + Z_{T0})}\left[\frac{Z_{T1}}{(Z_{eq} + Z_{SL})} - \frac{Z_{T0}}{(Z_{eq,0} + Z_{SL,0})}\right]$$

and

$$\mathbf{I}_{a''} = -\frac{\mathbf{E}_T}{(2Z_{T1} + Z_{T0})}\left[\frac{Z_{T1}}{(Z_{eq} + Z_{SL})}(2\cos\theta) + \frac{Z_{T0}}{(Z_{eq,0} + Z_{SL,0})}\right]$$

$$\mathbf{I}_{b''} = \frac{\mathbf{E}_T}{(2Z_{T1} + Z_{T0})}\left[\frac{Z_{T1}}{(Z_{eq} + Z_{SL})}\left(\cos\theta - \sqrt{3}\sin\theta\right) - \frac{Z_{T0}}{(Z_{eq,0} + Z_{SL,0})}\right]$$

$$\mathbf{I}_{c''} = \frac{\mathbf{E}_T}{(2Z_{T1} + Z_{T0})}\left[\frac{Z_{T1}}{(Z_{eq} + Z_{SL})}\left(\cos\theta + \sqrt{3}\sin\theta\right) - \frac{Z_{T0}}{(Z_{eq,0} + Z_{SL,0})}\right]$$

(13.114)

For a line to line fault:

$$\mathbf{I}_a = j\frac{2\mathbf{E}_T}{(Z_{eq} + Z_{SL})\sqrt{n^2+3}}\left(\sin\frac{\theta}{2}\right) \quad , \quad \mathbf{I}_b = -j\frac{\mathbf{E}_T}{(Z_{eq} + Z_{SL})\sqrt{n^2+3}}\left(\sqrt{3}\cos\frac{\theta}{2} + \sin\frac{\theta}{2}\right)$$

$$\mathbf{I}_c = j\frac{\mathbf{E}_T}{(Z_{eq} + Z_{SL})\sqrt{n^2+3}}\left(\sqrt{3}\cos\frac{\theta}{2} - \sin\frac{\theta}{2}\right)$$

(13.115)

and

$$\mathbf{I}_{a'} = 0 \quad , \quad \mathbf{I}_{b'} = j\frac{\sqrt{3}\mathbf{E}_T}{2\left(Z_{eq} + Z_{SL}\right)} \quad , \quad \mathbf{I}_{c'} = -j\frac{\sqrt{3}\mathbf{E}_T}{2\left(Z_{eq} + Z_{SL}\right)} \tag{13.116}$$

and

$$\mathbf{I}_{a''} = -j\frac{\mathbf{E}_T}{\left(Z_{eq} + Z_{SL}\right)}(\sin\theta)$$

$$\mathbf{I}_{b''} = 2j\frac{\mathbf{E}_T}{\left(Z_{eq} + Z_{SL}\right)}\left(\sqrt{3}\cos\theta + \sin\theta\right) \tag{13.117}$$

$$\mathbf{I}_{c''} = -2j\frac{\mathbf{E}_T}{\left(Z_{eq} + Z_{SL}\right)}\left(\sqrt{3}\cos\theta - \sin\theta\right)$$

For a double line to ground fault:

$$\mathbf{I}_a = \frac{2\mathbf{E}_T}{(Z_{T1} + 2Z_{T0})}\left\{\frac{1}{\left(Z_{eq} + Z_{SL}\right)\sqrt{n^2+3}}\left[Z_{T1}\left(\cos\frac{\theta}{2}\right) + j(Z_{T1} + 2Z_{T0})\left(\sin\frac{\theta}{2}\right)\right] - \frac{Z_{T0}}{n\left(Z_{eq,0} + Z_{SL,0}\right)}\right\}$$

$$\mathbf{I}_b = \frac{2\mathbf{E}_T}{(Z_{T1} + 2Z_{T0})}\left\{\frac{1}{\left(Z_{eq} + Z_{SL}\right)\sqrt{n^2+3}}\left[\frac{Z_{T1}}{2}\left(-\cos\frac{\theta}{2} + \sqrt{3}\sin\frac{\theta}{2}\right) - j\left(\frac{Z_{T1}}{2} + Z_{T0}\right)\left(\sqrt{3}\cos\frac{\theta}{2} + \sin\frac{\theta}{2}\right)\right] - \frac{Z_{T0}}{n\left(Z_{eq,0} + Z_{SL,0}\right)}\right\}$$

$$\mathbf{I}_c = \frac{2\mathbf{E}_T}{(Z_{T1} + 2Z_{T0})}\left\{\frac{1}{\left(Z_{eq} + Z_{SL}\right)\sqrt{n^2+3}}\left[-\frac{Z_{T1}}{2}\left(\cos\frac{\theta}{2} + \sqrt{3}\sin\frac{\theta}{2}\right) + j\left(\frac{Z_{T1}}{2} + Z_{T0}\right)\left(\sqrt{3}\cos\frac{\theta}{2} - \sin\frac{\theta}{2}\right)\right] - \frac{Z_{T0}}{n\left(Z_{eq,0} + Z_{SL,0}\right)}\right\}$$

(13.118)

and

$$\mathbf{I}_{a'} = -\frac{\mathbf{E}_T}{(Z_{T1}+2Z_{T0})}\left\{\frac{Z_{T1}}{(Z_{eq}+Z_{SL})}-\frac{Z_{T0}}{(Z_{eq,0}+Z_{SL,0})}\right\}$$

$$\mathbf{I}_{b'} = -\frac{\mathbf{E}_T}{(Z_{T1}+2Z_{T0})}\left\{\frac{1}{(Z_{eq}+Z_{SL})}\left[-\frac{Z_{T1}}{2}-j\sqrt{3}\left(\frac{Z_{T1}}{2}+Z_{T0}\right)\right]-\frac{Z_{T0}}{(Z_{eq,0}+Z_{SL,0})}\right\}$$

$$\mathbf{I}_{c'} = -\frac{\mathbf{E}_T}{(Z_{T1}+2Z_{T0})}\left\{\frac{1}{(Z_{eq}+Z_{SL})}\left[-\frac{Z_{T1}}{2}+j\sqrt{3}\left(\frac{Z_{T1}}{2}+Z_{T0}\right)\right]-\frac{Z_{T0}}{(Z_{eq,0}+Z_{SL,0})}\right\}$$

(13.119)

and

$$\mathbf{I}_{a''} = -\frac{\mathbf{E}_T}{(Z_{T1}+2Z_{T0})}\left\{\frac{1}{(Z_{eq}+Z_{SL})}\left[Z_{T1}(\cos\theta)+j(Z_{T1}+2Z_{T0})(\sin\theta)\right]-\frac{Z_{T0}}{(Z_{eq,0}+Z_{SL,0})}\right\}$$

$$\mathbf{I}_{b''} = -\frac{\mathbf{E}_T}{(Z_{T1}+2Z_{T0})}\left\{\frac{1}{(Z_{eq}+Z_{SL})}\left[\frac{Z_{T1}}{2}\left(-\cos\theta+\sqrt{3}\sin\theta\right)-j\left(\frac{Z_{T1}}{2}+Z_{T0}\right)\left(\sqrt{3}\cos\theta+\sin\theta\right)\right]-\frac{Z_{T0}}{(Z_{eq,0}+Z_{SL,0})}\right\}$$

$$\mathbf{I}_{c''} = -\frac{\mathbf{E}_T}{(Z_{T1}+2Z_{T0})}\left\{\frac{1}{(Z_{eq}+Z_{SL})}\left[-\frac{Z_{T1}}{2}\left(\cos\theta+\sqrt{3}\sin\theta\right)+j\left(\frac{Z_{T1}}{2}+Z_{T0}\right)\left(\sqrt{3}\cos\theta-\sin\theta\right)\right]-\frac{Z_{T0}}{(Z_{eq,0}+Z_{SL,0})}\right\}$$

(13.120)

Except for the double line to ground fault, the fault currents can be taken as real in force and stress calculations. In the case of normal currents for the units with three windings on a leg, the currents are complex. Leakage fields under normal conditions are needed for loss calculations. Thus for double line to ground faults and for leakage field calculations for the standard delta and 2 core design, calculations with complex currents must be performed.

14. COST MINIMIZATION

Summary Transformer design is primarily determined by minimizing the overall cost, including the cost of materials, labor, and losses. This minimization, however, must take into account constraints which may be imposed on the transferred power, the impedance, the flux density, the overall height of the tank, etc. Since the cost and constraints are generally non-linear functions of the design variables, a non-linear constrained optimization method is required. We examine several such methods, developing in greater detail the one which appears to be best suited to our needs. It is then applied to transformer design, considering for simplicity only major cost components and constraints.

14.1 INTRODUCTION

A transformer must perform certain functions such as transforming power from one voltage level to another without overheating or without damaging itself when certain abnormal events occur, such as lightning strikes or short circuits. Moreover, it must have a reasonable lifetime (> 20 years) if operated under rated conditions. Satisfying these basic requirements still leaves a wide latitude in possible designs. A transformer manufacturer will therefore find it in its best economic interest to choose, within the limitations imposed by the constraints, that combination of design parameters which results in the lowest cost unit. To the extent that the costs and constraints can be expressed analytically in terms of the design variables, the mathematical theory of optimization with constraints can be applied to this problem.

Optimization is a fairly large branch of mathematics with major specialized subdivisions such as linear programming, unconstrained optimization, and linear or non-linear equality or inequality constrained optimization. Transformer design optimization falls into the most general category of such methods, namely non-linear equality and inequality constrained optimization. In this area, there are no algorithms or iteration schemes which guarantee that a global optimum (in our case a minimum) will be found. Most of the algorithms proposed in the literature will converge to a local optimum with varying degrees of efficiency although even this is not guaranteed if one starts the iterations too far from a local optimum. Some insight is therefore usually required to find a suitable starting point for the iterations. Often past experience can serve as a guide.

There is however a branch of optimization theory called geometric programming which does guarantee convergence to a global minimum, provided the function to be minimized, called the objective or cost function, and the constraints are expressible in a certain way [Duf67]. These restricted functional forms, called posynomials, will be discussed later. This method is very powerful if the problem functions conform to this type. Approximating techniques have been developed to cast other types of functions into posynomial form. Later versions of this technique removed many of its restrictions but at the expense of no longer guaranteeing that it will converge to a global minimum [Wil67]. One of the earliest applications of geometric programming was to transformer design.

Whatever choice of optimization method is made, there is also the question of how much detail to include in the problem description. Although the goal is to find the lowest cost, one might wish that the solution should provide sufficient information so that an actual design could be produced with little additional work. In general, the more detail one can generate from the optimization process, the less work which will be required later on. However, it would be unrealistic to expect that the optimum cost design for a transformer, for example, would automatically satisfy all the mechanical, thermal, and electrical constraints that require sophisticated design codes to evaluate. Rather, these constraints can be included in an approximate (conservative) manner in the optimization process. When the design codes are subsequently employed, hopefully only minor adjustments would need to be made in the design parameters to produce a workable unit.

We will briefly describe geometric programming, presenting enough of the formalism to appreciate some of its strengths and weaknesses. Then we will present a more general approach which is developed in greater detail, since it is applied subsequently to the transformer cost minimization problem.

14.2 GEOMETRIC PROGRAMMING

Geometric programming requires that the function to be minimized, the objective or cost function, and all the constraints be expressed as posynomials. Using the notation of Ref. [Duf67], these are functions of the form

$$g = u_1 + u_2 + \cdots + u_n \tag{14.1}$$

with

$$u_i = c_i t_1^{a_{i1}} t_2^{a_{i2}} \cdots t_m^{a_{im}} \tag{14.2}$$

where the c_i are positive constants and the variables t_1 , t_2 , ..., t_m are positive. The exponents a_{ij} can be of either sign or zero. They can also be integer or non-integer.

The method makes use of the geometric inequality

$$\delta_1 U_1 + \delta_2 U_2 + \cdots \delta_n U_n \ge U_1^{\delta_1} U_2^{\delta_2} \cdots U_n^{\delta_n} \tag{14.3}$$

where the U_i are non-negative and the δ_i are positive weights that satisfy

$$\delta_1 + \delta_2 + \cdots \delta_n = 1 \tag{14.4}$$

called the normality condition. Equality obtains in (14.3) if and only if all U_i are equal. Letting $u_i = \delta_i U_i$, (14.3) can be rewritten

$$g = u_1 + u_2 + \cdots + u_n \ge \left(\frac{u_1}{\delta_1}\right)^{\delta_1} \left(\frac{u_2}{\delta_2}\right)^{\delta_2} \cdots \left(\frac{u_n}{\delta_n}\right)^{\delta_n} \tag{14.5}$$

Substituting (14.2) for the u_i in (14.5) , we obtain

$$g = u_1 + u_2 + \cdots + u_n \ge \left(\frac{c_1}{\delta_1}\right)^{\delta_1} \left(\frac{c_2}{\delta_2}\right)^{\delta_2} \cdots \left(\frac{c_n}{\delta_n}\right)^{\delta_n} t_1^{D_1} t_2^{D_2} \cdots t_m^{D_m} \tag{14.6}$$

where $D_j = \sum_{i=1}^{n} \delta_i a_{ij}$. If we can choose the weights δ_i in such a way that

$$D_j = \sum_{i=1}^{n} \delta_i a_{ij} = 0 \quad , \quad j = 1, \cdots m \tag{14.7}$$

then the right hand side of (14.6) will be independent of the variables t_j. (14.7) is called the orthogonality condition. With the D_j's so chosen, the right hand side of (14.6) is referred to as the dual function, $v(\delta)$, and is given by

$$v(\delta) = \left(\frac{c_1}{\delta_1}\right)^{\delta_1} \left(\frac{c_2}{\delta_2}\right)^{\delta_2} \cdots \left(\frac{c_n}{\delta_n}\right)^{\delta_n} \tag{14.8}$$

where the vector $\delta = (\delta_1, \delta_2, \ldots, \delta_n)$. Using vector notation for the variables t_j, $\mathbf{t} = (t_1, t_2, \ldots t_n)$, (14.6) can be rewritten

$$g(\mathbf{t}) \geq v(\delta) \tag{14.9}$$

subject to (14.7).

As (14.9) shows, the objective function $g(\mathbf{t})$ is bounded below by the dual function $v(\delta)$. Thus the global minimum of $g(\mathbf{t})$ cannot be less than $v(\delta)$. Similarly the global maximum of $v(\delta)$ cannot be greater than $g(\mathbf{t})$. We now show that the minimum of $g(\mathbf{t})$ and the maximum of $v(\delta)$ are equal. Let $\mathbf{t} = \mathbf{t'}$ at the minimum. At this point, the derivative of $g(\mathbf{t})$ with respect to each variable must vanish,

$$\frac{\partial g(\mathbf{t'})}{\partial t_j} = 0 \tag{14.10}$$

Since $t_j' > 0$ by assumption, (14.10) can be multiplied by t_j',

$$t_j' \frac{\partial g(\mathbf{t'})}{\partial t_j} = 0 = \sum_{i=1}^{n} t_j' \frac{\partial u_i(\mathbf{t'})}{\partial t_j} = \sum_{i=1}^{n} u_i(\mathbf{t'}) a_{ij} \tag{14.11}$$

using (14.1) and (14.2). Dividing by $g(\mathbf{t'})$ and letting

$$\delta_i' = \frac{u_i(\mathbf{t'})}{g(\mathbf{t'})} \tag{14.12}$$

we see that (14.11) becomes

$$\sum_{i=1}^{n} \delta_i' a_{ij} = 0 \tag{14.13}$$

Thus the δ_i' satisfy the orthogonality condition (14.7). Moreover, the δ_i' sum to 1 and thus satisfy the normality condition (14.4). Therefore we can write

$$g(\mathbf{t}') = \left(g(\mathbf{t}')\right)^{\delta_1+\delta_2+\cdots\delta_n} = \left(g(\mathbf{t}')\right)^{\delta_1}\left(g(\mathbf{t}')\right)^{\delta_2}\cdots\left(g(\mathbf{t}')\right)^{\delta_n} \tag{14.14}$$

But, using (14.12), (14.14) becomes

$$g(\mathbf{t}') = \left(\frac{u_1(\mathbf{t}')}{\delta_1'}\right)^{\delta_1'}\left(\frac{u_2(\mathbf{t}')}{\delta_2'}\right)^{\delta_2'}\cdots\left(\frac{u_n(\mathbf{t}')}{\delta_n'}\right)^{\delta_n'} \tag{14.15}$$

The right hand side of (14.15) is $v(\delta')$. Thus at the minimum,

$$g(\mathbf{t}') = v(\delta') \tag{14.16}$$

This equation, along with (14.9), shows that the minimum of $g(\mathbf{t})$ and the maximum of $v(\delta)$ are equal and are in fact a global minimum and a global maximum.

Note that at the minimum, according to (14.12),

$$u_i(\mathbf{t}') = g(\mathbf{t}')\delta_i' = v(\delta')\delta_i' \tag{14.17}$$

This is also a requirement for the geometric inequality to become an equality. In the geometric programming approach, one looks for a maximum of the dual function $v(\delta)$ subject to the normality and orthogonality conditions. Having found this $v(\delta')$ and the corresponding δ_i', (14.17) can be used to find the design variables t_j' at the minimum of g. (14.17) also shows that, at the minimum, the weights δ_i' give the relative importance (cost) of each term in the objective (cost) function.

In geometric programming, all the constraints must be expressed in the form

$$g_k(\mathbf{t}) \le 1 \quad , \quad k = 1,\cdots p \tag{14.18}$$

where the g_k are posynomials. Write

$$g_k = u_1 + u_2 + \cdots u_n \tag{14.19}$$

as before with the u_i given by (14.2). (We ignore labeling the u_i with k for simplicity.) In dealing with inequality constraints, the weights are allowed to be unnormalized. Letting λ denote their sum, we have

$$\lambda = \delta_1 + \delta_2 + \cdots + \delta_n \tag{14.20}$$

However, in applying the geometric inequality (14.5), we must use normalized weights given by δ_i / λ, so we get

$$g_k = u_1 + u_2 + \cdots + u_n \geq \left(\frac{u_1}{\delta_1}\right)^{\delta_1/\lambda} \left(\frac{u_2}{\delta_2}\right)^{\delta_2/\lambda} \cdots \left(\frac{u_n}{\delta_n}\right)^{\delta_n/\lambda} \lambda \tag{14.21}$$

Raising everything to the power λ, we obtain

$$1 \geq g_k \geq \left(\frac{u_1}{\delta_1}\right)^{\delta_1} \left(\frac{u_2}{\delta_2}\right)^{\delta_2} \cdots \left(\frac{u_n}{\delta_n}\right)^{\delta_n} \lambda^{\lambda} \tag{14.22}$$

There is an inequality like (14.22) for each constraint. We need to label these appropriately to avoid confusion. Let

$$J[k] = \{m_k, m_{k+1}, \cdots n_k\} \quad , \quad k = 0,1,\cdots,p \tag{14.23}$$

be a set of integers labeling the terms in constraint k when $k > 0$. We have $m_o = 1$, $m_1 = n_0+1$, $m_2 = n_1+1$, $n_p = n$ for a total of n terms in all of the functions. We have m positive variables t_i, $i = 1, \ldots, m$. Hence

$$g_k(t) = \sum_{i \in J[k]} c_i t_1^{a_{i1}} t_2^{a_{i2}} \cdots t_m^{a_{im}} \tag{14.24}$$

where $g_o(t) = g(t)$ is the objective function. We now multiply all of the inequalities together, i.e. the extreme left and right sides of (14.5) and (14.22) for each constraint, to get

$$g(\mathbf{t}) \geq \prod_{i=1}^{n}\left(\frac{u_i}{\delta_i}\right)^{\delta_i}\prod_{k=1}^{p}\lambda_k^{\lambda_k} \tag{14.25}$$

Using (14.24), this becomes

$$g(\mathbf{t}) \geq \prod_{i=1}^{n}\left(\frac{c_i}{\delta_i}\right)^{\delta_i}\prod_{k=1}^{p}\lambda_k^{\lambda_k} = v(\delta) \tag{14.26}$$

with the auxiliary conditions

$$D_j = \sum_{i=1}^{n}\delta_i a_{ij} = 0 \quad , \quad j = 1,\cdots m \tag{14.27}$$

which eliminates the t_j dependencies. In addition, we have the normality condition

$$\sum_{i \in J[0]}\delta_i = 1 \tag{14.28}$$

All the weights δ_i are required to be non-negative,

$$\delta_i \geq 0 \quad , \quad i = 1,\cdots n \tag{14.29}$$

The problem of minimizing the objective function subject to its constraints becomes one of maximizing the dual function $v(\delta)$ subject to the dual constraints (14.27), (14.28), and (14.29). These dual constraints, aside from the non-negativity condition, are linear constraints in contrast to the, in general, non-linear constraints of the original formulation. There are as many unknowns δ_i as terms in the objective plus constraint functions, namely n. However (14.27) and (14.28) must be satisfied by these unknowns and there are $m + 1$ of these, where m is the number of design variables. The degree of difficulty is defined as $n - m - 1$. When this equals 0, this means that there are as many unknowns as equations and the solution is found by solving the set of linear equations (14.27) and (14.28) simultaneously. When this is greater than 0, techniques for maximizing an, in general, non-linear function subject to linear constraints must be employed. Under very general conditions, it can be shown that the minimum of the objective function subject to its constraints equals the maximum of the dual function subject to the dual constraints [Duf67]. Furthermore at the maximizing point δ', the following equations hold

$$u_i = \begin{cases} \delta_i' v(\delta) & , \quad i \in J[0] \\ \dfrac{\delta_i'}{\lambda_k} & , \quad i \in J[k] \end{cases} \tag{14.30}$$

which permits the determination of the design variables t_j' at the minimizing point of the objective function via (14.2).

As a simple example, consider minimizing

$$g(\mathbf{t}) = 10t_1t_2^{-1} + 20t_2t_3^{-2}$$

subject to

$$g_1(\mathbf{t}) = 2t_1 + 0.5t_1^{-2}t_3 \le 1$$

There are four terms in total so that there are four δ_i. The dual function is

$$v(\delta) = \left(\frac{10}{\delta_1}\right)^{\delta_1}\left(\frac{20}{\delta_2}\right)^{\delta_2}\left(\frac{2}{\delta_3}\right)^{\delta_3}\left(\frac{0.5}{\delta_4}\right)^{\delta_4}(\delta_3 + \delta_4)^{\delta_3+\delta_4}$$

subject to

Normality	$\delta_1 + \delta_2 = 1$
t_1 exponent = 0	$\delta_1 + \delta_3 - 2\delta_4 = 0$
t_2 exponent = 0	$-\delta_1 + \delta_2 = 0$
t_3 exponent = 0	$-2\delta_2 + \delta_4 = 0$

This problem has degree of difficulty = 0, so the solution is determined by solving the above linear equations to obtain

$$\delta_1' = \frac{1}{2} \quad , \quad \delta_2' = \frac{1}{2} \quad , \quad \delta_3' = \frac{3}{2} \quad , \quad \delta_4' = 1$$

The maximum of the dual function is therefore

$$v(\delta') = \left(\frac{10}{1/2}\right)^{1/2}\left(\frac{20}{1/2}\right)^{1/2}\left(\frac{2}{3/2}\right)^{3/2}\left(\frac{0.5}{1}\right)^{1}\left(\frac{5}{2}\right)^{5/2} = 215.2$$

The design variables can be found from

$$10t_1t_2^{-1} = \frac{1}{2}(215.2)$$

$$20t_2t_3^{-2} = \frac{1}{2}(215.2)$$

$$2t_1 = \frac{3}{5}$$

$$\frac{1}{2}t_1^{-2}t_3 = \frac{2}{5}$$

These can be solved by taking logarithms or in this case by substitution. The solution is

$$t_1' = 0.3 \quad , \quad t_2' = 0.028 \quad , \quad t_3' = 0.072$$

This solution can be checked by substituting these values into the objective function $g(t')$ which should produce the same value as given by $v(\delta')$ above.

When the degree of difficulty is not zero, the method becomes one of maximizing a non-linear function subject to linear constraints for which various solution strategies are available. For problems formulated in the language of geometric programming, these strategies lead to the global maximum of the dual function (global minimum of the objective function. This desirable outcome is not, however, without a price. In the case of transformer design optimization, the number of terms in a realistic cost function, not to mention the constraints, is quite large whereas the number of design variables can be kept reasonably small. This means that the degree of difficulty is large. In addition, it can be quite awkward to express some of the constraints in posynomial form. A further difficulty is that the constraints must be expressed as inequalities whereas in some cases an equality constraint may be desired.

14.3 NON-LINEAR CONSTRAINED OPTIMIZATION

The general non-linear optimization problem with constraints can be formulated in the following way:

$$\text{minimize} \qquad f(\mathbf{x}) \quad , \quad \mathbf{x} \text{ is an } n-\text{vector}$$

subject to the equality constraints

$$c_i(\mathbf{x}) = 0 \quad , \quad i = 1, \cdots, t \tag{14.31}$$

and the inequality constraints

$$c_i(\mathbf{x}) \geq 0 \quad , \quad i = t+1, \cdots, m$$

where f and the c_i are non-linear functions in general. We will assume throughout that they are at least twice differentiable. Although we have expressed this optimization in terms of a minimum, maximizing a function f is equivalent to minimizing $-f$. Also any equality or inequality constraint involving analytic functions can be expressed in the above manner. For example, a constraint of the type $c_i(\mathbf{x}) \leq 0$ can be rewritten $-c_i(\mathbf{x}) \geq 0$. Thus the above formulation is quite general.

14.3.1 Characterization of the Minimum

We will be concerned with finding a relative minimum of $f(\mathbf{x})$ subject to the constraints. This may or may not be a global minimum for the particular problem considered. We first consider how to characterize such a minimum, i.e. how do we know we have reached a relative minimum while satisfying all the constraints. Let $\mathbf{x}^*$ denote the design variable vector at a minimum. In this section, minimum always means relative minimum. If an inequality constraint is not active at the minimum, i.e. $c_i(\mathbf{x}^*) > 0$, then it will remain positive in some neighborhood about the minimum. (An inequality constraint is active if $c_i(\mathbf{x}) = 0$ and it is said to be violated if $c_i(\mathbf{x}) < 0$.) Thus one could move some small distance away from the minimum without violating this constraint. In effect, near the minimum, this constraint places no restrictions on the design variables. It can therefore be ignored. Only those inequality constraints which are active play a role in characterizing the minimum. We therefore include them in the list of equality constraints in our formulation (14.31). Strategies to add or drop inequalities from the list of equality constraints will be discussed later.

Much of the methodology presented in this section is based on Ref. [Gil81]. We adopt their notation which is briefly reviewed here. First a distinction is made between column and row vectors. A column vector is denoted by

$$\mathbf{v} = \begin{pmatrix} v_1 \\ v_2 \\ \vdots \\ v_n \end{pmatrix}$$

where the v_i are its components. If there are n components, it can also be referred to as an n-vector. Its transpose is a row vector, i.e.

$$\mathbf{v}^T = \begin{pmatrix} v_1 & v_2 & \cdots & v_n \end{pmatrix}$$

where T denotes transpose. The dot or scalar product between two vectors $\mathbf{v}$ and $\mathbf{w}$ is given by

$$\mathbf{v}^T\mathbf{w} = v_1 w_1 + v_2 w_2 + \cdots + v_n w_n$$

It is essentially the matrix product between a row and column vector in that order. If the order were reversed, we would have

$$\mathbf{v}\mathbf{w}^T = \begin{pmatrix} v_1 w_1 & v_1 w_2 & \cdots & v_1 w_n \\ v_2 w_1 & v_2 w_2 & & \\ \vdots & & \ddots & \\ v_n w_1 & & & v_n w_n \end{pmatrix}$$

i.e. the matrix product between a column and row vector in that order. A matrix M is called an $m \times n$ matrix if it has m rows and n columns. The i , j th matrix element is denoted M_{ij} and these are organized into an array

$$M = \begin{pmatrix} M_{11} & M_{12} & \cdots & M_{1n} \\ M_{21} & M_{22} & & \\ \vdots & & \ddots & \\ M_{m1} & & & M_{mn} \end{pmatrix}$$

A matrix vector product between an $m \times n$ matrix and an n-vector, denoted $M\mathbf{v}$,'is an m-vector whose components are

$$(M\mathbf{v})_i = \sum_{j=1}^{n} M_{ij} v_j$$

If we multiply on the left by a row m-vector $\mathbf{w}^T$, we get the scalar product between $\mathbf{w}$ and the above vector

$$\mathbf{w}^T M\mathbf{v} = \sum_{i=1}^{m}\sum_{j=1}^{n} w_i M_{ij} v_j$$

Further notation will be introduced as it is encountered.

Using Taylor's theorem, expand the functions $f(\mathbf{x})$ and the $c_i(\mathbf{x})$ about the minimizing point $\mathbf{x}^*$. Let $\mathbf{x} = \mathbf{x}^* + \varepsilon\mathbf{p}$, where $\mathbf{p}$ is an n-dimensional vector and ε a positive constant. Then

$$f(\mathbf{x}^* + \varepsilon\mathbf{p}) = f(\mathbf{x}^*) + \varepsilon\mathbf{g}^T(\mathbf{x}^*)\mathbf{p} + \frac{1}{2}\varepsilon^2\mathbf{p}^T G(\mathbf{x}^*)\mathbf{p} + \cdots \qquad (14.32)$$

where $\mathbf{g}$ is the gradient vector and is given by

$$\mathbf{g}^T(\mathbf{x}) = \begin{pmatrix} \frac{\partial f}{\partial x_1} & \frac{\partial f}{\partial x_2} & \cdots & \frac{\partial f}{\partial x_n} \end{pmatrix} \qquad (14.33)$$

G is the Hessian matrix which is given by

$$G(\mathbf{x}) = \begin{pmatrix} \frac{\partial f}{\partial x_1 \partial x_1} & \frac{\partial f}{\partial x_1 \partial x_2} & \cdots & \frac{\partial f}{\partial x_1 \partial x_n} \\ \frac{\partial f}{\partial x_2 \partial x_1} & \frac{\partial f}{\partial x_2 \partial x_2} & & \\ \vdots & & \ddots & \\ \frac{\partial f}{\partial x_n \partial x_1} & & & \frac{\partial f}{\partial x_n \partial x_n} \end{pmatrix} \qquad (14.34)$$

The Hessian matrix is symmetric because of the equality of mixed partial derivatives. Vector dot products or matrix products are implied in (14.32). Higher order terms in the expansion (14.32) are assumed to be small for the values of ε considered.

Similarly we have for the equality constraints

$$c_i(\mathbf{x}^* + \varepsilon\mathbf{p}) = c_i(\mathbf{x}^*) + \varepsilon\mathbf{a}_i^T(\mathbf{x}^*)\mathbf{p} + \frac{1}{2}\varepsilon^2\mathbf{p}^T G_i(\mathbf{x}^*)\mathbf{p} + \cdots \tag{14.35}$$

where $\mathbf{a}_i$ is its gradient vector and G_i its Hessian matrix. Here $c_i(\mathbf{x}^*) = 0$. We would like to continue to satisfy the constraints as we move away from the minimum since they are assumed to apply for any choice of design parameters. This means that the first order term, i.e. the term in ε should vanish,

$$\mathbf{a}_i^T(\mathbf{x}^*)\mathbf{p} = 0 \tag{14.36}$$

There is an equation like this for each equality constraint. Let A denote a matrix whose i-th row is $\mathbf{a}_i^T$. Then (14.36) can be expressed compactly for all equality constraints as

$$A\mathbf{p} = 0 \tag{14.37}$$

A has t rows and n columns. It also depends on $\mathbf{x}$ or $\mathbf{x}^*$ but we omit this for clarity. We assume that $t \leq n$ and that A has full row rank, i.e. that the constraint gradient vectors $\mathbf{a}_i$ are linearly independent. If this is not true, then one or more of the constraints is redundant (to first order) and can be dropped from the list.

From (14.37), we see that $\mathbf{p}$ must be orthogonal to the row space of A. Thus $\mathbf{p}$ lies in the orthogonal complement of this row space. (Any vector space is decomposable into disjoint subspaces which are orthogonal complements of each other.) Let Z denote a matrix whose column vectors form a basis of this orthogonal subspace of dimension $n - t$. Thus Z has n rows and $n - t$ columns. This orthogonality can be expressed by

$$AZ = 0 \tag{14.38}$$

Since $\mathbf{p}$ lies in this orthogonal subspace, it can be expressed as a linear combination of its basis vectors, that is a linear combination of the column vectors of Z. Thus

$$\mathbf{p} = Z\mathbf{p}_Z \tag{14.39}$$

where $\mathbf{p}_Z$ is a vector of dimension $n - t$.

Probably the best way to obtain Z is via a triangularization of A using Hausholder transformation matrices. These are orthogonal matrices of the form

$$H = I - \frac{2\mathbf{w}\mathbf{w}^{\mathrm{T}}}{|\mathbf{w}|^2} \tag{14.40}$$

where $\mathbf{w}$ is a vector of magnitude $|\mathbf{w}|$ and I is the unit diagonal matrix. By suitably choosing $\mathbf{w}$, the columns of any matrix can be successively put in upper triangular form [Gol89]. In our case, we apply this procedure to A^{T} to obtain

$$H_{\mathrm{n}} \cdots H_2 H_1 A^{\mathrm{T}} = QA^{\mathrm{T}} = \begin{pmatrix} R \\ 0 \end{pmatrix} \tag{14.41}$$

where Q is an $\mathrm{n} \times \mathrm{n}$ orthogonal matrix since it is the product of orthogonal matrices, R is an upper triangular matrix of dimension $\mathrm{t} \times \mathrm{t}$, and 0 is a zero matrix of dimension $(\mathrm{n} - \mathrm{t}) \times \mathrm{t}$. Taking the transpose of (14.41), we get

$$AQ^{\mathrm{T}} = (L \quad 0) \tag{14.42}$$

where $L = R^{\mathrm{T}}$ is a lower triangular matrix and 0 is the $\mathrm{t} \times (\mathrm{n} - \mathrm{t})$ zero matrix. Thus we can take Z to be the last $\mathrm{n} - \mathrm{t}$ columns of Q^{T} since these are orthogonal to the rows of A. Since Q is an orthogonal matrix, its column vectors form an orthonormal basis for the whole space. Thus the first t columns form an orthonormal basis for the row space of A and the last $\mathrm{n} - \mathrm{t}$ columns form an orthonormal basis for its orthogonal complement. We group the first t columns into a matrix called Y so that

$$Q^{\mathrm{T}} = (Y \quad Z) \tag{14.43}$$

With $\mathbf{p}$ restricted to the form (14.39), the expression (14.32) becomes to first order

$$\mathrm{f}(\mathbf{x}^* + \varepsilon\mathbf{p}) = \mathrm{f}(\mathbf{x}^*) + \varepsilon\mathbf{g}^{\mathrm{T}}(\mathbf{x}^*) Z\mathbf{p}_Z + \cdots \tag{14.44}$$

For $\mathrm{f}(\mathbf{x}^*)$ to be a minimum, we must have

$$f\left(\mathbf{x}^* + \varepsilon\mathbf{p}\right) \geq f\left(\mathbf{x}^*\right) \tag{14.45}$$

In order for this to hold, the term in ε in (14.44) must be positive or zero. Since ε is positive, if this term is non-zero for some vector $\mathbf{p}_Z$, it can be made negative by possibly changing $\mathbf{p}_Z$ to $-\mathbf{p}_Z$ and thus violate (14.45). Hence we must require at the minimum

$$\mathbf{g}^T\left(\mathbf{x}^*\right)Z = 0 \tag{14.46}$$

This implies that the gradient is in the row space of A. Therefore

$$\mathbf{g}\left(\mathbf{x}^*\right) = A^T\boldsymbol{\lambda}^* \tag{14.47}$$

where $\boldsymbol{\lambda}^*$ is a vector of dimension t called the vector of Lagrange multipliers. (Note that the column space of A^T is the row space of A.

Let us now consider that one of our equality constraints at the minimum was originally an inequality constraint. Let this be the c_j th constraint. Then a displacement vector $\mathbf{p}$ could be chosen so that

$$c_j\left(\mathbf{x}^* + \varepsilon\mathbf{p}\right) \geq 0 \tag{14.48}$$

without violating the constraint. Noting that $c_j(\mathbf{x}^*) = 0$, from (14.35) with j replacing i, we see that to first order in ε, we get

$$\mathbf{a}_j^T\left(\mathbf{x}^*\right)\mathbf{p} \geq 0 \tag{14.49}$$

Assume that $\mathbf{p}$ satisfies the other constraints, i.e. (14.36) holds for $i \neq j$. Then rewriting (14.47), we have

$$\mathbf{g}\left(\mathbf{x}^*\right) = \lambda_1^*\mathbf{a}_1 + \lambda_2^*\mathbf{a}_2 + \cdots + \lambda_j^*\mathbf{a}_j + \cdots + \lambda_t^*\mathbf{a}_t \tag{14.50}$$

where we have omitted the dependence of the $\mathbf{a}$'s on $\mathbf{x}^*$. For the above choice of $\mathbf{p}$,

$$\mathbf{g}^T\left(\mathbf{x}^*\right)\mathbf{p} = \lambda_j^*\mathbf{a}_j^T\mathbf{p} \tag{14.51}$$

Substituting into (14.32), we get to first order in ε

$$f(\mathbf{x}^* + \varepsilon\mathbf{p}) = f(\mathbf{x}^*) + \varepsilon\lambda_j^* \mathbf{a}_j^T \mathbf{p} \tag{14.52}$$

If $f(\mathbf{x}^*)$ is a local minimum, then (14.45) shows that the term $\varepsilon\lambda_j^* \mathbf{a}_j^T \mathbf{p} \geq 0$. But $\varepsilon > 0$ together with (14.49), shows that we must have

$$\lambda_j^* \geq 0 \tag{14.53}$$

This says that the Lagrange multipliers associated with the inequality constraints that are active at the minimum must be non-negative. If the j - th Lagrange multiplier were negative then the direction $\mathbf{p}$ chosen above can be used to lower the value of f without violating the j - th constraint thus contradicting the assumption what we are at a minimum. In searching for a minimum, a negative Lagrange multiplier for an active inequality constraint is an indication that the constraint can be dropped (deactivated). This strategy is useful in deciding when to drop a constraint from the active set.

The discussion so far applies equally well to whether we are looking for a minimum or maximum of the objective function. The only difference is that the sign of the Lagrange multiplier of an active inequality constraint would be negative for a maximum. To distinguish a minimum from a maximum, we need to go to higher order. In the case of non-linear constraints, we can no longer move along a straight line with direction $\mathbf{p}$ when considering higher order effects because the constraints would eventually be violated. (With linear constraints, one could move freely along the direction $\mathbf{p}$ given by (14.39) without violating the active constraints.) For non-linear constraints, we need to move along a curve $\mathbf{x}(\theta)$ parametrized by θ chosen in such a way that the constraints are not violated to second order as we move a small amount $\Delta\theta$ away from the minimum. At the minimum, let $\theta = \theta^*$ and $x^* = x(\theta^*)$. Let $\mathbf{p}$ be the tangent vector to this curve at the minimum, i.e.

$$\mathbf{p} = \left.\frac{d\mathbf{x}}{d\theta}\right|_{\theta=\theta^*} \tag{14.54}$$

Using Taylor's theorem in one variable to second order for the active constraints, we have

$$c_i\left(\mathbf{x}\left(\theta^* + \Delta\theta\right)\right) = c_i\left(\mathbf{x}^*\right) + \frac{dc_i}{d\theta}\Delta\theta + \frac{1}{2}\frac{d^2c_i}{d\theta^2}\Delta\theta^2 + \cdots \tag{14.55}$$

All derivatives are evaluated at θ^*. Using the chain rule, we obtain

$$\frac{dc_i}{d\theta} = \sum_{j=1}^{n} \frac{\partial c_i}{\partial x_j}\frac{dx_j}{d\theta} = \mathbf{a}_i^T \mathbf{p} \tag{14.56}$$

and, using (14.56),

$$\begin{aligned}\frac{d^2c_i}{d\theta^2} &= \frac{d\mathbf{a}_i^T}{d\theta}\mathbf{p} + \mathbf{a}_i^T\frac{d\mathbf{p}}{d\theta} \\ &= \sum_{j=1}^{n}\left(\frac{d\mathbf{a}_i^T}{dx_j}\frac{dx_j}{d\theta}\mathbf{p}\right) + \mathbf{a}_i^T\frac{d\mathbf{p}}{d\theta} \\ &= \mathbf{p}^T G_i\left(\mathbf{x}^*\right)\mathbf{p} + \mathbf{a}_i^T\frac{d\mathbf{p}}{d\theta}\end{aligned} \tag{14.57}$$

where G_i is the Hessian of the i-th constraint defined previously. Since $c_i(\mathbf{x}^*) = 0$ at the minimum, in order for (14.55) to vanish to second order, we must have both (14.56) and (14.57) vanish. The vanishing of (14.56) yields the first order results such as (14.36) previously derived. The vanishing of (14.57) yields the new second order results. Just as we previously turned (14.56) into a matrix equation so it applied to all the constraints, equation (14.37), we apply the same procedure to (14.57). Organizing the $\mathbf{a}_i^T$ into rows of the A matrix and setting (14.57) to zero, we get

$$A\frac{d\mathbf{p}}{d\theta} = -\begin{pmatrix}\mathbf{p}^T G_1 \mathbf{p} \\ \mathbf{p}^T G_2 \mathbf{p} \\ \vdots \\ \mathbf{p}^T G_t \mathbf{p}\end{pmatrix} \tag{14.58}$$

Now consider the behavior of the objective function f along this curve which satisfies the constraints to second order. Using Taylor's theorem again in one variable,

$$f\left(\mathbf{x}\left(\theta^{*}+\Delta\theta\right)\right)=f\left(\mathbf{x}^{*}\right)+\frac{df}{d\theta}\Delta\theta+\frac{1}{2}\frac{d^{2}f}{d\theta^{2}}\Delta\theta^{2}+\cdots \tag{14.59}$$

Since, by assumption $f(\mathbf{x}^{*})$ is a minimum, and since $\Delta\theta$ can have either sign, we must have to first order

$$\frac{df}{d\theta}=\sum_{j=1}^{n}\frac{\partial f}{\partial x_{j}}\frac{dx_{j}}{d\theta}=\mathbf{g}^{T}\mathbf{p}=0 \tag{14.60}$$

This is the same first order condition we obtained previously and which led to formulas (14.46) and (14.47) which are still valid in the present context. As mentioned, they would be true whether we were at a minimum or maximum. Now since we are at a minimum,

$$f\left(\mathbf{x}\left(\theta^{*}+\Delta\theta\right)\right)\geq f\left(\mathbf{x}^{*}\right) \tag{14.61}$$

Since the first order term in (14.59) is zero, (14.61) implies that the second order term is non-negative. Using the same mathematics as was used for this term for the constraints, equation (14.57), we find using (14.60),

$$\begin{aligned}\frac{d^{2}f}{d\theta^{2}}&=\frac{d\mathbf{g}^{T}}{d\theta}\mathbf{p}+\mathbf{g}^{T}\frac{d\mathbf{p}}{d\theta}\\&=\sum_{j=1}^{n}\left(\frac{d\mathbf{g}^{T}}{dx_{j}}\frac{dx_{j}}{d\theta}\mathbf{p}\right)+\mathbf{g}^{T}\frac{d\mathbf{p}}{d\theta}\\&=\mathbf{p}^{T}G\left(\mathbf{x}^{*}\right)\mathbf{p}+\mathbf{g}^{T}\frac{d\mathbf{p}}{d\theta}\geq 0\end{aligned} \tag{14.62}$$

Substituting for $\mathbf{g}^{T}$ from (14.47) into (14.62), we obtain

$$\mathbf{p}^{T}G\left(\mathbf{x}^{*}\right)\mathbf{p}+\lambda^{*T}A\frac{d\mathbf{p}}{d\theta}\geq 0 \tag{14.63}$$

Substituting (14.58) into (14.63), we get

$$\mathbf{p}^{T}\left(G\left(\mathbf{x}^{*}\right)-\sum_{i=1}^{t}\lambda_{i}^{*}G_{i}\left(\mathbf{x}^{*}\right)\right)\mathbf{p}\geq 0 \tag{14.64}$$

Using (14.39) for $\mathbf{p}$, (14.64) becomes

$$\mathbf{p}_Z^T\left[Z^T\left(G(\mathbf{x}^*)-\sum_{i=1}^{t}\lambda_i^* G_i(\mathbf{x}^*)\right)Z\right]\mathbf{p}_Z \geq 0 \tag{14.65}$$

Since (14.65) must be true for any choice of $\mathbf{p}_Z$, this requires the matrix within the square brackets to be positive semi-definite. Often minimization with constraints is formulated in terms of a Lagrangian function

$$\mathcal{L}(\mathbf{x},\lambda) = f(\mathbf{x}) - \sum_{i=1}^{t}\lambda_i^T c_i(\mathbf{x}) \tag{14.66}$$

Then (14.47) follows from requiring the Lagrangian to vanish to first order at the minimum and (14.65) follows from requiring the projected Hessian of the Lagrangian to be positive semi-definite. (The Hessian is projected by means of the Z matrix.) The Lagrangian approach does not appear to be as well motivated as the one we have taken.

Condition (14.65) together with (14.47) are necessary conditions for a minimum subject to the constraints. They are not, however, sufficient. This is because if (14.65) were 0 for some non-zero vector $\mathbf{p}_Z$, the point $\mathbf{x}^*$ could be a saddle point. To eliminate this possibility, the inequality in (14.65) must be replaced by a strict inequality, $>$, so that the matrix is positive definite. This is sufficient to guarantee a minimum. We will assume this requirement to be met in the minimization problems we deal with.

14.3.2 Solution Search Strategy

Because the functions we are dealing with are, in general, non-linear, iterations are required in order to arrive at a minimum from some starting set of the design variables. Assuming the iterations converge, we can then check that the conditions derived in the previous section hold at the converged point to guarantee that it is a minimum. This may or may not be a global minimum. However, if suitable starting values are chosen based on experience, then for most practical problems, the minimum reached will be, with high probability, a global minimum.

One of the best convergence strategies is Newton-Raphson iteration. Provided one is near the minimum, this method converges very rapidly.

In fact, for a quadratic function, it converges in one step. Applying this method to the constraint functions, with $\mathbf{p} = \Delta\mathbf{x}$, we get

$$c_i(\mathbf{x}+\mathbf{p}) \approx c(\mathbf{x}) + \mathbf{a}_i^T\mathbf{p} = 0 \tag{14.67}$$

Collecting these into a vector of constraint functions, $\mathbf{c}(\mathbf{x}) = (c_1 \ c_2 \ldots c_t)$ and using the matrix A whose rows are a_i^T as defined previously, (14.67) can be written to include all the active constraints as

$$A(\mathbf{x})\mathbf{p} = -\mathbf{c}(\mathbf{x}) \tag{14.68}$$

where the dependence of A on the value of $\mathbf{x}$ at the current iterate is indicated. Since $\mathbf{p}$ is an n-vector in the space of design variables, we can express it in terms of the basis we obtained from the QR factorization of A. Thus, using (14.43)

$$\mathbf{p} = Y\mathbf{p}_Y + Z\mathbf{p}_Z \tag{14.69}$$

where $\mathbf{p}_Y$ is a t-vector in the row space of A and $\mathbf{p}_Z$ is an (n - t)-vector in its orthogonal complement. Using (14.38), (14.68) becomes

$$AY\mathbf{p}_Y = -\mathbf{c}(\mathbf{x}) \tag{14.70}$$

where the dependence of the matrices on $\mathbf{x}$ has been supressed for clarity. However, from (14.42) and (14.43), we see that $AY = L$. Thus (14.70) becomes

$$L\mathbf{p}_Y = -\mathbf{c}(\mathbf{x}) \tag{14.71}$$

Since L is a lower triangular matrix, (14.71) can be solved readily using forward substitution. Thus the QR factorization of A has an additional benefit in facilitating the solution of a matrix equation.

We now apply the Newton-Raphson technique to (14.47) which must be satisfied at the minimum

$$\begin{aligned} \mathbf{g}(\mathbf{x}+\mathbf{p}) - A^T(\mathbf{x}+\mathbf{p})\boldsymbol{\lambda} &= \mathbf{g}(\mathbf{x}+\mathbf{p}) - \sum_{i=1}^{t}\mathbf{a}_i(\mathbf{x}+\mathbf{p})\lambda_i \\ &\approx \mathbf{g}(\mathbf{x}) + G(\mathbf{x})\mathbf{p} - A^T\boldsymbol{\lambda} - \sum_{i=1}^{t}\lambda_i G_i(\mathbf{x})\mathbf{p} = 0 \end{aligned} \tag{14.72}$$

Rewriting (14.72),

$$\left(G(\mathbf{x}) - \sum_{i=1}^{t} \lambda_i G_i(\mathbf{x})\right)\mathbf{p} = -\left(\mathbf{g}(\mathbf{x}) - A^T\boldsymbol{\lambda}\right) \tag{14.73}$$

Define

$$W(\mathbf{x}) = G(\mathbf{x}) - \sum_{i=1}^{t} \lambda_i G_i(\mathbf{x}) \tag{14.74}$$

This is the Hessian of the Lagrangian function. Then (14.73) can be rewritten

$$W(\mathbf{x})\mathbf{p} = -\mathbf{g}(\mathbf{x}) + A^T\boldsymbol{\lambda} \tag{14.75}$$

Multiply this last expression by Z^T to obtain

$$Z^T W(\mathbf{x})\mathbf{p} = -Z^T \mathbf{g}(\mathbf{x}) \tag{14.76}$$

The term $Z^T A^T \boldsymbol{\lambda}$ vanishes by the transpose of (14.38). Using (14.69) for $\mathbf{p}$, (14.76) becomes

$$Z^T W(\mathbf{x}) Z \mathbf{p}_Z = -Z^T\left(\mathbf{g}(\mathbf{x}) + W(\mathbf{x}) Y \mathbf{p}_Y\right) \tag{14.77}$$

Having obtained $\mathbf{p}_Y$ from (14.71), this last equation can be used to solve for $\mathbf{p}_Z$.

The projected Hessian of the Lagrangian function occurs in (14.77). It was previously shown that this matrix must be positive definite to insure a minimum. However, during the iterations involved in finding a minimum, this matrix may not be positive definite. If this occurs, the Newton-Raphson iteration method may have difficulty converging. One method of circumventing this problem is to modify this matrix so that it is positive definite at each iteration step. The procedure must be such that when the matrix eventually becomes positive definite, no modification is made.

One way of producing a positive definite matrix that is not too different from the matrix to be altered is called a modified Cholesky factorization [Gil81]. In this method, a Cholesky factorization is begun as if the matrix were positive definite. For a positive definite matrix, M, a Cholesky factorization would result in the factorization

$$M = PDP^{\mathrm{T}} \tag{14.78}$$

where P is a lower triangular matrix and D a positive diagonal matrix, i.e. a diagonal matrix with all positive elements. If, during the course of the factorization of a not necessarily positive definite matrix, an element of D is calculated to be zero or negative, the corresponding diagonal element of M is increased until the element of D becomes positive. Limitations are also placed on how large the values of P can become during the factorization. These limitations can also be achieved by increasing the diagonal values of the original matrix. The result is a Cholesky factorization of a modified matrix M' which is related to the original matrix by

$$M' = PDP^{\mathrm{T}} = M + E \tag{14.79}$$

where E is a positive diagonal matrix. Using this approach, (14.77) is solved for $\mathbf{p}_Z$ at each iteration.

With $\mathbf{p}_Y$ and $\mathbf{p}_Z$ determined, the complete step given by (14.69) is known. This step may, however, be too large. Some of the equality constraints may be violated to too great an extent or the objective function may not decrease. The step may also be so large as to violate an inequality constraint that is not active. We would like to keep the direction of $\mathbf{p}$ but restrict its magnitude based on these considerations. Consider multiplying $\mathbf{p}$ by ε to get a step $\varepsilon\mathbf{p}$. By letting ε vary from 0 to 1, we can check these other criteria at each ε, stopping when they become too invalidated or when $\varepsilon = 1$. One of these criteria would be that we not violate any of the inactive inequality constraints. If this occurs at some value of $\varepsilon \leq 1$, then this restricts the step size and the constraint which is just violated is added to the list of active constraints on the next iteration. Another criteria to use to restrict the step size is to define a merit function which measures to what extent the constraints are satisfied and to what extent the objective function is decreasing in value along the step. One choice for this function is

$$\text{merit function} = f(\mathbf{x}) + \rho\sum_{i=1}^{t} c_i(\mathbf{x})^2 \tag{14.80}$$

where ρ is a positive constant. The values of ε are increased towards 1 so long as this merit function decreases or until one of the inactive constraints is violated. When this function starts increasing or an inactive constraint is violated, that value of ε is used to determine the step size $\varepsilon\mathbf{p}$. In this approach, units must be chosen so that the relative magnitudes of f and the c_i are close. This means that, for values of $\mathbf{x}$ which can be expected to violate the constraints during the iterations, the values of f and the c_i are of the same order of magnitude. Then ρ can really be used to weigh the importance of satisfying the constraints relative to achieving a minimum of the objective function. In our work, we have chosen $\rho = 20$. Thus we gradually increase ε until either the merit function starts increasing or an inactive constraint is violated or $\varepsilon = 1$, whichever comes first.

Having chosen ε, we solve for new values of the Lagrange multipliers using (14.75) with $\mathbf{p}$ replaced by the actual step taken, $\varepsilon\mathbf{p}$. Rearranging (14.75), we obtain

$$A^{\mathrm{T}}\boldsymbol{\lambda} = \mathbf{g}(\mathbf{x}) + W(\mathbf{x})\varepsilon\mathbf{p} \tag{14.81}$$

which is an equation for the unknown $\boldsymbol{\lambda}$, since all other quantities are known. The solution of this equation is also facilitated by the results of the QR factorization of A. From (14.41), we can write

$$A^{\mathrm{T}} = Q^{\mathrm{T}}\begin{pmatrix} R \\ 0 \end{pmatrix} \tag{14.82}$$

since $Q^{\mathrm{T}}Q = QQ^{\mathrm{T}} = I$ for orthogonal matrices. Here I is the unit diagonal matrix. Substituting into (14.81) and multiplying by Q, we obtain

$$\begin{pmatrix} R \\ 0 \end{pmatrix}\boldsymbol{\lambda} = Q\left(\mathbf{g}(\mathbf{x}) + W(\mathbf{x})\varepsilon\mathbf{p}\right) \tag{14.83}$$

But, using (14.43), this becomes

$$\begin{pmatrix} R \\ 0 \end{pmatrix}\boldsymbol{\lambda} = \begin{pmatrix} Y^{\mathrm{T}} \\ Z^{\mathrm{T}} \end{pmatrix}\left(\mathbf{g}(\mathbf{x}) + W(\mathbf{x})\varepsilon\mathbf{p}\right) \tag{14.84}$$

Only the top t equations in (14.84) are needed to determine $\boldsymbol{\lambda}$ and these are

$$R\lambda = Y^{\mathrm{T}}\left(\mathbf{g}(\mathbf{x}) + W(\mathbf{x})\varepsilon\mathbf{p}\right) \tag{14.85}$$

Since $R = L^{\mathrm{T}}$ is upper triangular, (14.85) can be directly solved by back substitution. This value of λ is used in the next iterate. At the start, all the λ_i are set to 1.

At the next iterate, the values of the design variables become

$$\mathbf{x}_{\text{new}} = \mathbf{x}_{\text{old}} + \varepsilon\mathbf{p} \tag{14.86}$$

where $_{\text{old}}$ labels the present values and $_{\text{new}}$ the new values for the design variables. This process is continued until successive iterations produce negligible changes in the design variables and the Lagrange multipliers. At this point, it is necessary to check that the conditions which should hold at a minimum are satisfied. Thus the projected Hessian of the Lagrangian should be positive definite and the gradient of the objective function should satisfy (14.47) to some level of accuracy. The equality constraints should also equal 0 to some level of accuracy. The Lagrange multipliers of the inequality constraints which are active at the minimum should be positive. None of the inequality constraints should be violated, i.e. < 0.

During the course of the iterations, the number of equality constraints may change as inequality constraints are added or dropped from the list of active constraints. They are added when a step size threatens to violate an inactive constraint and they are dropped when a Lagrange multiplier becomes negative. Strictly speaking, the Lagrange multiplier of an active inequality constraint only needs to be positive at the minimum. This does not have to hold away from the minimum as occurs during the iterations. However, there will be some neighborhood about the minimum where they will remain positive if the constraint is active at the minimum. We also noted previously that the gradients of the active constraints are assumed to be linearly independent. If this is not true, then constraints are dropped until a linearly independent set is obtained. This is determined during the QR factorization of A since the process requires the rows of A to be linearly independent.

For certain types of problems, it may be necessary for all or some of the variables to be positive. This can be treated as a special type of inequality constraint. Thus if variable x_j must be positive, we add to the list of constraints

$$c_k(\mathbf{x}) = x_j \geq 0 \tag{14.87}$$

where k labels the inequality constraint. The gradient vector for this constraint is given by

$$\mathbf{a}_k^T = (0 \ \cdots \ 0 \ 1 \ 0 \ \cdots \ 0) \tag{14.88}$$

where the 1 is in the j-th position. Its Hessian matrix is identically zero. Thus, this type of restriction can be handled like an ordinary inequality constraint. It has no influence on the minimization unless x_j becomes 0 or negative during the course of the iterations.

14.3.3 Practical Considerations

It is useful to choose units so that all the design variables are comparable in magnitude. It is also desirable to keep the objective function and the constraints (violated to some degree or away from 0 if inequalities) of comparable magnitude. A good choice for this magnitude is unity. For example, if the objective function is cost, one could express it in $, kilo$, or mega$, whichever produces a value of order of magnitude unity at the minimum. Similarly an equality or inequality constraint of the form

$$A + B + C + D + \cdots \geq 0$$

could be divided by the absolute value of the maximum term so as to bring it closer to 1. This may not be possible in all cases so some other strategy may be necessary. This procedure will prevent the merit function from favoring one constraint more that another.

Although the method developed here applies to any non-linear, twice differentiable objective or constraint functions, in practice it is found that these functions are often of the form

$$F(\mathbf{x}) = \sum_{i=1}^{m} b_i x_1^{a_{i1}} x_2^{a_{i2}} \cdots x_n^{a_{in}} = \sum_{i=1}^{m} b_i \prod_{j \neq k}^{n} x_j^{a_{ij}} \tag{14.89}$$

where the coefficients b_i and the exponents a_{ij} are arbitrary constants. This differs from the posynomials considered earlier where the coefficients b_i all had to be positive. The k-th component of the gradient vector of this function is given by

$$\frac{\partial F(\mathbf{x})}{\partial x_k} = \sum_{i=1}^{m} b_i a_{ik} x_k^{a_{ik}-1} \prod_{j \neq k}^{n} x_j^{a_{ij}} \tag{14.90}$$

The k , q -th entry of its Hessian matrix, where $k \neq q$, is given by

$$\frac{\partial F(\mathbf{x})}{\partial x_k \partial x_q} = \sum_{i=1}^{m} b_i a_{ik} a_{iq} x_k^{a_{ik}-1} x_q^{a_{iq}-1} \prod_{\substack{j \neq k \\ j \neq q}}^{n} x_j^{a_{ij}} \tag{14.91}$$

When $k = q$, the diagonal entry of the Hessian matrix is

$$\frac{\partial^2 F(\mathbf{x})}{\partial x_k^2} = \sum_{i=1}^{m} b_i a_{ik} (a_{ik} - 1) x_k^{a_{ik}-2} \prod_{j \neq k}^{n} x_j^{a_{ij}} \tag{14.92}$$

Some care must be exercised in using the above formulas when a variable or an exponent vanishes. Also when a variable is negative and the corresponding exponent in non-integer, the expression becomes imaginary. In the latter case, which could occur during the iterations even if the variables are constrained to be non-negative, the exponent is simply rounded off to the nearest integer. This allows the iterations to proceed. Eventually as the minimum is approached, the appropriate variables will become non-negative so that this rounding becomes unnecessary. If a variable is expected to be negative at the minimum, then it should occur with an integer exponent in all the functions.

14.4 APPLICATION TO TRANSFORMER DESIGN

In designing a transformer, we normally wish to minimize the cost. Our objective function $f(\mathbf{x})$ is therefore a cost function. It will, in general, have many terms. These will include material costs, labor costs, the cost of losses to the customer, and overhead costs. These component costs, as well as the constraint functions, must be expressed in terms of a basic set of design variables. Although the choice of design variables is somewhat arbitrary, they should be chosen in such a way that the cost and constraint functions can be easily expressed in terms of them. Here we consider a 2 winding, 3 phase , core-form power transformer. We will simplify the details in order to focus more on the method.

14.4.1 Design Variables

Our basic design variables are:

(1) B Core flux density in Tesla
(2) J_s OA current density in the secondary or LV winding in kAmps/in^2
(3) R_c Core radius in inches
(4) g HV-LV gap in inches
(5) R_s Mean radius of the secondary or LV winding in inches
(6) R_p Mean radius of the primary or HV winding in inches
(7) h_s Height of the secondary winding in inches
(8) t_s Thickness (radial build) of the secondary winding in inches
(9) t_p Thickness (radial build) of the primary winding in inches
(10) M_c Weight of the core steel in kilo-pounds
(11) M_t Weight of the tank in kilo-pounds

Note that the last two weights can be expressed in terms of the other design variables. However, since some of the material and labor costs and losses are easily expressed in terms of them, we find it convenient to include them in the set of basic design variables. Their dependence on the other variables will be expressed in terms of equality constraints. The units chosen for the above variables are such that their magnitudes are all in the range of about 1 to 100. These units are used internally in the computer optimization program. As far as input and output is concerned, i.e. what the user deals with, the units are a matter of familiarity and can differ from the above.

Figure 14.1 illustrates some of the design variables geometrically. We have not considered the height of the primary winding a design variable since, in our designs, it is usually taken to be an inch shorter then the secondary winding. We express this as $h_p = \alpha h_s$, where h_p is the height of the primary winding and α is a fraction ≈ 0.95. g_c and g_o are gaps which are fixed and inputted by the user. g_c depends on whether a tertiary or tap winding is present under the LV winding and g_o depends on the phase to phase voltages. H is the window height and T the window width. X is the maximum stack width $\approx 2R_c$. These are expressible in terms of the other variables.

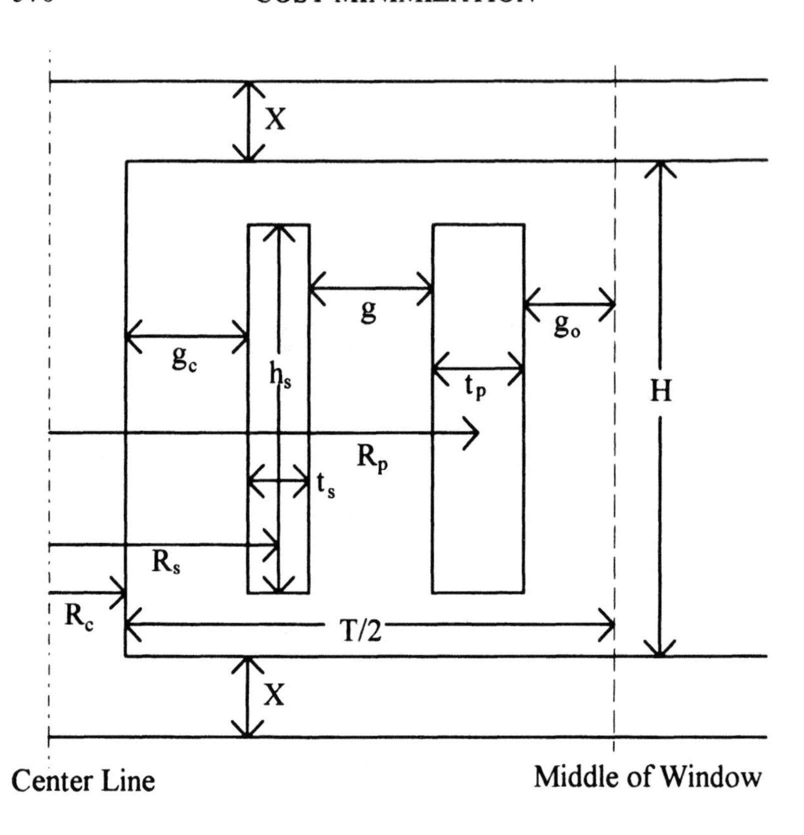

Figure 14.1 Geometry of a 2 winding core-form power transformer illustrating some of the design variables

14.4.2 Cost Function

One of the cost components is the cost of the copper. To obtain this, we need the copper weight. In terms of the design variables, the weight of the 3 secondary windings (for 3 phases) is

$$M_s = 3d_{Cu}\eta_s(2\pi R_s h_s t_s) \tag{14.93}$$

where d_{Cu} is the copper density and η_s is the fill factor, i.e. the fraction of the coil's cross-sectional area which is copper. If this is in pounds, it must then be multiplied by a \$/lb cost to arrive at the cost to include in the objective function. This \$/lb cost must reflect the cost of the raw copper plus the add-on cost of forming it into an insulated wire or cable plus any overhead or storage costs. The fill factor will be determined separately by selecting a cable size which meets constraints on the coil's cross sectional dimensions and the number of turns. If a tertiary or tap winding is present, its cost is determined as some fraction of the secondary winding's cost. The weight of the primary winding is similarly

$$M_p = 3d_{Cu}\eta_p\alpha\left(2\pi R_p h_s t_p\right) \tag{14.94}$$

where η_p is the primary winding's fill factor. This also gets multiplied by a $/lb which will differ from that of the secondary winding because it will generally be made of a different type of wire or cable.

The weight of the core in pounds is simply $1000M_c$. This is multiplied by a $/lb for the core steel before inclusion in the cost function. This $/lb also includes overhead and storage costs. The tank cost is similarly $1000M_t$ times the appropriate $/lb. M_c and M_t will be expressed in terms of the other variables by means of equality constraints.

Other material costs are the cost of the oil, the insulation such as cylinders, key-spacers, and lead support structure, the cost of leads, clamping, bushings, tap-gear, conservator if present, radiators, and auxiliary reactors or series transformers. The details are omitted here to simplify the discussion.

Load and no-load losses are part of the overall cost of operating a transformer and are included in the cost function. A customer buys a transformer not on just its initial manufacturing cost but also on the cost of operating it over many years. These latter costs depend on how efficient the unit is, i.e. how high are its losses. The load losses include the Joule heating in the windings (copper losses) and the losses due to the stray flux impinging on tank walls, clamps, etc. (stray losses). The losses in the secondary winding under OA conditions can be expressed as

$$W_s = \rho_{Cu}\left(1+ecf_s\right)J_s^2 V_s \tag{14.95}$$

where ρ_{Cu} is the copper resistivity which is evaluated at the appropriate temperature. ecf_s is the eddy current factor which is due to stray flux and depends on the type of wire or cable making up the winding. V_s is the copper volume which can be expressed

$$V_s = 3\eta_s\left(2\pi R_s h_s t_s\right) \tag{14.96}$$

Since J_s is in kA/in^2, ρ_s needs to be in appropriate units and an overall multiplying factor is necessary to get W_s in kW. This must then be multiplied by the cost of load losses in \$/kW. Similarly, the primary winding's losses are given by

$$W_p = \rho_{Cu}\left(1+ecf_p\right)J_p^2 V_p \tag{14.97}$$

where the copper volume is

$$V_p = 3\eta_p\alpha\left(2\pi R_p h_s t_p\right) \tag{14.98}$$

and where J_p is the current density in the primary winding under OA conditions. Because the ampere-turns of the primary and secondary are equal under balanced conditions, we can obtain J_p in terms of J_s by

$$J_p = \frac{J_s \eta_s t_s}{\alpha \eta_p t_p} \tag{14.99}$$

This is then substituted into (14.97). We then multiply W_p by the \$/kW cost of load losses before adding it to the cost function. The stray losses are also part of the load losses and must be added to the cost function but we omit discussing them here for simplicity. If the transformer includes a reactor, step voltage regulator, or series transformer, then the cost of losses associated with these components must be added to the cost function.

The no-load losses are essentially the core losses. Generally the core steel manufacturer provides a curve or polynomial expression for the core losses in W/lb in terms of the core flux density, e.g.

$$\text{Core Loss}\left(\text{W/lb}\right) = a_0 + a_1 B + a_2 B^2 + \cdots a_k B^k$$

Multiplying this by the core weight in pounds, $1000M_c$, gives a core loss for an ideal core. Actual cores have higher losses due to non-uniform flux in the corners, due to building stresses, and other factors. These must be accounted for in the expression for the no-load loss. Having done this, the core loss in kW is then multiplied by the cost of the no-load losses in \$/kW before adding to the cost function.

Labor costs are an important part of the cost of a transformer since power transformers are usually custom made. The various specialized types of labor must be tied to the appropriate design variables. For instance, coil winding labor costs should be expressed in terms of the coil weight or coil dimensions while core stacking labor costs should be expressed in terms of the core weight or core dimensions. Labor associated with tank welding or painting should be tied to the tank weight or tank dimensions. Correlations need to be established between the hours required to perform a certain job function and the appropriate design variables. These correlations will differ from manufacturer to manufacturer and are omitted in this discussion. Establishing meaningful correlations requires much effort. Such correlations should be revised periodically as conditions warrant. Once established, they should be multiplied by a labor cost in \$/hr , which includes labor overhead, before adding to the cost function.

As can be seen, the cost function can get rather lengthy. However, all the terms mentioned above are expressions of the form given by formula (14.89). Labor hour correlations can also be expressed this way. Therefore, gradients and Hessians can be obtained by means of the formulas given in Section 3.3. Since we wish the overall cost to be close to unity in magnitude and since our transformers cost in the neighborhood of a million dollars, we express all costs in mega-dollars.

14.4.3 Equality Constraints

Equality constraints are those which must be satisfied exactly in the final design. Probably the most important such constraint is on the total transferred power or MVA which is specified by the customer. We need to express it in terms of the chosen design variables. In the following, we work in terms of phase quantities. The power transferred per phase, P, is given by $P = MVA/3$. This is also given by

$$P = V_s I_s \tag{14.100}$$

where V_s is the rms secondary phase voltage in kV and I_s is the rms secondary phase current in kA. Using Faraday's law,

$$V_s = \frac{10^{-3}}{\sqrt{2}} N_s A_{Fe} 2\pi f B \tag{14.101}$$

where B is the peak flux density in Tesla, N_s is the number of secondary turns, A_{Fe} is the area of the iron in a cross-section of the core steel in m^2, and f is the frequency in Hz. Expressing A_{Fe} in terms of our design variables, we have

$$A_{Fe} = (0.0254)^2 \eta_c \pi R_c^2 \tag{14.102}$$

where 0.0254 converts inches to meters and η_c is a fill factor for the core steel, i.e. the fraction of actual steel in a circle of radius R_c. The current I_s is given in terms of the current density by

$$I_s = \frac{J_s \eta_s h_s t_s}{N_s} \tag{14.103}$$

Substituting these into the expression for power, we see that N_s cancels out and we find

$$P = \left[10^{-3}\sqrt{2}\pi^2 (0.0254)^2 f\right] \eta_c \eta_s B R_c^2 J_s h_s t_s \tag{14.104}$$

Divide by P to get this to be or order 1 and express it in the form of our standard equality constraints, $c_i(x) = 0$,

$$\left[\frac{10^{-3}\sqrt{2}\pi^2 (0.0254)^2 f}{P}\right] \eta_c \eta_s B R_c^2 J_s h_s t_s - 1 = 0 \tag{14.105}$$

This is an expression of the form (14.89). Note that -1 is of this form with all the $a_{ij} = 0$, $b_l = -1$, and $m = 1$. We have kept the numerical constants explicit in (14.105) in order to reduce errors.

Perhaps the next most important equality constraint is on the per unit reactance between the primary and secondary windings which is specified by the customer. A simplified expression for this which works well in practice is

$$X(\text{p.u.}) = 7.5606 \times 10^{-5} \frac{(VI)_b}{(V/N)_b^2} \frac{r^2}{(h+s)} \left[\frac{R_s t_s}{3} + \frac{R_p t_p}{3} + R_g g\right] \tag{14.106}$$

where r is the co-ratio (= 1 for a two winding transformer), R_g is the mean radius of the gap between the primary and secondary windings and is given by

$$R_g = \frac{R_s + R_p}{2} + \frac{t_s - t_p}{4} \tag{14.107}$$

h is the average height of the two windings,

$$h = \left(\frac{1+\alpha}{2}\right)h_s \tag{14.108}$$

s is a correction factor for fringing flux,

$$s = 0.32\left(R_p + \frac{t_p}{2} - R_c\right) \tag{14.109}$$

$(VI)_b$ is the base MVA/phase = P and $(V/N)_b$ is the base kV/turn which is obtained from (14.101),

$$\frac{V_s}{N_s} = \left[10^{-3}\sqrt{2}\pi^2(0.0254)^2 f\right]\eta_c R_c^2 B \tag{14.110}$$

Substituting into (14.106) and converting to standard form, we obtain

$$\left[\frac{7.5606\times10^{-5}\,Pr^2}{\left[10^{-3}\sqrt{2}\pi^2(0.0254)^2 f\right]^2 X(p.u.)}\right]\frac{1}{\eta_c^2 R_c^4 B^2 h_s}\times$$

$$\left\{\frac{R_s t_s}{3} + \frac{R_p t_p}{3} + \frac{R_s g}{2} + \frac{R_p g}{2} + \frac{t_s g}{4} - \frac{t_p g}{4}\right\} - \left(\frac{1+\alpha}{2}\right) - 0.032\frac{R_p}{h_s} - 0.16\frac{t_p}{h_s} + 0.32\frac{R_c}{h_s} = 0 \tag{14.111}$$

This is in the form of equation (14.89) and has been suitably normalized to order of magnitude 1. The constants are again kept explicit to reduce errors.

We treated R_p as an independent variable thus far since it appears in many formulas. However, it can really be expressed in terms of other design variables as

$$R_p = R_s + \frac{t_s}{2} + g + \frac{t_p}{2} \tag{14.112}$$

Converting to standard form and suitably normalizing, we get the equality constraint

$$\frac{R_s}{R_p} + \frac{t_s}{2R_p} + \frac{g}{R_p} + \frac{t_p}{2R_p} - 1 = 0 \tag{14.113}$$

An equality constraint is needed for the core weight in kilo-lbs. In terms of the window height H, window width T, and the maximum sheet width X which make up the core stacks shown in Fig. 14.1, this is given by

$$M_c = 10^{-3} d_{Fe} \eta_c \pi R_c^2 \left[3H + 4T + 6X + 0.30235 \frac{R_c}{\eta_c} \right] \tag{14.114}$$

The last term is a correction factor for the joints. d_{Fe} is the density of the core iron in lbs/in^3. In terms of the basic design variables,

$$H = h_s + slack_s \tag{14.115}$$

where $slack_s$ is a slack distance in the window which depends on the voltage or BIL of the winding and is a constant for the unit under consideration,

$$T = 2\left(R_p + \frac{t_p}{2} + g_o - R_c \right) \tag{14.116}$$

and X will be taken here as $2R_c$ although a more exact formula can be used for greater accuracy. Thus we obtain, in standard normalized form,

$$\left[10^{-3} d_{Fe} \pi\right] \frac{\eta_c R_c^2}{M_c} \left\{ 3h_s + 8R_p + 4t_p + 4R_c + 0.30235 \frac{R_c}{\eta_c} + 3slack_s + 8g_o \right\} - 1 = 0 \tag{14.117}$$

There is an equality expressing the tank weight in terms of the other variables. The tank dimensions depend, not only on the size of the 3 phase core and coils but also on clearances based on voltage considerations and on the presence of reactors, tap leads, etc. The expression is fairly complicated and is omitted here.

14.4.4 Inequality Constraints

We place an inequality constraint on the mean radius of the LV winding since it must not drop below a minimum value given by

$$R_s \geq R_c + g_c + \frac{t_s}{2} \tag{14.118}$$

Expressing this in standard form, we have

$$\frac{R_s}{R_c} - \frac{t_s}{2R_c} - \frac{g_c}{R_c} - 1 \geq 0 \tag{14.119}$$

The HV-LV gap g must not fall below a minimum value given by voltage or BIL considerations. Calling this minimum gap g_{min}, leads to the inequality

$$g \geq g_{min} \tag{14.120}$$

or, in standard form

$$\frac{g}{g_{min}} - 1 \geq 0 \tag{14.121}$$

The flux density B is limited above by the saturation of iron or by a lower value determined by overvoltage or sound level considerations. Calling the maximum value B_{max} leads to the inequality in standard form

$$\frac{B_{max}}{B} - 1 \geq 0 \tag{14.122}$$

There is also a reasonable limit on the OA current density, J_s, which we call J_{max}. This is based on cooling considerations which are most severe under FA conditions. But since the FA current density is related to the OA current density by the ratio of FA and OA MVA's, the current limitation can be placed on the OA current density. Thus we have, in standard form

$$\frac{J_{max}}{J_s} - 1 \geq 0 \tag{14.123}$$

It may be necessary to limit the tank height for shipping purposes. Referring to Fig. 14.1, the tank height can be expressed as

$$H + 2X + slack_t \leq H_{max} \tag{14.124}$$

where H_{max} is the maximum tank height and $slack_t$ is the vertical slack of the core in the tank and will depend on the clamping structure, the presence of leads, etc. $slack_t$ will be a constant for a given unit. H is given by (14.115) and, taking $X \approx 2R_c$, we get

$$h_s + 4R_c + slack_s + slack_t \leq H_{max} \tag{14.125}$$

or, in standard form,

$$\left(\frac{H_{max} - slack_s - slack_t}{H_{max}}\right) - \frac{h_s}{H_{max}} - \frac{4R_c}{H_{max}} \geq 0 \tag{14.126}$$

The quantity in parentheses in the above formula is a constant for the unit under consideration.

There are other inequalities involving the radial forces on the windings which are limited by the tensile strength of the copper used, the cooling capacity of the windings which depends on the winding current density and fill factor, and the impulse strength which depends on the cable type, voltage level, etc. These inequalities are fairly inexact since detailed design codes are ultimately used to determine winding stresses, temperatures, and impulse voltages. Nevertheless, they place some restrictions on the initial design so that later on, in the detailed design phase, difficulties are not encountered. In addition, all the design variables are positive so that a positivity condition on these variables must be included among the inequality constraints.

14.4.5 Optimization Strategy

A transformer cost minimization program has been developed based on the formulation described above. It uses the 11 basic design variables discussed earlier. Initial values must be selected for these variables to start the iteration process. Because of the equality constraints, of which there are 5 , we can only select 6 variables independently. For these, we choose an initial value for B , $B_{init} = 0.9B_{max}$, where B_{max} is the maximum flux density specified in the input, an initial value for J_s , $J_{s,init}$

$= 0.75J_{max}$, where J_{max} is the inputted maximum OA current density, an initial value for g , $g_{init} = g_{min}$, with g_{min} the inputted minimum high-low gap, an initial value for h_s , $h_{s,init} = 60$, i.e. a secondary coil height of 60 inches is typical for many of our units, an initial value for t_s, $t_{s,init} = 3$, i.e. a secondary coil width of 3 inches, and an initial value for t_p , $t_{p,init} = 4$, i.e. a primary coil width of 4 inches, both coil widths typical for our units. We also choose , for starting values, $\eta_s = \eta_p = 0.5$, $\eta_c = 0.88$, and $ecf_s = ecf_p = 0.1$. The remainder of the starting values for the design variables are determined by satisfying the equality constraints or equations such as Faraday's law which was used in their derivation or by other means based on convenience. For greater flexibility, the user can override the starting values for the 6 variables mentioned above.

The iterations are then started. The cost and constraint functions are evaluated, together with their gradients and Hessians. The Lagrange multipliers are initially set to 1 for the equality constraints. The Hessian of the Lagrangian is then formed. The matrix A is formed and subjected to a QR factorization. During this process, if a row vector of A is found to be linearly dependent on the other row vectors, the corresponding constraint is dropped from the active set. This process produces the Y , Z , L , and $R = L^T$ matrices. The vector $\mathbf{p}_Y$ is then obtained via equation (14.71). Equation (14.77) is then solved for $\mathbf{p}_Z$ using a modified Cholesky factorization method. The new step direction is then determined , $\mathbf{p} = Y\mathbf{p}_Y + Z\mathbf{p}_Z$. The merit function (14.80) is then evaluated for steps of size $\varepsilon\mathbf{p}$ where ε increases from 0 to 1 in small increments. This process continues until (1) the merit function starts increasing, (2) one of the inequality constraints is violated, or (3) $\varepsilon = 1$. If (2) is true, then the violated constraint is included among the active constraints on the next iteration. For the value of ε determined by the above procedure, new Lagrange multipliers are determined by means of (14.85). If one of these is negative for an active inequality constraint, that constraint is dropped from the active set on the next iteration. A new set of design variables in then determined, using (14.86), and the process repeated until convergence is achieved, i.e. until the changes in the design variables and Lagrange multipliers are negligible. At convergence, the projected Hessian of the Lagrangian must be positive definite, the equality constraints must be satisfied to a given level of accuracy, the inequality constraints must not be violated, and (14.47) must be satisfied to a given level of accuracy.

At this stage, details concerning the wire or cable types for the two windings are worked out based on the optimum dimensions h_s , t_s , $h_p = \alpha h_s$, and t_p . From the formula for the given power per phase, P ,

$$P = V_s I_s = V_p I_p \tag{14.127}$$

where V_s and V_p are known (inputted) secondary and primary voltages, we can determine the phase currents I_s and I_p for the secondary and primary windings respectively. From these phase currents and the optimized current densities J_s and J_p, we obtain secondary and primary turn areas $A_s = I_s/J_s$ and $A_p = I_p/J_p$. We use (14.110) to get Volts/turn in terms of the optimized variables and, using the known secondary and primary voltages, we obtain the number of secondary and primary turns N_s and N_p. In addition, from the BIL levels of the windings which are inputted, we obtain, via tables, the required paper and key spacer thicknesses for the two windings. Staying within these constraints, a wire size (magnet wire or cable) is selected by a search procedure from those that are available. In arriving at a suitable wire type, eddy current loss and impulse strength limitations are considered. The possibility of needing to use several parallel wires or cables is also considered in the search process. If a wire (cable) size or parallel package of wires (cables) cannot be found which would meet all the requirements, the wire area is then allowed to change. This would change the current density on subsequent iterations but as long as it remains below J_{max}, this is acceptable.

Having arrived at a suitable wire size, from the paper and key spacer thicknesses along with other allowances such as for cross-overs and thinning, the fill factors of the two windings η_s and η_p can be determined. Also from the optimized core radius, a better value for the core fill factor η_c can be determined. These fill factors are then used in subsequent iterations.

With optimized design variables as starting values and the newly determined fill factors, iterations are resumed until convergence is achieved. Wire sizes are then determined for these new optimized parameters and new fill factors determined and the process continued until wire sizes and fill factors no longer change between sets of iterations.

At this point, the core radius may not correspond to one of our standard radii. It is therefore set to the nearest standard value and an equality constraint is imposed on the core radius, $R_c = R_0$, or in standard form

$$\frac{R_c}{R_0} - 1 = 0 \tag{14.128}$$

where R_0 is a standard radius. The entire iteration process is then restarted with this new equality constraint. Since the starting values for the design variables and the fill factors are the values arrived at by the optimization process up to this point, the iterations with the fixed core radius usually converge very quickly.

On output, the program prints out the total cost of the optimized unit and a detailed breakdown of this cost such as the cost of each coil, the core cost, the tank cost, the oil cost, the cost of radiators, the cost of load and no-load losses, labor costs itemized by each job type, and a summary of the total material, labor, loss, and overhead costs. In addition, the physical values associated with these costs such as pounds of material, kW of loss, or labor hours are also printed out. The optimum values of the design variables are printed out as well as additional design information such as the tank dimensions, details of the magnet wire or cable used for each winding, the number of turns and the turns/disk for each winding, the paper and key spacer thicknesses for the windings, the number of radiators and their height, and the size of the conservator if needed.

REFERENCES

[Abr72] M. Abramowitz and I. A. Stegun, Editors, "Handbook of Mathematical Functions", Dover Publications, Inc., New York, 1972.

[Ame57] "American Institute of Physics Handbook", McGraw-Hill book Co., Inc., NY, 1957.

[Ansoft] MAXWELL® is a product of Ansoft Corp., Four Station Square, Suite 660, Pittsburgh, PA 15219-1119.

[Aub92] J. Aubin and Y. Langhame, "Effect of Oil Viscosity on Transformer Loading Capability at Low Ambient Temperatures", IEEE Trans. on Power Delivery, Vol. 7, No. 2, pp. 516-524, April 1992.

[Bel77] W. R. Bell, "Influence of Specimen Size on the Dielectric Strength of Transformer Oil", Trans. Elect. Insul., Vol. EI-12, No. 4, Aug. 1977, pp. 281-292.

[Blu51] L. F. Blume, A. Boyajian, G. Camilli, T. C. Lennox, S. Minneci, and V. M. Montsinger, "Transformer Engineering", 2nd Edition, John Wiley & Sons, Inc., New York, 1951.

[Bos72] A. K. Bose, "Dynamic Response of Windings Under Short Circuit", CIGRE International Conference on Large High Tension Electric Systems, 1972 Session, Aug. 28-Sept. 6, 1972.

[Chu60] R. V. Churchill, "Complex Variables and Applications", 2nd Edition, McGraw-Hill Book Co., Inc., New York, 1960.

[Cla62] F. M. Clark, "Insulating Materials for Design and Engineering Practice", John Wiley & Sons, Inc., New York, 1962.

[Cle39] J. E. Clem, "Equivalent Circuit Impedance of Regulating Transformers", A.I.E.E. Transactions, Vol. 58, 1939, pp. 871-873.

[Cyg87] S. Cygan and J. R. Laghari, "Dependence of the Electric Strength on Thickness Area and Volume of Polypropylene", IEEE Trans. on Elect. Insul., Vol. EI-22, No. 6, Dec. 1987, pp. 835-837.

[Dai73] J. W. Daily and D. R. F. Harleman, "Fluid Dynamics", Addison-Wesley Publishing Co., Inc., Reading, Mass., 1973.

[Dan90] M. G. Danikas, "Breakdown of Transformer Oil", IEEE Electrical Insulation Magazine, Sept./Oct. 1990, Vol. 6, No. 5, pp. 27-34.

[Del94] R. M. Del Vecchio, "Magnetostatics", Encyclopedia of Applied Physics, Vol. 9, VCH Publishers, Inc., 1994, pp. 207-227.

[Del98] R. M. Del Vecchio, B. Poulin, and R. Ahuja, "Calculation and Measurement of Winding Disk Capacitances with Wound-in-Shields", IEEE Trans. on Power Delivery, Vol. 13, No. 2, April 1998, pp. 503-509.

[Del99] R. M. Del Vecchio and P. Feghali, "Thermal Model of a Disk Coil with Directed Oil Flow", 1999 IEEE Transmission and Distribution Conference, Now Orleans, LA, 11-16 April, 1999, pp. 914-919.

[Duf67] R. J. Duffin, E. L. Peterson, and C. Zener, "Geometric Programming", Wiley, NY,1967.

[Dwi61] H. B. Dwight, "Tables of Integrals and Other Mathematical Data", 4th Edition, The MacMillan Co., New York, 1961.

[Eas65] C. Eastgate, "Simplified steady-state thermal calculations for naturally cooled transformers", Proc. IEE, Vol. 112, No. 6, pp. 1127-1134, June 1965.

[End57] H. S. Endicott and K. H. Weber, "Electrode Area Effect for the Impulse Breakdown of Transformer Oil", Trans. IEEE, Aug. 1957, pp. 393-398.

[For69] J. A. C. Forrest and R. E. James, Patent no. 1,158,325, "Improvements in or relating to Windings for Inductive Apparatus", The patent Office, London, July 16,1969.

[Franc] M. A. Franchek and T. A. Prevost, EHV Weidmann, private communication.

[Gil81] P. E. Gill, W. Murray, and M. H. Wright, "Practical Optimization", Academic Press, NY, 1981.

[Gol89] G. H. Golub and C. F. Van Loan, "Matrix Computations", Second Edition, The Johns Hopkins University Press, Baltimore and London, 1989.

[Gro73] F. W. Grover, "Inductance Calculations", Dover Publications, Inc., New York, N.Y., Special edition prepared for the Instrument Society of America, 1973.

[Gum58] E. J. Gumbel, "Statistics of Extremes", Columbia University Press, New York, 1958.

[Hig75] H. Higake, K. Endou, Y. Kamata, and M. Hoshi, "Flashover Characteristics of Transformer Oil in Large Cylinder-Plane Electrodes up to Gap Length of 200 mm, and Their Application to the Insulation from Outer Winding to Tank in UHV Transformers", IEEE PES winter meeting, New York, NY, Jan. 26-31, 1975.

[Hir71] K. Hiraishi, Y. Hore, and S. Shida, "Mechanical Strength of Transformer Windings Under Short-Circuit Conditions", Paper 71 TP 8-PWR, paper presented at the IEEE Winter Power Meeting, New York, N.Y., Jan 31-Feb. 5, 1971

[Hob39] J. E. Hobson and W. A. Lewis, "Regulating Transformers in Power-System Analysis", A.I.E.E. Transactions, Vol. 58, 1939, pp. 874-886.

[Hue72] Lawrence P. Huelsman, "Basic Circuit Theory with Digital Computations", Prentice-Hall, Inc., Englewood Cliffs, NJ, 1972.

[IEE81] "Guide for Loading Mineral Oil Immersed Transformers Up to and Including 100 MVA with 55 °C or 65 °C Average Winding Rise", ANSI / IEEE C57.92 - 1981.

[IEE93] IEEE Standard General Requirements for Liquid-Immersed Distribution, Power, and Regulating Transformers, IEEE Std. C57.12.00-1993, IEEE, Aug. 27, 1993.

[Jes96] Sandor Jeszenszky, "History of Transformers", IEEE Power Engineering Review, Vol. 16, No. 12, Dec. 1996, pp. 9 - 12.

[Kau68] R. B. Kaufman and J. R. Meador, "Dielectric Tests for EHV Transformers", IEEE Trans. on PAS, Vol. PAS-87, No. 1, Jan. 1968, pp. 135-145.

[Kok61] J. A. Kok, "Electrical Breakdown of Insulating Liquids", Interscience Publishers, Inc., New York, 1961.

[Kra98] A. Kramer and J. Ruff, "Transformers for Phase Angle Regulation Considering the Selection of On-Load Tap-Changers", IEEE Transactions on Power Delivery, Vol. 13, No. 2, April 1998, pp. 518-525.

[Kre80] F. Kreith and W. Z. Black, "Basic Heat Transfer",Harper & Row Publishers, NY,1980

[Kuf88] E. Kuffel and W. S. Zaengl, "High Voltage Engineering", Pergamon Press, New York, 1988.

[Lam66] J. Lammeraner and M. Stafl, "Eddy Currents", Iliffe Books, Ltd., London, 1966.

[Lyo37] W. V. Lyon, "Applications of the Method of Symmetrical Components", Mc Graw-Hill Book Co., Inc., N.Y., 1937.

[McN91] W. J. McNutt, "Insulation Thermal Life Considerations for Transformer Loading Guides", paper presented at the IEEE/PES Summer Meeting, San Diege, CA, July 28 - Aug. 1, 1991.

[Meh97] Sam P. Mehta, Nicola Aversa, and Michael S. Walker, "Transforming transformers", IEEE Spectrum, July, 1997, pp. 43 - 49.

[Mik78] A. Miki, T. Hosoya, and K. Okuyama, "A calculation Method for Impulse Voltage Distribution and Transferred Voltage in Transformer Windings", IEEE Trans. on PAS, Vol. PAS-97, No. 3, May/June, 1978, pp. 930-939.

[MIT43] Members of the staff of the Dept. of Electrical Engineering of the Massachusetts Institute of Technology, "Magnetic Circuits and Transformers", The M.I.T. Press, Massachusetts Institure of Technology, Cambridge Mass., 1943.

[Mos79] H. P. Moser, "Transformerboard", Special print of Scientia Electrica, 1979.

[Mos87] H. P. Moser and V. Dahinden, "Transformerboard II", H. Weidmann AG, Ch-8640, Rapperswil, Switzerland, 1987.

[Nel89] J. K. Nelson, "An Assessment of the Physical Basis for the Application of Design Criteria for Dielectric Structures", IEEE Trans. on Elect. Insul., Vol. 24, No. 5, Oct. 1989, pp. 835-847.

[Nor48] E. T. Norris, "The Lightening strength of Power Transformers", J. Inst. Elect. Engr., Vol. 95, 1948, pp. 389-406.

[Nuy78] R. van Nuys, "Interleaved High-Voltage Transformer Windings", IEEE Trans. on Power Apparatus and Systems, Vol. PAS-97, No. 5, Sept./Oct. 1978, pp 1946-1954.

[Oli80] A. J. Oliver, " Estimation of transformer winding temperatures and coolant flows using a general network method", IEE Proc., Vol. 127, Pt. C, No. 6, pp. 395-405, Nov. 1980.

[Ols80] R. M. Olson, "Essentials of Engineering Fluid Mechanics", 4th Edition, Harper & Row Publishers, N. Y., 1980.

[Pal69] S. Palmer and W. A. Sharpley, "Electric Strength of Transformer Insulation", Proc. IEE, Vol. 116, 1969, pp. 1965-1973.

[Pat80] M. R. Patel, "Dynamic stability of helical and barrel coils in transformers against axial short-circuit forces", IEE Proc., Vol. 127, Pt. C, No. 5, Sept., 1980, pp 281-284.

[Pav93] D. Pavlik, D. C. Johnson, and R. S. Girgis, "Calculation and Reduction of Stray and Eddy Losses in Core-Form Transformers Using a Highly Accurate Finite Element Technique", IEEE Trans. on Power Delivery, Vol. 8, No. 1, January, 1993.

[Pie92] L. W. Pierce, "An Investigation of the Thermal Performance of an Oil Filled Transformer Winding", IEEE Trans. on Power Delivery, Vol. 7, No. 3, pp. 1347-1358, July 1992.

[Pie92a] L. W. Pierce, "Predicting Liquid Filled Transformer Loading Capability", IEEE / IAS Petroleum and Chemical Industry Technical Conference, San Antonio, TX, Sept. 28-30, 1992.

[Pry58] R. H. Pry and C. P. Bean, "Calculation of the Energy Loss in Magnetic Sheet Materials Using a Domain Model", J. Appl. Phys., Vol 29, No. 3, 1958, pp. 532-533.

[Pug62] E. M. Pugh and E. W. Pugh, "Principles of Electricity and Magnetism", Addison-Wesley Publishing Co., Inc., Reading, Mass., 1962.

[Rab56] L. Rabins, "Transformer Reactance Calculations with Digital Computers", AIEE Trans., Vol. 75 Pt. I, July 1956, pp. 261-267.

[Roh85] W. M. Rohsenow, J. P. Hartnett, and E. N. Ganic, Editors, "Handbook of Heat Transfer Fundamentals", 2nd Edition, McGraw-Hill Book Co., N. Y., 1985.

[Rud40] R. Rudenberg, "Performance of Traveling Waves in Coils and Windings", Trans. of the A.I.E.E., Vol. 59, 1940, pp. 1031-1040.

[Rud68] R. Rudenberg, "Electrical Shock Waves in Power Systems", Harvard University Press, Cambridge, Mass., 1968.

[Smy68] W. R. Smythe, "Static and Dynamic Electricity", 3rd Ed., McGraw-Hill Book Co., New York, N.Y., 1968.

[Ste62], [9] E. Stenkvist and L. Torseke, "Short-circuit Problems in Large Transformers", CIGRE, 1962, Report No. 142, Appendix II.

[Ste62a], 1) W. D. Stevenson, Jr., "Elements of Power System Analysis", McGraw-Hill Book Co., Inc., N.Y., 1962.

[Ste64] G. M. Stein, "A Study of the Initial Serge Distribution in Concentric Transformer Windings", IEEE Trans. on PAS, Vol. PAS-83, Sept. 1964, pp. 877-893.

[Ste72] R. B. Steel, W. M. Johnson, J. J. Narbus, M. R. Patel, and R. A. Nelson, "Dynamic Measurements in Power Transformers Under Short-Circuit Conditions", International Conference on Large High Tension Electric Systems, CIGRE, 1972 Session, Aug. 28-Sept. 6, 1972.

[Tay58] E. D. Taylor, B. Berger, and B. E. Western, "An Experimental Approach to the Cooling of Transformer Coils by Natural Convection", IEE, Paper No. 2505 S, April 1958.

[Tho79] H. A. Thompson, F. Tillery, and D. U. von Rosenberg, "The Dynamic Response of Low Voltage, High Current, Disk Type Transformer Windings to Through Fault Loads", IEEE Trans. on Power Apparatus and Systems, Vol. PAS-98, No. 3, May/June 1979, pp. 1091-1098.

[Tim55] S. Timoshenko, "Strength of Materials", Parts I and II, 3rd Edition, D. Van Nostrand Co., Inc., Princeton, N.J., 1955 (Part II 1956).

[Tim70] S. P. Timoshenko and J. N. Goodier, "Theory of Elasticity", 3rd Edition, Mc Graw-Hill Book Co., N.Y., 1970.

[Tri82] N. Giao Trinh, C. Vincent, and J. Regis, "Statistical Dielectric Degradation of Large-Volume Oil-Insulation", IEEE Trans. on Power Apparatus and Systems (PAS), Vol. PAS-101, No. 10, Oct. 1982, pp. 3712-3721.

[Wat66] M. Waters, "The Short-Circuit Strength of Power Transformers", Macdonald & Co. (Publishers) Ltd., London, 1966.

[Web56] K. H. Weber and H. S. Endicott, "Area Effect and its Extremal Basis for the Electric Breakdown of Transformer Oil", AIEE Transactions, June, 1956, pp. 371-381.

[Wes64] Central Station Engineers of the Westinghouse Electric Corp., E. Pittsburgh, PA, "Electrical Transmission and Distribution Reference Book", Westinghouse Electric Corp., E. Pittsburgh, PA, 1964.

[Wil53] W. R. Wilson, "A Fundamental Factor Controlling the Unit Dielectric Strength of Oil", AIEE Trans. on PAS, Vol. 72, Feb. 1953, pp. 68-74.

[Wil67] D. J. Wilde and C. S. Beightler, "Foundations of Optimization", Prentice-Hall, Inc., Englewood Cliffs, N. J., 1967.

INDEX

—R—

—V—

—W—

—X—

—Y—

—Z—